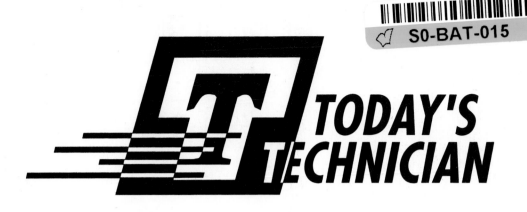

TODAY'S TECHNICIAN

Shop Manual for

Automatic Transmissions and Transaxles

Third Edition

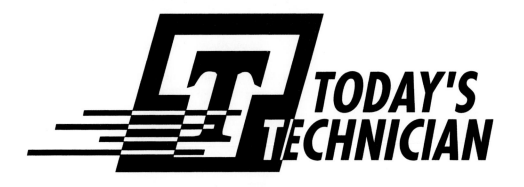

Shop Manual for
Automatic Transmissions and Transaxles

Third Edition

■

Jack Erjavec
Professor Emeritus, Columbus State Community College
Columbus, Ohio

THOMSON
™
DELMAR LEARNING

Australia   Canada   Mexico   Singapore   Spain   United Kingdom   United States

THOMSON
DELMAR LEARNING

Today's Technician: Automatic Transmissions and Transaxles Shop Manual
Third Edition
Jack Erjavec

Business Unit Director: Alar Elken	**Executive Marketing Manager:** Maura Theriault	**Production Editor:** Barbara L. Diaz
Executive Editor: Sandy Clark	**Channel Manager:** Fair Huntoon	**Art/Design Coordinator:** Rachel Baker
Acquisitions Editor: Sanjeev Rao	**Marketing Coordinator:** Brian McGrath	**Editorial Assistant:** Jill Carnahan
Developmental Editor: Alison Weintraub	**Executive Production Manager:** Mary Ellen Black	

NOTICE TO THE READER

Contents

Photo Sequences

Job Sheets

Preface

Thanks to the support the Today's Technician Series has received from those who teach automotive technology, Delmar Learning, a division of Thomson Learning, Inc., is able to live up to its promise to provide new editions every three years. We have listened to our critics and our fans and present this new revised edition. By revising our series every three years, we can and will respond to changes in the industry and in the certification process and to the ever-changing needs of those who teach automotive technology.

The *Today's Technician* series, by Delmar Learning, features textbooks that cover all mechanical and electrical systems of automobiles and light trucks. Principal titles correspond with the eight major areas of the National Institute for Automotive Service Excellence (ASE) certification. Additional titles include remedial skills and theories common to all of the certification areas and advanced or specialized subject areas that reflect the latest technological trends.

Each title is divided into two manuals: a Classroom Manual and a Shop Manual. Dividing the material into two manuals provides the reader with the information needed to begin a successful career as an automotive technician without interrupting the learning process by mixing cognitive and performance-based learning objectives.

Each Classroom Manual contains the principles of operation for each system and subsystem. It also discusses the design variations used by different manufacturers. The Classroom Manual is organized to build upon basic facts and theories. The primary objective of this manual is to allow the reader to gain an understanding of how each system and subsystem operates. This understanding is necessary to diagnose the complex automobile systems.

The understanding acquired by using the Classroom Manual is required for competence in the skill areas covered in the Shop Manual. All of the high-priority skills, as identified by ASE, are explained in the Shop Manual. The Shop Manual also includes step-by-step instructions for diagnostic and repair procedures. Photo Sequences are used to illustrate many of the common service procedures. Other common procedures are listed and are accompanied with fine-line drawings and photographs that allow the reader to visualize and conceptualize the finest details of the procedure. The Shop Manual also contains the reasons for performing the procedures, as well as when that particular service is appropriate.

The two manuals are designed to be used together and are arranged in corresponding chapters. Not only are the chapters in the manual linked together, the contents of the chapters are also linked. Both manuals contain clear and thoughtfully selected illustrations. Many of the illustrations are original drawings or photos prepared for inclusion in the series. This means that the art is a vital part of each manual.

The page layout is designed to include information that would otherwise break up the flow of information presented to the reader. The main body of the text includes all of the "need-to-know" information and illustrations. In the side margins are many of the special features of the series. Items such as definition of new terms, common trade jargon, tools list, and cross-referencing are placed in the margin, out of the normal flow of information so as not to interrupt the thought process of the reader.

Highlights of This Edition—Classroom Manual

The notable change in this new edition is the organization of topics. The reviewers of the previous edition had many suggestions for reorganizing the information. The order of the topics and the new chapter titles reflect the most common reviewer suggestions. In addition to the reorganization, this edition has been thoroughly updated. Much of the updating includes current electronic applications, especially OBD-II systems. Current model transmissions are used as examples throughout the text. Some are discussed in detail. This includes five-speed and constantly variable transmissions. This new edition also has more information on nearly all automatic transmission-related topics. In addition to more coverage of electronics, there is more detailed coverage on hydraulics, torque converter clutches, pumps, multiple-friction disc packs, bands, one-way clutches, valves, pressure controls, and final drive units. Also, a new feature, the "Author's Note," has been added. This feature allows me to add simple explanations or examples to the coverage of something complex. It also allows me the opportunity to encourage the students in their studies. Finally, the art has been updated throughout the text to enhance comprehension and improve visual interest.

Highlights of This Edition—Shop Manual

This text was reorganized to correlate with the new organization of the Classroom Manual. This change created new chapters, not just new chapter titles, and allowed for more coverage of all topics. Chapters 1 and 2 are new and cover the need-to-know, transmission-related information about tools, safety, and typical service procedures. The rest of the chapters are new or have been thoroughly updated. Much of the updating focuses on the diagnosis and service of electronic systems, especially OBD-II systems. Currently accepted service procedures are used as examples throughout the text. These procedures also served as the basis for new job sheets that are included in the text. Finally, the art has been updated throughout the text to enhance comprehension and improve visual interest. The new art includes new and updated photo sequences.

Jack Erjavec
Series Advisor

Classroom Manual

To stress the importance of safe work habits, the Classroom Manual dedicates one full chapter to safety. Included in this chapter are common safety practices, safety equipment, and safe handling of hazardous materials and wastes. This includes information on MSDS and OSHA regulations. Other features of this manual include:

Cognitive Objectives

These objectives define the contents of the chapter and define what the student should have learned upon completion of the chapter.

Each topic is divided into small units to promote easier understanding and learning.

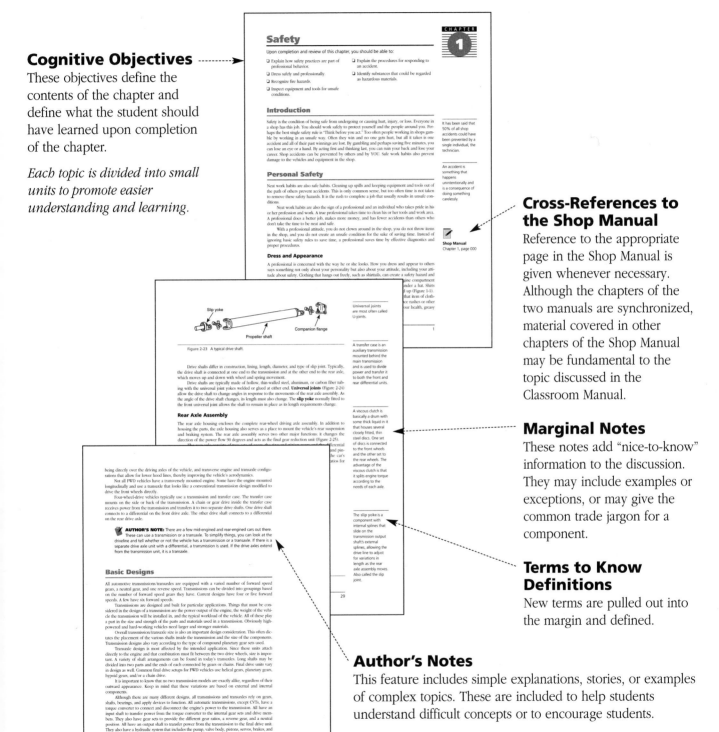

Cross-References to the Shop Manual

Reference to the appropriate page in the Shop Manual is given whenever necessary. Although the chapters of the two manuals are synchronized, material covered in other chapters of the Shop Manual may be fundamental to the topic discussed in the Classroom Manual.

Marginal Notes

These notes add "nice-to-know" information to the discussion. They may include examples or exceptions, or may give the common trade jargon for a component.

Terms to Know Definitions

New terms are pulled out into the margin and defined.

Author's Notes

This feature includes simple explanations, stories, or examples of complex topics. These are included to help students understand difficult concepts or to encourage students.

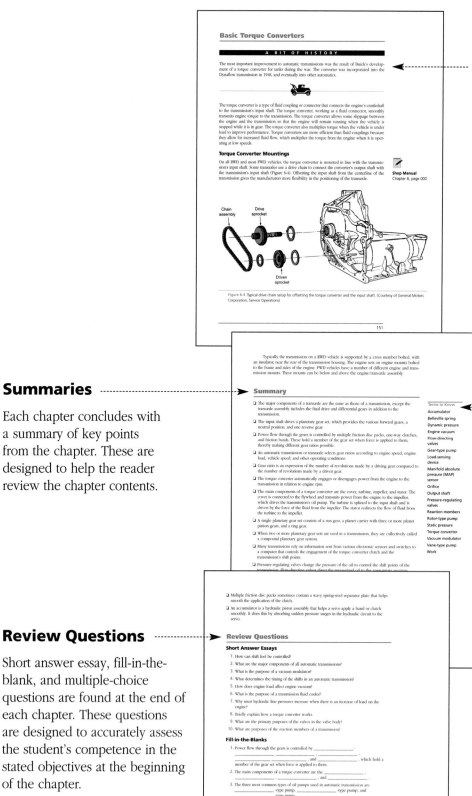

A Bit of History

This feature gives the student a sense of the evolution of the automobile. This feature not only contains nice-to-know information, but also should spark some interest in the subject matter.

Summaries

Each chapter concludes with a summary of key points from the chapter. These are designed to help the reader review the chapter contents.

Terms to Know List

A list of new terms appears next to the Summary.

Review Questions

Short answer essay, fill-in-the-blank, and multiple-choice questions are found at the end of each chapter. These questions are designed to accurately assess the student's competence in the stated objectives at the beginning of the chapter.

Shop Manual

To stress the importance of safe work habits, the Shop Manual also dedicates one full chapter to safety. Other important features of this manual include:

Performance-Based Objectives

These objectives define the contents of the chapter and define what the student should have learned upon completion of the chapter. These objectives also correspond to the list of required tasks for ASE certification.

Although this textbook is not designed to simply prepare someone for the certification exams, it is organized around the ASE task list. These tasks are defined generically when the procedure is commonly followed and specifically when the procedure is unique for specific vehicle models. Imported and domestic model automobiles and light trucks are included in the procedures.

Marginal Notes

These notes add "nice-to-know" information to the discussion. They may include examples or exceptions, or may give the common trade jargon for a component.

Special Tools Lists

Whenever a Special Tool is required to complete a task, it is listed in the margin next to the procedure.

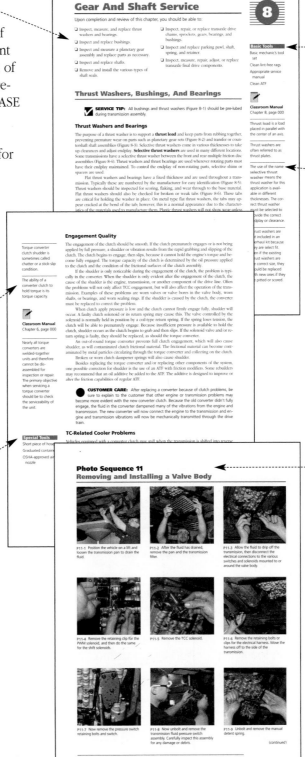

Basic Tools Lists

Each chapter begins with a list of the Basic Tools needed to perform the tasks included in the chapter.

Terms to Know Definitions

New terms are pulled out into the margin and defined.

Photo Sequences

Many procedures are illustrated in detailed Photo Sequences. These detailed photographs show the students what to expect when they perform particular procedures. They also can provide for the student a familiarity with a system or type of equipment, which the school may not have.

Customer Care
This feature highlights those little things a technician can do or say to enhance customer relations.

Service Tips
Whenever a short-cut or special procedure is appropriate, it is described in the text. These tips are generally those things commonly done by experienced technicians.

Cross-References to the Classroom Manual
Reference to the appropriate page in the Classroom Manual is given whenever necessary. Although the chapters of the two manuals are synchronized, material covered in other chapters of the Classroom Manual may be fundamental to the topic discussed in the Shop Manual.

Cautions and Warnings
Throughout the text, warnings are given to alert the reader to potentially hazardous materials or unsafe conditions. Cautions are given to advise the student of things that can go wrong if instructions are not followed or if a nonacceptable part or tool is used.

Job Sheets
Located at the end of each chapter, the Job Sheets provide a format for students to perform procedures covered in the chapter. A reference to the ASE Task addressed by the procedure is referenced on the Job Sheet.

Case Studies

Case Studies concentrate on the ability to properly diagnose the systems. Beginning with Chapter 3, each chapter ends with a case study in which a vehicle has a problem, and the logic used by a technician to solve the problem is explained.

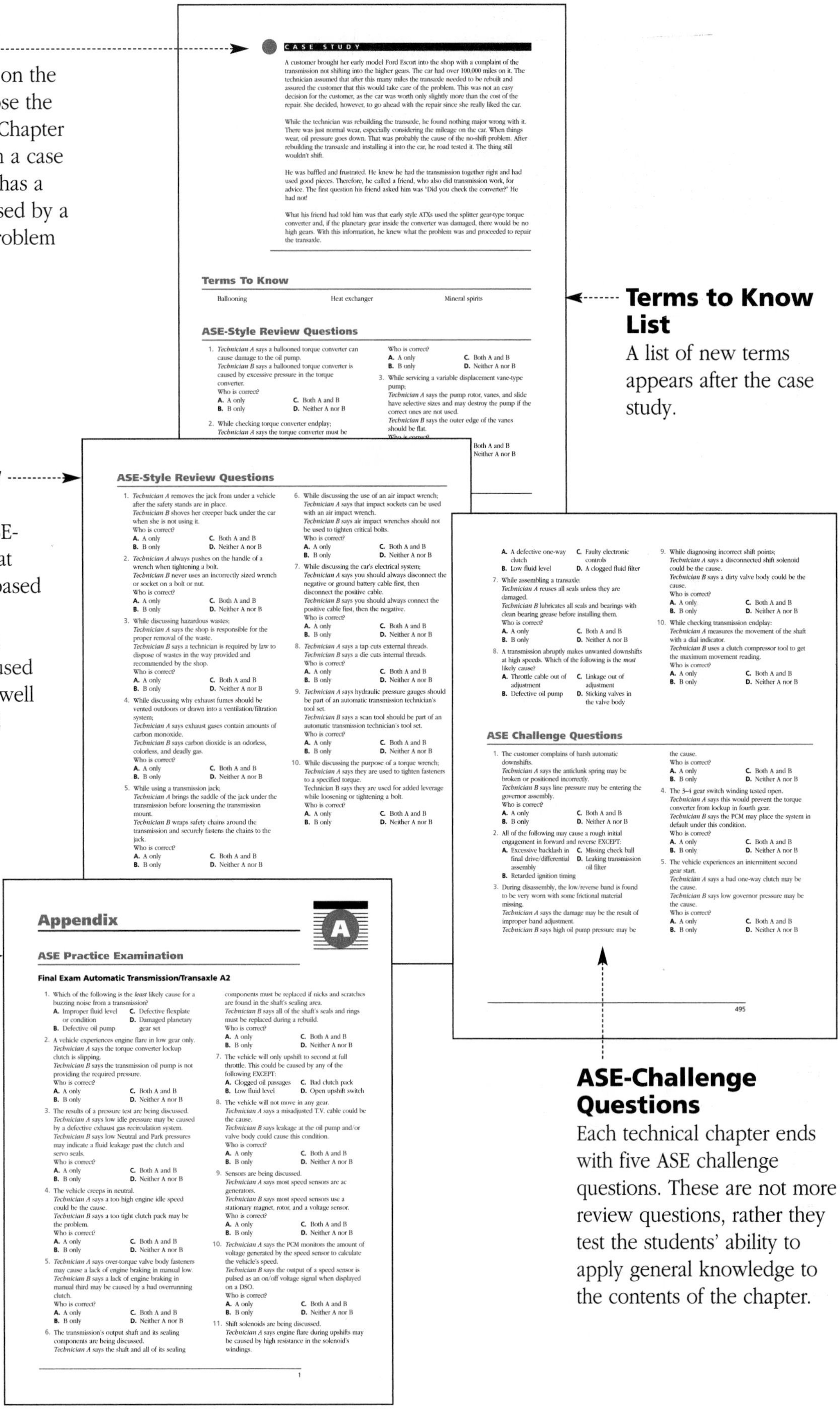

ASE-Style Review Questions

Each chapter contains ASE-style review questions that reflect the performance-based objectives listed at the beginning of the chapter. These questions can be used to review the chapter as well as to prepare for the ASE certification exam.

ASE Practice Examination

A 50 question ASE practice exam, located in the appendix, is included to test students on the contents of the Shop Manual.

Terms to Know List

A list of new terms appears after the case study.

ASE-Challenge Questions

Each technical chapter ends with five ASE challenge questions. These are not more review questions, rather they test the students' ability to apply general knowledge to the contents of the chapter.

CASE STUDY

A customer brought her early model Ford Escort into the shop with a complaint of the transmission not shifting into the higher gears. The car had over 100,000 miles on it. The technician assumed that after this many miles the transaxle needed to be rebuilt and assured the customer that this would take care of the problem. This was not an easy decision for the customer, as the car was worth only slightly more than the cost of the repair. She decided, however, to go ahead with the repair since she really liked the car.

While the technician was rebuilding the transaxle, he found nothing major wrong with it. There was just normal wear, especially considering the mileage on the car. When things wear, oil pressure goes down. That was probably the cause of the no-shift problem. After rebuilding the transaxle and installing it into the car, he road tested it. The thing still wouldn't shift.

He was baffled and frustrated. He knew he had the transmission together right and had used good pieces. Therefore, he called a friend, who also did transmission work, for advice. The first question his friend asked him was 'Did you check the converter?' He had not!

What his friend had told him was that early style ATXs used the splitter gear-type torque converter and, if the planetary gear inside the converter was damaged, there would be no high gears. With this information, he knew what the problem was and proceeded to repair the transaxle.

Terms To Know

Ballooning   Heat exchanger   Mineral spirits

ASE-Style Review Questions

1. *Technician A* says a ballooned torque converter can cause damage to the oil pump.
   Technician B says a ballooned torque converter is caused by excessive pressure in the torque converter.
   Who is correct?
   A. A only   C. Both A and B
   B. B only   D. Neither A nor B

2. While checking torque converter endplay; *Technician A* says the torque converter must be

3. While servicing a variable displacement vane-type pump;
   Technician A says the pump rotor, vanes, and slide have selective sizes and may destroy the pump if the correct ones are not used.
   Technician B says the outer edge of the vanes should be flat.

ASE-Style Review Questions

1. *Technician A* removes the jack from under a vehicle after the safety stands are in place.
   Technician B shoves her creeper back under the car when she is not using it.
   Who is correct?
   A. A only   C. Both A and B
   B. B only   D. Neither A nor B

2. *Technician A* always pushes on the handle of a wrench when tightening a bolt.
   Technician B never uses an incorrectly sized wrench or socket on a bolt or nut.
   Who is correct?
   A. A only   C. Both A and B
   B. B only   D. Neither A nor B

3. While discussing hazardous wastes;
   Technician A says the shop is responsible for the proper removal of the waste.
   Technician B says a technician is required by law to dispose of wastes in the way provided and recommended by the shop.
   Who is correct?
   A. A only   C. Both A and B
   B. B only   D. Neither A nor B

4. While discussing why exhaust fumes should be vented outdoors or drawn into a ventilation/filtration system;
   Technician A says exhaust gases contain amounts of carbon monoxide.
   Technician B says carbon dioxide is an odorless, colorless, and deadly gas.
   Who is correct?
   A. A only   C. Both A and B
   B. B only   D. Neither A nor B

5. While using a transmission jack;
   Technician A brings the saddle of the jack under the transmission before loosening the transmission mount.
   Technician B wraps safety chains around the transmission and securely fastens the chains to the jack.
   Who is correct?
   A. A only   C. Both A and B
   B. B only   D. Neither A nor B

6. While discussing the use of an air impact wrench;
   Technician A says that impact sockets can be used with an air impact wrench.
   Technician B says air impact wrenches should not be used to tighten critical bolts.
   Who is correct?
   A. A only   C. Both A and B
   B. B only   D. Neither A nor B

7. While discussing the car's electrical system;
   Technician A says you should always disconnect the negative or ground battery cable first, then disconnect the positive cable.
   Technician B says you should always connect the positive cable first, then the negative.
   Who is correct?
   A. A only   C. Both A and B
   B. B only   D. Neither A nor B

8. *Technician A* says a tap cuts external threads.
   Technician B says a die cuts internal threads.
   Who is correct?
   A. A only   C. Both A and B
   B. B only   D. Neither A nor B

9. *Technician A* says hydraulic pressure gauges should be part of an automatic transmission technician's tool set.
   Technician B says a scan tool should be part of an automatic transmission technician's tool set.
   Who is correct?
   A. A only   C. Both A and B
   B. B only   D. Neither A nor B

10. While discussing the purpose of a torque wrench;
    Technician A says they are used to tighten fasteners to a specified torque.
    Technician B says they are used for added leverage while loosening or tightening a bolt.
    Who is correct?
    A. A only   C. Both A and B
    B. B only   D. Neither A nor B

A. A defective one-way clutch   C. Faulty electronic controls
B. Low fluid level   D. A clogged fluid filter

7. While assembling a transaxle;
   Technician A reuses all seals unless they are damaged.
   Technician B lubricates all seals and bearings with clean bearing grease before installing them.
   Who is correct?
   A. A only   C. Both A and B
   B. B only   D. Neither A nor B

8. A transmission abruptly makes unwanted downshifts at high speeds. Which of the following is the *most* likely cause?
   A. Throttle cable out of adjustment   C. Linkage out of adjustment
   B. Defective oil pump   D. Sticking valves in the valve body

9. While diagnosing incorrect shift points;
   Technician A says a disconnected shift solenoid could be the cause.
   Technician B says a dirty valve body could be the cause.
   Who is correct?
   A. A only   C. Both A and B
   B. B only   D. Neither A nor B

10. While checking transmission endplay;
    Technician A measures the movement of the shaft with a dial indicator.
    Technician B uses a clutch compressor tool to get the maximum movement reading.
    Who is correct?
    A. A only   C. Both A and B
    B. B only   D. Neither A nor B

ASE Challenge Questions

1. The customer complains of harsh automatic downshifts.
   Technician A says the anticlunk spring may be broken or positioned incorrectly.
   Technician B says line pressure may be entering the governor assembly.
   Who is correct?
   A. A only   C. Both A and B
   B. B only   D. Neither A nor B

2. All of the following may cause a rough initial engagement in forward and reverse EXCEPT:
   A. Clogged oil passages   C. Missing check ball
   B. Retarded ignition timing   final drive/differential assembly   D. Leaking transmission oil filter

3. During disassembly, the low/reverse band is found to be very worn with some frictional material missing.
   Technician A says the damage may be the result of improper band adjustment.
   Technician B says high oil pump pressure may be

the cause.
   Who is correct?
   A. A only   C. Both A and B
   B. B only   D. Neither A nor B

4. The 3-4 gear switch winding tested open.
   Technician A says this would prevent the torque converter from lockup in fourth gear.
   Technician B says the PCM may place the system in default under this condition.
   Who is correct?
   A. A only   C. Both A and B
   B. B only   D. Neither A nor B

5. The vehicle experiences an intermittent second gear start.
   Technician A says a bad one-way clutch may be the cause.
   Technician B says low governor pressure may be the cause.
   Who is correct?
   A. A only   C. Both A and B
   B. B only   D. Neither A nor B

495

Appendix

ASE Practice Examination

Final Exam Automatic Transmission/Transaxle A2

1. Which of the following is the *least* likely cause for a buzzing noise from a transmission?
   A. Improper fluid level or condition   C. Defective flexplate
   B. Defective oil pump   D. Damaged planetary gear set

2. A vehicle experiences engine flare in low gear only.
   Technician A says the torque converter lockup clutch is slipping.
   Technician B says the transmission oil pump is not providing the required pressure.
   Who is correct?
   A. A only   C. Both A and B
   B. B only   D. Neither A nor B

3. The results of a pressure test are being discussed.
   Technician A says low idle pressure may be caused by a defective exhaust gas recirculation system.
   Technician B says Neutral and Park pressures may indicate a fluid leakage past the clutch and servo seals.
   Who is correct?
   A. A only   C. Both A and B
   B. B only   D. Neither A nor B

4. The vehicle creeps in neutral.
   Technician A says a too high engine idle speed could be the cause.
   Technician B says a too tight clutch pack may be the problem.
   Who is correct?
   A. A only   C. Both A and B
   B. B only   D. Neither A nor B

5. *Technician A* says over-torque valve body fasteners may cause a lack of engine braking in manual low.
   Technician B says a lack of engine braking in manual third may be caused by a bad overrunning clutch.
   Who is correct?
   A. A only   C. Both A and B
   B. B only   D. Neither A nor B

6. The transmission's output shaft and its sealing components are being discussed.
   Technician A says the shaft and all of its sealing

components must be replaced if nicks and scratches are found in the shaft's sealing area.
   Technician B says all of the shaft's seals and rings must be replaced during a rebuild.
   Who is correct?
   A. A only   C. Both A and B
   B. B only   D. Neither A nor B

7. The vehicle will only upshift to second at full throttle. This could be caused by any of the following EXCEPT:
   A. Clogged oil passages   C. Bad clutch pack
   B. Low fluid level   D. Open upshift switch

8. The vehicle will not move in any gear.
   Technician A says a misadjusted T.V. cable could be the cause.
   Technician B says leakage at the oil pump and/or valve body could cause this condition.
   Who is correct?
   A. A only   C. Both A and B
   B. B only   D. Neither A nor B

9. Sensors are being discussed.
   Technician A says most speed sensors are ac generators.
   Technician B says most speed sensors use a stationary magnet, rotor, and a voltage sensor.
   Who is correct?
   A. A only   C. Both A and B
   B. B only   D. Neither A nor B

10. *Technician A* says the PCM monitors the amount of voltage generated by the speed sensor to calculate the vehicle's speed.
    Technician B says the output of a speed sensor is pulsed as an on/off voltage signal when displayed on a DSO.
    Who is correct?
    A. A only   C. Both A and B
    B. B only   D. Neither A nor B

11. Shift solenoids are being discussed.
    Technician A says engine flare during upshifts may be caused by high resistance in the solenoid's windings.

1

Reviewers

I would like to extend a special thanks to those who saw things I overlooked and for their contributions to this text:

Travis DeClerk
Black River Technical College
Pocahontas, AR

Clifton Owen
Griffin Technical College
Griffin, GA

Richard Hausmann
Owens Community College
Toledo, OH

Mario Schwarz
Santa Fe Community College
Gainesville, FL

John L. Linden
Pittsburgh, PA

Contributing Companies

I would also like to thank these companies who provided technical art for this edition:

Automotive Lift Institute

CRC Industries, Inc.

General Motors Corporation

Lincoln Automotive

National Automotive Technicians Education Foundation (NATEF)

National Institute for Automotive Service Excellence (ASE)

OTC Tool and Equipment, a division of SPX Corporation

Snap-on Tools Company

Portions of materials contained herein have been reprinted with permission of General Motors Corporation, Service Operations.

Tools and Safety

CHAPTER 1

Upon completion and review of this chapter, you should be able to:

❏ Identify and describe the purpose of hand tools commonly found in a basic automotive technician's tool set.

❏ Describe the use of common pneumatic, electrical, and hydraulic power tools found in an automotive service department.

❏ Describe some of the special tools used to service automatic transmissions.

❏ Explain the importance of safety and accident prevention in an automotive shop.

❏ Explain the basic principles of personal safety, including the use of protective eye

wear, clothing, gloves, shoes, and hearing protection.

❏ Explain the procedures and precautions for safely using tools and equipment.

❏ Explain the precautions that need to be followed to safely raise a vehicle on a lift.

❏ Properly lift heavy objects.

❏ Extinguish the common variety of fires.

❏ Safely work around and with batteries.

❏ Describe the purpose of the laws concerning hazardous wastes and materials, including the right-to-know laws.

Common Hand Tools

Automatic transmission technicians use a variety of tools (Figure 1-1). As you progress through the chapters of this manual, you will be introduced to the special tools required to complete a particular task. These tools will be listed in the margin, next to the procedure, and included in the procedure. At the beginning of each chapter, a list of basic tools is given in the margin. It is those tools that will be discussed in the remainder of this chapter. Some of these are basic hand tools, while others are special and power tools.

Although every technician's toolbox contains many different tools, there are certain hand tools that are a must. These are described in the following paragraphs.

Wrenches

A basic tool set should include a set of box- and open-end wrenches. A **box-end wrench** completely encircles a nut or the head of a bolt and is less apt to slip and cause damage or injury. Often, it is difficult to place a box-end wrench around a nut or bolt because of its surroundings. An **open-end wrench** has an open squared end and often can be used where a box wrench will not fit. However, an open-end wrench is more likely to slip under force. Open- and box-end wrenches have different sizes at either end.

Box-end wrenches are available in either 6- or 12-point ends. 12-point box-end wrenches allow you to work in tighter areas than do 6-point wrenches. An open-end wrench is normally the best tool for turning the nut down or holding a bolt head.

You should have a complete set of both open- and box-end wrenches in your tool set. However, to reduce cost and storage space, you may want to have a set of combination wrenches that have an open-end wrench on one end and a box wrench on the other. Both ends of these wrenches are sized the same and can be used interchangeably on the same nut or bolt. Most older domestic cars are built with bolts and nuts that require wrenches of SAE (USCS) sizes. These wrenches are most commonly found in increments of 1/16 of an inch. Most imported and newer domestic cars require metric-sized wrenches. These wrenches are made in increments of 1 mm. To be able to work on today's transmissions, you should have complete sets of both SAE and metric wrenches.

A **box-end wrench** has ends that surround the bolt head or nut.

An **open-end wrench** grips the bolt head or nut only on two sides.

The USCS and metric systems are two of the most common measuring standards. The USCS is what we use in the United States, whereas the metric system is used in most other parts of the world. The next chapter covers these measuring systems in great detail.

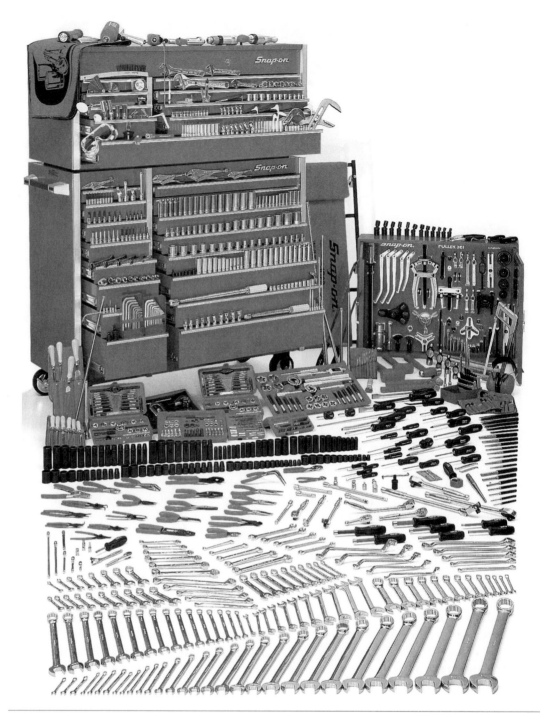

Figure 1-1 A driveline technician needs a wide variety of tools. (Courtesy of Snap-on Tools Company)

WARNING: Metric and USCS wrenches are not interchangeable. For example, a 9/16-inch wrench is 0.02 inches larger than a 14-millimeter nut. If the 9/16-inch wrench is used to turn or hold a 14-mm nut, the wrench will probably slip. This may cause the points of the bolt head or nut to round off and can possibly cause skinned knuckles.

An **Allen wrench** is no more than a hex-shaped rod that fits into a matching recess in a bolt head.

Allen wrenches or hex-head wrenches are used to tighten and loosen setscrews, which are commonly used in transmissions and transaxles. The appropriately sized wrench fits into a machined hex-shaped recess in the bolt or screw.

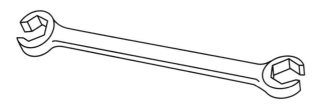

Figure 1-2 A six-point flare nut wrench. (Courtesy of Matco Tools)

Flare nut wrenches are commonly called line wrenches.

To loosen or tighten line or tubing fittings, flare nut wrenches (Figure 1-2) should be used rather than open-end wrenches. Using an open-end wrench will tend to round the corners of the nut; these nuts are typically made of soft metal and can distort easily. Line wrenches surround the nut and provide a better grip on the fitting.

Ratchets and Sockets

A set of SAE and metric sockets combined with a ratchet handle and a few extensions should also be included in your tool set. These sockets should be 3/8-inch drive, although 1/4- and 1/2-inch drive sets are also handy. The ratchet allows you to turn the socket with force in one direction and without force in the other direction. This allows you to tighten or loosen a bolt without removing and resetting the wrench after you have turned it. You may also want a long breaker bar to fit your sockets. Breaker bars offer increased leverage when loosening very tight bolts.

There are many designs of sockets for most sizes. A six-point socket has stronger walls and improved grip on a bolt compared to a normal 12-point socket. However, 6-point sockets have half the positions of a 12-point socket. Six-point sockets are mostly used on fasteners that are rusted or rounded. Eight-point sockets are available to use on square nuts or square-headed bolts. Some axle and transmission assemblies use square-headed plugs in the fluid reservoir. Sockets are also available as deep-well sockets that are used to reach a nut when it is on a bolt or stud with long threads.

WARNING: Deep-well sockets are also good for reaching nuts and bolts in limited access areas. Deep-well sockets should not be used when a regular size socket will work. The longer socket develops more twist torque and tends to slip off the fastener.

Extensions allow you to put the handle in the best position while working. They range from 1 inch to 3 feet long. Also available are universal joints that allow a technician to turn a bolt or nut at a slight angle.

Torque Wrenches

A **torque wrench** is used to measure the tightness of a fastener.

Torque wrenches measure how tight a nut or bolt is. Many of a car's nuts and bolts should be tightened to a certain amount and have a torque specification that is expressed in foot-pounds (SAE) or Newton-meters (metric). A foot-pound is the work or pressure accomplished by a force of 1 pound through a distance of 1 foot. A Newton-meter is the work or pressure accomplished by a force of 1 kilogram through a distance of 1 meter.

Torque wrenches come with drives that correspond with sockets: 1/4 inch, 3/8 inch, and 1/2 inch. There are four types of torque wrenches: dial-type, breakover-type, torsion bar-type, and the digital readout-type. For most drivetrain work, a dial, torsion bar, or digital readout type is recommended (Figure 1-3). These have a scale that can be read and can be used to measure turning effort, as well as tightening bolts. With the breakover-type, you must dial in the desired torque. The wrench makes an audible click when you have reached the correct force.

Quite often when working on automatic transmissions a inch-pound torque wrench (Figure 1-4) is required. These are smaller in size and measure torque in finer increments.

Classroom Manual
Chapter 2, page 17

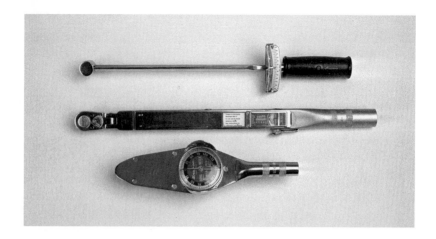

Figure 1-3 The three basic types of torque indicating wrenches.

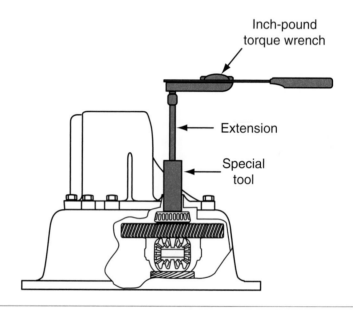

Figure 1-4 A inch-pound torque wrench measuring preload.

Screwdrivers

There are many styles of screwdrivers available. The commonly used ones are the standard tip or blade style and Phillips types. A blade style fits into a straight slot in the head of the screw. A **Phillips screwdriver** has a cross point that allows for more gripping power and stability. The cross point has four surfaces that insert into a like pattern in the head of the screw, making the screwdriver less likely to slip out of the screw. Your tool set should include both blade and Phillips drivers in a variety of lengths from 2-inch "stubbies" to 12-inch screwdrivers. Some vehicles may require special screwdrivers, such as those with a Torx head design. In fact, Torx-type fasteners are becoming common in automatic transmissions.

Phillips screwdrivers have four gripping surfaces.

Pliers

The two most commonly used pliers are a pair of interlocking jaw pliers that are about eight or nine inches long and a pair of diagonal cutters about seven inches long that can be used to cut wire and remove cotter pins. Other designs are also used while servicing a vehicle's drivetrain. These include slip-joint, needle nose, duckbill, adjustable joint, and offset needle nose pliers. It is also recommended that you have a pair of vise-grip pliers to hold parts while grinding or to use as a "third hand."

Hammers and Mallets

Your tool set should include at least three types of hammers: two ball-peen hammers, one 8-ounce and one 12- to 16-ounce, and a small sledgehammer. You should also have a plastic and lead or brass-faced mallet. Hammers are used with punches and chisels, and mallets are used for tapping parts apart or aligning parts together. A soft-faced mallet will not harm the part it is hitting against, whereas a hammer will (Figure 1-5).

Adjustable joint pliers are commonly called water pump pliers or channel locks.

Punches and Chisels

A variety of punches and chisels is used by an automatic transmission technician. Your tool set should include a variety of drift punches and starter punches. Drift punches are used to remove drift and roll pins. Some drifts are made of brass; these should be used whenever you are concerned about possible damage to the pin or surface surrounding the pin. Tapered punches are used to line up boltholes. Starter or center punches are used to make an indent before drilling to prevent the drill bit from wandering. A variety of chisels is also recommended; it should include flat, cape, round-nose cape, and diamond point chisels.

Files

A set of files should also be included in your tool set. Files are used to remove metal and to de-burr parts. Files are available in many different shapes: round, half-round, flat, crossing, knife, square, and triangular. The most commonly used files are the half-round and flat with either single-cut or double-cut designs. A single-cut file has its cutting grooves lined up diagonally across the face of the file. The cutting grooves of a double-cut file run diagonally in both directions across the face. Double-cut files are considered first cut or roughening files because they can remove large amounts of metal. Single-cut files are considered finishing files because they remove small amounts of metal. Thread files are often used to clean the external threads of a fastener.

Taps and Dies

Often problems are caused by defective fasteners or damaged threads in the bore of an assembly. Fasteners can be replaced or their threads restored with a die. A tap can cut and restore the threads in a bore. It is recommended that you have two sets of taps and dies: one USCS and one metric.

A die is used to cut external threads and a tap is used on internal threads.

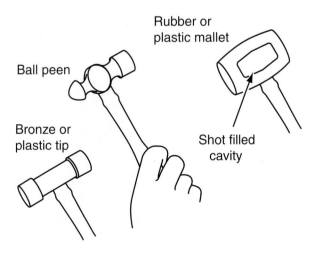

Figure 1-5 A variety of hammers and mallets.

Power Tools

Power tools make a technician's job easier. They operate faster and with more torque than hand tools. However, power tools require greater safety measures. Power tools do not stop unless they are turned off. Power is furnished by air (pneumatic), electricity, or hydraulic fluid. Pneumatic tools are typically used by technicians because they have more torque, weigh less, and require less maintenance than electric power tools. However, electric power tools tend to cost less than the pneumatics. Electric power tools can be plugged into most electric wall sockets, but to use a pneumatic tool you must have an air compressor and an air storage tank.

▲ **WARNING:** Carelessness or mishandling of power tools can cause serious injury. Make sure you know how to operate a tool before using it.

Air Wrenches

An impact wrench uses compressed air or electricity to hammer or impact a nut or bolt loose or tight. Light-duty impact wrenches are available in three drive sizes: 1/4 inch, 3/8 inch, and 1/2 inch, and two heavy-duty sizes: 3/4 inch and 1 inch.

■ **CAUTION:** Impact wrenches should not be used to tighten critical parts or parts that may be damaged by the hammering force of the wrench.

▲ **WARNING:** The sockets designed for impact wrenches are constructed of thicker, softer steel to withstand the force of the impact. Ordinary sockets must not be used with impact wrenches; they will crack or shatter because of the force and can cause injury.

Air ratchets are often used during disassembly or reassembly work to save time. Because the ratchet turns the socket without an impact force, these wrenches can be used on most parts and with ordinary sockets. Air ratchets usually have a 3/8-inch drive. Air ratchets are not torque sensitive; therefore a torque wrench should be used on all fasteners after snugging them up with an air ratchet.

Blowgun

A **blowgun** is an air nozzle used to control the release of compressed air.

Blowguns are used for blowing off parts during cleaning. Never point a blowgun at yourself or someone else. A blowgun (Figure 1-6) snaps into one end of an air hose and directs airflow when

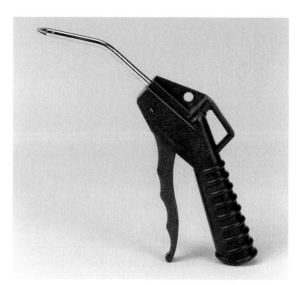

Figure 1-6 An OSHA-approved air blowgun. (Courtesy of DeVilbiss Co.)

a button is pressed. Always use an Occupational Safety and Health Administration (OSHA)-approved air blowgun. Before using a blowgun, be sure it has not been modified to eliminate air-bleed holes on the side.

Bench Grinder

This electric power tool is generally bolted to a workbench. The grinder should have safety shields and guards. Always wear face protection when using a grinder. Make sure you follow the correct procedure and only perform operations suited for a grinding wheel, because the wheel may shatter or explode if it is used improperly.

A bench grinder is classified by wheel size. Six- to ten-inch wheels are the most common in auto repair shops. Three types of wheels are available with this bench tool.

- ❏ *Grinding wheel*. Used for a wide variety of grinding jobs from sharpening cutting tools to deburring.
- ❏ *Wire wheel brush*. Used for general cleaning and buffing, removing rust and scale, paint removal, deburring, and so forth.
- ❏ *Buffing wheel*. Used for general purpose buffing, polishing, and light cutting.

Trouble Light

Adequate light is necessary when working under and around automobiles. A trouble light can be battery powered (like a flashlight) or need to be plugged into a wall socket. Some shops have trouble lights that pull down from a reel suspended from the ceiling. Trouble lights use either an incandescent bulb or fluorescent tube. Because incandescent bulbs can pop and burn, it is highly recommended that you only use fluorescent bulbs. Take extra care when using a trouble light. Make sure the cord doesn't get caught in a rotating object. The bulb or tube is surrounded by a cage or enclosed in clear plastic to prevent accidental breaking and burning.

Presses

Many automotive jobs require the use of a powerful force to assemble or disassemble parts that are **press-fit** together. Removing and installing piston pins, servicing rear axle bearings, pressing brake drum and rotor studs, and transmission assembly work are just a few of the examples. Presses can be hydraulic, electric, air, or hand driven. Capacities range up to 150 tons of pressing force, depending on the size and design of the press. Smaller arbor and C-frame presses can be bench or pedestal mounted, while high-capacity units are freestanding or floor mounted (Figure 1-7).

When a part is forced into an opening that is slightly smaller than the part itself, it is **press-fit**.

 WARNING: Always wear safety glasses when using a press.

Figure 1-7 A floor-mounted hydraulic press. (Courtesy of Snap-on Tools Company)

Lifting Tools

Lifting tools are necessary tools for most drivetrain repair procedures. Typically these tools are provided for by the shop and are not the property of a technician. Correct operating and safety procedures should always be followed when using lifting tools.

Jacks

Jacks are used to raise a vehicle off the ground and are available in two basic designs and in a variety of sizes. The most common jack is the hydraulic floor jack (Figure 1-8), which is classified by the weights it can lift: 1-1/2, 2, and 2-1/2 tons, and larger sizes for heavier vehicles. These jacks are controlled by moving the handle up and down. The other design of portable floor jack uses compressed air. Pneumatic jacks are operated by controlling air pressure at the jack.

Safety Stands

Safety stands are commonly called jack stands.

When a vehicle is raised by a jack, it should be supported by *safety stands* (Figure 1-9). Never work under a car with only a jack supporting it. Always use safety stands. Hydraulic seals in the jack can let go and allow the vehicle to drop. Service manuals note the proper locations for jacking and supporting a vehicle while it is raised from the ground. Always follow the guidelines that are provided.

Hydraulic Lift

The hydraulic floor lift is the safest lifting tool and is able to raise the vehicle high enough to allow you to walk and work under it. Various safety features prevent a hydraulic lift from dropping if a seal does leak or if air pressure is lost. Before lifting a vehicle, make sure the lift is correctly positioned.

Transmission Jacks

Transmission jacks (Figure 1-10) are designed to help you while removing a transmission from under the vehicle. The weight of the transmission makes it difficult and unsafe to remove it

Figure 1-8 Typical hydraulic jack. (Courtesy of Lincoln Automotive)

Figure 1-9 Safety stands are used to support a vehicle after it has been jacked up.

Figure 1-10 Typical hydraulic transmission jack.

Figure 1-11 An engine holding fixture for a FWD vehicle. (Courtesy of OTC Tool and Equipment, Division of SPX Corporation)

without much assistance and/or a transmission jack. These jacks fit under the transmission and are typically equipped with hold down chains. These chains are used to secure the transmission to the jack. The transmission's weight rests on the jack's saddle.

Transmission jacks are available in two basic styles. One is used when the vehicle is raised by a hydraulic jack and is sitting on jack stands. The other style is used when the vehicle is raised on a lift.

Engine hoists are often referred to as "cherry pickers."

Engine Hoist

The engine hoist allows you to lift an engine from a car. Hydraulic pressure converts power to a mechanical advantage and lifts the engine from the car. After the engine has been removed, the engine should be mounted on an engine stand and not left dangling on the engine hoist.

FWD Engine Support Fixtures

Often the transmission and engine are removed from front-wheel-drive (FWD) vehicles as a unit. Other times, the engine is left in the vehicle and the transaxle removed as a separate piece. When the latter is the case, the engine must be supported while it is in the vehicle before, during, and after transaxle removal. Special fixtures mount to the vehicle's upper frame or suspension parts (Figure 1-11). This support has a bracket that is attached to the engine. With the bracket in place, the engine's weight is then on the support fixture and the transmission can be removed.

Transmission/Transaxle Holding Fixtures

Special holding fixtures should be used to support the transmission or transaxle after it has been removed from the vehicle. These holding fixtures may be stand-alone units (Figure 1-12) or may be bench mounted and allow the transmission to be easily repositioned during repair work.

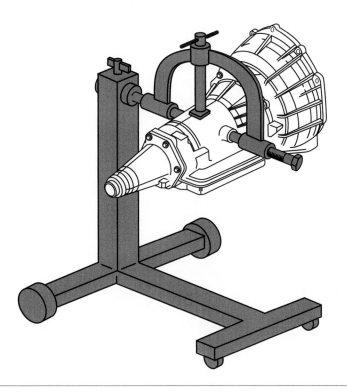

Figure 1-12 A stand-alone transmission holding stand.

Special Tools

Scan Tools

The introduction of computer-controlled systems brought with it the need for tools capable of troubleshooting electronic control systems. There are a variety of computer scan tools available today that do just that. A **scan tool** (Figure 1-13) is a microprocessor designed to communicate with the vehicle's computer. Connected to the computer through diagnostic connectors, a scan tool can access trouble codes, run tests to check system operations, and monitor the activity of the system. Trouble codes and test results are displayed on an LED screen, or printed out on the scanner printer.

A **scan tool** is a computer designed to interface with another computer to allow for easier system diagnosis.

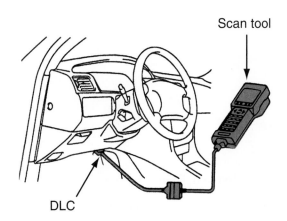

Scan tool

DLC

Figure 1-13 An OBD-II compliant scan tool connected to the vehicle's DLC.

Scan tools will retrieve fault codes from a computer's memory and digitally display these codes on the tool. A scan tool may also perform many other diagnostic functions depending on the year and make of the vehicle. Most aftermarket scan tools have removable modules that are updated each year. These modules are designed to test the computer systems on various makes of vehicles. For example, some scan testers have a 3-in-1 module that tests the computer systems on Chrysler, Ford, and General Motors vehicles. A 10-in-1 module is also available to diagnose computer systems on vehicles imported by 10 different manufacturers. These modules plug into the scan tool.

Scan tools are capable of testing many onboard computer systems such as transmission controls, engine computers, antilock brake computers, air bag computers, and suspension computers, depending on the year and make of the vehicle and the type of scan tester. In many cases, the technician must select the computer system to be tested with the scanner after it has been connected to the vehicle.

The scan tool is connected to specific diagnostic connectors on various vehicles. Most manufacturers have one diagnostic connector. This connects the data wire from each onboard computer to a specific terminal in this connector. Other vehicle manufacturers have several different diagnostic connectors on each vehicle, and each of these connectors may be connected to one or more onboard computers. A set of connectors is supplied with the scanner to allow tester connection to various diagnostic connectors on different vehicles.

The scanner must be programmed for the model year, make of vehicle, and type of engine. With some scan tools this selection is made by pressing the appropriate buttons on the tester, as directed by the digital tester display. On other scan testers, the appropriate memory card must be installed in the tester for the vehicle being tested. Some scan testers have a built-in printer to print test results, while other scan testers may be connected to an external printer.

As automotive computer systems become more complex, the diagnostic capabilities of scan testers continue to expand. Many scan testers now have the capability to store, or "freeze," data into the tester during a road test (Figure 1-14), and then play back this data when the vehicle is returned to the shop.

Some scan testers now display diagnostic information based on the fault code in the computer memory. Service bulletins published by the scan tester manufacturer may be indexed by the tester after the vehicle information is entered in the tester. Other scan testers will display sensor specifications for the vehicle being tested.

Trouble codes are only set by the vehicle's computer when an input signal is entirely out of its normal range. The codes help technicians identify the cause of the problem when this is the

Figure 1-14 Using a scan tool during a road test. (Courtesy of OTC Tool and Equipment, Division of SPX Corporation)

Figure 1-15 Using a breakout box to check an individual circuit.

case. If a signal is within its normal range but is still not correct, the vehicle's computer may not display a trouble code. However, a problem will still exist. As an aid to identify this type of problem, most manufacturers recommend that the signals to and from the computer be carefully looked at. This is done through the use of a scan tool or **breakout box** (Figure 1-15). A breakout box allows the technician to check voltage and resistance readings between specific points within the computer's wiring harness.

With OBD-II, the diagnostic connectors look the same on all vehicles. Also, any scan tool designed for OBD-II will work on all OBD-II systems; therefore, the need to have designated scan tools or cartridges is eliminated. The OBD-II scan tool has the ability to run diagnostic tests on all emission-related systems and has "freeze frame" capabilities.

A **breakout box** allows access to all of the wires in the designated wiring harness.

Gear and Bearing Pullers

Many tools are designed for a specific purpose. An example of a special tool is a gear and universal bearing puller (Figure 1-16). Many gears and bearings have a slight interference fit (press-fit) when they are installed on a shaft or in a housing. Something that has a press-fit has an interference fit. For example, the inside diameter of a bore is 0.001 inch smaller than the outside

Figure 1-16 A final drive's side bearing being removed with a universal bearing puller.

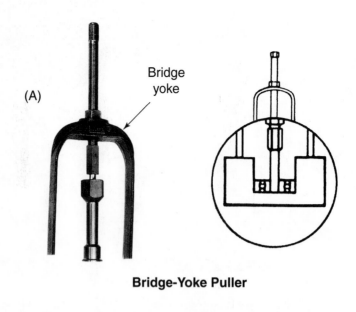

Bridge-Yoke Puller

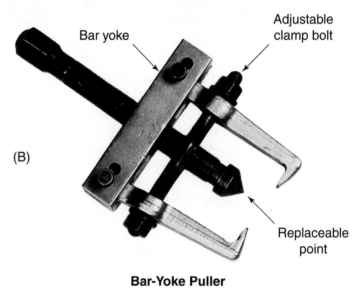

Bar-Yoke Puller

Figure 1-17 (A) A bridge-yoke puller. (B) A bar-yoke puller.

diameter of a shaft. When the shaft is fitted into the bore it must be pressed in to overcome the 0.001 inch interference fit. This press-fit prevents the parts from moving on each other. The removal of these gears and bearings must be done carefully to prevent damage to the gears, bearings, or shafts. Prying or hammering can break or bind the parts. A puller with the proper jaws and adapters should be used to remove gears and bearings. Using the proper puller, the force required to remove a gear or bearing can be applied with a slight and steady motion. Various gear and bearing puller styles and sizes are shown in Figure 1-17.

Bushing and Seal Pullers and Drivers

Another commonly used group of special tools are the various designs of bushing and seal drivers (Figure 1-18) and pullers (Figure 1-19). Pullers are either a threaded or slide hammer type tool. Always make sure you use the correct tool for the job; bushings and seals are easily damaged

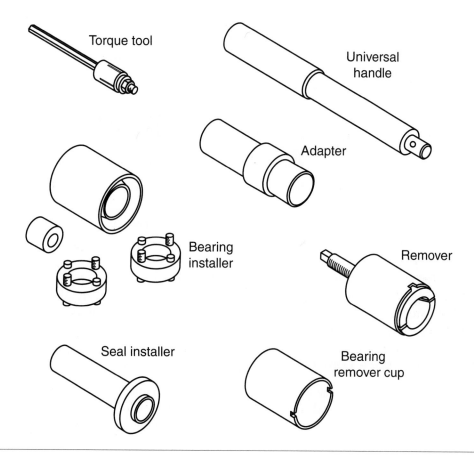

Figure 1-18 An assortment of bushing and seal drivers.

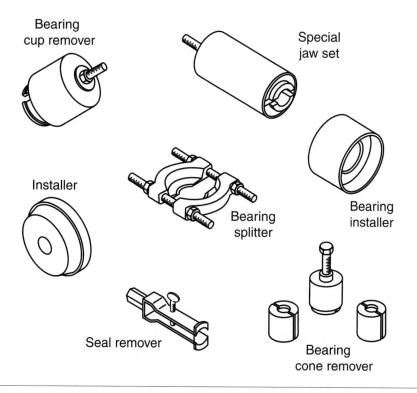

Figure 1-19 An assortment of pullers and removal tools.

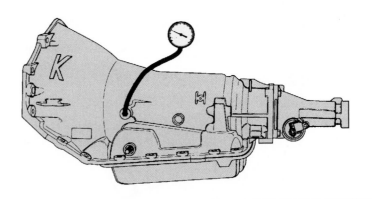

Figure 1-20 A hydraulic pressure gauge connected to a transmission.

if the wrong tool or procedure is used. Car manufacturers and specialty tool companies work closely together to design and manufacture special tools required to repair cars. Most of these special tools are listed in the appropriate service manuals. Likewise, all special tools needed for the procedures discussed in this manual will be displayed in the margin.

Hydraulic Pressure Gauge Set

A common diagnostic tool for automatic transmissions is a hydraulic pressure gauge. A pressure gauge measures pressure in pounds per square inch (psi) and/or kilopascals (kPa). The gauge is normally part of a kit that contains various fittings and adapters (Figure 1-20).

Special Tool Sets

Often, an automatic transmission technician will run into many different styles and sizes of retaining rings that hold subassemblies together or keep them in a fixed location. Using the correct tool to remove and install these rings is the only safe way to work with them. All automatic transmission technicians should have an assortment of retaining ring pliers (Figure 1-21).

Vehicle manufacturers and specialty tool companies work closely together to design and manufacture special tools required to repair transmissions. Most of these special tools are listed in the appropriate service manuals and are part of each manufacturer's Essential Tool Kit. All special tools required for the procedures discussed in this manual will be displayed in the margin.

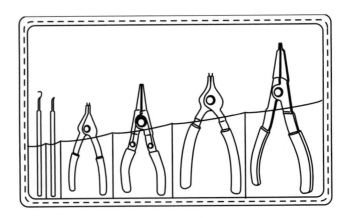

Figure 1-21 Retaining ring pliers set for transmission service

Safe Work Practices

Working on automobiles can be dangerous. It can also be fun and very rewarding. To keep the fun and rewards rolling in, you need to try to prevent accidents by working safely. In an automotive repair shop, there is great potential for serious accidents, simply because of the nature of the business and the equipment used. Through carelessness, the automotive repair industry can be one of the most dangerous occupations.

Unless you want to get hurt or want your fellow students or employees to get hurt, you should strive to work safely. Shop accidents can cause serious injury, temporary or permanent disability, and death. Think about these facts:

❑ Vehicles, equipment, and many parts are very heavy; their weight can cause severe injuries.
❑ Many parts of a car become very hot and can cause severe burns.
❑ High fluid pressures can build up inside the cooling system, fuel system, or battery; these can spray dangerous fluids on you or into your eyes.
❑ Batteries contain highly corrosive and potentially explosive acids; these can cause bad skin burns or blindness.
❑ Fuels and commonly used cleaning solvents are flammable.
❑ Exhaust fumes are poisonous and can be deadly.
❑ During some repairs, technicians can be exposed to harmful dust particles and vapors that can cause permanent and terminal diseases.

All of these are enough to scare you away from working on cars. But, the chances of your being injured while working on a car are close to nil if you learn to work safely and use common sense. Shop safety is the responsibility of everyone in the shop—you, your fellow students or technicians, and your employer or instructor. Everyone must work together to protect the health and welfare of all who work in the shop.

Personal Safety

Personal safety simply involves those precautions you take to protect yourself from injury. This includes wearing protective gear, dressing for safety, working professionally, and correctly handling tools and equipment.

Eye Protection

Your eyes can become infected or permanently damaged by many things in a shop. Some procedures, such as grinding, result in tiny particles of hot metal and dust that are thrown off at very high speeds. These metal and dirt particles can easily get into your eyes, causing scratches or cuts on your eyeball. Pressurized gases and liquids escaping from a ruptured hose or hose fitting can spray a great distance. If these chemicals get into your eyes, they can cause blindness. Dirt and sharp bits of corroded metal can easily fall down into your eyes while you are working under a vehicle.

Eye protection should be worn whenever you are exposed to these risks. To be safe, you should wear it whenever you are working in the shop. There are many types of eye protection available. To provide adequate eye protection, safety glasses have lenses made of safety glass. They also offer some sort of side protection. Regular prescription glasses do not offer sufficient protection and, therefore, should not be worn as a substitute for safety glasses. When prescription glasses are worn in a shop, they should be fitted with side shields.

Wearing safety glasses at all times is a good habit to get into. To help develop this habit wear safety glasses that fit well and feel comfortable. Driveline and transmission work is either performed on the vehicle or at a bench. For nearly all services performed on the vehicle, eye protection should be worn. This is especially true while working under the vehicle.

Some procedures may require that you wear other eye protection in addition to safety glasses. For example, when you are cleaning parts with a pressurized spray, you should wear a face shield. The face shield not only gives added protection to your eyes, but also protects the rest of your face.

If chemicals such as battery acid, fuel, or solvents get into your eyes, flush them continuously with clean water. Have someone call a doctor and get medical help immediately.

Many shops have eye wash stations or safety showers that should be used whenever you or someone else has been sprayed or splashed with a chemical.

Clothing

Your clothing should be well fitted and comfortable but made with strong material. Loose, baggy clothing can easily get caught in moving parts and machinery. Neckties should not be worn. Some technicians prefer to wear coveralls or shop coats to protect their personal clothing. Your work clothing should offer you some protection but should not restrict your movement.

Hair and Jewelry

Long hair and loose, hanging jewelry can create the same type of hazard as loose-fitting clothing. They can get caught in moving engine parts and machinery. If you have long hair, tie it back or tuck it under a cap.

Never wear rings, watches, bracelets, or neck chains. These can easily get caught in moving parts and cause serious injury.

Shoes

Automotive work involves the handling of many heavy objects that can be accidentally dropped on your feet or toes. Always wear leather or similar material shoes or boots with no-slip soles. Steel-tipped safety shoes can give added protection to your feet. Tennis or basketball shoes, street shoes, and sandals are inappropriate in the shop.

Gloves

Classroom Manual
Chapter 1, page 4

Good hand protection is often overlooked. A scrape, cut, or burn can limit your effectiveness at work for many days. A well-fitted pair of heavy work gloves should be worn during operations such as grinding and welding or when handling hot components. Always wear approved rubber gloves when handling strong and dangerous caustic chemicals. They can easily burn your skin. Be very careful when handling these types of chemicals.

Many technicians wear thin, surgical-type latex gloves whenever they are working on vehicles. These offer little protection against cuts but do offer protection against disease and grease buildup under and around your fingernails. These gloves are comfortable and are quite inexpensive.

Ear Protection

Exposure to very loud noise levels for extended periods of time can lead to a loss of hearing. Air wrenches, engines running under a load, and vehicles running in enclosed areas can all generate annoying and harmful levels of noise. Simple earplugs or earphone-type protectors should be worn in environments that are constantly noisy.

Respiratory Protection

It is not uncommon for a technician to work with chemicals that have toxic fumes. Air or respiratory masks should be worn whenever you will be exposed to toxic fumes. Cleaning parts with solvents and painting are the most common activities that call for respiratory masks to be worn. Masks should also be worn when handling parts (such as clutch discs on older vehicles) that have asbestos dust on them, or when handling hazardous materials.

Professional Behavior

Accidents can be prevented simply by the way you act. The following are some guidelines to follow while working in a shop. This list does not include everything you should or shouldn't do; it merely gives you some things to think about.

- ❏ Never smoke while working on a vehicle or while working with any machine in the shop.
- ❏ Playing around is not fun when it sends someone to the hospital. Such things as air nozzle fights, creeper races, and practical jokes have no place in the shop.
- ❏ To prevent serious burns, keep your skin away from hot metal parts such as the radiator, exhaust manifold, tailpipe, catalytic converter, and muffler.
- ❏ Always disconnect electric engine cooling fans when working around the radiator. Many of these will turn on without warning and can easily chop off a finger or hand. Make sure you reconnect the fan after you have completed your repairs.
- ❏ When working with a hydraulic press, make sure the pressure is applied in a safe manner. It is generally wise to stand to the side when operating the press.
- ❏ Properly store all parts and tools by putting them away in a place where people will not trip over them. This practice not only cuts down on injuries, but also reduces time wasted looking for a misplaced part or tool.

Tool and Equipment Safety

Hand Tool Safety

Hand tools should always be kept clean and be used only for the purpose for which they were designed. Use the right tool for the right job. If you use the wrong tool, you will damage the part, the tool, or yourself. Oily hand tools can slip out of your hand while you are using them, causing broken fingers or at least cut or skinned knuckles. Your tools should also be inspected for cracks, broken parts, or other dangerous conditions before you use them. The following list includes examples of what to do when using certain tools:

1. Always pull on a wrench handle.
2. Always use the correct size of wrench.
3. Use a box-end or socket wrench whenever possible.
4. Use an adjustable wrench only when it is absolutely necessary; pull the wrench so that the force of the pull is on the non-adjustable jaw.
5. When using an air impact wrench, always use impact sockets.
6. Never use wrenches or sockets that have cracks or breaks.
7. Never use a wrench or pliers as a hammer.
8. Never use pliers to loosen or tighten a nut; use the correct wrench.
9. Always be sure to strike an object with the full face of the hammerhead.
10. Always wear safety glasses when using a hammer and/or chisel.
11. Never strike two hammer heads together.
12. Never use screwdrivers as chisels.

If the tool that you are using is pointed or sharp, always aim it away from yourself. Knives, chisels, and scrapers must be used in a motion that will keep the point or blade moving away from your body. Always hand a pointed or sharp tool to someone else with the handle toward the person you are handing the tool to.

Power Tool Safety

Power tools are operated by an outside source of power, such as electricity, compressed air, or hydraulic pressure. Safety around power tools is very important. Serious injury can result from carelessness. Always wear safety glasses when using power tools.

If the tool is electrically powered, make sure it is properly grounded. Check the wiring for cracks in the insulation, as well as for bare wires, before using it. Also, when using electrical power tools, never stand on a wet or damp floor. Disconnect the power source before doing any work on the machine or tool. Before plugging in any electric tool, make sure its switch is in the OFF position. When you are done using the tool, turn it off and unplug it. Never leave a running power tool unattended.

When using power equipment on a small part, never hold the part in your hand. Always mount the part in a bench vise or use vise grip pliers. Never try to use a machine or tool beyond its stated capacity or for operations requiring more than the rated power of the tool.

When working with larger power tools, such as bench or floor grinding wheels, check the machine and the grinding wheels for signs of damage before using them. If the wheels are damaged, they should be replaced and not used. Check the speed rating of the wheel and make sure it matches the speed of the machine. Never spin a grinding wheel at a speed higher than it is rated for. Be sure to place all safety guards in position. A safety guard is a protective cover over a moving part. Although the safety guards are designed to prevent injury, you should still wear safety glasses and/or a face shield while using the machine. Make sure there are no people or parts around the machine before starting it. Keep your hands and clothing away from the moving parts. Maintain a balanced stance while using the machine.

Classroom Manual
Chapter 1, page 5

Compressed Air Equipment Safety

Compressed air is used to inflate tires, apply paint, and drive tools. Compressed air can be dangerous when it is not used properly. When using compressed air, safety glasses and/or a face shield should be worn. Particles of dirt and pieces of metal, blown by the high-pressure air, can penetrate your skin or get into your eyes.

Before using a compressed air tool, check all hose connections and for signs of bulging along the air hose. Pneumatic tools must always be operated at the pressure recommended by the manufacturer.

Always hold an air nozzle or air control device securely when starting or shutting off the compressed air. A loose nozzle can whip suddenly and cause serious injury. Never point an air nozzle at anyone. Never use compressed air to blow dirt from your clothes or hair. Never use compressed air to clean the floor or workbench. Also, never spin bearings with compressed air; one of the steel balls or rollers might fly out and cause serious injury.

Lift Safety

Always be careful when raising a vehicle on a lift or a hoist. Adapters and hoist plates must be positioned correctly on twin post and rail-type lifts to prevent damage to the underbody of the vehicle. There are specific lift points. These points allow the weight of the vehicle to be evenly supported by the adapters or hoist plates. The correct lift points can be found in the vehicle's service manual. Figure 1-22 shows typical locations for unibody and frame cars. These diagrams are for illustration only. Always follow the manufacturer's instructions. Before operating any lift or hoist, carefully read the operating manual and follow the operating instructions.

Once you feel the lift supports are properly positioned under the vehicle, raise the lift until the supports contact the vehicle. Then, check the supports to make sure they are in full contact with the vehicle. Shake the vehicle to make sure it is securely balanced on the lift, and then raise the lift to the desired working height.

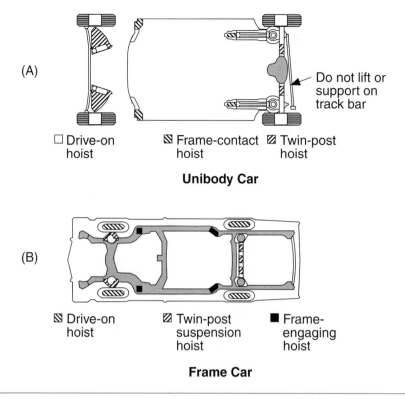

Figure 1-22 Typical lift points for (A) unibody and (B) frame/body vehicles.

The Automotive Lift Institute (ALI) is an association concerned with the design, construction, installation, operation, maintenance, and repair of automotive lifts. Their primary concern is safety. Every lift approved by ALI has the label shown in Figure 1-23. It is a good idea to read through the safety and maintenance tips included on this label before using a lift.

Transmission Jack Safety

Transmission jacks are designed to help you, not hurt you. Special care must be taken whenever using a transmission jack. It is important that the weight of the transmission is on the saddle of the jack and that the hold down chains or other transmission holding devices are secure before moving the transmission. Also you need to make sure another jack or support supports the weight of the engine.

Once the transmission is on the jack's saddle, lower the saddle to the lowest point you can. Then move the transmission from under the vehicle. If you fail to lower the jack before you move it, the jack and transmission will be very unstable and it will be unsafe to roll it.

Jack and Safety Stand Safety

A vehicle can be raised off the ground by a hydraulic jack. A handle on the jack is moved up and down to raise part of a vehicle and a valve is turned to release the hydraulic pressure in the jack to lower the part. At the end of the jack is a lifting pad. The pad must be positioned under an area of the vehicle's frame or at one of the manufacturer's recommended lift points. Never place the pad under the floor pan or under steering and suspension components; these are easily damaged by the weight of the vehicle. Always make sure the transmission is in neutral and position the jack so the wheels of the vehicle can roll as the vehicle is being raised.

AUTOMOTIVE LIFT

SAFETY TIPS

Post these safety tips where they will be a constant reminder to your lift operator. For information specific to the lift, always refer to the lift manufacturer's manual.

1. Inspect your lift daily. Never operate if it malfunctions or if it has broken or damaged parts. Repairs should be made with original equipment parts.

2. Operating controls are designed to close when released. Do not block open or override them.

3. Never overload your lift. Manufacturer's rated capacity is shown on nameplate affixed to the lift.

4. Positioning of vehicle and operation of the lift should be done only by trained and authorized personnel.

5. Never raise vehicle with anyone inside it. Customers or by-standers should not be in the lift area during operation.

6. Always keep lift area free of obstructions, grease, oil, trash and other debris.

7. Before driving vehicle over lift, position arms and supports to provide unobstructed clearance. Do not hit or run over lift arms, adapters, or axle supports. This could damage lift or vehicle.

8. Load vehicle on lift carefully. Position lift supports to contact at the vehicle manufacturer's recommended lifting points. Raise lift until supports contact vehicle. Check supports for secure contact with vehicle. Raise lift to desired working height. CAUTION: If you are working under vehicle, lift should be raised high enough for locking device to be engaged.

9. Note that with some vehicles, the removal (or installation) of components may cause a critical shift in the center of gravity, and result in raised vehicle instability. Refer to the vehicle manufacturer's service manual for recommended procedures when vehicle components are removed.

10. Before lowering lift, be sure tool trays, stands, etc. are removed from under vehicle. Release locking devices before attempting to lower lift.

11. Before removing vehicle from lift area, position lift arms and supports to provide an unobstructed exit (See Item #7).

These "Safety Tips," along with "Lifting it Right," a general lift safety manual, are presented as an industry service by the Automotive Lift Institute. For more information on this material, write to: ALI, P.O. Box 1519, New York, NY 10101.

Look For This Label on all Automotive Service Lifts.

AUTOMOTIVE LIFT INSTITUTE, INC.

Figure 1-23 Automotive lift safety tips. (Courtesy of Automotive Lift Institute)

⚠️ **WARNING:** Never use a lift or jack to move something heavier than it is designed for. Always check the rating before using a lift or jack. If a jack is rated for 2 tons, do not attempt to use it for a job weighing 5 tons. To do so would be dangerous for you and the vehicle.

Safety stands are supports of different heights that sit on the floor. They are placed under a sturdy chassis member, such as the frame or axle housing, to support the vehicle. Once the safety stands are in position, the hydraulic pressure in the jack should be slowly released until the weight of the vehicle is on the stands. Like jacks, safety stands also have a capacity rating. Always use a safety stand that has the correct rating.

Never move under a vehicle when it is supported only by a hydraulic jack; rest the vehicle on the safety stands before moving under the vehicle. The jack should be removed after the safety stands are set in place. This eliminates a possible hazard, such as a jack handle sticking out into a walkway. A jack handle that is bumped or kicked can cause a tripping accident or cause the vehicle to fall.

Chain Hoist and Crane Safety

Heavy parts of automobiles, such as engines, are removed by using chain hoists or cranes. Another term used for a chain hoist is chain fall. To prevent serious injury, chain hoists and cranes must be properly attached to the parts being lifted (Figure 1-24). Always use bolts with enough strength to support the object being lifted. After you have attached the lifting chain or cable to the part that is being removed, have your instructor check it. Place the chain hoist or crane directly over the assembly. Then, attach the chain or cable to the hoist.

Cleaning Equipment Safety

Parts cleaning is a necessary step in most repair procedures. Cleaning automotive parts can be divided into four basic categories: chemical cleaning, thermal cleaning, abrasive cleaning, and steam cleaning. Regardless of what method you use, you should always follow the precautions given by the manufacturer and you should always wear the recommended protective gear.

Battery Safety

When possible, you should disconnect the battery of a car before you disconnect any electrical wire or component. This prevents the possibility of a fire or electrical shock. It also eliminates the

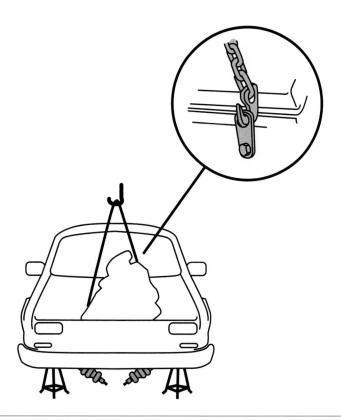

Figure 1-24 When using a chain hoist to pull an engine, the attachments to the engine should be secure and strong.

possibility of an accidental short, which can ruin the car's electrical system. To properly disconnect the battery, you should disconnect the negative or ground cable first, then disconnect the positive cable. Since electrical circuits require a ground to be complete, by removing the ground cable you eliminate the possibility of a circuit accidentally becoming completed. When reconnecting the battery, connect the positive cable first, then the negative. Remember that some late-model transmissions may not function normally and may need to relearn the vehicle and driver.

WARNING: Never smoke or cause sparks around a battery. The slightest increase in heat can cause the battery to explode.

CUSTOMER CARE: To keep your customer happy always record the stations set on the radio before disconnecting the battery. The absence of power will remove the stations from the radio's memory. Also be sure to reset those stations and the clock after the battery has been reconnected. It is very wise to use a memory saver, which is a battery pack, so that you will not need to reset everything.

Vehicle Operation

When the customer brings a vehicle in for service, certain driving rules should be followed to ensure your safety and the safety of those working around you. For example, before moving a car into the shop, buckle your safety belt. Make sure no one is near, the way is clear, and there are no tools or parts under the car before you start the engine. Check the brakes before putting the vehicle in gear. Then, drive slowly and carefully in and around the shop.

When road testing the car, obey all traffic laws. Drive only as far as is necessary to check the automobile and verify the customer's complaint. Never make excessively quick starts, turn corners too quickly, or drive faster than conditions allow.

If the engine must be running while you are working on the car, block the wheels to prevent the car from moving. Place the transmission into park for automatic transmissions or in neutral for manual transmissions. Set the parking (emergency) brake. Never stand directly in front of or behind a running vehicle.

Run the engine only in a well-ventilated area to avoid the danger of poisonous **carbon monoxide (CO)** in the engine exhaust. CO is an odorless but deadly gas. Most shops have an exhaust ventilation system, and you should always use it. Connect the hose from the vehicle's tailpipe to the intake for the vent system. Make sure the vent system is turned on before running the engine. If the work area does not have an exhaust venting system, use a hose to direct the exhaust out of the building.

Exhaust contains an odorless, colorless, and deadly gas: **carbon monoxide**. This poisonous gas gives very little warning to the victim and can kill in just a few minutes.

Lifting and Carrying

When lifting a heavy object like a transmission, use a hoist or have someone else help you. If you must work alone, *always* lift heavy objects with your legs, not your back. Bend down with your legs, not your back, and securely hold the object you are lifting; then stand up keeping the object close to you (Figure 1-25). Trying to "muscle" something with your arms or back can result in severe damage to your back. It might even end your career and limit what you do the rest of your life!

You should also use back-protection devices when you are lifting a heavy object. Always lift and work within your ability and ask others to help when you are not sure whether or not you can handle the size or weight of an object. Even small, compact parts can be surprisingly heavy or unbalanced. Think about how you are going to lift something before beginning.

Figure 1-25 When lifting a heavy object, always use your legs and keep the object close to your body.

Fire Hazards and Prevention

In case of a fire you should know the location of the fire extinguishers and fire alarms in the shop and should also know how to use them. You should also be aware of the different types of fires and the fire extinguishers used to put out those types of fires.

Basically, there are four types of fires. **Class A fires** are those in which wood, paper, and other ordinary materials are burning. **Class B fires** are those involving flammable liquids, such as gasoline, diesel fuel, paint, grease, oil, and other similar liquids. **Class C fires** are electrical fires. **Class D fires** are unique types of fires, for the material burning is a metal. An example of this is a burning "mag" wheel; the magnesium used in the construction of the wheel is a flammable metal and will burn brightly when subjected to high heat.

Fires are classified by what is burning: wood and paper = **Class A**, liquids = **Class B**, electrical fires = **Class C**, and metals = **Class D**.

Fire Extinguishers

Using the wrong type of extinguisher may cause the fire to grow, instead being put out. All extinguishers are marked with a symbol or letter to signify what class of fire they were intended for (Table 1-1). Typically, fire extinguishers that use pressurized water should only be used on Class A fires. Soda acid extinguishers are also used to put out Class A fires. Foam type extinguishers are used on Class A and B fires; the foam should be sprayed directly on the fire. Those extinguishers filled with a dry chemical should be aimed at the base of a Class B or Class C fire, so as to suffocate the fire. Some extinguishers that use a multipurpose dry chemical are also suitable for three classes of fires—Classes A, B, and C. Carbon dioxide is also used in some fire extinguishers that are designed to put out Class B and Class C type fires. Halon gas extinguishers can also be used on Class B and Class C fires. Class D fires require the use of special dry powder type extinguisher. Some extinguishers are designed to fight more than one class of fire and will have multiple ratings, such as A and B, B and C, or A, B, C and D. You should know exactly the location of each fire extinguisher in the shop and what the rating of each extinguisher is before you need one. You need to be able to react immediately when a fire starts.

Table 1-1 GUIDE TO EXTINGUISHER SELECTION

	Class of Fire	Typical Fuel Involved	Type of Extinguisher
Class A Fires (green)	**For Ordinary Combustibles** Put out a Class A fire by lowering its temperature or by coating the burning combustibles.	Wood Paper Cloth Rubber Plastics Rubbish Upholstery	Water[*1] Foam[*] Multipurpose dry chemical[4]
Class B Fires (red)	**For Flammable Liquids** Put out a Class B fire by smoldering it. Use an extinguisher that gives a blanketed flame-interrupting effect; cover whole flaming liquid surface.	Gasoline Oil Grease Paint Lighter fluid	Foam[*] Carbon dioxide[5] Halogenated agent[6] Standard dry chemical[2] Purple K dry chemical[3] Multipurpose dry chemical[4]
Class C Fires (blue)	**For Electrical Equipment** Put out a Class C fire by shutting off power as quickly as possible and by always using a nonconducting extinguishing agent to prevent electric shock.	Motors Appliances Wiring Fuse boxes Switchboards	Carbon dioxide[5] Halogenated agent[6] Standard dry chemical[2] Purple K dry chemical[3] Multipurpose dry chemical[4]
Class D Fires (yellow)	**For Combustible Metals** Put out a Class D fire of metal chips, turnings, or shavings by smothering or coating with a specially designed extinguishing agent.	Aluminum Magnesium Potassium Sodium Titanium Zirconium	Dry powder extinguishers and agents only

*Cartridge-operated water, foam, and soda-acid types of extinguishers are no longer manufactured. Those extinguishers should be removed from service when they become due for their next hydrostatic pressure test.

Notes:

(1) Freezes in low temperatures unless treated with antifreeze solution, usually weighs over 20 pounds, and is heavier than any other extinguisher mentioned.

(2) Also called ordinary or regular dry chemical. (solution bicarbonate)

(3) Has the greatest initial fire-stopping power of the extinguishers mentioned for Class B fires. Be sure to clean residue immediately after using the extinguisher so sprayed surfaces will not be damaged. (potassium bicarbonate)

(4) The only extinguisher that fights Class A, B, and C fires. However, it should not be used on fires in liquefied fat or oil of appreciable depth. Be sure to clean residue immediately after using the extinguisher so sprayed surfaces will not be damaged. (ammonium phosphates)

(5) Use with caution when in unventilated, confined spaces.

(6) May cause injury to the operator if the operator inhales extinguisher agent (a gas) or the gases produced when the agent is applied to a fire.

Using a Fire Extinguisher

Remember, during a fire never open doors or windows unless it is absolutely necessary; the extra draft will only make the fire worse. Make sure the fire department and others in the shop are contacted before or during your attempt to extinguish a fire. To extinguish a fire (see Photo Sequence 1), stand about 8 feet from the fire. Hold the extinguisher firmly in an upright position. Aim the

Photo Sequence 1
Using a Dry Chemical Fire Extinguisher

P1-1 Multipurpose dry chemical fire extinguisher.

P1-2 Hold the fire extinguisher in an upright position.

P1-3 Pull the safety pin from the handle.

P1-4 Stand 8 feet from the fire. Do not go any close to the fire.

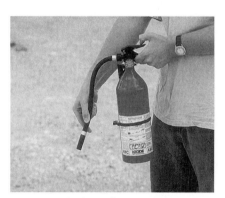

P1-5 Free the hose from its retainer and aim it at the base of the fire.

P1-6 Squeeze the lever while sweeping the hose from side to side. Keep the hose aimed at the *base* of the fire.

nozzle at the base of the fire and with a side-to-side motion sweep the entire width of the fire. Stay low to avoid inhaling the smoke. If it gets too hot or too smoky, get out. Never go back into a burning building for anything. As a guide to how to use an extinguisher, remember the word "PASS."

Pull the pin from the handle of the extinguisher.

Aim the extinguisher's nozzle at the base of the fire.

Squeeze the handle.

Sweep the entire width of the fire with the contents of the extinguisher.

If there is not a fire extinguisher handy, a blanket or fender cover may be used to smother the flames. You must be careful when doing this because the heat of the fire may burn you and the blanket. If the fire is too great to smother, move everyone away from the fire and call the local fire department. A simple under-the-hood fire can cause the total destruction of the car and the building and even take lives. You must be able to respond quickly and precisely to avoid a disaster.

Work Area Safety

Your work area should be kept clean and safe. The floor and bench tops should be kept clean, dry, and orderly. Any oil, coolant, or grease on the floor can make it slippery. Slips can result in serious injuries. To clean up oil, use commercial oil absorbent. Keep all water off the floor. Water can cause smooth floors to become slippery, and also poses a danger because electricity flows well through water. Aisles and walkways should be kept clean and open wide enough so that movement through them is easy. Make sure the work areas around machines are large enough to allow the machines to be operated safely.

Make sure all drain covers are snugly in place. Open drains or covers that are not flush to the floor can cause toe, ankle, and leg injuries.

Keep an up-to-date list of emergency telephone numbers clearly posted next to the telephone. These numbers should include a doctor, hospital, and fire and police departments. Also, the work area should have a first-aid kit for treating minor injuries and eye-flushing kits readily available. You should know where these items are kept.

Gasoline is a highly flammable volatile liquid. Something that is flammable catches fire and burns easily. A volatile liquid is one that vaporizes very quickly. Flammable volatile liquids are potential firebombs. Always keep gasoline or diesel fuel in an approved safety can (Figure 1-26), and never use gasoline to clean your hands or tools.

Handle all solvents (or any liquids) with care to avoid spillage. Keep all solvent containers closed, except when pouring. Proper ventilation is very important in areas where volatile solvents and chemicals are used. Solvents and other combustible materials must be stored in approved and designated storage cabinets or rooms. Storage rooms should have adequate ventilation.

Be extra careful when transferring flammable materials from bulk storage. Static electricity can build up enough to create a spark that could cause an explosion. Discard or clean all empty solvent containers. Solvent fumes in the bottom of these containers are very flammable. Never light matches or smoke near flammable solvents and chemicals, including battery acids.

Figure 1-26 Flammable liquids should be stored in safety-approved containers. (Courtesy of Gutman Advertising Agency)

Oily rags should also be stored in an approved metal container. When these oily, greasy, or paint-soaked rags are left lying about or are not stored properly, they can cause spontaneous combustion. Spontaneous combustion results in a fire that starts by itself, without the use of a match or some other external source of ignition.

Accidents

Make sure you are aware of the location and contents of the shop's first aid kit. There should be an eye wash station in the shop so that you can rinse your eyes thoroughly should you get acid or some other irritant into them. Find out if there is a resident nurse in the shop or at the school, and find out where the nurse's office is. If there are specific first aid rules in your school or shop, make sure you are aware of them and follow them. Some first aid rules apply to all circumstances and are normally included in everyone's rules. If someone is overcome by carbon monoxide, get him or her fresh air immediately. Burns should be cooled immediately by rinsing them with water. Whenever there is severe bleeding from a wound, try to stop the bleeding by applying pressure with clean gauze on or around the wound, and get medical help. Never move someone who may have broken bones unless the person's life is otherwise endangered. Moving that person may cause additional injury. Call for medical assistance.

Classroom Manual
Chapter 1, page 6

Hazardous Materials

A typical shop contains many potential health hazards for those working in it. These hazards can cause injury, sickness, health impairments, discomfort, and even death. These hazards can be classified as:

❏ Chemical hazards—caused by high concentrations of vapors, gases, or solids in the form of dust.
❏ Hazardous wastes—those substances that are the result of a service being performed.
❏ Physical hazards—excessive noise, vibration, pressures, and temperatures.
❏ Ergonomic hazards—conditions that impede normal and/or proper body position and motion.

There are many government agencies charged with ensuring safe work environments for all workers. These include the **Occupational Safety and Health Administration (OSHA)**, Mine Safety and Health Administration (MSHA), and National Institute for Occupational Safety and Health (NIOSH). These, in addition to state and local governments, have instituted regulations that must be understood and followed. Everyone in a shop has the responsibility for adhering to these regulations.

OSHA

In 1970, OSHA was formed by the federal government to "assure safe and healthful working conditions for working men and women; by authorizing enforcement of the standards developed under the Act; by assisting and encouraging the States in their efforts to assure safe and healthful working conditions by providing research, information, education, and training in the field of occupational safety and health."

Safety standards have been established that are to be applied consistently across the country. It is the employer's responsibility to provide a place of employment that is free from all recognized hazards and that will be inspected by government agents knowledgeable in the law of working conditions. OSHA controls all safety and health issues of the automotive industry.

The **Occupational Safety and Health Administration (OSHA)** was formed by the federal government in 1970 to assure safe and healthful working conditions for working men and women.

Right-To-Know Laws

According to the **right-to-know laws**, employees have the right to know what hazardous materials they will be working with.

Classroom Manual
Chapter 1, page 7

An important part of a safe work environment is the employee's knowledge of potential hazards. Every employee in the shop is protected by **right-to-know laws** concerning all chemicals. These laws started with OSHA's Hazard Communication Standard, which was published in 1983. This document was originally intended for chemical companies and manufacturers that require employees to handle potentially hazardous materials in the workshop. Since then, the majority of states have enacted their own right-to-know laws. The federal courts have decided that these regulations should apply to all companies, including auto repair shops.

The general intent of right-to-know laws is that employers should provide their employees with a safe working place as it relates to hazardous materials. Specifically, there are three areas of employer responsibility.

Primarily, all employees must be trained about their rights under the legislation, the nature of the hazardous chemicals in their workplace, and the contents of the labels on the chemicals. All of the information about each chemical must be posted on material safety data sheets (MSDS) and must be accessible. The manufacturer of the chemical must give these sheets to its customers, if they are requested to do so. They detail the chemical composition and precautionary information for all products that can present a health or safety hazard. The Canadian equivalents to MSDS are called workplace hazardous materials information systems (WHMIS).

An MSDS must include the following information about the product:

- ❏ The trade and chemical name of the product.
- ❏ The manufacturer of the product.
- ❏ All of the ingredients of the product.
- ❏ Health hazards such as headaches, skin rashes, nausea, and dizziness.
- ❏ The product's physical description. This information may include the product's color, odor, permissible exposure limit (PEL), threshold limit value (TLV), specific gravity, boiling point, freezing point, evaporation data, and volatility rating.
- ❏ The product's explosion and fire data, such as the flash point.
- ❏ The reactivity and stability data.
- ❏ The product's weight compared to air.
- ❏ Protection data including first aid and proper handling.

Employees must become familiar with the general uses, protective equipment, accident or spill procedures, and any other information regarding the safe handling of the hazardous material. This training must be given to employees annually and provided to new employees as part of their job orientation.

Secondly, each hazardous material must be properly labeled, indicating what health, fire, or reactivity hazard it poses and what protective equipment is necessary when handling each chemical. The manufacturer of hazardous waste materials must provide all warnings and precautionary information, which must be read and understood by the user before application. Attention to all label precautions is essential for the proper use of the chemical and for prevention of hazardous conditions. A list of all hazardous materials used in the shop must be posted for the employees to see.

Finally, shops must maintain documentation on the hazardous chemicals in the workplace, proof of training programs, records of accidents or spill incidents, satisfaction of employee requests for specific chemical information via the MSDS, and a general right-to-know compliance procedure manual utilized within the shop.

⚠️ **WARNING:** When handling any hazardous material, be sure to wear the proper safety equipment that is covered under the right-to-know law. Follow all required procedures correctly. This includes the use of approved respirator equipment.

As mentioned before, some of the materials used in auto repair shops can be dangerous. The solvents and other chemical products used in an auto shop carry warnings and caution information that must be read and understood by all who use them. Some of the common hazardous materials that automotive technicians use include cleaning chemicals, fuels, paints and thinners, battery electrolyte, refrigerants, and engine coolant.

Hazardous Waste

In the United States, OSHA regulates the use of many hazardous materials. The Environmental Protection Agency (EPA) regulates the disposal of **hazardous waste**. A summary of these regulations follows:

❑ All businesses that generate hazardous waste must develop a hazardous waste policy.
❑ Each hazardous waste generator must have an EPA identification number.
❑ When waste is transported for disposal, a licensed waste hauler must be used to transport and dispose of the waste. A copy of a written manifest (an EPA form) must be kept by the shop.
❑ Hazardous material manufacturers must publish information related to hazardous materials they produce and/or distribute.
❑ All hazardous materials must be labeled. The label must include the product's trade name, the manufacturer of the product, the chemical name of the product, and safety information about the product.
❑ A MSDS must be available for all hazardous materials.

Regulations on hazardous waste handling and generation have led to the development of equipment that is now commonly found in shops. Examples of these are thermal cleaning units, close-loop steam cleaners, waste oil furnaces, oil filter crushers, refrigerant recycling machines, engine coolant recycling machines, and highly absorbent cloths.

Many repair and service procedures generate what are known as hazardous wastes. Dirty solvents and cleaners are good examples of hazardous wastes. Not only are the waste materials of concern but also whether the shop uses the different chemicals. Something is classified as a hazardous waste by the Environmental Protection Agency if it is on the EPA list of known harmful materials or has one or more of the following characteristics.

Ignitability. If it is a liquid with a flash point below 140°F or a solid that can spontaneously ignite.

Corrosivity. If the chemical dissolves metals and other materials or burns the skin, it is corrosive.

Reactivity. Any material that reacts violently with water or other materials or releases cyanide gas, hydrogen sulfide gas, or similar gases when exposed to low pH acid solutions. This also includes material that generates toxic mists, fumes, vapors, and flammable gases.

EP toxicity. Materials that leach one or more of eight heavy metals in concentrations greater than 100 times primary drinking water standard concentrations.

A complete EPA list of hazardous wastes can be found in the Code of Federal Regulations. It should be noted that no material is considered hazardous waste until the shop is finished using it and is ready to dispose of it.

> **CAUTION:** The shop is ultimately responsible for the safe disposal of hazardous wastes, even after the waste leaves the shop. Only licensed waste removal companies should be used to dispose of the waste. Make sure you know what the company is planning to do with the waste. Make sure you have a written contract stating what is supposed to happen with the waste. Leave nothing to chance. In the event of an emergency hazardous waste spill, contact the National Response Center (1-800-424-8802) immediately. Failure to do so can result in a $10,000 fine, a year in jail, or both.

A **hazardous waste** is any used material or any by-product that can be classified as potentially hazardous to one's health and/or the environment.

OSHA and the EPA have other strict rules and regulations that help to promote safety in the auto shop. These are described throughout this text whenever they are applicable. Maintaining a vehicle involves handling and managing a wide variety of materials and wastes. Some of these wastes can be toxic to fish, wildlife, and humans when improperly managed. No matter how much waste is produced, it is to the shop's legal and financial advantage to manage the wastes properly and, even more importantly, to prevent pollution.

Handling Shop Wastes

Oil

Recycle oil. Set up equipment, such as a drip table or screen table with a used oil collection bucket, to collect oils dripping off parts. Place drip pans underneath vehicles that are leaking fluids onto the storage area. Do not mix other wastes with used oil, except as allowed by your recycler. Used oil generated by a shop (and/or oil received from household "do-it-yourself" generators) may be burned on site in a commercial space heater. Also, used oil may be burned for energy recovery. Contact state and local authorities to determine requirements and to obtain necessary permits.

Oil Filters

Drain for at least 24 hours, crush, and recycle used oil filters.

Transmission Fluid

Like other fluids, automatic transmission fluid should be recycled. Normally, used transmission fluid is stored in a designated container, although some recyclers may allow the fluid to be poured in with used motor oil. Always check with the recycler prior to mixing transmission fluid with motor oil.

Batteries

Recycle batteries by sending them to a reclaimer or back to the distributor. Keeping shipping receipts can demonstrate that you have done the recycling. Store batteries in watertight, acid-resistant containers. Inspect batteries for cracks and leaks when they come in. Treat a dropped battery as if it were cracked. Acid residue is hazardous because it is corrosive and may contain lead and other toxics. Neutralize spilled acid by using baking soda or lime, and dispose of it as hazardous material.

Metal Residue from Machining

Collect metal filings when machining metal parts. Keep them separate and recycle them if possible. Prevent metal filings from falling into a storm sewer drain.

Refrigerants

Recover and/or recycle refrigerants during the service and disposal of motor vehicle air conditioners and refrigeration equipment. It is not allowable to knowingly vent refrigerants to the atmosphere. Recovery and/or recycling during service must be performed by an EPA-certified technician using certified equipment and following specified procedures.

Solvents

Replace hazardous chemicals with less toxic alternatives that perform as well. For example, substitute water-based cleaning solvents for petroleum-based solvent degreasers. To reduce the amount of solvent used when cleaning parts, use a two-stage process: dirty solvent followed by fresh solvent. Hire a hazardous waste management service to clean and recycle solvents. (Some spent solvents must be disposed of as hazardous waste, unless recycled properly). Store solvents in closed containers to prevent evaporation. Evaporation of solvents contributes to ozone depletion and smog formation. In addition, the residue from evaporation must be treated as a hazardous waste. Properly label spent solvents; store them on drip pans or in diked areas and only with compatible materials.

Containers
Cap, label, cover, and properly store above-ground outdoor liquid containers and small tanks within a diked area and on a paved impermeable surface to prevent spills from running into surface or ground water.

Other Solids
Store materials such as scrap metal, old machine parts, and worn tires under a roof or tarpaulin to protect them from the elements and to prevent the possibility of creating contaminated runoff. Consider recycling tires by retreading them.

Liquid Recycling
Collect and recycle coolants from radiators. Store transmission fluids, brake fluids, and solvents containing chlorinated hydrocarbons separately, and recycle or dispose of them properly.

Shop Towels/Rags
Keep waste towels in a closed container marked "contaminated shop towels only." To reduce costs and liabilities associated with disposal of used towels, which can be classified as hazardous wastes, investigate using a laundry service that is able to treat the wastewater generated from cleaning the towels.

Hazardous Waste Disposal

Hazardous wastes must be properly stored and disposed of. It is the shop's responsibility to hire a reputable, financially stable, and state-approved hauler, who will dispose of the shop wastes legally. Select a licensed hazardous waste hauler after seeking recommendations and reviewing the firm's permits and authorizations. If hazardous waste is dumped illegally, your shop may be held responsible.

Always keep hazardous waste separate, properly labeled, and sealed in the recommended containers. The storage area should be covered and may need to be fenced and locked if vandalism could be a problem.

Summary

- ❏ The three common types of wrenches in a technician's tool set are the open-end, box-end, and combination wrenches.
- ❏ Sockets are available in different drives and sizes. Special type sockets are also available, such as deep-well and impact sockets.
- ❏ Metric and SAE size wrenches are not interchangeable. An auto technician should have a variety of both types.
- ❏ Basic tool sets should include a variety of standard and Phillips screwdrivers.
- ❏ The hand tap is used for hand cutting internal threads and for cleaning and restoring previously cut threads. Hand-threading dies cut external threads and fit into holders called die stocks.

- ❏ Torque wrenches are used to tighten fasteners to a specified torque.

- ❏ Always use the correct tool, in the correct way, for the job.

- ❏ Many special tools are available for special purposes. These tools are not normally part of a basic tool set but are purchased on an as-needed basis.

- ❏ It is everyone's job to periodically check for safety hazards in a shop.

- ❏ Heavy objects should be lifted with your legs not your back.

- ❏ There are two areas of housekeeping you are responsible for: your work area and the rest of the shop.

- ❏ Special care should be taken whenever using power tools, such as impact wrenches, gear pullers, and jacks.

- ❏ Dressing safely for work is very important. This includes snug-fitting clothing, eye and ear protection, protective gloves, strong shoes, and caps to cover long hair.

- ❏ When shop noise exceeds safe levels, protect your ears by wearing earplugs or earmuffs.

- ❏ Attention to safety while using any tool is a must, particularly when using power tools. Before plugging in a power tool, make sure the power switch is off. Disconnect the power before servicing the tool.

- ❏ Carelessness or mishandling of power tools can cause serious injury. Safety measures are needed when working with tools such as impact and air ratchet wrenches, blowguns, bench grinders, lifts, hoists, and hydraulic presses.

- ❏ Always observe all relevant safety rules when operating a vehicle lift or hoist. Jacks, jack stands, chain hoists, and cranes can also cause injury if not operated safely.

- ❏ Make sure the transmission is secure on the transmission jack before lowering it and lower the jack before moving the jack.

- ❏ Use care whenever it is necessary to remove a vehicle in the shop. Carelessness and playing around can lead to a damaged vehicle and serious injury.

- ❏ Carbon monoxide gas is a poisonous gas present in engine exhaust fumes. It must be properly vented from the shop using tailpipe hoses or other reliable methods.

- ❏ Adequate ventilation is also necessary when working with any volatile solvent or material.

- ❏ Gasoline and diesel fuel are highly flammable and should be kept in approved safety cans.

- ❏ Never light matches near any combustible materials.

- ❏ It is important to know when to use each of the various types of fire extinguishers. When fighting a fire, aim the nozzle at the base of the fire and use a side-to-side sweeping motion.

- ❏ Material safety data sheets contain important chemical information and must be furnished to all employees annually. New employees should be given the sheets as part of their job orientation.

- ❏ All asbestos waste should be disposed of according to OSHA and EPA regulations.

- ❏ Right-to-know laws were introduced in 1983 and are designed to protect employees who must handle hazardous materials and wastes on the job.

- ❏ The EPA lists many materials as hazardous, provided they have one or more of the following characteristics: ignitability, corrosivity, reactivity, and EP toxicity.

Terms to Know

Allen wrench	Class B fire	Open-end wrench
Blowgun	Class C fire	Phillips screwdriver
Box-end wrench	Class D fire	Press-fit
Breakout box	Hazardous waste	Right-to-know laws
Carbon monoxide (CO)	Occupational Safety and Health Association (OSHA)	Scan tool
Class A fire		Torque wrench

ASE-Style Review Questions

1. *Technician A* removes the jack from under a vehicle after the safety stands are in place.
   Technician B shoves her creeper back under the car when she is not using it.
   Who is correct?
   A. A only **C.** Both A and B
   B. B only **D.** Neither A nor B

2. *Technician A* always pushes on the handle of a wrench when tightening a bolt.
   Technician B never uses an incorrectly sized wrench or socket on a bolt or nut.
   Who is correct?
   A. A only **C.** Both A and B
   B. B only **D.** Neither A nor B

3. While discussing hazardous wastes:
   Technician A says the shop is responsible for the proper removal of the waste.
   Technician B says a technician is required by law to dispose of wastes in the way provided and recommended by the shop.
   Who is correct?
   A. A only **C.** Both A and B
   B. B only **D.** Neither A nor B

4. While discussing why exhaust fumes should be vented outdoors or drawn into a ventilation/filtration system:
   Technician A says exhaust gases contain amounts of carbon monoxide.
   Technician B says carbon dioxide is an odorless, colorless, and deadly gas.
   Who is correct?
   A. A only **C.** Both A and B
   B. B only **D.** Neither A nor B

5. While using a transmission jack:
   Technician A brings the saddle of the jack under the transmission before loosening the transmission mount.
   Technician B wraps safety chains around the transmission and securely fastens the chains to the jack.
   Who is correct?
   A. A only **C.** Both A and B
   B. B only **D.** Neither A nor B

6. While discussing the use of an air impact wrench;
   Technician A says that impact sockets can be used with an air impact wrench.
   Technician B says air impact wrenches should not be used to tighten critical bolts.
   Who is correct?
   A. A only **C.** Both A and B
   B. B only **D.** Neither A nor B

7. While discussing the car's electrical system:
   Technician A says you should always disconnect the negative or ground battery cable first, then disconnect the positive cable.
   Technician B says you should always connect the positive cable first, then the negative.
   Who is correct?
   A. A only **C.** Both A and B
   B. B only **D.** Neither A nor B

8. *Technician A* says a tap cuts external threads.
   Technician B says a die cuts internal threads.
   Who is correct?
   A. A only **C.** Both A and B
   B. B only **D.** Neither A nor B

9. *Technician A* says hydraulic pressure gauges should be part of an automatic transmission technician's tool set.

Technician B says a scan tool should be part of an automatic transmission technician's tool set.

Who is correct?

A. A only

B. B only

C. Both A and B

D. Neither A nor B

10. While discussing the purpose of a torque wrench: *Technician A* says they are used to tighten fasteners to a specified torque.

Technician B says they are used for added leverage while loosening or tightening a bolt.

Who is correct?

A. A only

B. B only

C. Both A and B

D. Neither A nor B

Job Sheet 1

Name _____ Date _____

Shop Safety Survey

As a professional technician, safety should be one of your first concerns. This job sheet will increase your awareness of shop safety rules and equipment. As you survey your shop area and answer the questions, you will learn how to evaluate the safeness of your workplace.

Procedure

Your instructor will review your progress throughout this job sheet and should sign off on the sheet when you complete it.

1. Before you begin to evaluate your work area, evaluate yourself. Are you dressed to work safely? ☐ Yes ☐ No

If no, what is wrong?

_____

2. Are your safety glasses OSHA approved? ☐ Yes ☐ No

Do they have protection shields? ☐ Yes ☐ No

3. Walk around the shop and note any area that poses a potential safety hazard or is an area that you should be aware of.

_____

_____

Any true hazards should be brought to the attention of the instructor immediately.

4. Are there safety areas marked around grinders and other machinery? ☐ Yes ☐ No

5. What is the line air pressure in the shop? _____ psi

What should it be? _____ psi

6. Where are the tools stored in the shop?

_____

_____

7. If you could, how would you improve the tool storage area?

_____

_____

8. What types of hoists are used in the shop?

_____

_____

9. Ask your instructor to demonstrate the proper use of a hoist.

10. Where is the first aid kit(s) kept in the work area?

_____

_____

11. What is the shop's procedure for dealing with an accident?

_____

_____

12. Have your instructor supply you with a vehicle make, model, and year. Using the appropriate service manual, find the location of the correct lift points for that vehicle. On the rear of this sheet, draw a simple figure showing where these lift point are.

Instructor's Response _____

_____

_____

_____

Job Sheet 2

Name _____ Date _____

Working Safely Around Air Bags

Upon completion of this job sheet, you should be able to work safely around and with air bag systems.

Tools and Materials

A vehicle with air bags
Service manual for this vehicle
Component locator for this vehicle
Safety glasses
A DMM

Describe the vehicle being worked on.

Year _____ Make _____ VIN _____

Model _____

Procedure

1. Locate the information about the air bag system in the service manual. How are the critical parts of the system identified in the vehicle?

   _____

   _____

   _____

   _____

2. List the main components of the air bag system and describe their locations.

   _____

   _____

   _____

   _____

   _____

   _____

   _____

   _____

3. There are some very important guidelines to follow when working with and around air bag systems. These are listed below with some key words left out. Read through these guidelines and fill in the blanks with the correct words.

   a. Wear _____ _____ when servicing an air bag system and handling an air bag module.

b. Wait at least _____ minutes after disconnecting the battery before beginning any service. The reserve _____ module is capable of storing enough energy to deploy the air bag for up to _____ minutes after battery voltage is lost.

c. Always handle all _____ and other components with extreme care. Never strike or jar a sensor, especially when the battery is connected. This can cause deployment of the air bag.

d. Never carry an air bag by its _____ or _____ , and, when carrying it, always face the trim and air bag _____ from your body. When placing a module on a bench, always face the trim and air bag _____ .

e. Deployed air bags may have a powdery residue on them. _____ is produced by the deployment reaction and is converted to _____ when it comes in contact with the moisture in the atmosphere. Although it is unlikely that chemicals will still be on the bag, it is wise to wear _____ _____ and _____ when handling a deployed air bag. Immediately wash your hands after handling a deployed air bag.

f. A live air bag must be _____ before it is disposed of. A deployed air bag should be disposed of in a manner consistent with the _____ and manufacturer's procedures.

g. Never use a battery- or AC-powered _____ , _____ , or any other type of test equipment in the system unless the manufacturer specifically says to. Never probe with a _____ _____ for voltage.

Instructor's Response _____

_____

_____

_____

Job Sheet 3

Name _____ Date _____

Identification of Special Tools

Upon completion of this job sheet, you should be able to use the service manual to determine what special tools are required to properly service a particular transmission.

Tools and Materials

Appropriate service manual

Procedure

Your instructor will assign you a transmission type. You will use the service manual to identify and describe the special tools recommended for servicing this transmission. You will also identify any tools your shop has that can be used in place of the factory-specified tools.

1. The transmission assigned to you is:

   Transmission Model _____

   from: year _____ make _____ model _____

   The vehicle's VIN is: _____

2. The manual you will use to find the information:

   _____

3. List the special tools referenced in the overhaul section of the service manual for this transmission (include in your list what part or service each of the tools is used for):

   _____

   _____

   _____

   _____

   _____

   _____

   _____

   _____

4. List the special tools that are available for your use in the shop:

   _____

   _____

   _____

   _____

   _____

   _____

   _____

   _____

5. List the available tools that can be used in place of the specified tools:

_____

_____

_____

_____

_____

_____

Instructor's Response _____

_____

_____

_____

Typical Shop Procedures

Upon completion and review of this chapter, you should be able to:

- List the basic units of measure for length, volume, and mass in the two measuring systems.
- Convert measurements between the SAE and the metric systems of units.
- Describe the different types of fasteners used in the automotive industry.
- Identify the major measuring instruments and devices used by technicians.
- Explain what these instruments and devices measure and how to use them.
- Describe the proper procedure for measuring with a micrometer.

- Read a vernier scale.
- Describe the measurements normally taken by an automatic transmission technician.
- Using math and measuring instruments, determine the corrective action to take when end clearances are not within specifications.
- Describe the different sources for service information that are available to technicians.
- Describe the requirements for ASE certification as an automotive technician and a master auto technician.

Measuring Systems

Servicing modern automatic transmissions requires the use of various tools. Covered in this chapter are the commonly used measuring instruments. In order to use these tools, you must have an understanding of measuring systems. This knowledge also applies to the various hand and power tools discussed in the previous chapter. Two different systems of weights and measures are currently used in the United States—the **United States Customary System** (**USCS**), which is often referred to as the Society of Automotive Engineers (SAE) system and the **International** or metric **System** (**SI**). The original English settlers brought the USCS units of measurement with them. The United States is slowly changing over to the metric system; during the changeover, cars produced in the United States are being made with both English and metric fasteners and specifications. Most of the world outside the United States uses the metric system.

United States Customary System

In the United States Customary System, linear measurements are measured by inches, feet, and yards. The inch can be broken down into fractions, such as 1/64, 1/32, 1/16, 1/8, 1/4, and 1/2 of an inch (Figure 2-1). The inch is also commonly broken down into decimals. When an inch is divided into tenths, each part is a tenth of an inch (0.1 in.). Tenths of an inch can be further divided by ten into hundredths of an inch (0.01 in.). The division after hundredths is thousandths (0.001 in.); these are followed by ten–thousandths (0.0001 in.).

Metrics

In the metric system, the basic measurement unit of length is the **meter**. For exact measurements, the meter is divided into units of ten. The first division is called a decimeter (dm). The second division is the centimeter (cm) and the third and most commonly used division is the **millimeter** (**mm**). One dm is equal to 0.1 meters, 1 cm equals 0.01 m, and 1 mm equals 0.001 m.

Because some vehicles have metric fasteners, some have USCS fasteners, and others have both, automotive technicians must have both English and metric tools. Vehicle specifications are normally listed in both meters and inches; therefore, measuring tools such as micrometers and dial

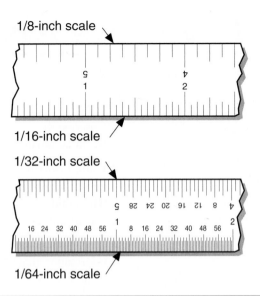

Figure 2-1 An inch is normally broken down into these fractional increments.

indicators are available in both measuring systems. The following are some common equivalents between the two systems.

Linear Measurements

    1 meter (m) = 39.37 inches (in.)
    1 millimeter (mm) = 0.03937 inch
    1 inch = 25.4 millimeters
    1 mile = 1.6093 kilometers

Volume Measurements

    1 cubic inch = 16.387 cubic centimeters
    1000 cubic centimeters = 1 liter (l)
    1 liter (l) = 61.02 cubic inches

Temperature Measurements

    1 degree Fahrenheit (F) = 9/5(C + 32 degrees)
    1 degree Celsius (C) = 5/9(F − 32 degrees)

Pressure Measurements

    1 pound per square inch (psi) = 0.07031 kilograms (kg) per square centimeter
    1 kilogram per square centimeter = 14.22334 pounds per square inch

Torque Measurements

    10 foot-pound = 13.558 Newton (N) meters
    1 N-m = 0.7375 ft.-lb.

Fasteners

Fasteners are used to secure or hold parts of something together. Many types and sizes of fasteners are used by the automotive industry. Each fastener is designed for a specific purpose and condition. One of the most commonly used types of fastener is the threaded fastener. Threaded fasteners include bolts, nuts, screws, and similar items that allow a technician to install or remove parts easily.

Threaded fasteners are available in many sizes, designs, and threads. The threads can be either cut or rolled into the fastener. Rolled threads are 30% stronger than cut threads. They also offer better fatigue resistance because there are no sharp notches to create stress points. Fasteners are made to Imperial or metric measurements. There are four classifications for the threads of Imperial fasteners: Unified National Coarse (UNC), Unified National Fine (UNF), Unified National Extrafine (UNEF), and Unified National Pipe Thread (UNPT or NPT). Metric fasteners are also available in fine and coarse threads.

Coarse threads are used for general-purpose work, especially when rapid assembly and disassembly is required. Fine-threaded fasteners are used when greater holding force is necessary. They are also used when greater resistance to vibration is desired.

Bolts have a head on one end and threads on the other. Bolts are identified by defining the head size, shank diameter, thread pitch, length, (Figure 2-2) and grade. Many bolts have a shoulder below the head and the threads do not travel all the way from the head to the end of the bolt.

Studs are rods with threads on both ends. Most often, the threads on one end are coarse and the other end is fine thread. One end of the stud is screwed into a threaded bore. A hole in the part to be secured is fitted over the stud and held in place with a nut that is screwed over the stud. Studs are used when the clamping pressures of a fine thread are needed and a bolt will not work. If the material the stud is being screwed into is soft (such as aluminum) or granular (such as cast iron), fine threads will not withstand a great amount of pulling force on the stud. Therefore, a coarse thread is used to secure the stud in the workpiece and a fine-threaded nut is used to secure the other part to it. Doing this allows the clamping force of fine threads and the holding power of coarse threads.

Nuts are used with other threaded fasteners when the fastener is not threaded into a piece of work. Many different designs of nuts are found on today's cars. The most common one is the hex nut, which is used with studs and bolts and is tightened with a wrench.

Setscrews are used to prevent rotary motion between two parts, such as a pulley and shaft. Setscrews are either headless and require an Allen wrench or screwdriver to loosen and tighten them, or they have a square head.

Bolt Identification

The **bolt head** is used to loosen and tighten the bolt; a socket or wrench fits over the head and is used to screw the bolt in or out. The size of the bolt head varies with the diameter of the bolt and is available in Imperial and metric wrench sizes. Many confuse the size of the head with the

The **bolt head** is used to loosen and tighten the bolt.

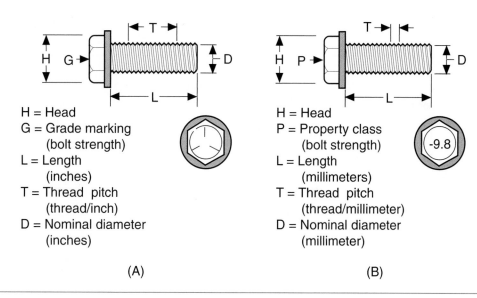

H = Head
G = Grade marking
   (bolt strength)
L = Length
   (inches)
T = Thread pitch
   (thread/inch)
D = Nominal diameter
   (inches)

(A)

H = Head
P = Property class
   (bolt strength)
L = Length
   (millimeters)
T = Thread pitch
   (thread/millimeter)
D = Nominal diameter
   (millimeter)

(B)

Figure 2-2 (A) English and (B) metric bolt terminology.

size of the bolt. The size of a bolt is determined by the diameter of its shank. The size of the bolt head determines what size wrench is required to screw it.

Bolt diameter is the measurement across the major diameter of the threaded area or across the **bolt shank**. The length of a bolt is measured from the bottom surface of the head to the end of the threads.

The **thread pitch** of a bolt in the Imperial system is determined by the number of threads that are in one inch of the threaded bolt length and is expressed in number of threads per inch. An UNF bolt with a 3/8-inch diameter would be a 3/8 × 24 bolt. It would have 24 threads per inch. Likewise a 3/8-inch UNC bolt would be called a 3/8 × 16.

The distance, in millimeters, between two adjacent threads determines the thread pitch in the metric system. This distance will vary between 1.0 and 2.0 and depends on the diameter of the bolt. The lower the number, the closer the threads are placed and the finer the threads are.

The bolt's tensile strength, or grade, is the amount of stress or stretch it is able to withstand before it breaks. The type of material the bolt is made of and the diameter of the bolt determines its grade. In the Imperial system, the tensile strength of a bolt is identified by the number of radial lines (**grade marks**) on the bolt's head. More lines mean higher tensile strength (Table 2-1). Count the number of lines and add two to determine the grade of a bolt.

A property class number on the bolt head identifies the grade of metric bolts. This numerical identification is comprised of two numbers. The first number represents the tensile strength of the bolt. The higher the number, the greater the tensile strength. The second number represents the yield strength of the bolt. This number represents how much stress the bolt can take before it is not able to return to its original shape without damage. The second number represents a percentage rating. For example, a 10.9 bolt has a tensile strength of 1000 MPa (145,000 psi) and a yield strength of 900 MPa (90% of 1000). A 10.9 metric bolt is similar in strength to a SAE grade 8 bolt.

Nuts are graded to match their respective bolts. For example, a grade 8 nut must be used with a grade 8 bolt. If a grade 5 nut were used, a grade 5 connection would result. Grade 8 and critical applications require the use of fully hardened flat washers. Unlike soft washers, these will not dish out when torqued.

Bolt heads can pop off because of **fillet** damage. The fillet is the smooth curve where the shank flows into the bolt head (Figure 2-3). Scratches in this area introduce stress to the bolt head,

---

The **bolt shank** is the smooth area measured from the bottom surface of the head to the start of the threads.

The **thread pitch** of a bolt in the Imperial system is expressed in number of threads per inch. The distance in millimeters between two adjacent threads determines the thread pitch in the metric system.

In the Imperial system, the tensile strength of a bolt is identified by the number of radial lines (**grade marks**) on the bolt's head.

Table 2-1 STANDARD BOLT STRENGTH MARKINGS

SAE Grade Markings					
DEFINITION	No lines: un-marked indeterminate quality SAE grades 0-1-2	3 Lines: common commercial quality automotive and AN bolts SAE grade 5	4 Lines: medium commercial quality automotive and AN bolts SAE grade 6	5 Lines: rarely used SAE grade 7	6 Lines: best commercial quality NAS and aircraft screws SAE grade 8
MATERIAL	Low carbon steel	Medium carbon steel, tempered	Medium carbon steel, quenched and tempered	Medium carbon alloy steel	Medium carbon alloy steel, quenched and tempered
TENSILE STRENGTH	65,000 psi	120,000 psi	140,000 psi	140,000 psi	150,000 psi

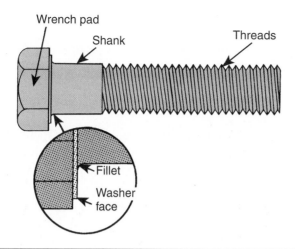

Figure 2-3 Bolt fillet detail.

causing failure. Removing any burrs around the edges of holes can protect the bolt head. Also place flat washers with their rounded, punched side against the bolt head and their sharp side against the work surface.

Fatigue breaks are the most common type of bolt failure. A bolt becomes fatigued from working back and forth when it is too loose. Under tightening the bolt causes this problem. Bolts can also be broken or damaged by over tightening, being forced into a nonmatching thread, or bottoming out, which happens when the bolt is too long.

Tightening Bolts

Any fastener is nearly worthless if it is not as tight as it should be. When a bolt is properly tightened, it will be "spring loaded" against the part it is holding. This spring effect is caused by the stretch of the bolt when it is tightened. Normally a properly tightened bolt is stretched to 70% of its elastic limit. The elastic limit of a bolt is that point of stretch in which the bolt will not return to its original shape when it is loosened. Not only will an over tightened or stretched bolt not have sufficient clamping force, it will also have distorted threads. The stretched threads will make it more difficult to screw and unscrew the bolt or a nut on the bolt. Always check the service manual to see if there is a torque specification for a bolt before tightening it. If there is, use a torque wrench and tighten the bolt properly.

Measuring Tools

Many of the procedures discussed in this manual require exact measurements of parts and clearances. Accurate measurements require the use of precision measuring devices that are designed to measure things in very small increments. Measuring tools are delicate instruments and should be handled with great care. Never strike, pry, drop, or force these tools. Also make sure you clean them before and after every use.

> ✔ **SERVICE TIP:** It is a good idea to check all measuring tools against known good equipment or tool standards. This will ensure that they are operating properly and are capable of accurate measurement.

There are many different measuring devices used by automotive technicians. This chapter will only cover those that are commonly used to service transmissions and other driveline components.

Bolt heads can pop off because of **fillet** damage. The fillet is the smooth curve where the shank flows into the bolt head.

Machinist's Rule

A machinist's rule is divided into increments based on different scales: USCS, metric, or decimal.

The **machinist's rule** looks very much like an ordinary ruler. Each edge of this basic measuring tool is divided into increments based on a different scale. A typical machinist's rule based on the Imperial system of measurement may have scales based on 1/8-, 1/16-, 1/32-, and 1/64-inch intervals. Of course, metric machinist's rules are also available. Metric rules are usually divided into 0.5-mm and 1-mm increments.

Some machinist's rules may be based on decimal intervals. These are typically divided into 1/10-, 1/50-, and 1/1,000-inch (0.1, 0.03, and 0.01) increments. Decimal machinist's rules are very helpful when measuring dimensions that are specified in decimals; they make such measurements much easier.

Machinist's rules are commonly used to make adjustments on transmissions. In Figure 2–4, a machinist rule is used during the adjustment of line pressure. The procedure calls for a specific pressure regulator spring height.

Vernier Caliper

A vernier caliper is a precision measuring instrument capable of measuring outside and inside diameters, as well as depth.

A **vernier caliper** has a movable scale that is parallel to a fixed scale (Figure 2-5). These precision measuring instruments are capable of measuring outside and inside diameters and most will even measure depth (Figure 2-6). Vernier calipers are available in both Imperial and metric scales. The main scale of the caliper is divided into inches; most measure up to six inches. Each inch is divided into 10 parts, each equal to 0.100 inch. The area between the 0.100 marks is divided into four. Each of these divisions is equal to 0.025 inches.

The vernier scale has 25 divisions, each one representing 0.001 inch. Measurement readings are taken by combining the main and vernier scales. At all times, only one division line on the main scale will line up with a line on the vernier scale. This is the basis for accurate measurements.

To read the caliper, locate the line on the main scale that lines up with the zero (0) on the vernier scale. In Figure 2-7, the correct reading is 0.025 inches. If the zero lined up with the 1 on the main scale, the reading would be 0.100 inches. If the zero on the vernier scale does not line up exactly with a line on the main scale, then look for a line on the vernier scale that does line up with a line on the main scale.

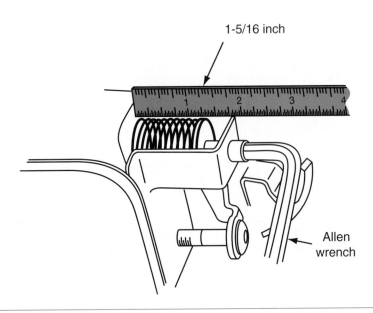

Figure 2-4 Using a machinist's rule to make an adjustment to line pressure.

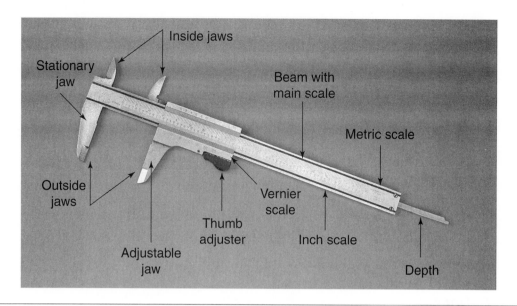

Figure 2-5 A vernier caliper.

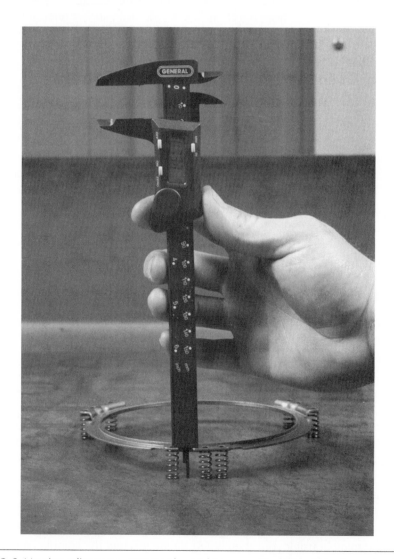

Figure 2-6 Vernier calipers are commonly used to measure depth. This dimension is read in the same way as thickness or width.

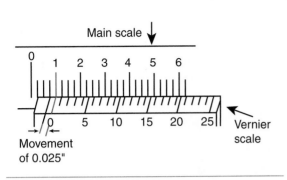

Figure 2-7 Measuring 0.025 on the vernier scale.

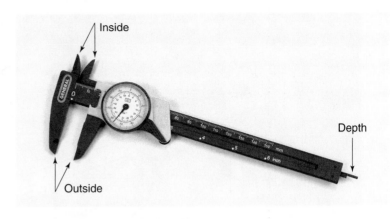

Figure 2-8 A dial vernier caliper.

Dial Caliper

The **dial caliper** (Figure 2-8) is an easier-to-use version of the vernier caliper. Imperial calipers commonly measure dimensions from 0 to 6 inches. Metric dial calipers typically measure from 0 to 150 mm in increments of 0.02 mm. The dial caliper features a depth scale, bar scale, dial indicator, inside measurement jaws, and outside measurement jaws.

The main scale of an Imperial dial caliper is divided into one-tenth (0.1) inch graduations. The dial indicator is divided into one-thousandth (0.001) inch graduations. Therefore, one revolution of the dial indicator needle equals one-tenth inch on the bar scale.

A metric dial caliper is similar in appearance but the bar scale is divided into 2-mm increments. Additionally, on a metric dial caliper, one revolution of the dial indicator needle equals 2 mm.

Both English and metric dial calipers use a thumb-operated roll knob for fine adjustment. When you use a dial caliper, always move the measuring jaws backward and forward to center the jaws on the object being measured. Make sure the caliper jaws lay flat on or around the object. If the jaws are tilted in any way, you will not obtain an accurate measurement.

Although dial calipers are precision measuring instruments, they are only accurate to plus or minus two-thousandths (±0.002) of an inch. Micrometers are preferred when extremely precise measurements are desired.

Micrometers

The **micrometer** is used to measure linear outside and inside dimensions. Both outside and inside micrometers are calibrated and read in the same manner. Measurements on both are taken with the measuring points in contact with the surfaces being measured.

The major components and markings of a micrometer include the frame, anvil, spindle, locknut, sleeve, sleeve numbers, sleeve long line, thimble marks, thimble, and ratchet (Figure 2-9). Micrometers are calibrated in either inch or metric graduations and are available in a range of sizes.

Micrometers can be used to measure the diameter and/or thickness of an object. The common objects measured with a micrometer by automatic transmission technicians are washers. Critical clearances throughout the transmission require that correctly sized washers are used.

Reading an Outside Micrometer

To measure a small object, such as a thrust washer, with an outside micrometer, open the jaws of the tool and slip the object between the spindle and the anvil. While holding the object against the anvil, turn the thimble using your thumb and forefinger until the spindle contacts the object. Use only enough pressure on the thimble to allow the object to just fit between the tips of the

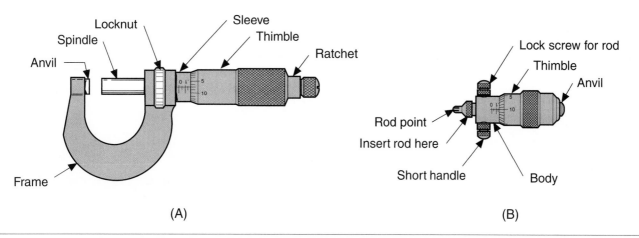

Figure 2-9 Major components of (A) an outside and (B) inside micrometer.

anvil and spindle. The object should slip through with only a very slight resistance. When a satisfactory feel is reached, lock the micrometer.

Since each graduation on the sleeve represents 0.025 inch (Figure 2-10), begin reading the measurement by counting the visible lines on the sleeve and multiply that number by 0.025. The graduations on the thimble assembly define the area between the lines on the sleeve; therefore, the number indicated on the thimble should be added to the measurement shown on the sleeve. The sum is the outside diameter of the object.

To measure a larger object, hold the frame of the micrometer and slip it over the object. Turn the thimble while continuing to slip the micrometer over the object until you feel a very slight resistance. Rock the micrometer from side to side while doing this to make sure the spindle cannot be closed any farther. Then, lock the micrometer and take a measurement reading. Photo Sequence 2 guides you through the correct procedure for measuring with and reading a micrometer.

SERVICE TIP: Like all tools, measuring tools should only be used for the purpose they were designed for. Some instruments are not accurate enough for very precise measurements; others are too accurate to be practical for less critical measurements.

Some technicians use a digital micrometer, which is easier to read. These tools do not have the various scales; rather, the measurement is displayed and read directly off the micrometer.

To measure a larger object such as a clutch piston, select a micrometer of the proper size. Micrometers are available in a number of different sizes. The size is dictated by the smallest measurement it can make to the largest. Examples of these sizes are the 0-to-1-inch, 1-to-2-inch, 2-to-3-inch, and 3-to-4-inch micrometers.

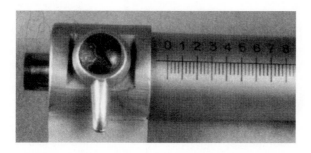

Figure 2-10 The graduations of a micrometer sleeve represent 0.025 inch.

Photo Sequence 2
Typical Procedure for Using a Micrometer

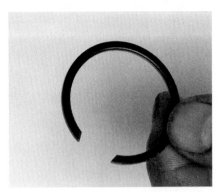

P2-1 Micrometers can be used to measure the diameter and thickness of many objects. A common example for a transmission tech is a selective snap ring.

P2-2 Because the thickness of the snap ring is less than one inch, a 0- to 1-inch micrometer will be used to measure the snap ring.

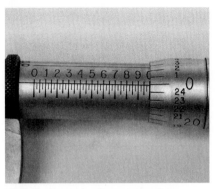

P2-3 Each graduation on the sleeve represents 0.025 inch. To read a measurement on a micrometer, begin by counting the visible lines on the sleeve, then multiple 0.025 by that number.

P2-4 The graduations on the thimble assembly define the area between the lines on the sleeve. The number indicated on the thimble is added to the measurement shown on the sleeve.

P2-5 Micrometer reading of 0.375 inch.

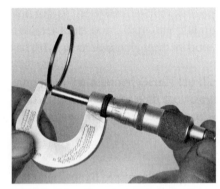

P2-6 Position the micrometer over the snap ring and slowly close the micrometer around the snap ring until a slight drag is felt while moving the micrometer back and forth over the snap ring.

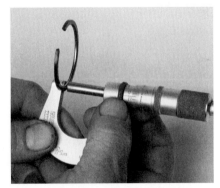

P2-7 To prevent the reading from changing while you move the micrometer away from the stem, use your thumb to activate the lock lever.

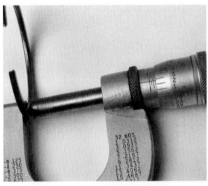

P2-8 This reading (0.078) represents the thickness of the snap ring.

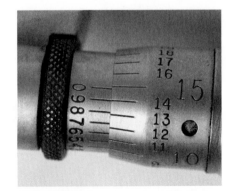

P2-9 Some micrometers are able to measure in 0.0001-inch increments. Use this type of micrometer when the specifications call for this degree of accuracy.

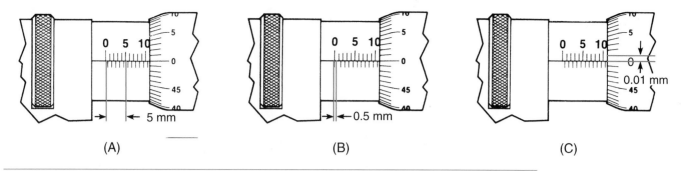

Figure 2-11 Reading a metric micrometer: (A) 5 mm plus (B) 0.5 mm plus (C) 0.01 mm equals 5.51 mm.

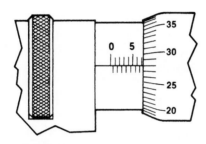

Figure 2-12 The total reading on this micrometer is 7.28 mm.

Reading a Metric Outside Micrometer

The metric micrometer is read in the same manner as the inch-graduated micrometer, except the graduations are expressed in the metric system of measurement. Readings are obtained as follows.

❑ Each number on the sleeve of the micrometer represents 5 millimeters (mm) or 0.005 meter (m) (Figure 2-11A).

❑ Each of the ten equal spaces between each number, with index lines alternating above and below the horizontal line, represents 0.5 mm or five tenths of a millimeter. One revolution of the thimble changes the reading one space on the sleeve scale or 0.5 mm (Figure 2-11B).

❑ The beveled edge of the thimble is divided into 50 equal divisions with every fifth line numbered: 0, 5, 10, . . .45. Since one complete revolution of the thimble advances the spindle 0.5 mm, each graduation on the thimble is equal to one hundredth of a millimeter (Figure 2-11C).

❑ As with the inch-graduated micrometer, the three separate readings are added together to obtain the total reading (Figure 2-12).

Feeler Gauge

A **feeler gauge** is a thin strip of metal or plastic of known and closely controlled thickness. Several of these metal strips are often assembled together as a feeler gauge set that looks like a pocket knife (Figure 2-13). The desired thickness gauge can be pivoted away from the others for convenient use. A steel feeler gauge pack usually contains strips or leaves of 0.002- to 0.010-inch

A **feeler gauge** is a device made up of metal strips or blades finished to an accurate thickness and is used for measuring the clearance between two parts.

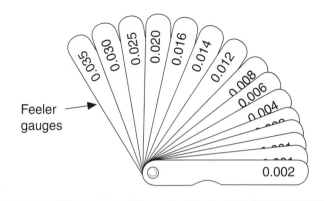

Feeler gauges → 0.035 0.030 0.025 0.020 0.016 0.014 0.012 0.008 0.006 0.004 0.002

Figure 2-13 Typical feeler gauge pack.

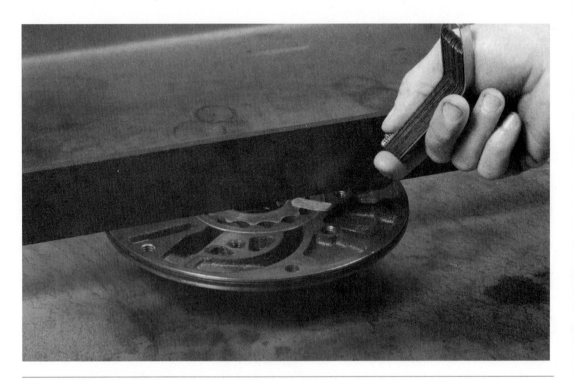

Figure 2-14 Using a feeler gauge and straight edge to check the flatness of a transmission pump.

thickness (in steps of 0.001 inch) and leaves of 0.012- to 0.024-inch thickness (in steps of 0.002 inch). Metric feeler gauges are also available.

A feeler gauge can be used by itself to measure clearances and gaps or it can be used with a precision straightedge to measure alignment and surface warpage (Figure 2-14).

Screw Pitch Gauge

The use of a **screw pitch gauge** (Figure 2-15) provides a quick and accurate method of checking the thread pitch of a fastener. The leaves of this measuring tool are marked with the various pitches. To check the pitch of threads, simply match the teeth of the gauge with the threads of the fastener. Then, read the pitch from the leaf.

Screw pitch gauges are available for the various types of fastener threads used by the automotive industry: Unified National Coarse and Fine threads, metric threads, International Standard threads, and Whitworth threads.

A **screw pitch gauge** is a tool used to measure the threads per inch on a threaded fastener.

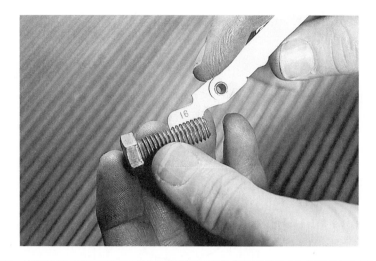

Figure 2-15 Using a screw pitch gauge to check the thread pitch on a bolt.

Dial Indicator

The **dial indicator** (Figure 2-16) is calibrated in 0.001-inch (one-thousandth inch) increments. Metric dial indicators are also available. Both types are used to measure movement. Common uses of the dial indicator include measuring valve lift, journal concentricity, flywheel or brake rotor runout, gear backlash, and crankshaft endplay. Dial indicators are available with various face markings and measurement ranges to accommodate many measuring tasks.

To use a dial indicator, position the indicator rod against the object to be measured. Then, push the indicator toward the work until the indicator needle travels far enough around the gauge face to permit movement to be read in either direction (Figure 2-17). Zero the indicator needle on the gauge. Always be sure the range of the dial indicator is sufficient to allow the amount of movement required by the measuring procedure. For example, never use a 1-inch indicator on a component that will move 2 inches.

A **dial indicator** is a measuring instrument that measures changes in dimension as the indicator is moved across an object or when the object is moved over the stem of the indicator.

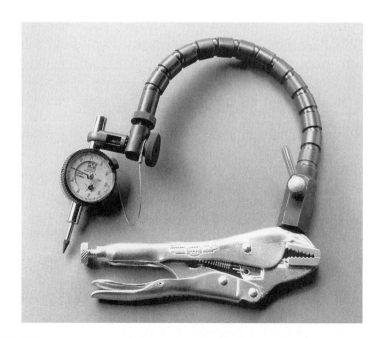

Figure 2-16 A dial indicator with a highly adaptive holding fixture.

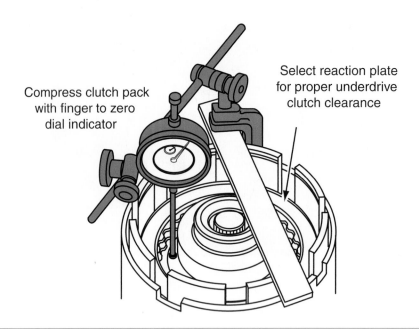

Compress clutch pack with finger to zero dial indicator

Select reaction plate for proper underdrive clutch clearance

Figure 2-17 A dial indicator is often used in automatic transmission service. This setup is measuring the clearance of a multiple friction disc pack.

Basic Gear Adjustments

While a drivetrain is in operation, the gears, shafts, and bearings are subjected to loads and vibrations. Because of this, the drivetrain must normally be adjusted for the proper fit between parts. These adjustments require the use of precision measuring tools. There are three basic adjustments that are made when reassembling a unit or when a problem suggests that readjustment is necessary. Adjusting the clearance or play between two gears in mesh is referred to as adjusting the backlash. Endplay adjustments limit the amount of end to end movement of a gear shaft. Preload is an adjustment made to put a load on an assembly to offset the loads the assembly will face during operation.

Backlash in Gears

Backlash is the clearance between two gears in mesh.

Backlash is the clearance between two gears in mesh (Figure 2-18). Excessive backlash can be caused by worn gear teeth, improper meshing of teeth, or bearings that do not support the gears properly. Excessive backlash can result in a severe impact on the gear teeth from sudden stops or direction changes of the gears, which can cause broken gear teeth and gears. Insufficient backlash causes excessive overload wear on the gear teeth and could cause premature gear failure.

Backlash is measured with a dial indicator mounted so that its stem is in line with the rotation of the gear and perpendicular to the angle of the teeth (Figure 2-19). The gear is moved in both directions while the other gear it meshes with is held. The amount of movement on the dial indicator equals the amount of backlash present. The proper placement of shims on a gear shaft is the normal procedure for making backlash adjustments.

Endplay in Gears and Shafts

End clearance is often referred to as **endplay** and is the measurable axial or end-to-end looseness of a bearing, assembly, or shaft.

Endplay refers to the measurable axial or end–to–end looseness of a bearing. Endplay is always measured in an unloaded condition. To check endplay, a dial indicator is mounted against the side of a gear or the end of a shaft (Figure 2-20). The gear shaft is then pried in both directions and the readings noted. The difference between the two readings is the amount of endplay. Shims or adjusting nuts are widely used to adjust endplay.

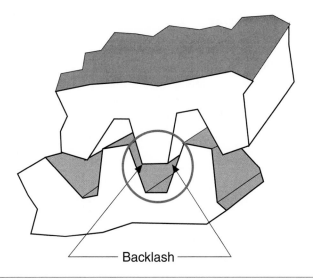

Backlash

Figure 2-18 Backlash is the clearance between two gears in mesh.

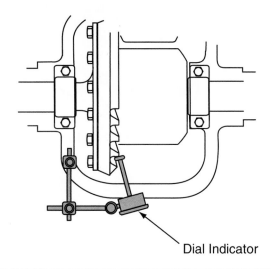

Dial Indicator

Figure 2-19 A dial indicator is normally used to measure the backlash of a gearset. The plunger tip should be in line with the gear tooth movement in order to get the most accurate reading.

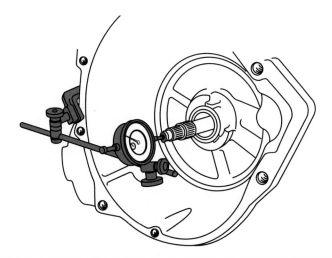

Figure 2-20 This dial indicator setup will measure the endplay of the input shaft.

Choosing Selective Washers

A common procedure is the checking and correcting of subassembly end clearances. When these dimensions are correct, the transmission will have good shift characteristics and will wear normally. Look at the forward clutch assembly in Figure 2-21. The clearances in this clutch pack are critical and are controlled by the retaining snap ring. The specifications for this clearance is 0.046 to 0.068 inch (1.17 to 1.72 mm) and the snap rings are available in four different thicknesses: 0.060 to 0.064 inch (1.52 to 1.62 mm), 0.074 to 0.078 inch (1.87 to 1.98 mm), 0.086 to 0.106 inch (2.23

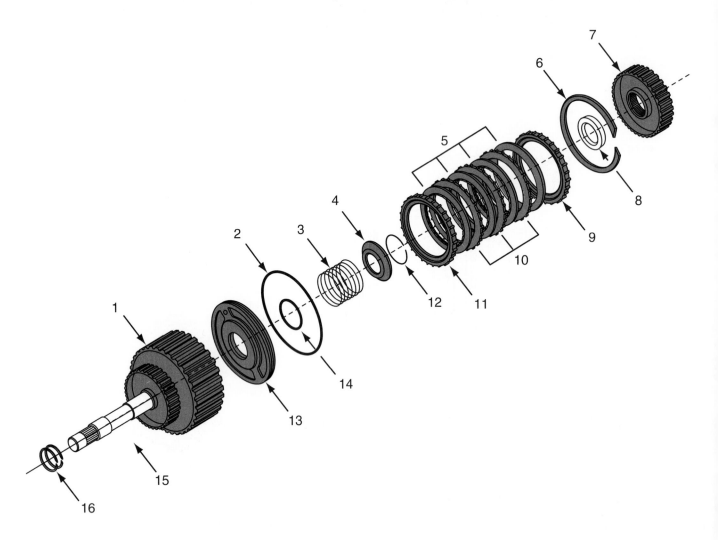

LEGEND

1. Clutch cylinder and shaft
2. Clutch piston outer seal
3. Clutch piston return spring
4. Clutch retainer return spring
5. Clutch steel plates
6. Selective retaining snap ring
7. Clutch hub
8. Clutch hub front bearing

9. Clutch pressure plate
10. Clutch friction plates
11. Clutch pressure spring
12. Retaining snap ring
13. Clutch piston
14. Clutch piston inner spring
15. Clutch bearing
16. Input shaft seal

Figure 2-21 A front clutch assembly: note the location of the retaining snap ring; the thickness of this ring controls the end clearance of the pack.

to 2.69 mm), and 0.102 to 0.106 inch (2.59 to 2.69 mm). To determine what size snap ring should be used, the end clearance of the pack needs to be measured.

To do this, assemble the pack with the existing snap ring. Position a dial indicator on the clutch pack's pressure plate (Figure 2-22). Then zero the dial indicator. Now push down on the clutch pack, release the pressure, and zero the indicator again.

Using the required special tool, lift up on the clutch pack until the pressure plate is fully seated against the retaining snap ring. Now read the dial indicator. The reading is the end clearance of the pack with the existing snap ring. If the measurement was not within specifications, a snap ring of a different thickness must be used. For example, if the end clearance measurement was 0.074 inches, a thicker snap ring must be installed. To determine how thick, remove the existing snap ring and measure its thickness. With a micrometer the thickness was found to be 0.060 inches. By adding the measured clearance to the snap ring thickness, we know the clearance of the pack. In this case that clearance is 0.134 inches. To determine what thickness snap ring is best to use, subtract the specification from this total clearance figure. Since the specification is a range, subtract the extremes of the range from the clearance (0.134 – 0.046 = 0.078 and 0.134 – 0.068 = 0.066). What we find is the correct snap ring is also within a range. Since the snap rings are available in four different sizes, the thickness that best meets the need is the one that should be used. In this case, a snap ring with a thickness of 0.074 to 0.078 inch is the best choice.

If the part number on a selective snap ring or washer can be read, its thickness can be determined by referring the number to specifications in the service manual.

Preloading of Geartrains

When normal operating loads are great, geartrains are often preloaded to reduce the deflection of parts. The amount of **preload** is specified in shop manuals and must be correct for the design of the bearings and the strength of the parts. If bearings are excessively preloaded, they will heat up and fail. When bearings are set too loose, the gears or shaft will wear rapidly due to the great amounts of deflection they will experience. Geartrains are preloaded by shims, thrust washers, adjusting nuts, or by using double-race bearings. Preload adjustments are normally checked by measuring turning effort or torque with a torque wrench (Figure 2-23).

Preload is an affixed amount of pressure constantly applied to a component.

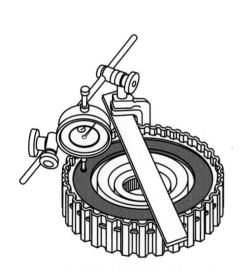

Figure 2-24 To measure end clearance on this clutch pack, the dial indicator is placed on the pressure plate and zeroed, then the pressure plate is depressed and a reading taken.

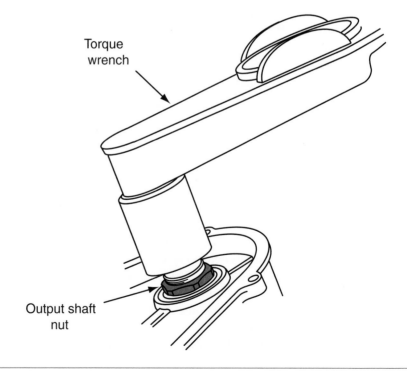

Figure 2-23 Turning torque is checked with a torque wrench, normally an inch-pound wrench.

Service Manuals

Service manuals are sometimes simply called shop manuals.

Service manuals are a necessary part of transmission and driveline service. There is no way a technician can remember all of the procedures and specifications needed to repair an automobile correctly. Thus, a good technician relies on service manuals and other information sources for this information. Good information plus knowledge allows a technician to fix a problem with the least bit of frustration and at the lowest cost to the customer. Service manuals also provide drawings and photographs that show where and how to perform certain procedures on the particular car you are working on. Special tools or instruments are listed and shown, when they are required. Precautions are also given to prevent injury or damage to parts. Perhaps the most important tools you will use are service manuals.

Most automobile manufacturers publish a service manual or set of manuals for each model and year of their cars. These manuals provide the best and most complete information for those cars. The most commonly used specifications and procedures are compiled in comprehensive service manuals. Various editions are available, covering different ranges of model years for both domestic and imported cars. Specialized shop manuals are also produced for special areas, such as transmissions. Similar to manufacturers' manuals in many ways, they do not provide as much information or detail as do the manufacturers' manuals.

Although the manuals from different publishers vary in presentation and arrangement of topics, all service manuals are easy to read and follow after you become familiar with their organization. Most shop manuals are divided into a number of sections, each covering different aspects of the vehicle (Figure 2-24). The beginning sections commonly provide vehicle identification and basic maintenance information. The remaining sections deal with each different vehicle system in detail, including diagnostic, service, and overhaul procedures. Each section has an index indicating more specific areas of information.

To use a service manual:

1. Select the appropriate manual for the vehicle being serviced.
2. Use the table of contents to locate the section that applies to the work being done.
3. Use the index at the front of that section to locate the required information.
4. Carefully read the information and study the applicable illustrations and diagrams.
5. Follow all of the required steps and procedures given for that service operation.
6. Adhere to all of the given specifications and perform all measurement and adjustment procedures with accuracy and precision.

Throughout this book, you will be told to refer to the appropriate shop manual to find the correct procedures and specifications. Although the various systems of all automobiles function in much the same way, there are many variations in design. Each design has its own set of repair and diagnostic procedures. Therefore, it is important that you always follow the recommendations of the manufacturer to identify and repair problems.

Since many technical changes occur on specific vehicles each year, manufacturers' service manuals need to be constantly updated. Updates are published as service bulletins (often referred to as Technical Service Bulletins or TSBs) that show the changes in specifications and repair procedures during the model year. These changes do not appear in the service manual until the next year. The car manufacturer provides these bulletins to dealers and repair facilities on a regular basis.

Automotive manufacturers also publish a series of technician reference books. The publications provide general instructions on the service and repair of their vehicles with their recommended techniques.

SECTION	SUBJECT	CONTENTS	
00	GENERAL INFORMATION	0A GENERAL INFORMATION 0B MAINTENANCE	0C VIBRATION AND NOISE DIAGNOSIS
01	FRAME AND SHEET METAL	1A FRAME AND BUMPERS	1B SHEET METAL
02	ENGINE	2 ENGINE 2A ENGINE, DRIVEABILITY, AND DIAGNOSIS 2B1 ENGINE COOLING 2B2 RADIATOR 2C ENGINE ELECTRICAL 2C1 BATTERY	2C2 CRANKING SYSTEM 2C3 CHARGING SYSTEM 2C4 ENGINE WIRING 2D IGNITION SYSTEM 3E EXHAUST
03	HEATING AND AIR CONDITIONING	3A HEATING AND VENTILATION	3B AIR CONDITIONING
04	TRANSMISSION	4A AUTOMATIC TRANSMISSION 4B MANUAL TRANSMISSION	4C CLUTCH 4D TRANSFER CASE
05	PROPELLER SHAFT AND AXLES	5A PROPELLER SHAFT 5B FRONT AXLE	5C REAR AXLE
06	STEERING, SUSPENSION, TIRES AND WHEELS	6A TIRES AND WHEELS 6B FRONT END ALIGNMENT 6C FRONT SUSPENSION 6D REAR SUSPENSION	6E1 STEERING COLUMN-STANDARD 6E2 STEERING COLUMN-TILT 6F1 POWER STEERING 6F2 STEERING LINKAGE
07	BRAKES	7A HYDRAULIC BRAKES 7B HYDRAULIC FOUNDATION BRAKES 7C HYDRAULIC BRAKE BOOSTER SYSTEMS	7D PARKING BRAKES 7E1 FOUR-WHEEL ANTILOCK BRAKE SYSTEM 7E2 REAR-WHEEL ANTILOCK BRAKE SYSTEM
08	ELECTRICAL	8A ELECTRICAL CIRCUITS 8B LIGHTING SYSTEMS 8C INSTRUMENT PANEL AND GAUGES	8D CHASSIS ELECTRICAL 8E WIPER/WASHER SYSTEMS
09	ACCESSORIES	9A AUDIO SYSTEMS	9B CRUISE CONTROL
10	BODY	10A1 DOORS 10A2 SEATS 10A3 WINDOWS	10A4 INTERIOR TRIM 10B MAINTENANCE 10C EXTERIOR TRIM
	INDEX	ALPHABETICAL INDEX	

Figure 2-24 Service manuals are divided into sections, each covering different systems of the vehicle.

Aftermarket Suppliers' Guides and Catalogs

Many of the larger parts manufacturers have excellent guides on the various parts they manufacture or supply. They also provide updated service bulletins on their products. Other sources for up-to-date technical information are trade magazines and trade associations.

General and Specialty Repair Manuals

Service manuals are also published by independent companies rather than by the manufacturers. However, they pay for and get most of their information from the carmakers. They contain component information, diagnostic steps, repair procedures, and specifications for several car makes in one book. Information is usually condensed and is more general in nature than the manufacturer's manuals. The condensed format allows for more coverage in less space and, therefore, is not always specific. They may also contain several years of models as well as several car makes in one book.

Flat-Rate Manuals

Flat-rate manuals contain standards for the length of time a specific repair is supposed to require. Normally, they also contain a parts list with approximate or exact prices of parts. They are excellent for making cost estimates and are published by manufacturers and independents.

Computer-based Information

The same information that is available in service manuals and bulletins is also available on compact disk-read only memory (CD-ROM) systems (Figure 2-25). A single compact disk can hold a quarter million pages of text, eliminating the need to maintain a huge library containing all of the printed manuals. Using a CD-ROM to find information is also easier and quicker. The disks are updated periodically and not only contain the most recent service bulletins but also engineering and field service fixes. Service procedures and specifications are also available through the Internet. Using electronics to store and sort data allows for quick access to information.

Hotline Services

There are many companies that provide on-line help to technicians. As the complexity of the automobile grows, so does the popularity of these services. One of the most commonly used hotlines is Autoline Telediagnosis, which has over twenty "experts" available for inquiries. These

Figure 2-25 The use of CD-ROMs and computers makes accessing information quick and easy.

experts are technicians familiar with the different systems of certain manufacturers. Armed with various factory service manuals, general service manuals, service manuals, and electronic data sources, these brand experts give information to technicians to help them through diagnostic and repair procedures.

Logical Diagnostics

The word "**diagnosis**" is commonly used in automotive manuals. However, it is a term that is seldom defined. Diagnosis is not guessing and it is more than following a series of interrelated steps in order to find the solution to a specific problem. Diagnosis is a way of looking at systems that are not functioning the way they should and finding out why. It is knowing how the system should work and deciding if it is working correctly. Through an understanding of the purpose and operation of the system, you can accurately diagnose problems.

Most good technicians use the same basic diagnostic approach. Simply because this is a logical approach, it can quickly lead to discovering the cause of a problem. Logical diagnosis follows these steps:

1. Gather information about the problem.
2. Verify that the problem exists.
3. Thoroughly define what the problem is and when it occurs.
4. Research all available information and knowledge to determine the possible causes of the problem.
5. Isolate the problem by testing. Most often this testing will include the use of a scan tool.
6. Follow the test procedures outlined by the manufacturer for the symptom and/or diagnostic codes retrieved from the computer.
7. Continue testing to pinpoint the cause of the problem.
8. Locate and repair the problem, then verify the repair.

Diagnosis is a systematic study of something to determine the cause of improper operation or failure.

Working as an Auto Tech

To be a successful automotive technician, you need to have good training, a desire to succeed, and be committed to become a good technician and a good employee. A good employee works well with others and strives to make the business successful. The required training is not just in the field of automotive technology. Good technicians need to have good reading, writing, and math skills. These skills will allow you to better understand and use the material found in service manuals and textbooks, as well as provide you with the basics for good communications with customers and others.

Compensation

Technicians are typically paid according to their abilities. Most often, new or apprentice technicians are paid by the hour. While being paid they are learning the trade and the business. Time is usually spent working with a master technician or doing low-skilled jobs. As an apprentice learns more, he or she can earn more and take on more complex jobs. Once technicians have demonstrated a satisfactory level of skills, they can go on flat rate.

Flat rate is a pay system in which a technician is paid for the amount of work he or she does. Each job has a flat rate time. Pay is based on that time, regardless of how long it took to complete the job. As an example of how this system works, consider a technician who is paid $15.00 per flat rate. If a job has a flat rate time of 3 hours, the technician will be paid $45.00 for the job, regardless of how long it took to complete it. Experienced technicians beat the flat rate time nearly all of the time. Their weekly pay is based on the time "turned in," not on the time spent. If the technician turns in 60 hours of work in a 40-hour workweek, he or she actually earned $22.50 each hour worked. However, if he or she turned in only 30 hours in the 40-hour week, the hourly pay is $11.25.

The flat rate system favors good technicians who work in a shop that has a large volume of work. The use of flat rate times makes it possible to give more accurate repair estimates to customers. It also rewards skilled and productive technicians.

Employer-Employee Relationships

Being a good employee requires more than learning the skills for the job. When you begin a job, you enter into a business agreement with your employer. When you become an employee, you sell your time, skills, and efforts. In return, your employer pays you for these resources.

As part of the employment agreement, your employer also has certain responsibilities:

- ❏ *Instruction and supervision*—You should be told what is expected of you. A supervisor should observe your work and tell you if it is satisfactory and offer ways to improve your performance.
- ❏ *Clean, safe place to work*—An employer should provide a clean and safe work area as well as a place for personal cleanup.
- ❏ *Wages*—You should know how much you are to be paid, what your pay will be based on, and when you will be paid before accepting a job.
- ❏ *Fringe benefits*—When you are hired, you should be told what benefits, such as paid vacations and employer contributions to health insurance and retirement plans, you can expect.
- ❏ *Opportunity*—This means you should be given a chance to succeed and possibly advance within the company.
- ❏ *Fair treatment*—All employees should be treated equally, without prejudice or favoritism.

On the other side of this business transaction, employees have responsibilities to their employers. Your obligations as an employee include the following:

- ❏ *Regular attendance*—A good employee is reliable. Businesses cannot operate successfully unless their workers are on the job.
- ❏ *Following directions*—As an employee, you are part of a team and doing things your way may not serve the best interests of the company.
- ❏ *Responsibility*—You must be willing to answer for your behavior and work. You also need to realize that you are legally responsible for the work you do.
- ❏ *Productivity*—Remember, you are paid for your time as well as your skills and effort.
- ❏ *Loyalty*—Loyalty is expected and by being loyal you will act in the best interests of your employer, both on and off the job.

Another responsibility you have as an employee is good customer relations. Learn to listen and communicate clearly. Be polite and organized, particularly when dealing with customers. Always be as honest as you possibly can be.

Look and present yourself as a professional, which is what automotive technicians are. Professionals are proud of what they do and they show it. Always dress and act appropriately and watch your language, even when you think no one is near.

Respect the vehicles on which you work. They are important to the lives of your customers. Always return the vehicle to the owner in a clean, undamaged condition. Remember, a car is usually the second largest expense a customer has. Treat it that way. It doesn't matter if you like the car. It belongs to the customer; treat it respectfully.

Explain the repair process to the customer in understandable terms. Whenever you are explaining something to a customer, make sure you do so in a simple way without making the customer feel stupid. Always show the customers respect and be courteous to them. Not only is this the right thing to do but it also leads to loyal customers.

ASE Certification

The National Institute for **Automotive Service Excellence** (**ASE**) has established a voluntary certification program for automotive, heavy-duty truck, auto body repair, and engine machine shop technicians. In addition to these programs, ASE also offers individual testing in the areas of parts, alternate fuels, and advanced engine performance. This certification system combines voluntary testing with on-the-job experience to confirm that technicians have the skills needed to work on today's more complex vehicles. ASE recognizes two distinct levels of service capability—the automotive technician and the master automotive technician. The master automotive technician is certified by ASE in all eight major automotive systems. A certified automotive technician may have certification in one or more specialty areas.

To become ASE certified, a technician must pass one or more tests that stress diagnostic and repair problems. One of these areas is Automatic Transmissions and Transaxles. This test covers the diagnosis and repair of transmissions, transaxles, hydraulic circuits, electronic circuits, and torque converters. You will find content on all of these topics in this manual, as well as the Classroom Manual.

After passing at least one exam and providing proof of two years of hands-on work experience, the technician becomes ASE certified. Retesting is necessary every five years to remain certified. A technician who passes one examination receives an automotive technician shoulder patch. The master automotive technician patch is awarded to technicians who pass all eight of the basic automotive certification exams (Figure 2-26).

You may receive credit for one of the two years by substituting relevant formal training in one, or a combination, of the following:

- ❏ *High school training.* Three years of training may be substituted for one year of experience.
- ❏ *Post-high school training.* Two years of post-high school training in a public or private trade school, technical institute, community or four-year college, or in an apprenticeship program may be counted as one year of work experience.

The National Institute for **Automotive Service Excellence** (**ASE**) has established a voluntary certification program for technicians employed in the many related fields of the automotive industry.

Figure 2-26 The ASE certification shoulder patch worn by automotive technicians certified in one or more areas is shown on the left. When they are certified in all eight areas, they become Master Automotive Technicians and can wear the patch shown on the right.

- *Short courses.* For shorter periods of post-high school training, you may substitute one month of training for one month of work experience.
- You may receive full credit for the experience requirement by satisfactorily completing a three- or four-year apprenticeship program.

Each certification test consists of 40 to 80 multiple-choice questions. The questions are written by a panel of technical service experts, including domestic and import vehicle manufacturers, repair and test equipment and parts manufacturers, working automotive technicians, and automotive instructors. All questions are pretested and quality-checked on a national sample of technicians before they are included in the actual test. Many test questions force the student to choose between two distinct repair or diagnostic methods. The knowledge and skills needed to pass the tests follows.

- *Basic Technical Knowledge*—What is it? How does it work? This requires knowing what is in a system and how the system works. It also calls for knowing the procedures and precautions to be followed in making repairs and adjustments.
- *Repair Knowledge and Skill*—What is a likely source of a problem? How do you fix it? This requires you to understand and to apply generally accepted procedures and precautions for inspecting, disassembling, rebuilding, replacing, and/or adjusting components within a particular system.
- *Testing and Diagnostic Knowledge and Skill*—How do you find what is wrong? How do you know you corrected a problem? This requires that you are able to recognize that a problem does exist and to know what steps should be taken to identify the cause of the problem.

For further information on the ASE certification program, write: National Institute for Auto Service Excellence (ASE), 13505 Dulles Technology Drive, Herndon, VA 22071 or www.asecert.org.

Summary

- The USCS measuring system is based on the inch, and fine measurements are expressed in fractions of an inch or decimals.
- The metric system is based on the meter, and fine measurements are expressed in tenths, hundredths, and thousandths.
- Whenever bolts are replaced, they should be replaced with exactly the same size, grade, and type as those that were installed by the manufacturer.
- Dial indicators are used to measure movement and are commonly used to measure the backlash and endplay of a set of gears.
- A micrometer can be used to measure the outside diameter of shafts and the inside diameter of holes. It is calibrated in either inch or metric graduations.
- The screw pitch gauge provides a fast and accurate method of measuring the threads per inch (pitch) of fasteners. This is done by matching the teeth of the gauge with the fastener's threads and reading the pitch directly from the leaf of the gauge.
- Gear backlash is a statement of how tightly the teeth of two gears mesh.
- The primary source of repair and specification information for any vehicle is the manufacturer's service manual. Updates are published as service bulletins. These include changes made during the model year, which will not appear in the manual until the following year.
- Flat-rate manuals are ideal for making cost estimates. Published by manufacturers and independent companies, they contain figures showing how long specific repairs should take to complete, as well as a list of the necessary parts and their prices.

❏ Diagnosis means finding the cause or causes of a problem. It requires a thorough understanding of the purpose and operation of the various automotive systems. Diagnostic charts found in service manuals can aid in diagnostics.

❏ Besides learning technical and mechanical skills, service technicians must learn to work as part of a team. As an employee, you will have certain responsibilities to both your employer and customers.

❏ Customer relations is an extremely important part of doing business. Professional, courteous treatment of customers and their vehicles is a must.

❏ The National Institute for Automotive Service Excellence (ASE) actively promotes professionalism within the industry. Its voluntary certification program for automotive technicians and master auto technicians helps guarantee a high level of quality service.

Terms to Know

Automotive Service Excellence (ASE)	Feeler gauge	Micrometer
Backlash	Fillet	Millimeter
Bolt head	Flat-rate	Preload
Bolt shank	Grade marks	Screw pitch gauge
Diagnosis	International System (SI)	Thread pitch
Dial caliper	Machinist's rule	United States Customary Service (USCS)
Dial indicator	Meter	Vernier caliper
Endplay		

ASE-Style Review Questions

1. *Technician A* says the USCS measuring system is based on the inch and fine measurements are expressed in fractions of an inch or decimals.
   Technician B says the metric system is based on the meter and fine measurements are expressed in tenths, hundredths, and thousandths.
   Who is correct?
   A. A only   **C.** Both A and B
   B. B only   **D.** Neither A nor B

2. *Technician A* says one millimeter is equal to 0.001 meters.
   Technician B says one meter is equal to 0.03937 inches.
   Who is correct?
   A. A only   **C.** Both A and B
   B. B only   **D.** Neither A nor B

3. While discussing automotive fasteners:
   Technician A says bolt sizes are listed by their appropriate wrench size.

Technician B says whenever bolts are replaced, they should be replaced with exactly the same size and type that was installed by the manufacturer.
Who is correct?
   A. A only   **C.** Both A and B
   B. B only   **D.** Neither A nor B

4. While adjusting the end clearance of a clutch with a specified clearance of 0.044 to 0.051 inches and a measured clearance of 0.075 inches with the existing thrust washer thickness of 0.055 inches:
   Technician A uses a washer with a thickness between 0.079 and 0.086 inches to correct the end clearance.
   Technician B says the proper thickness for the washer is determined by adding the measured clearance to the thickness of the existing washer, then subtracting the specifications from that total.
   Who is correct?
   A. A only   **C.** Both A and B
   B. B only   **D.** Neither A nor B

5. *Technician A* says dial indicators are commonly used to measure the backlash of a set of gears.
Technician B says dial indicators are commonly used to measure the endplay of a set of gears.
Who is correct?
   A. A only
   B. B only
   C. Both A and B
   D. Neither A nor B

6. *Technician A* says that gear backlash is a statement of how much radial movement a gear shaft has.
Technician B says gear backlash is a statement of how tightly the teeth of two gears mesh.
Who is correct?
   A. A only
   B. B only
   C. Both A and B
   D. Neither A nor B

7. *Technician A* says one of the most important tools for a technician is a shop or service manual.
Technician B says technicians must constantly update their tools in order to work on newer vehicles.
Who is correct?
   A. A only
   B. B only
   C. Both A and B
   D. Neither A nor B

8. *Technician A* says a vernier caliper can be used to measure the outside diameter of something.

Technician B says a vernier caliper can be used to measure the inside diameter of a bore.
Who is correct?
   A. Technician A
   B. Technician B
   C. Both A and B
   D. Neither A nor B

9. *Technician A* uses a dial caliper to take inside and outside measurements.
Technician B uses a dial caliper to take depth measurements.
Who is correct?
   A. Technician A
   B. Technician B
   C. Both A and B
   D. Neither A nor B

10. *Technician A* says that after an individual passes a particular ASE certification exam, he or she is certified in that test area.
Technician B says all of the questions on an ASE certification exam are written as Technician A and Technician B questions.
Who is correct?
   A. Technician A
   B. Technician B
   C. Both A and B
   D. Neither A nor B

Job Sheet 4

Name _____ Date _____

Checking End Clearance of a Multiple-Friction Disc Pack

Upon completion of this job sheet, you should be able to use a feeler gauge and a service manual to check end clearance of a multiple-friction disc pack.

ASE Correlation

This job sheet is related to the ASE Automatic Transmission and Transaxle Test's content area: *Off-Vehicle Transmission and Transaxle Repair: Friction and Reaction Units.*
Task: Measure clutch pack clearance; adjust as needed.

Tools and Materials

Feeler gauge set
A multiple-friction disc pack
Service manual
Goggles or safety glasses with side shields

Describe the transmission the pack is from:

Model and type of transmission: _____

Vehicle the transmission is from:

Year _____ Make _____ Model _____

VIN _____ Engine type and size _____

Procedure

1. Insert a feeler gauge between the pressure plate of the pack and the snap ring. What is the measured clearance? _____

2. What is the specified clearance? _____

3. If the measured clearance is not the same as the specified clearance, what should be done?

   _____

   _____

4. Tip the clutch assembly on its side and insert a feeler gauge between the pressure plate and the adjacent friction plate. What is the measured clearance? _____

5. What is the specified clearance? _____

6. If the measured clearance is not the same as what was specified, what should be done?

   _____

   _____

7. What other measuring instrument can be used to measure the end clearance of this pack?

_____

_____

Instructor's Response _____

_____

_____

_____

Diagnosis, Maintenance, and Basic Adjustments

Upon completion and review of this chapter, you should be able to:

❏ Listen to the driver's complaint, road test the vehicle, then determine the needed repairs.

❏ Diagnose unusual fluid usage, level, and condition problems.

❏ Replace automatic transmission fluid and filters.

❏ Diagnose noise and vibration problems.

❏ Diagnose electronic, mechanical, and vacuum control systems.

❏ Inspect, replace, and align powertrain mounts.

❏ Inspect, adjust, and replace vacuum modulator, lines, and hoses.

❏ Diagnose mechanical and vacuum control systems; determine needed repairs.

❏ Perform oil pressure tests; determine needed repairs.

❏ Inspect, adjust, and replace manual valve shift linkage.

❏ Inspect, adjust, and replace cables or linkages for throttle valve (T.V.) kickdown and accelerator pedal.

❏ Inspect and replace external seals and gaskets while the transmission is in the vehicle.

❏ Inspect, repair, and replace extension housing.

❏ Inspect and replace speedometer drive gear, driven gear, and retainers while the transmission is in the vehicle.

❏ Inspect and replace parking pawl, shaft, spring, and retainer while the transmission is in the vehicle.

❏ Adjust bands.

Basic Tools

Technician's basic tool set

Appropriate service manual

Automatic transmissions are used in many rear-wheel-drive and four-wheel-drive vehicles. Automatic transaxles are most used on front-wheel-drive vehicles. The major components of a transaxle are the same as those in a transmission, except the transaxle assembly includes the final drive and differential gears. An automatic transmission/transaxle shifts automatically without the driver moving a gearshift or pressing a clutch pedal.

Because of the many similarities between a transmission and a transaxle, most of the diagnostic and service procedures are similar. Therefore, all references to a transmission apply equally to a transaxle unless otherwise noted. This rule also holds true for the questions on the certification test for automatic transmissions and transaxles.

 SERVICE TIP: Whenever you are diagnosing or repairing a transaxle or transmission, make sure you refer to the appropriate service manual before you begin.

Transmissions are strong and typically trouble-free units that require little maintenance. Maintenance normally includes fluid checks and scheduled linkage adjustments and oil changes.

Nearly all current automatic transmissions have electronic controls. These controls work with the hydraulic and mechanical systems of the transmission to provide reliable and efficient operation. Because of the variety of different electronic systems used by the manufacturers and the complexity of some of these systems, electrical and electronic transmission systems are covered in their own chapter (Chapter 4). The content of this chapter applies to all transmissions, whether they have electronic controls or not.

Some of the procedures noted in this chapter are for both in-vehicle and out-of-vehicle service.

Identification

Prior to beginning any service or repair work, be sure you know exactly which transmission you are working on. This will ensure that you are following the correct procedures and specifications and installing the correct parts. Proper identification can be difficult because transmissions cannot be accurately identified by the way they look. The only exception to this is the shape of the oil pan, which sometimes can be used as identification for some transmissions.

The only positive way to identify the exact design of the transmission is by its identification numbers. *Transmission identification numbers* are found on stickers on the transmission (Figure 3-1), as stamped numbers in the case or on a metal tag held by a bolt head. Use a service manual to decipher the identification number. Most identification numbers include the model, manufacturer, and assembly date. Whenever you work with a transmission with a metal ID tag, make sure the tag is put back on the transmission so that the next technician will be able to properly identify the transmission.

Some examples of the location of and the information contained on the identification tags of common transmissions follow.

On Chrysler FWD vehicles, the transaxle can be identified by a transaxle identification number (TIN) stamped on a boss, located on the transaxle housing just above the oil pan flange. On RWD vehicles, the TIN is stamped on a pad on the left side of the transmission case's oil pan flange. In addition to the TIN, each transmission carries an assembly part number that must be referenced when ordering transaxle replacement parts. Transmission operation requirements are different for each vehicle and engine combination. Some internal parts will differ between models. Always refer to the seven-digit part number for positive transmission identification when replacing parts.

Late-model Ford RWD vehicles can be identified by an identification code letter found on the lower line of the vehicle certificate label under "TR." This label is attached to the left (driver's) side door lock post. A RWD transmission can be identified by a metal tag attached to the transmission by the lower extension housing retaining bolt. The tag on a transaxle is attached to the valve body cover and gives the transmission model number, line shift code, build date code, and assembly and serial numbers (Figure 3-2).

Early Ford RWD models can be identified by an identification tag located under the lower intermediate servo cover bolt or attached to the lower extension housing retaining bolt. A number appearing after the suffix indicates internal parts in the transmission have been changed after initial production startup. The top line of the tag shows the transmission model number and build date code.

The transaxle in General Motors FWD vehicles can be identified by an identification number stamped on a plate attached to the rear face of the transaxle (Figure 3-3). The transmission/transaxle model is printed on the Service Parts Identification label located inside the vehicle.

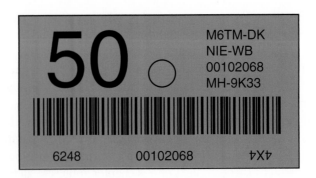

Figure 3-1 An example of a transmission identification sticker. These are typically stuck to the transmission in an easily viewed spot.

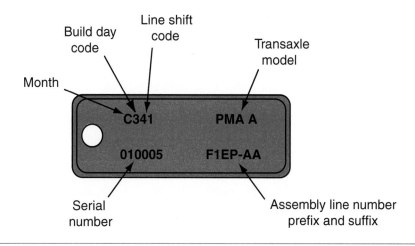

Figure 3-2 A typical identification tag for a Ford Motor Co. automatic transmission.

The transmission in early RWD General Motors products can be identified by the production number, which is located on the ID plate attached to the right side of the case near the modulator. The production number consists of a year code, a two-character model code, and a build-date code.

The transmission in late-model RWD GM products can be identified by a letter code contained in the identification number. The ID number is stamped on the transmission case above the oil pan rail on the right rear side. The identification number contains information that must be used when ordering replacement parts.

The transaxle in many Hondas and Acuras can be identified by an identification number stamped on a metal pad on top of the transmission. The first two characters indicate the transmission model.

Aisin Warner transmissions, which are used by Jeep, Isuzu, and Volvo, can be identified by a plate attached to a side of the transmission case. The plate shows the transmission model number and serial number.

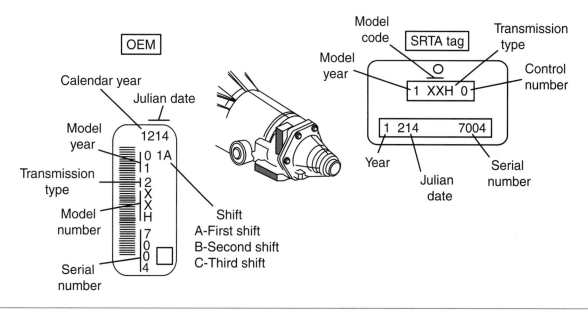

Figure 3-3 Location of and information contained on a General Motors transaxle identification plate.

The transmission used in some models from Audi, Porsche, and Volkswagen can be identified by numbers cast into the top rear of the case. The transaxle model code is identified by figures stamped into the torque converter housing. These figures consist of the model code and build-date code. Some models have code letters and the date of manufacture stamped into the machined flat on the bellhousing rim. The valve body may also be stamped on its machined boss. The valve body identification tag is secured with valve body mounting screws. A torque converter code letter is stamped on the side of the attaching lug. Some of these model vehicles use a Mercedes transmission. These units have their model stamped on the right pan rail of the case.

Models that use Borg Warner transmissions have an identification number stamped on a plate attached to the torque converter housing near the throttle cable and distributor.

Jatco transmissions, which are manufactured by the Japan Automatic Transmission Company, are used by Chrysler, Mitsubishi, Mazda, and Nissan. The model may be identified by a stamped metal plate attached to the right side of the transmission case. The model code appears on the second line and the serial number is on the bottom line.

Some Mazda, Mercedes Benz, Toyota, and Subaru (Gunma) transmissions are best identified by the eleventh character in the VIN, which is located at the top left of the instrument panel and on the transaxle flange on the exhaust side of the engine or the driver's door post. Most Mercedes Benz transmissions have the identification number stamped into the pan rail on the right side of the case.

BMW, Peugeot, and some models of Volvo use ZF transmissions, which have an identification plate fixed to the left side of the transmission case. The lower left series of numbers on the plate indicate the number of gears, type of controls, type of gears, and torque capacity.

Diagnostics

An automatic transmission receives engine power through a torque converter, which is indirectly attached to the engine's crankshaft. Hydraulic pressure in the converter allows power to flow from the torque converter to the transmission's input shaft. The only time a torque converter is mechanically connected to the input shaft of the transmission is during torque converter clutch engagement.

The input shaft drives a planetary gearset, which provides the different forward gears, a neutral position, and reverse. Power flow through the gears is controlled by multiple friction disc packs, one-way clutches, and friction bands (Figure 3-4). These hold a member of the gearset when hydraulic pressure is applied to them. By holding different members of the planetary gearset, different gear ratios are possible.

Hydraulic pressure is routed to the correct holding element by the transmission's valve body, which contains many hydraulic valves and controls the pressure and direction of the hydraulic fluid.

Because of the complexities involved in the operation of an automatic transmission, diagnostics can be quite complicated. This is especially true if the technician does not have a thorough understanding of the operation of a normal working transmission.

Before beginning to rebuild or repair a transmission, make sure it has a problem and requires repairs. Gathering as much information as you can to describe the problem will help you decide if the problem is a transmission problem.

Automatic transmission problems are usually caused by one or more of the following conditions: poor engine performance, problems in the hydraulic system, abuse resulting in overheating, mechanical malfunctions, electronic failures, or improper adjustments. Begin to diagnose these problems by checking the condition of the fluid and its level, conducting a thorough visual inspection, and by checking the various linkage adjustments. Although electronics problems are quite common, these will be covered in detail in the next chapter.

Poor engine performance can have a drastic effect on transmission operation. Low engine vacuum will cause a vacuum modulator to sense a load condition when it is actually not present.

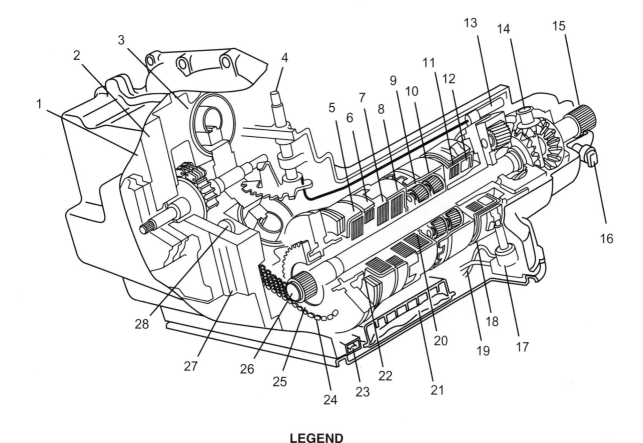

LEGEND

1. TFP manual valve position switch assembly
2. Control valve body assembly
3. Torque converter assembly
4. Manual shaft and detent lever assembly
5. 2nd clutch
6. Reverse clutch
7. Coast clutch
8. Direct clutch
9. Reaction planetary gearset
10. Input planetary gearset
11. Forward clutch
12. Lo roller clutch
13. Parking lock actuator assembly
14. Differential and final drive assembly

15. Output stub shaft
16. Output speed sensor
17. Lo/Reverse servo assembly
18. Lo/Reverse band
19. Oil feed tube assembly
20. Input sprag clutch assembly
21. Oil filter assembly
22. 2nd roller clutch
23. Oil level control
24. Drive link assembly
25. Driven sprocket
26. Output shaft
27. Channel plate assembly
28. Input speed sensor

Figure 3-4 Basic components of a 4T45-E transaxle. Note the location of planetary gearset and apply and holding devices. (Courtesy of General Motors Corporation, Service Operations)

This will cause delayed and harsh shifts. Delayed shifts can also result from the action of the throttle valve (T.V.) assembly, if the engine runs so badly that the throttle pedal must be frequently pushed to the floor to keep it running or to get it going.

Engine performance can also affect torque converter clutch operation. If the engine is running too poorly to maintain a constant speed, the converter clutch will engage and disengage at higher speeds. The customer complaint may be that the converter chatters; however, the problem may be the result of engine misses.

At higher elevations, the atmospheric pressure—and thus engine vacuum—is lower. Normally, measured vacuum will decrease one inch for each 1000 feet of elevation.

Classroom Manual
Chapter 3, page 60

If the dipstick has a HOT level marked on it, this mark should be used only when the dipstick is hot to the touch.

If the vehicle has an engine performance problem, the cause should be found and corrected before any conclusions on the transmission are made. A quick way to identify if the engine is causing shifting problems is to connect a vacuum gauge to the engine and take a reading while the engine is running. The gauge should be connected to the intake manifold vacuum, anywhere below the throttle plates. A normal vacuum gauge reading is steady and at 17 in. Hg. The rougher the engine runs, the more the gauge readings will fluctuate. The lower the vacuum readings, the more severe the problem.

Diagnosing a problem should follow a systematic procedure to eliminate every possible cause that can be corrected without removing the transmission.

In order to properly diagnose a problem, you must totally understand the customer's concern or complaint. Make sure you know all you can about the conditions that exist when the problem occurs. The following are conditions that will help you in your diagnosis, so get as much detail as you can before proceeding with your troubleshooting.

- ❏ Cold or hot vehicle operating temperatures
- ❏ Cold or hot outside temperatures
- ❏ Loaded or unloaded vehicle
- ❏ City or highway driving
- ❏ Type of terrain
- ❏ Upshifting
- ❏ Downshifting
- ❏ Particular gear ranges
- ❏ Coasting
- ❏ Braking
- ❏ Transmission maintenance

Fluid Check

Your diagnosis should begin with a fluid level check. To check the ATF level, make sure the vehicle is on a level surface. Before removing the dipstick, wipe all dirt off the protective disc and the dipstick handle. On most automobiles, the ATF level can be checked accurately only when the transmission is at operating temperature and the engine is running. Remove the dipstick and wipe it clean with a lintfree white cloth or paper towel. Reinsert the dipstick, remove it again, and note the reading. Markings on a dipstick indicate ADD levels, and on some models, FULL levels for cool, warm, or hot fluid (Figures 3-5 and 3-6).

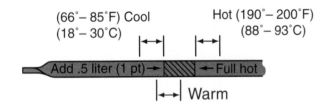

Note: Do not overfill. It takes only one pint to raise level from "Add" to "Full" with a hot transmission.

Figure 3-5 Typical dipstick markings for an automatic transmission. (Courtesy of General Motors Corporation, Service Operations)

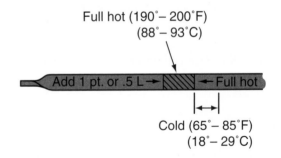

Note: "Cold" reading is ABOVE "Full" mark.

Figure 3-6 Typical dipstick markings for an automatic transaxle. (Courtesy of General Motors Corporation, Service Operations)

✔ SERVICE TIP: The temperature of the transmission or transaxle is a vital factor when checking ATF levels. Fluid level checking can be misleading because some transmissions have bimetallic elements that block fluid flow to the transmission cooler until a predetermined temperature is reached. The bimetal element keeps the fluid inside the transmission hot.

If the fluid level is low or off the crosshatch section of the dipstick, the problem could be external fluid leaks. Check the transmission case, oil pan, and cooler lines for evidence of leaks.

Low fluid levels may also be caused by poor maintenance. The level needs to be checked on a regular basis. If the transmission is equipped with a vacuum modulator, low fluid level may result from a bad diaphragm in the modulator. Engine vacuum will draw the fluid into the cylinders where the engine will burn it. This problem may be found quickly by looking for fluid in the vacuum hose at the modulator. There should be none.

Low fluid levels can cause a variety of problems. Air can be drawn into the oil pump's inlet circuit and mixed with the fluid. This will result in aerated fluid, which causes slow pressure buildup and low pressures, which will cause slippage between shifts. Air in the pressure regulator valve will cause a buzzing noise when the valve tries to regulate pump pressure.

Excessively high fluid levels can also cause **aeration.** As the planetary gears rotate in high fluid levels, air can be forced into the fluid. Aerated fluid can foam, overheat, and oxidize. All of these problems can interfere with normal valve, clutch, and servo operation. Foaming may be evident by fluid leakage from the transmission's vent.

> **Aeration** is the process of mixing air with a liquid.

● CUSTOMER CARE: Customers should be made aware that fluid level and condition should be checked at least every six months. Temperature fluctuations from summer to winter can cause a thermal breakdown of ATF. Even high-quality fluids can experience breakdown as a result of these frequent and extreme temperature changes. It is said that nearly 90% of all transmission failures are caused by fluid breakdown or **oxidation.**

The condition of the fluid should be checked while checking the fluid level. Examine the fluid carefully. The normal color of ATF is pink or red. If the fluid has a dark brownish or blackish color and/or a burned odor, the fluid has been overheated. A milky color indicates that water is mixed with the fluid possibly by engine coolant leaking into the transmission's cooler in the radiator. If there is any question about the condition of the fluid, drain out a sample for closer inspection.

After checking the ATF level and color, wipe the dipstick on absorbent white paper and look at the stain left by the fluid. Dark particles are normally band or clutch material, while silvery metal particles are normally caused by the wearing of the transmission's metal parts. If the dipstick cannot be wiped clean, it is probably covered with varnish, which results from fluid oxidation. Varnish will cause the spool valves to stick, leading to improper shifting speeds. Varnish or other heavy deposits indicate the need to change the transmission's fluid and filter.

> **Oxidation** occurs when something is mixed with oxygen to produce an oxygen-containing compound. Oxidation is also the term given to the chemical breakdown of a substance or compound caused by its combination with oxygen.

✔ SERVICE TIP: Contaminated fluid can sometimes be felt better than seen. Place a few drops of fluid between two fingers and rub them together. If the fluid feels dirty or gritty, it is contaminated with burned frictional material.

Fluid Changes

● CUSTOMER CARE: Abusive driving can overheat a transmission and cause fluid oxidation and breakdown. Inform your customers that they should always stay within the recommended towing load for their vehicle. Also tell them to avoid excessive rocking back and forth when they are stuck in snow or mud.

The transmission's fluid and filter should be changed whenever there is an indication of oxidation or contamination. Photo Sequence 3 shows the correct procedure for performing an

> **Special Tools**
> Large drain pan
> Clean white rags
> Inch-pound torque wrench

Photo Sequence 3
Typical Procedure for Performing an Automatic Transmission Fluid and Filter Change

P3-1 To begin the procedure for changing automatic transmission fluid and filter, raise the car on a lift. Before working under the car, make sure it is safely positioned on the lift.

P3-2 Locate the transmission pan. Remove any part that may interfere with the removal of the pan.

P3-3 Position the oil drain pan under the transmission pan. A large diameter drain pan helps prevent spills.

P3-4 Loosen all of the pan bolts and remove all but three at one end. This will cause fluid to flow out around the pan into the drain pan.

P3-5 Supporting the pan with one hand, remove the remaining bolts and pour the fluid from the transmission pan into the drain pan.

P3-6 Carefully inspect the transmission pan and the residue in it. The condition of the fluid is often an indication of the condition of the transmission and serves as a valuable diagnostic tool.

P3-7 Remove the old pan gasket and wipe the transmission pan clean with a lint-free rag. Check the magnet; metal particles may indicate component damage.

P3-8 Unbolt the fluid filter from the transmission's valve body. Keep the drain pan under the transmission while doing this. Additional fluid may leak out of the filter or the valve body.

P3-9 Before continuing, compare the old filter and pan gasket with the new ones.

Photo Sequence 3 (cont'd.)
Typical Procedure for Performing an Automatic Transmission Fluid and Filter Change

P3-10 Install the new filter onto the valve body and tighten the attaching bolts to proper specifications. Then lay the new pan gasket over the sealing surface of the pan and move the pan into place.

P3-11 Install the attaching bolts and tighten them to the recommended specifications. Carefully read the specifications for installing the filter and pan. Some transmissions require torque specifications of inch-pounds rather than the typical foot-pounds.

P3-12 With the pan tightened, lower the car. Pour ATF into the transmission. Start the engine. With the parking brake applied and your foot on the brake pedal, move the shift lever through the gears. This allows the fluid to circulate throughout the entire transmission. After the engine reaches normal operating temperature, check the transmission fluid level and correct it if necessary.

automatic transmission fluid and filter change. Periodic fluid and filter changes are also part of the preventive program for most vehicles. The frequency of this service depends on the conditions that the transmission normally operates under. Severe usage requires that the fluid and filter be changed every 15,000 miles. Severe usage is defined as:

1. More than 50% operation in heavy city traffic during hot weather (above 90°F).
2. Police, taxi, commercial-type operation, and trailer towing.

The mileage interval that a manufacturer will recommend for a fluid and filter change will also depend on the type of transmission. For example, some General Motors transmissions use aluminum valves in their valve body. Since aluminum is softer than steel, aluminum valves are less tolerant of abrasives and dirt in the fluid. Therefore, to prolong the life of the valves, more frequent fluid changes are recommended by General Motors.

CUSTOMER CARE: Older transmissions did not require as frequent fluid and filter changes as do the later-model vehicles. Customers should be made aware of the recommended frequency and the reason for this change. In older vehicles, both the engine and transmission ran cooler, which extended the life of transmission fluid. These transmissions operated at 175° F and the transmission fluid lasted about 100,000 miles before oxidizing. However, fluid life is halved with every 20° its temperature rises above 175° F. For example, fluid life expectancy drops to 50,000 miles at 195° F and just 25,000 miles at 215° F, which is the temperature at which most newer transmissions run.

WARNING: Be careful when draining the transmission fluid. It can be very hot and will tend to adhere to your skin, causing severe burns.

Change the fluid only when the engine and transmission are at normal operating temperatures. On most transmissions, you must remove the oil pan to drain the fluid. Some transmission

pans on recent vehicles include a drain plug. A filter or screen is normally attached to the bottom of the valve body. Filters are made of paper or fabric and are held in place by screws, clips, or bolts (Figure 3-7). Filters should be replaced, not cleaned.

To drain and refill a typical transmission, the vehicle must be raised on a hoist. After the vehicle is safely in position, place a drain pan with a large opening under the transmission's oil pan. Then loosen the pan bolts and tap the pan at one corner to break it loose. Fluid will begin to drain from around the pan. When all fluid is drained, remove the oil pan.

After draining, carefully remove the pan. There will be some fluid left in the pan. Be prepared to dump it into the drain pan. Check the bottom of the pan for deposits and metal particles. Slight contamination, blackish deposits from clutches and bands, is normal. Other contaminants should be of concern. Clean the oil pan and its magnet.

Remove the filter and inspect it. Use a magnet to determine if metal particles are steel or aluminum. Steel particles indicate severe internal transmission wear or damage. If the metal particles are aluminum, they may be part of the torque-converter stator. Some torque converters use phenolic plastic stators; therefore, metal particles found in these transmissions must be from the transmission itself. Filters are always replaced, whereas screens are cleaned. Screens are removed in the same way as filters. Clean a screen with fresh solvent and a stiff brush. Remove any traces of the old pan gasket on the case housing. Then, install a new filter and gasket on the bottom of the valve body and tighten the retaining bolts to the specified torque. If the filter is sealed with an O-ring, make sure it is properly installed.

CAUTION: Make sure you check the specifications. The required torque is often given in inch-pounds. You can easily break the bolts or damage something if you tighten the bolts to foot-pounds.

Remove any traces of the old pan gasket from the oil pan. Make sure the mounting flange of the oil pan is not distorted or bent. Then, reinstall the pan using the gasket or sealant recommended by the manufacturer. Normally RTV sealant is recommended. Tighten the pan retaining bolts to the specified torque; over tightening can cause pan leaks.

Pour a little less than the required amount of fluid into the transaxle through the dipstick tube. Always use the recommended type of ATF. The wrong fluid will alter the shifting characteristics of the transmission. For example, if type F fluid is used in a transmission designed for Dexron-type fluid, the shifting will be harsher.

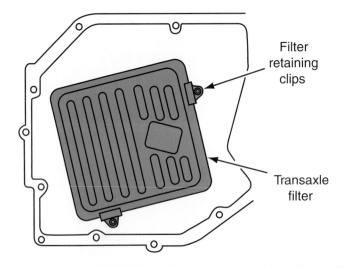

Figure 3-7 Transmission fluid filters are attached to the transmission case by screws, bolts, and/or retaining clips.

Start the engine and allow it to idle for at least one minute. Then, with the parking and service brakes applied, move the gear selector lever momentarily to each position, ending in PARK. Recheck the fluid level and add a sufficient amount of fluid to bring the level to about 1/8-inch below the ADD mark.

 SERVICE TIP: On transmissions with a governor, after starting the engine, quickly move the gear selector to reverse. Many transmissions do not feed fluid to the governor when they are in reverse. By quickly placing the transmission in reverse, the chance of dirt entering the governor is greatly decreased.

Run the engine until it reaches normal operating temperature. Then recheck the fluid level. It should be in the HOT region on the dipstick. Make sure the dipstick is fully seated into the dipstick tube opening. This will prevent dirt from entering into the transmission.

Some transaxles vent through the dipstick handle or tube.

Case porosity is caused by tiny holes that are formed by trapped air bubbles during the casting process.

Some transmissions do not use a gasket; instead the pan is sealed with RTV.

Fluid Leaks

Continue your diagnostics by conducting a quick and careful visual inspection. Check all drivetrain parts for looseness and leaks. If the transmission fluid was low or if there was no fluid, raise the vehicle and carefully inspect the transmission for signs of leakage. Leaks are often caused by defective gaskets or seals. Common sources of leaks are the oil pan seal, rear cover, and final drive cover (on transaxles), extension housing, speedometer drive gear assembly, and electrical switches mounted into the housing (Figure 3-8). The housing itself may have a porosity problem, allowing fluid to seep through the metal. Case porosity may be repaired using an epoxy-type sealer.

Oil Pan

A common cause of fluid leakage is the seal of the oil pan to the transmission housing. If there are signs of leakage around the rim of the pan, retorquing the pan bolts may correct the problem. If tightening the pan does not correct the problem, the pan must be removed and a new gasket installed. Make sure the sealing surface of the pan's rim is flat and capable of providing a seal before reinstalling it.

Classroom Manual
Chapter 3, page 72

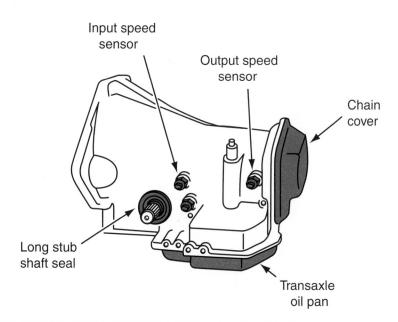

Figure 3-8 Possible sources of fluid leaks on the transaxle.

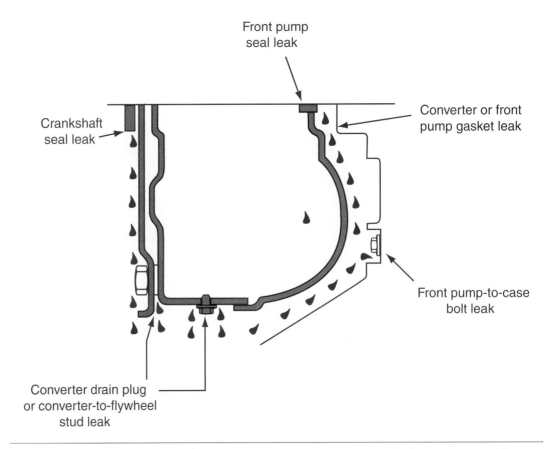

Figure 3-9 By determining the direction of fluid travel, the cause of a fluid leak around the torque converter can be identified.

Torque Converter

Torque converter problems can be caused by a leaking converter (Figure 3-9). This type of problem may be the cause of customer complaints of slippage and a lack of power. To check the converter for leaks, remove the converter access cover and examine the area around the torque converter shell. An engine oil leak may be falsely diagnosed as a converter leak. The color of engine oil is different from transmission fluid and may help identify the true source of the leak. However, if the oil or fluid has absorbed much dirt, both will look the same. An engine leak typically leaves an oil film on the front of the converter shell, whereas a converter leak will cause the entire shell to be wet. If the transmission's oil pump seal is leaking, only the back side of the shell will be wet. If the converter is leaking or damaged, it should be replaced.

Extension Housing

An oil leak stemming from the mating surfaces of the extension housing and the transmission case may be caused by loose bolts. To correct this problem, tighten the bolts to the specified torque. Also check for signs of leakage at the rear of the extension housing. Fluid leaks from the seal of the extension housing can be corrected with the transmission in the car. Often, the cause for the leakage is a worn extension housing bushing, which supports the sliding yoke of the drive shaft. Normally, bushing wear or damage is caused by faulty Universal joints. When the drive shaft is installed, the clearance between the sliding yoke and the bushing should be minimal. If the clearance is satisfactory, a new oil seal will correct the leak. If the clearance is excessive, the repair requires that a new seal and a new bushing be installed. If the seal is faulty, the transmission vent should be checked for blockage.

Classroom Manual
Chapter 3, page 72

Rear Oil Seal and Bushing Replacement

Procedures for the replacement of the rear oil seal and bushing vary little with each car model. General procedures for the replacement of the oil seals and bushings follow.

To replace the rear seal:

1. Remove the drive shaft.
2. Remove the old seal from the extension housing (Figure 3-10).
3. Lubricate the lip of the seal, then install the new seal in the extension housing.
4. Install the drive shaft.

To replace the rear bushing and seal:

1. Remove the drive shaft from the car. Inspect the yoke for wear and replace it if necessary.
2. Insert the appropriate puller tool into the extension housing until it grips the front side of the bushing.
3. Pull the seal and bushing from the housing.
4. Using the correct tools, drive a new bushing into the extension housing.
5. Install a new seal in the housing (Figure 3-11).
6. Install the drive shaft.

Speedometer Drive

The vehicle's speedometer can be purely electronic, which requires no mechanical hookup to the transmission, or it can be driven off the output shaft via a cable. If the transmission is equipped with a vehicle speed sensor, the bore and sensor can be a source of leaks. The sensor is retained

Special Tools

Slide hammer

Seal remover tool

Bushing removing tool

Seal driver

Bushing driver

Classroom Manual
Chapter 3, page 75

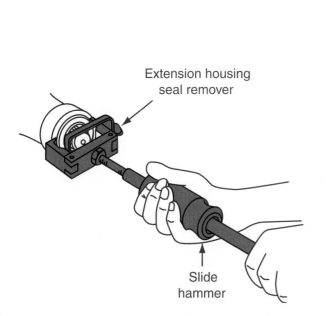

Figure 3-10 Extension housing rear seals can be removed with a slide hammer and a special removal tool.

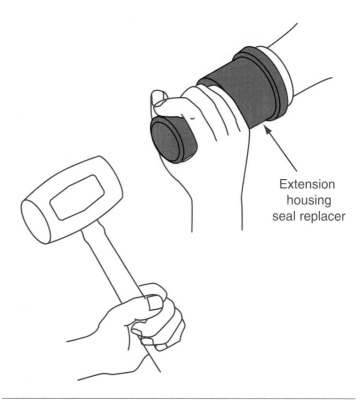

Figure 3-11 Rear seals can be installed easily into the extension housing with the proper driver and a hammer.

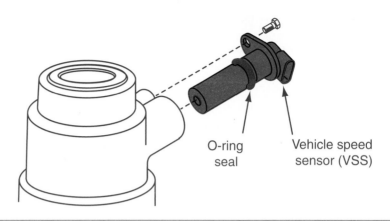

O-ring
seal

Vehicle speed
sensor (VSS)

Figure 3-12 A VSS and its position in a transaxle.

in the bore with a retaining nut or bolt (Figure 3-12). An oil leak at the speedometer cable or vehicle speed sensor (VSS) can be corrected by replacing the O-ring seal. While replacing the seal, inspect the drive gear for chips and missing teeth. Always lubricate the O-ring and gear prior to installation.

When replacing the speedometer drive seal, first clean the top of the speedometer cable retainer. Then remove the hold-down screw, which keeps the retainer in its bore. Carefully pull up on the speedometer cable, pulling the speedometer retainer and drive gear assembly from its bore. Carefully remove the old seal.

To reinstall the retainer, lightly grease and install the O-ring onto the retainer. Gently tap the retainer and gear assembly into its bore while lining up the groove in the retainer with the screw hole in the side of the clutch housing case. Install the hold-down screw and tighten it in place.

Electrical Connections

Shop Manual
Chapter 3, page 52

A visual inspection of the transmission should include a careful check of all electrical wires, connectors, and components. This inspection is especially important for electronically controlled transmissions and for transmissions that have a torque converter clutch. All faulty connectors, wires, and components should be repaired or replaced before continuing your diagnosis of the transmission.

Check all electrical connections to the transmission. Faulty connectors or wires can cause harsh or delayed and missed shifts. On transaxles, the connectors can normally be inspected through the engine compartment, whereas they can only be seen from under the vehicle on longitudinally mounted transmissions. To check the connectors, release the locking tabs and disconnect them, one at a time, from the transmission. Carefully examine them for signs of corrosion, distortion, moisture, and transmission fluid. A connector or wiring harness may deteriorate if ATF reaches it. Also check the connector at the transmission. Using a small mirror and flashlight may help you get a good look at the inside of the connectors. Inspect the entire transmission wiring harness for tears and other damage. Road debris can damage the wiring and connectors mounted underneath the vehicle.

Even the slightest amount of corrosion can affect the output of a sensor. Increased resistance will always change the voltage signal of a circuit.

A **throttle position sensor** is most often referred to as a TP sensor.

Because the operation of the engine and transmission is integrated through the control computer, a faulty engine sensor or connector may affect the operation of both the engine and the transmission. The various sensors and their locations can be identified by referring to the appropriate service manual. The engine control sensors that are the most likely to cause shifting problems are the throttle position sensor, MAP sensor, and vehicle speed sensor.

Throttle position sensors (Figure 3-13) are typically located on the side of the fuel injection throttle plate assembly. Remove the electrical connector from the sensor and inspect both

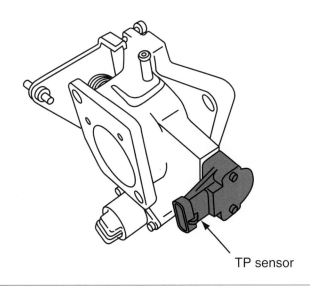

TP sensor

Figure 3-13 Typical throttle position (TP) sensor.

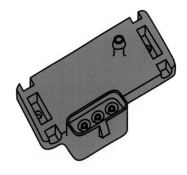

Figure 3-14 Typical manifold absolute pressure (MAP) sensor.

ends for signs of corrosion and damage. Poor contact can cause the transmission to shift erratically. Also inspect the wiring harness to the TP sensor for evidence of damage.

A **manifold absolute pressure (MAP) sensor** (Figure 3-14) usually is located on top of the intake manifold or is mounted to the firewall, where a vacuum hose connects it to the intake manifold. Check both ends of its three-pronged connector and wiring for corrosion and damage. Also check the condition of the vacuum hose. A vacuum leak can cause harsh shifting.

A **vehicle speed sensor** is often used on late-model cars to monitor speed. The function of this sensor is similar to a speedometer cable except it operates electronically rather than mechanically. Proper shift points cannot occur if the signal from the speed sensor is faulty. The sensor's connections and wiring should be inspected for signs of damage and corrosion. These sensors are located near the output shaft of the transmission.

Checking the Transaxle Mounts

The engine and transmission mounts on FWD cars are important to the operation of the transaxle. Any engine movement may change the effective length of the shift and throttle cables and therefore may affect the engagement of the gears. Delayed or missed shifts may result from linkage changes as the engine pivots on its mounts. Problems with transmission mounts may also affect the operation of a transmission on a RWD vehicle, but this type of problem will be less detrimental than the same type of problem on a FWD vehicle. Many shifting and vibration problems can be caused by worn, loose, or broken engine and transmission mounts. Visually inspect the mounts for looseness and cracks. To get a better look at the condition of the mounts, pull up and push down on the transaxle case while watching the mount. If the mount's rubber separates from the metal plate or if the case moves up but not down, replace the mount. If there is movement between the metal plate and its attaching point on the frame, tighten the attaching bolts to an appropriate torque.

Then, from the driver's seat, apply the foot brake, set the parking brake, and start the engine. Put the transmission into a gear and gradually increase the engine speed to about 1500–2000 rpm. Watch the torque reaction of the engine on its mounts. If the engine's reaction to the torque appears to be excessive, broken or worn drivetrain mounts may be the cause.

If it is necessary to replace the transaxle mount, make sure you follow the manufacturer's recommendations for maintaining the alignment of the drive line. Failure to do this may result in poor gear shifting, vibrations, and/or broken cables. Some manufacturers recommend that a holding fixture or special bolt be used to keep the unit in its proper location.

Classroom Manual
Chapter 3, page 75

A **manifold absolute pressure (MAP) sensor**, senses engine vacuum.

Vehicle speed sensors are used to track the current speed and the total miles traveled by a vehicle. This input is used by many computer systems in the vehicle.

Some manufacturers require that the mount bolts be discarded and new ones installed during reassembly.

Classroom Manual
Chapter 3, page 76

When removing the transaxle mount, begin by disconnecting the battery's negative cable. Disconnect any electrical connectors that may be located around the mount. It may be necessary to move some accessories, such as the horn, in order to service the mount without damaging some other assembly. Be sure to label any wires you remove to facilitate reassembly.

Install the engine support fixture (Figure 3-15) and attach it to an engine hoist. Lift the engine just enough to take the pressure off of the mounts. Remove the bolts attaching the transaxle mount to the frame and the mounting bracket, then remove the mount.

To install the new mount, position the transaxle mount in its correct location on the frame and tighten its attaching bolts to the proper torque. Install the bolts that attach the mount to the transaxle bracket. Prior to tightening these bolts, check the alignment of the mount. Once you have confirmed that the alignment is correct, tighten all loosened bolts to their specified torque. Remove the engine hoist fixture from the engine and reinstall all accessories and wires that may have been removed earlier.

> **✔ SERVICE TIP:** In the procedure for removing the mount, some manufacturers recommend the use of a special alignment bolt, which is installed in an engine mount. This bolt serves as an indicator of powertrain alignment. If excessive effort is required to remove the alignment bolt, the powertrain must be shifted to allow for proper alignment.

Classroom Manual
Chapter 3, page 61

Transmission Cooler and Line Inspection

Transmission coolers are a possible source of fluid leaks. The efficiency of the coolers is also critical to the operation and longevity of the transmission. The vehicle may be equipped with the standard type cooler in the radiator or may also have an auxiliary cooler. Both should be visually inspected and tested in the same way.

Follow these steps when inspecting the transmission cooler and associated lines and fittings:

1. Check the engine cooling system. The transmission cooler cannot be efficient if the engine's cooling system is defective. Repair all engine cooling system problems before continuing to check the transmission cooler.

2. Inspect the fluid lines and fitting between the cooler and the transmission (Figure 3-16). Check these for looseness, damage, signs of leakage, and wear. Replace any damaged lines. Replace or tighten any leaking fittings.

3. Inspect the engine's coolant for traces of transmission fluid. If ATF is present in the coolant, the transmission cooler leaks.

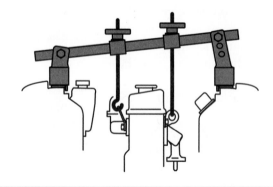

Figure 3-15 A typical engine support fixture in place to remove and/or install engine and transaxle mounts. (Courtesy of General Motors Corporation, Service Operations)

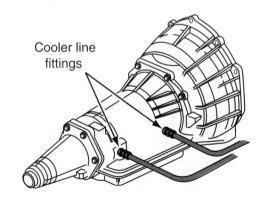

Cooler line fittings

Figure 3-16 Typical location of cooler line connections on a transmission.

4. Check the transmission's fluid for signs of engine coolant. Water or coolant will cause the fluid to appear milky with a pink tint. This is also an indication that the transmission cooler leaks and is allowing engine coolant to enter into the transmission fluid.

The cooler can be checked for leaks by disconnecting and plugging the transmission-to-cooler lines, at the radiator. Then remove the radiator cap to relieve any pressure in the system. Tightly plug one of the ATF line fittings at the radiator. Using the shop air supply with a pressure regulator, apply 50 to 70 psi of air pressure into the cooler at the other cooler line fitting. Look into the radiator. If you see bubbles, the cooler leaks.

The cooler system should also be checked for proper flow. To do this, make sure there are no fluid leaks. Start the engine and allow the transmission to reach normal operating temperature. Then turn the engine off and check the fluid level. Remove the transmission dipstick and install a funnel into the dipstick tube. Raise the vehicle up on a hoist. Then disconnect the cooler return line at the transmission. Securely fasten a rubber hose to the end of the cooler line and place the other end of the hose into the funnel. Now lower the vehicle and start the engine. Allow the engine to run at a fast idle and observe the flow of fluid into the funnel.

Another way to do this test is with the help of an assistant. Have the assistant pour the same amount of fluid into the filler tube as you are getting to flow into a bucket. If you get one pint, have the assistant pour in one pint. By replenishing the fluid, your assistant is ensuring that you will not run out of fluid. Make sure the assistant keeps track of what is being poured in. Continue the test for 30 seconds.

The preferred way to check flow is with a flow meter. A flow meter is preferred because it has restrictions on the return side, which is closer to reality than an open hose. To use a flow meter, connect it into the cooler lines, start the engine, and observe.

Most manufacturers list specifications for transmission cooler flow rates. This is normally expressed in pints per second (such as 1 pint in 15–20 seconds). You should observe a steady flow of fluid. If the flow is erratic or weak, the cooler may be restricted, the transmission's pump may be weak, or there is an internal leak in the transmission.

Road Testing the Vehicle

All transmission complaints should be verified by road testing the vehicle and attempting to duplicate the customer's complaint. A knowledge of the exact conditions that cause the symptom and a thorough understanding of transmissions will allow you to accurately diagnose problems. Many problems that appear to be transmission problems may in fact be problems with the engine, the drive shaft, the U-joint or CV joint, the wheel bearings, wheel/tire imbalance, or other conditions.

Make sure these are not the cause of the problem before you begin to diagnose and repair a transmission. Diagnosis becomes easy if you think about what is happening in the transmission when the problem occurs. If there is a shifting problem, think about the parts that are being engaged and what these parts are attempting to do.

SERVICE TIP: Always refer to your service manual to identify the particulars of the transmission you are diagnosing. It is also helpful to check for any Technical Service Bulletins that may be related to the customer's complaint.

Noises

Abnormal transmission noises and vibrations can be caused by faulty bearings, damaged gears, worn or damaged clutches and bands, or a bad oil pump as well as by contaminated fluid or improper fluid levels. Vibrations can also be caused by torque converter problems. All noises and vibrations that occur during the road test should be noted, as well as when they occur.

Transaxles have unique noises associated with them because of their construction. These noises can result from problems in the differential or drive axles. Use the following guide to determine if a noise is caused by the transaxle or by other parts in the drivetrain.

A knock at low speeds:

1. Worn drive axle CV joints or U-joints

2. Worn side gear hub counterbore

Noise most pronounced on turns:

1. Differential gear noise

2. Dry, worn CV joints

Clunk on acceleration or deceleration:

1. Loose engine mounts

2. Worn differential pinion shaft or side gear hub counterbore in the case

3. Worn or damaged drive axle inboard CV joints or U-joints

Clicking noise in turns:

1. Worn or damaged outboard CV joint

Road testing a vehicle to exactly duplicate the symptoms is essential to proper diagnosis. During the road test, the transmission should be operated in all possible modes and its operation noted. The customer's complaint should be verified and any other transmission malfunctions should be noted. Your observations during the road test, your understanding of the operation of a transmission, and the information given in service manuals will help you identify the cause of any transmission problem. If conducted properly, road testing is diagnosis through a process of elimination.

Before beginning your road test, find and duplicate from a service manual the chart (Figure 3-17) that shows the band and clutch application for different gear selector positions. Using these charts will greatly simplify your diagnosis of automatic transmission problems. It is also wise to have a notebook or piece of paper to jot down notes about the operation of the transmission. If the transmission is electronically controlled, take a scan tool on the road test with you. It will be an added convenience if the scan tool has memory or a printer. By doing this, you won't have to worry about remembering all of the details, nor will you have to decide if you remembered them correctly. Some manufacturers recommend the use of a symptoms chart or checklist that is provided in their service manuals (Figure 3-18).

Prior to road testing a vehicle, always check the transmission fluid level and condition of oil and correct any engine performance problems. Inspect the transmission for signs of fluid leakage. If leakage is evident, wipe off the leaking oil and dust. Then, note the location of the leaks in your notebook or on the symptoms chart.

Begin the road test with a drive at normal speeds to warm the engine and transmission. If a problem appears only when starting or when the engine and transmission are cold, record this symptom on the chart or in your notebook.

After the engine and transmission are warmed up, place the shift selector into the DRIVE position and allow the transmission to shift through all of its normal shifts. Any questionable transmission behavior should be noted.

During a road test, check for proper gear engagement as you move the selector lever to each gear position, including PARK. There should be no hesitation or roughness as the gears are engaging. Check for proper operation in all forward ranges, especially the 1–2, 2–3, 3–4 upshifts, and

Classroom Manual
Chapter 3, page 47

Range Reference

Range	Park/Neutral	Reverse	D				3			2			1	
Gear	N	R	1st	2nd	3rd	4th	1st	2nd	3rd	1st	2nd	3rd**	1st	2nd***
1-2 Shift Solenoid	ON	ON	ON	OFF	OFF	ON	ON	OFF	OFF	ON	OFF	OFF	ON	OFF
2-3 Shift Solenoid	OFF	OFF	OFF	OFF	ON	ON	OFF	OFF	ON	OFF	OFF	ON	OFF	OFF
2nd Clutch	—	—	—	A	A*	A*	—	A	A*	—	A	A*	—	A
2nd Roller Clutch	—	—	—	H	O	—	—	H	O	—	H	O	—	H
Int./4th Band	—	—	—	—	—	A	—	—	—	—	A	—	—	A
Reverse Clutch	—	A	—	—	—	—	—	—	—	—	—	—	—	—
Coast Clutch	—	—	—	—	—	A	A	A	A	A	A	A	A	A
Input Sprag	—	—	H	H	H	O	H	H	H	H	H	H	H	H
Direct Clutch	—	—	—	—	A	A	—	—	A	—	A	—	—	—
Forward Clutch	—	—	A	A	A	A*	A	A	A	A	A	A	A	A
Lo/Rev Band	A	A	—	—	—	—	—	—	—	—	—	—	A	—
Lo Roller Clutch	—	—	H	O	O	O	H	O	O	H	O	O	H	O

A=Applied. H=Holding. O=Overrunning. ON=The solenoid is energized. OFF= The solenoid is de-energized.
*=Applied with no load. **Manual SECOND/THIRD gear is only available above approximately 100 km/k (62 mph).
***=Manual FIRST/SECOND gear is only available above approximately 60 km/h (37 mph). Manual FIRST/THIRD gear is also possible at high vehicle speed as safety feature.

Figure 3-17 Typical application chart for the brakes, clutches, and solenoids of a transmission. This chart should be referred to during a road test and when determining the cause of any shifting problems. (Courtesy of General Motors Corporation, Service Operations)

converter clutch engagement during light throttle operation. These shifts should be smooth and positive and should occur at the correct speeds. These same shifts should feel firmer under medium-to-heavy throttle pressures. Transmissions equipped with a torque converter clutch should be brought to the specified apply speed and their engagement noted. Again, record the operation of the transmission in these different modes in your notebook or on the diagnostic chart.

TRANSMISSION/TRANSAXLE-CONCERN CHECK SHEET

Date ____/____/____

TRANSMISSION/TRANSAXLE-CONCERN CHECK SHEET
(*information required for technical assistance)

*VIN _____ * Mileage_____ R.O.# ____

*Model year _____ *Vehicle Model_____ *Engine ____

* Trans. Model _____ *Trans. Serial _____

S E R V I C E   A D V I S O R

CUSTOMER'S CONCERN

_____
_____
_____
_____

Check the items that describe the concern:

WHAT:	WHEN:	OCCURS:	USUALLY NOTICED:
_no power	_vehicle warm	_always	_idling
_shifting	_vehicle cold	_intermittent	_accelerating
_slips	_always	_seldom	_coasting
_noise	_not sure	_first time	_braking
_shudder			at ____MPH

T E C H N I C I A N

Preliminary Check Procedures   NOTE FINDINGS

Inspect
*fluid level and condition   _____
*engine performance-vacuum & EMC codes   _____
*TV cable and/or modulator vacuum   _____
*manual linkage adjustment   *Simulate the conditions
road test to verify concern   under which the customer's
        concern was observed

* PROPOSED OR COMPLETED REPAIRS

_____
_____
_____

TRANS. TEMPERATURE   _____HOT   _____COLD

*VACUUM READINGS AT MODULATOR
(180C, 250C, 350C, 400, 410-T4)
Readings at Modulator_____IN. HG.(Engine at Hot Idle, Trans. in Drive)
Check for Vacuum Response During Accelerator Movement

P R E S S U R E   T E S T

PRN / D / O21

	MINIMUM		MAXIMUM	
	SPEC. (from manual)	ACTUAL	SPEC. (from manual)	ACTUAL
	_____	_____	_____	_____
	_____	_____	_____	_____
	_____	_____	_____	_____

*FINDINGS BASED ON ROAD TEST

R O A D   T E S T

Check items on road test that describe customer's comments about:

Garage Shift	Upshifts	Downshifts	Torque Converter
Feel	_early	_busyness	Clutch
_engine stops	_harsh	_harsh	_busyness
_harsh	_delayed	_delayed	_harsh
_delayed	_slips	_slips	_no release
_no drive	_no upshift	_no downshift	_shudder
			_early apply
			_no apply
			_late apply

Concerns Occur When/During:

Check	Gear	Range	During
		P-N	_1-2 upshift
	1st		_2-3 upshift
	2nd	D	_3-4 upshift
	3rd		_4-3 downshift
	4th		_3-2 downshift
	1st		_2-1 downshift
	2nd	D	
	3rd		
	1st	2	Throttle Position
	2nd		_light
	1st	1	_medium
	Reverse	R	_heavy
			_W.O.T.

Noise
Type   When Noticed
_buzz   _always
_whine   _load sensitive
_clunk   _steering sensitive
        at _____MPH
        in _____gear

Pitch   Level
_low   _light
_medium   _medium
_high   _heavy

Figure 3-18 Typical road test checklist recommended by a manufacturer to identify the cause of shifting problems. (Courtesy of General Motors Corporation, Service Operations)

Force the transmission to kick down and record the quality of this shift and the speed at which it downshifts. These will be compared later to the specifications. Manual downshifts should also be made at a variety of speeds. The reaction of the transmission should be noted as should all abnormal noises, and the gears and speeds at which they occur.

After the road test, check the transmission for signs of leakage. Any new leaks and their probable cause should be noted. Then compare your written notes from the road test to the information given in the service manual to identify the cause of the malfunction. The service manual usually has a diagnostic chart to aid you in this process.

☑ **SERVICE TIP:** The following problems and their causes are given as examples. The actual causes of these types of problems will vary with the different models of transmissions. Always refer to the appropriate band and clutch application chart while diagnosing shifting problems. Using the chart will allow you to identify the cause of the shifting problems through the process of elimination.

These are the typical causes of common problems:

❏ If the shift for all forward gears is delayed, a slipping front or forward clutch is indicated.

❏ If there is a delay or slip when the transmission shifts into any two forward gears, a slipping rear clutch is indicated.

❏ If there is a shift delay while the transmission was in DRIVE and it shifted into third gear, either the front or rear clutch may be slipping. To determine which clutch is defective, look at the behavior of the transmission when one of the two clutches is not applied. For example, if the rear clutch is not engaged while the transmission is in reverse and there was no slippage in reverse, the problem of the slippage in third gear is caused by the rear clutch. If there is a delay only during shifts into reverse and from second to third gear, the Reverse/High clutch is slipping.

❏ If there is delayed shifting from first to second gear, there are intermediate clutch or band problems.

❏ If there is slippage in first gear when the gear selector is in the DRIVE position, but not when first gear is manually selected, the one-way or overrunning clutch may be the cause.

It is important to remember that delayed shifts or slippage may also be caused by leaking hydraulic circuits or sticking spool valves in the valve body. Since the application of bands and clutches is controlled by the hydraulic system, improper pressures will cause shifting problems. Other components of the transmission can also contribute to shifting problems. For example, on transmissions equipped with a vacuum modulator, if upshifts do not occur at the specified speeds or do not occur at all, the modulator may be faulty or the vacuum supply line may be leaking.

The reverse/high clutch is sometimes called the direct clutch.

Vacuum Modulator

Diagnosing a vacuum modulator begins with checking the vacuum at the line or hose to the modulator (Figure 3-19). The modulator should be receiving engine manifold vacuum. If it does, there are no vacuum leaks in the line to the modulator. Check the modulator itself for leaks with a handheld vacuum pump (Figure 3-20). The modulator should be able to hold approximately 18 in. Hg.

Classroom Manual
Chapter 3, page 67

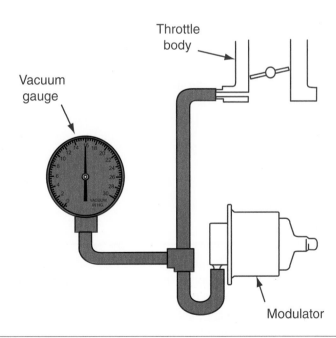

Figure 3-19 The correct hook-up for connecting a vacuum gauge to vacuum modulators.

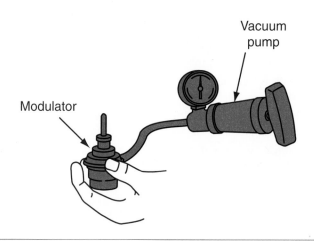

Vacuum pump

Modulator

Figure 3-20 The vacuum modulator can be checked for leakage and action by activating it with a hand-held vacuum pump and observing the vacuum gauge and the action of the modulator.

If transmission fluid is found when you disconnect the line at the modulator, the vacuum diaphragm in the modulator is leaking and the modulator should be replaced. If the vacuum source, vacuum lines, and vacuum modulator are in good condition but shift characteristics indicate a vacuum modulator problem, the modulator may need adjustment.

If no engine vacuum was found at the modulator, check the vacuum line from the engine to the modulator. If the engine is running it is very unlikely that it would have zero vacuum, so some vacuum should be present unless the line or fittings leak.

If the vacuum to modulator was low, check engine vacuum. Run the normal engine vacuum tests to determine where the problem is. A common problem for low vacuum is a restricted exhaust.

A vacuum drop test is a very important test. It checks the soundness of the vacuum sources and connecting lines and hoses. With the engine running, foot on the brake pedal, and the transmission in gear, quickly press the gas pedal to the floor then release it. The reading on the vacuum gauge should fall to zero and return to about 17 in. Hg. immediately. If it falls slowly or goes down only to about 5 in. there is a restriction. A restriction is also noted by slow return to 17 in.

Most modulators must be removed to be adjusted, it if is adjustable. However, there are some that have an external adjustment. This adjustment allows for fine tuning of modulator action. To remove a vacuum modulator from the transmission, loosen the retaining clamp and bolt. Some units are screwed into the transmission case. While pulling the modulator out of the housing, be careful not to lose the modulator actuating pin, which may fall out as the modulator is removed. Use a hand-held vacuum pump with a vacuum gauge and the recommended gauge pins to adjust the modulator according to specifications.

Diagnose Hydraulic and Vacuum Control Systems

A transmission's oil circuit chart is often referred to as the transmission's flow chart.

The best way to identify the exact cause of the problem is to use the results of the road test, logic, and the oil circuit charts for the transmission being worked on. Before doing this, however, you should always check all sources for information about the symptom. Also, make sure you check the basics: trouble codes in the computer, fluid level and condition, leaks, and mechanical and electrical connections. Using the oil circuits, you can trace problems to specific valves, servos, clutches, and bands.

The basic oil flow is the same for all transmissions. The oil pump supplies the fluid flow that is used throughout the transmission. Fluid from the pump always goes to the pressure-regulating valve. From there, the fluid is directed to the manual shift valve. When the gear selector is moved, the manual valve directs the fluid to other valves and to the apply devices. By following the flow of the fluid on the oil circuit chart, you can identify which valves and apply devices should be operating in each particular gear selector position. Through a process of elimination, you can identify the most probable cause of the problem.

In most cases, the transmission or transaxle will need to be removed to repair or replace the items causing the problem. However, some transmissions allow for a limited amount of service to the apply devices and control valves.

SERVICE TIP: Whenever there is a shifting problem that is not readily diagnosed, check the service manual and technical service bulletins before making assumptions about the problem or getting frustrated. Sometimes the cause of a problem is something that you would least expect. For example: A single, and hard to diagnose, problem in a 4L30 transmission can cause it to have no movement, forward movement in neutral, no forward movement, or no reverse gear. All of these complaints may be caused by a defective center support. The center support is made of two pieces, the aluminum housing and the steel sleeve, which is pressed into the center of the support. Grooves machined into the sleeve provide channels for several oil paths through the support. If the sleeve rotates, the potential for several problems exists. The most common complaint is a bind-up in reverse. It is not recommended that the sleeve be removed and a new one installed; rather, the entire center support assembly should be replaced.

Mechanical and/or vacuum controls can also contribute to shifting problems. The condition and adjustment of the various linkages and cables should be checked whenever there is a shifting problem. If all checks indicate that the problem is either an apply device or in the valving, an air pressure test can help identify the exact problem. Air pressure tests are also performed during disassembly to locate leaking seals and during reassembly to check the operation of the clutches and servos.

An air pressure test is conducted by applying clean, moisture-free air, at approximately 40 psi, through a rubber-tipped air nozzle. With the valve body removed, direct air pressure to the case holes that lead to the servo, accumulator, and clutch apply passages is possible (Figure 3-21). Cover the vent hole of the circuit being tested with a clean, lint-free shop towel; this will catch any

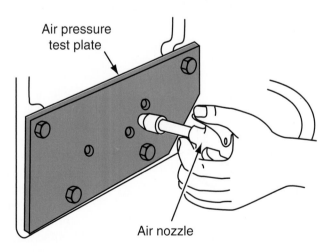

Figure 3-21 Using air pressure through a test plate to test hydraulic circuits.

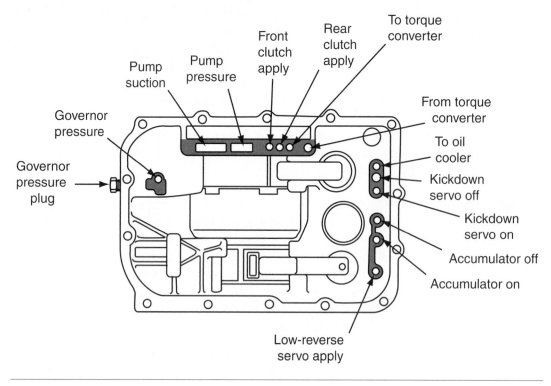

Figure 3-22 Air testing points on a typical transmission.

fluid that may spray out. You should clearly hear or feel a dull thud that indicates the action of a holding device. If a hissing noise is heard, a seal is probably leaking in that circuit (Figure 3-22). If you cannot hear the action of the servo or clutch, apply air pressure again and watch the assembly to see if it is reacting to the air. A servo or accumulator should react immediately to the release of the apply air. If it does not, something is making it stick. Repair or replace the apply devices if they do not operate normally.

Air pressure can normally be directed to the following circuits, through the appropriate hole for each: front clutch, rear clutch, kick down servo, low servo, and reverse servo. Some manufacturers recommend the use of a specially drilled plate and gasket, which is bolted to the transmission case (Figure 3-23). This plate not only clearly identifies which passages to test but also seals off the other passages. Air is applied directly through the holes in the plate.

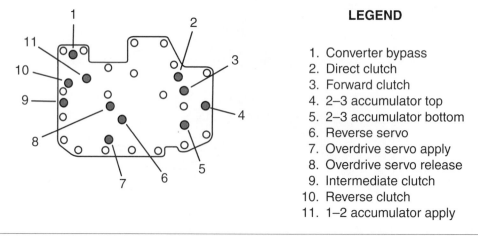

LEGEND

1. Converter bypass
2. Direct clutch
3. Forward clutch
4. 2–3 accumulator top
5. 2–3 accumulator bottom
6. Reverse servo
7. Overdrive servo apply
8. Overdrive servo release
9. Intermediate clutch
10. Reverse clutch
11. 1–2 accumulator apply

Figure 3-23 An example of a transmission air test plate.

Pressure Tests

If you cannot identify the cause of a transmission problem from your inspection or road test, a pressure test should be conducted. This test measures the fluid pressure of the different transmission circuits during the various operating gears and gear selector positions. Refer to the illustrations shown (Figures 3-24, 3-25, 3-26, and 3-27) for a quick review of power flow in the various gears. The number of hydraulic circuits that can be tested varies with the different makes and models of transmissions. However, most transmissions are equipped with pressure taps that allow the pressure test equipment to be connected to the transmission's hydraulic circuits (Figures 3-28 and 3-29). Pressure testing checks the operation of the oil pump, pressure regulator valve, throttle valve, and vacuum modulator system (if the vehicle is equipped with one), plus the governor assembly. Pressure tests can be conducted on all transmissions whether their pressures are regulated by a **variable force motor (VFM)**, vacuum modulator, or through conventional valving.

Classroom Manual
Chapter 3, page 65

Variable force motor (VFM), which is used to change and maintain line pressure.

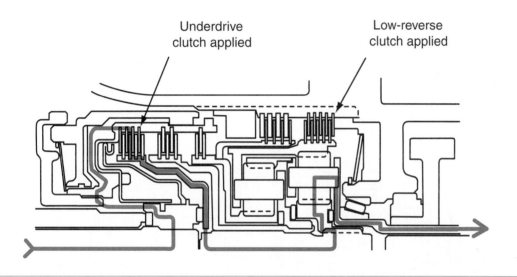

Figure 3-24 Power flow in low gear.

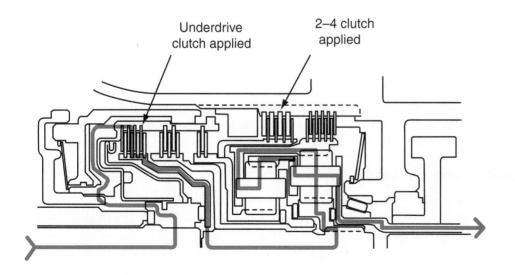

Figure 3-25 Power flow in second gear.

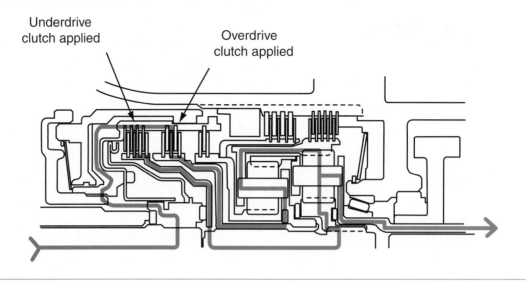

Figure 3-26 Power flow in third gear.

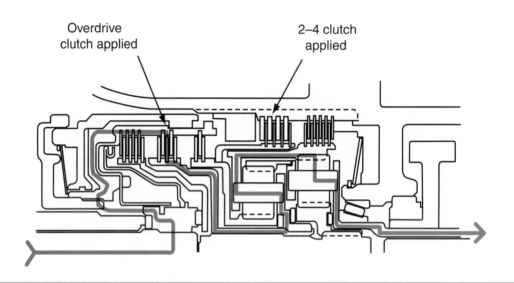

Figure 3-27 Power flow in fourth gear.

A pressure test has its greatest value when the transmission does not shift smoothly or when the shift timing is wrong. Both of these problems may be caused by excessive line pressure, which can be verified by a pressure test.

Conducting a Pressure Test

Before conducting a pressure test on an electronic automatic transmission, correct and clear all trouble codes retrieved from the system. Also make sure the transmission fluid level and condition is okay and that the shift linkage is in good order and properly adjusted. To conduct a pressure test, use a tachometer, two pressure gauges (electronic gauges if available), the correct manufacturer specifications (Figure 3-30), and (for vehicles equipped with a vacuum modulator) a vacuum gauge and a hand-held vacuum pump.

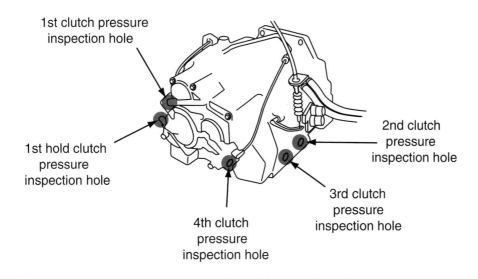

1st clutch pressure inspection hole

1st hold clutch pressure inspection hole

4th clutch pressure inspection hole

3rd clutch pressure inspection hole

2nd clutch pressure inspection hole

Figure 3-28 Pressure taps on the outside of a typical Honda transaxle.

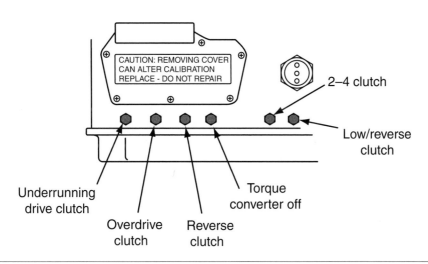

CAUTION: REMOVING COVER CAN ALTER CALIBRATION REPLACE - DO NOT REPAIR

2–4 clutch

Low/reverse clutch

Underrunning drive clutch

Overdrive clutch

Reverse clutch

Torque converter off

Figure 3-29 Pressure taps on the outside of a typical Chrysler transaxle.

The pressure is best conducted with three pressure gauges, but two will work. Two of the gauges should read up to 400 psi and the other to 100 psi. The two 400-psi gauges are usually used to check mainline and an individual circuit, such as mainline and direct or forward circuits. If a circuit is 10 psi lower than mainline pressure when they are both tested at exactly the same time, a leak is indicated. This is why it is best to use two 400-psi gauges. A 100-psi gauge may be used on TV and governor circuits. This ensures an observation of the critical pressures from these two circuits. Next to the scan tool, a pressure gauge is the most valuable tool for automatic transmission diagnostics.

✓ **SERVICE TIP:** When using two or more pressure gauges, be certain to put them on a "Tee" fitting at the same time to calibrate them. The gauges should be calibrated to read the same at 50, 75, 100, 125, and 150 psi.

The pressure gauges are connected to the pressure taps (Figure 3-31) in the transmission housing and routed so that the driver can see the gauges. The vehicle is then road tested and the

Gear Selector Position	Actual Gear	PRESSURE TAPS					
		Underdrive Clutch	Overdrive Clutch	Reverse Clutch	Torque Converter Clutch Off	2/4 Clutch	Low/Reverse Clutch
PARK * 0 mph	PARK	0–2	0–5	0–2	60–110	0–2	115–145
REVERSE * 0 mph	REVERSE	0–2	0–7	165–235	50–100	0–2	165–235
NEUTRAL * 0 mph	NEUTRAL	0–2	0–5	0–2	60–110	0–2	115–145
L # 20 mph	FIRST	110–145	0–5	0–2	60–110	0–2	115–145
3 # 30 mph	SECOND	110–145	0–5	0–2	60–110	115–145	0–2
3 # 45 mph	DIRECT	75–95	75–95	0–2	60–90	0–2	0–2
OD # 30 mph	OVERDRIVE	0–2	75–95	0–2	60–90	75–95	0–2
OD # 50 mph	OVERDRIVE WITH TCC	0–2	75–95	0–2	0–5	75–95	0–2

* Engine speed at 1500 rpm.
# CAUTION: Both front wheels must be turning at the same speed.

Figure 3-30 An air pressure chart. These pressures are for an engine speed of 1500 rpm.

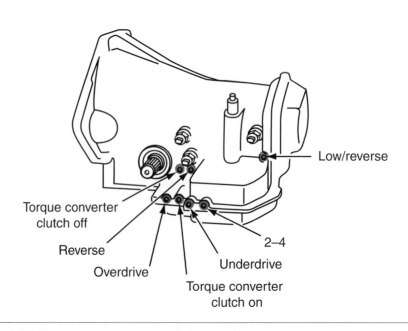

Figure 3-31 The pressure taps on a typical transaxle.

gauge readings observed during the following operational modes: slow idle, fast idle, and wide-open throttle (WOT).

WOT is a common acronym for wide-open throttle.

⚠ **WARNING:** Always stop the engine when connecting and disconnecting the pressure gauges at the transmission. The fluid is under pressure and can easily spray at you and get in your eyes if a hose, fitting, or gauge leaks, so always wear safety goggles when using a pressure gauge.

During the road test, observe the starting pressures and the steadiness of the increases that should occur with slight increases in load. The amount the pressure drops as the transmission shifts from one gear to another should also be noted. The pressure should not drop more than 15 psi between shifts.

Any pressure reading not within the specifications indicates a problem (Figure 3-32). Typically when the fluid pressures are low, there is an internal leak, clogged filter, low oil pump output, or faulty pressure regulator valve. If the fluid pressure increased at the wrong time or the pressure was not high enough, sticking valves or leaking seals are indicated. If the pressure drop between shifts was greater than approximately 15 psi, an internal leak at a servo or clutch seal is indicated. Always check the manufacturer's specifications for maximum drop off before jumping to any conclusions.

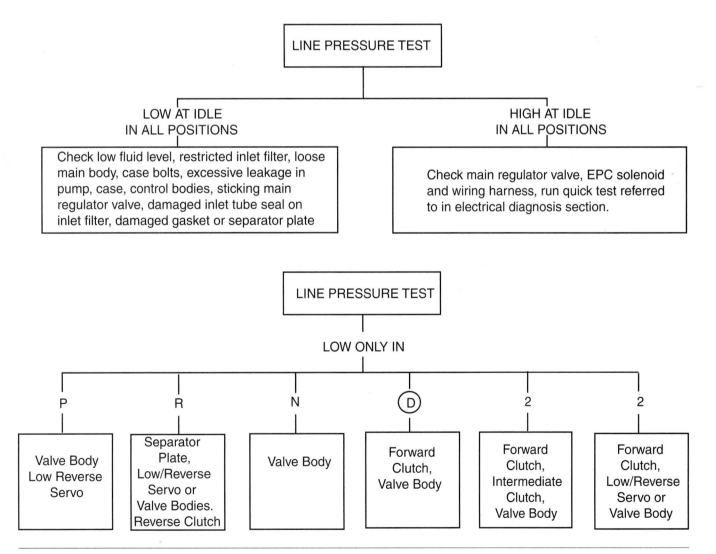

Figure 3-32 A troubleshooting chart for abnormal line pressure.

To maximize the usefulness of a pressure test and to be better able to identify specific problems, begin the test by measuring line pressure. Mainline pressure should be checked in all gear ranges and at the three basic engine speeds. It is very helpful for you to make a quick chart such as the one shown in Figure 3-33 to record the pressures during the test.

If the pressure in all operating gears is within specifications at slow idle, the pump and pressure regulator are working fine. If all pressures are low at slow idle, it is likely that there is a problem in the pump, a stuck open pressure regulator, clogged filter, improper fluid level, or there is an internal pressure leak. To further identify the cause of the problem, check the pressure in the various gears while the engine is at a fast idle.

If the pressures at fast idle are within specifications, the cause of the problem is normally a worn oil pump; however, the problem may be an internal leak or a pressure regulator valve that is stuck open. Internal leaks typically are more evident in a particular gear range because that is when ATF is being sent to a particular device through a particular set of valves and passages. If any of these components leaks, the pressure will drop when that gear is selected or when the transmission is operating in that gear.

Further diagnostics can be made by observing the pressure change when the engine is operated at WOT in each gear range. A clogged oil filter will normally cause a gradual pressure drop at higher engine speeds because the fluid cannot pass through the filter fast enough to meet the needs of the transmission and the faster turning oil pump. If the fluid pressure changed with the increase in engine speed, a stuck pressure regulator is the most probable cause of the problem. A stuck open pressure regulator may still allow the pressure to build with increases in engine speed, but it will not provide the necessary boost pressures.

If the pressures are high at slow idle, a stuck closed pressure regulator, a misassembled or stuck open boost valve, a poorly adjusted or damaged shift cable, or a stuck open throttle valve is indicated. If all of the pressures are low at WOT, pull on the TV cable or disconnect the vacuum hose to the vacuum modulator. If this causes the pressures to be in the normal range, the low pressure is caused by a faulty cable or there is a problem in the vacuum modulator or vacuum lines. If the pressures stayed below specifications, the most likely causes of the problem are the pump or the control system.

If all pressures are high at WOT, compare the readings to those taken at slow idle. If they were high at slow idle and WOT, a faulty pressure regulator is indicated. If the pressures were normal at slow idle and high at WOT, the throttle system is faulty. To verify that the low pressures are caused by a weak or worn oil pump, conduct a *reverse stall test*. If the pressures are low during this test but are normal during all other tests, a weak oil pump is indicated.

	Slow idle	Fast idle	WOT
P			///////
R			
N			///////
D			
3			
2			
1			

Figure 3-33 Pressure chart to aid in diagnostics.

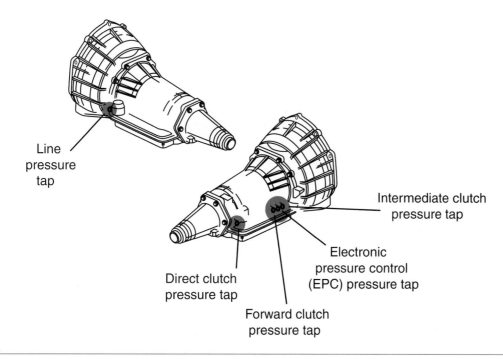

Line pressure tap

Intermediate clutch pressure tap

Direct clutch pressure tap

Electronic pressure control (EPC) pressure tap

Forward clutch pressure tap

Figure 3-34 Location of EPC pressure tap.

Transmission Pressure with TP at 1.5 Volts and Vehicle Speed above 8 Km/h (5 mph)					
Gear	EPC Tap	Line Pressure Tap	Forward Clutch Tap	Intermediate Clutch Tap	Direct Clutch Tap
1	276–345 kPa (40–50 Psi)	689–814 kPa (100–118 Psi)	620–745 kPa (90–108 Psi)	641–779 kPa (93–113 Psi)	0–34 kPa (0–5 Psi)
2	310–345 kPa (45–50 Psi)	731–869 kPa (106–126 Psi)	662–800 kPa (96–116 Psi)	689–827 kPa (100–120 Psi)	655–800 kPa (95–116 Psi)
3	341–310 kPa (35–45 Psi)	620–758 kPa (90–110 Psi)	0–34 kPa (0–5 Psi)	586–724 kPa (85–105 Psi)	551–689 kPa (80–100 Psi)

Figure 3-35 A pressure chart for a transmission equipped with an EPC.

On transmissions equipped with an electronic pressure control (PC) solenoid, if the line pressure is not within specifications the PC pressure needs to be checked. To do this, connect the pressure gauge to the PC tap (Figure 3-34). Start the engine and check PC pressure. Use the appropriate pressure chart for specifications (Figure 3-35). If the PC pressure is not within specifications, follow the recommended procedures for testing the PC. If the pressure is okay, be concerned with mainline pressure problems.

Governors

If the pressure tests suggested that there was a governor problem, it should be removed, disassembled, cleaned, and inspected. Some governors are mounted internally and the transmission

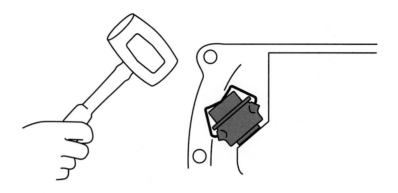

Figure 3-36 Some governor assemblies are contained in a separate housing and retained by a cover and external retaining clamp.

must be removed to service the governor, whereas others can be serviced by removing the extension housing or oil pan, or by detaching an external retaining clamp and then removing the unit (Figure 3-36).

A faulty governor or governor drive gear system typically causes improper shift points. However, some electronically controlled transmissions do not rely on the hydraulic signals from a governor; rather, they rely on the electrical signals from these sensors. Sensors, such as speed and load sensors, signal to the transmission's computer when gears should be shifted. Faulty electrical components and/or loose connections can also cause improper shift points.

Linkages

Many transmission problems are caused by improper adjustment of the linkages. All transmissions have either a cable or a rod-type gear selector linkage. Some transmissions also have a throttle valve linkage, while others use an electric switch connected to the throttle to control forced downshifts.

Normal operation of a *neutral safety switch* provides a quick check for the adjustment of the gear selector linkage. To do this, move the selector lever slowly until it clicks into the PARK position. Turn the ignition key to the START position. If the starter operates, the PARK position is correct. After checking the PARK position, move the lever slowly toward the NEUTRAL position until the lever drops at the end of the N stop in the selector gate. If the starter also operates at this point, the gearshift linkage is properly adjusted. This quick test also tests the adjustment of the neutral safety switch. If the engine does not start in either or both of these positions, the neutral safety switch or the gear selector linkage needs adjustment or repair. Also, the starter should not operate in any other gear selector position. If it does, adjust the linkage.

> ⚠️ **WARNING:** Since you must work under the vehicle to adjust most shift linkages, make sure you properly raise and support the vehicle before working under it. Also, wear safety glasses or goggles while working under the vehicle.

Neutral Safety Switch

Most neutral safety switches are combinations of a neutral safety switch and a backup lamp switch. Others have separate units for these two distinct functions. When the reverse light switch is not part of the neutral safety switch, it seldom needs adjustment, as it is directly activated by the transmission (Figure 3-37) or the gear selector linkage. A neutral safety switch allows current to pass from the starter switch to the starter when the lever is placed in the P or N position. Some neutral safety switches are nonadjustable. If these prevent starting in P or N and the gear selector linkage is correctly adjusted, they should be replaced.

The gear selector linkage is often referred to as the manual linkage because it controls the manual shift valve.

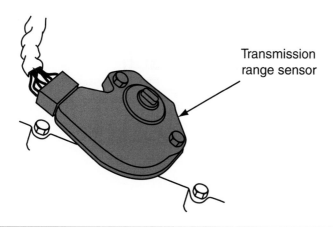

Transmission range sensor

Figure 3-37 A digital transmission range (TR) sensor that serves as a neutral start switch, back-up lamp switch, and as an input to the computer. It is located on the side of the transmission at the manual lever.

A voltmeter can be used to check the switch for voltage when the ignition key is turned to the START position with the shift lever in P or N. If there is no voltage, the switch should be adjusted or replaced.

To adjust a typical neutral safety switch:

1. Place the shift lever in neutral.
2. Loosen the attaching bolts for the switch.
3. Using an aligning pin, move the switch until the pin falls into the hole in its rotor.
4. Tighten the attaching bolts.
5. Recheck the switch for continuity. If no voltage is present, replace the switch.

Gear Selector Linkage

A worn or misadjusted gear selector linkage will affect transmission operation. The transmission's manual shift valve must completely engage the selected gear (Figure 3-38). Partial manual shift valve engagement will not allow the proper amount of fluid pressure to reach the rest of the valve body. If the linkage is misadjusted, poor gear engagement, slipping, and excessive wear can result. The gear selector linkage should be adjusted so the manual shift valve detent position in the transmission matches the selector level detent and position indicator.

To check the adjustment of the linkage, move the shift lever from the PARK position to the lowest DRIVE gear. Detents should be felt at each of these positions. If the detent cannot be felt in either of these positions, the linkage needs to be adjusted. While moving the shift lever, pay attention to the gear position indicator. Although the indicator will move with an adjustment of the linkage, the pointer may need to be adjusted so that it shows the exact gear after the linkage has been adjusted.

Linkage Adjustment

⚠️ **WARNING:** Always set the parking brake and block the wheels before moving the gear selector through its positions. This will prevent the vehicle from rolling over you or someone else.

To adjust a typical floor-mounted gear selector linkage:

1. Place the shift lever into the DRIVE position.
2. Loosen the locknuts and move the shift lever until DRIVE is properly aligned and the vehicle is in the D range.

Classroom Manual
Chapter 3, page 48

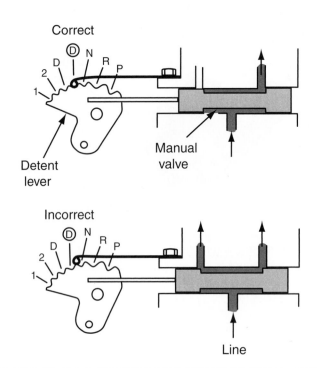

Figure 3-38 Incorrect linkage adjustments may cause the manual shift valve to be positioned improperly in its bore and cause slipping during gear changes. (Courtesy of General Motors Corporation, Service Operations)

3. Tighten the locknut (Figure 3-39).

To adjust a typical cable-type linkage:

1. Place the shift lever into the PARK position.

2. Loosen the clamp bolt on the shift cable bracket.

3. Make sure the preload adjustment spring engages the fork on the transaxle bracket.

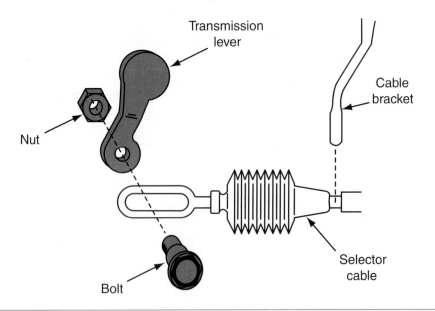

Figure 3-39 Gearshift linkages normally attach directly to the shift lever at the transmission and are retained by a locknut.

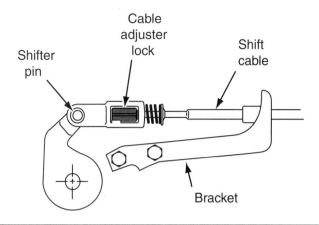

Figure 3-40 A shift cable adjuster lock assembly.

4. By hand, pull the shift lever to the front detent position (PARK), then tighten the clamp bolt. The shift linkage should now be properly adjusted.

5. If the cable assembly is equipped with an adjuster (Figure 3-40), move the shift lever into the PARK position and adjust the cable by rotating the adjuster into the lock position. Typically the adjuster will click when the lock is fully adjusted.

To adjust a typical rod-type linkage:

1. Loosen or disconnect the shift rod at the shift lever bracket.

2. Place the gear selector into PARK and the manual shift valve lever into the PARK detent position.

3. With both levers in position, tighten the clamp on the sliding adjustment to maintain their relationship. On the threaded type of linkage adjustment, lengthen or shorten the connection as needed. On some vehicles, you may need to adjust the neutral safety switch after resetting the linkage.

After adjusting any type of shift linkage, recheck it for detents throughout its range. Make sure a positive detent is felt when the shift lever is placed into the PARK position, as a safety measure. If you are unable to make an adjustment, the levers' grommets may be badly worn or damaged and should be replaced (Figure 3-41). When it is necessary to disassemble the linkage from the levers, the plastic grommets used to retain the cable or rod should be replaced. Use a prying tool to force the cable or rod from the grommet, then cut out the old grommet. Pliers can be used to snap the new grommets into the levers and the cable or rod into the levers.

> ✓ **SERVICE TIP:** If the proper prying tool is not used, the shift lever in some transmissions can easily break. If this tool is not available, apply some heat to the grommet. This should allow for easy separation of the cable or rod from the lever.

Throttle Valve Linkages

The throttle valve cable connects the throttle pedal movement to the throttle valve in the transmission's valve body. On some transmissions, the *throttle valve (TV) linkage* may control both the downshift valve and the throttle valve. Others use a vacuum modulator to control the throttle valve and a throttle linkage to control the downshift valve. Late-model transmissions may not have a throttle cable. Instead, they rely on electronic sensors and switches to monitor engine load and throttle plate opening. The action of the throttle valve produces throttle pressure (Figure 3-42). Throttle pressure is used as an indication of engine load and influences the speed at which automatic shifts will take place.

The throttle valve linkage is called the TV linkage.

Classroom Manual
Chapter 3, page 68

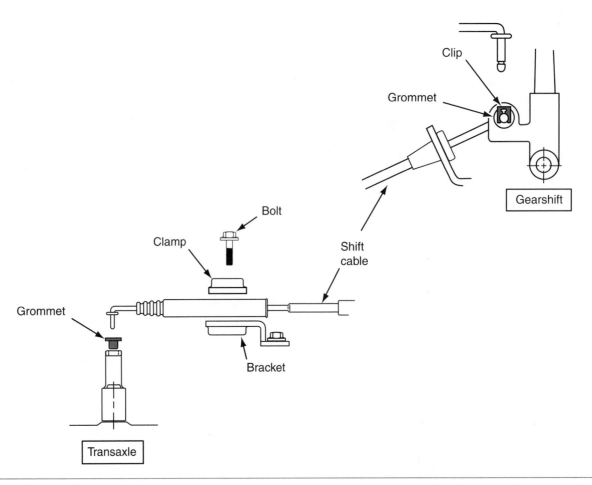

Figure 3-41 If the shift cable cannot be properly adjusted, the cable is distorted or some of the various grommets or brackets are worn and should be replaced.

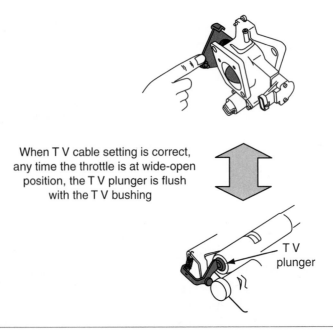

When T V cable setting is correct, any time the throttle is at wide-open position, the T V plunger is flush with the T V bushing

Figure 3-42 The movement of the throttle plate causes the T V valve to move. (Courtesy of General Motors Corporation, Service Operations)

A misadjusted TV linkage may result in a throttle pressure that is too low in relation to the amount the throttle plates are open, causing early upshifts or slipping. Throttle pressure that is too high can cause harsh and delayed upshifts, and part and wide-open throttle downshifts will occur earlier than normal. When adjusting the TV and downshift linkages, always follow the manufacturer's recommended procedures. An adjustment as small as a half-turn can make a big difference in shift timing and feel.

Some GM transmissions use a pull–push plastic lock or a metal push-release tab on the TV cable instead of a threaded adjustment. Follow the procedure given in the service manual to adjust this type of cable. If the adjustment is off by a single click of the adjusting mechanism, shift timing will be wrong.

CAUTION: It is important that you check the service manual before making adjustments to the throttle or downshift linkages. Some of these linkage systems are not adjustable and if you loosen them in an attempt to adjust them, you may break them.

To adjust a typical throttle cable:

1. Run the engine until it has reached normal operating temperature.
2. Loosen the cable mounting bracket or swivel lock screw.
3. Position the bracket so that both bracket alignment tabs are touching the transaxle, then tighten the lock screw to the recommended torque.
4. Release the readjust tab on the cable assembly.
5. Make sure the cable is free to slide all the way toward the engine against its stop, after the readjust or locking tab is released (Figure 3-43).
6. Move the throttle control lever clockwise against its internal stop, then press the readjust tab downward into its locked position.
7. At this point, the cable is adjusted and the backlash in the cable is removed.
8. Check the cable for free movement by moving the throttle lever counterclockwise and slowly releasing it to confirm it will return fully clockwise.
9. No lubrication is required for any component of the throttle cable system.
10. Check the adjustment of the cable by conducting a system pressure test.

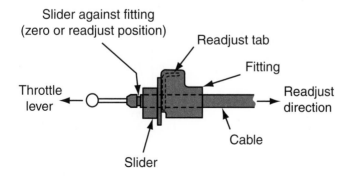

Figure 3-43 Typical adjusting mechanism for a T V cable. (Courtesy of General Motors Corporation, Service Operations)

To adjust a typical throttle lever rod:

1. Run the engine until it has reached normal operating temperature.
2. Loosen the linkage's swivel lock screw.
3. Make sure the swivel is free to slide along the flat end of the throttle rod. Disassemble and clean or repair parts to assure free movement.
4. Hold the transaxle's throttle lever firmly toward the engine and against its internal stop, then tighten the swivel lock screw.
5. The rod is now adjusted and any backlash in the linkage should be taken up by the preload spring.
6. Check the adjustment of the cable by conducting a system pressure test.

To adjust a typical downshift linkage:

1. Run the engine until it has reached normal operating temperature.
2. Put the transmission in neutral with the parking brake set and allow the engine to run at its normal idle speed.
3. Using the specified amount of pressure, press down on the downshift rod.
4. Rotate the adjustment screw to obtain the specified clearance between the screw and the throttle arm.

Kickdown Switch Adjustment

Classroom Manual
Chapter 3, page 71

Most late-model transmissions are not equipped with downshift linkages, instead they may use a *kickdown switch* typically located at the upper post of the throttle pedal (Figure 3-44). Some kickdown circuits are part of the computer control circuit, based on signals from a throttle position sensor. Movement of the throttle pedal to the wide-open position signals to the transmission that the driver desires a forced downshift.

To check the operation of the switch, fully depress the throttle pedal and listen for a click that should be heard just before the pedal reaches its travel stop. If the click is not heard, loosen the locknut and extend the switch until the pedal lever makes contact with the switch. If the pedal contacts the switch too early, the transmission may downshift during part-throttle operation.

If you feel and hear the click of the switch but the transmission still doesn't kick down, use an ohmmeter to check the continuity of the switch when it is depressed. (Make sure the ignition switch is off before connecting the meter.) An open switch will prevent forced downshifting, whereas a shorted switch can cause upshift problems. Defective switches should be replaced.

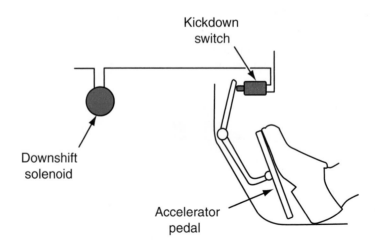

Figure 3-44 Action of a kickdown switch.

Other Adjustments

Band Adjustment

If a transmission problem still exists after the shift linkage and throttle pressure cable and rod have been adjusted and all electrical switches and sensors checked, the bands of the transmission may need adjustment. Photo Sequence 4 shows a typical procedure for adjusting a transmission band.

On some transmissions, slippage during shifting can be corrected by adjusting the holding bands. To help identify if a band adjustment will correct the problem, refer to the written results of your road test. Compare your results with the Clutch and Band Application Chart in the service manual. If slippage occurs when there is a gear change that requires holding by a band, the problem may be corrected by tightening the band.

On some vehicles, the bands can be adjusted externally with a torque wrench. On others, the transmission fluid must be drained and the oil pan removed. Locate the band-adjusting nut, and then clean off all dirt on and around the nut. Now, loosen the band-adjusting bolt locknut and back it off approximately five turns (Figure 3-45). Use a calibrated inch-pound torque wrench to tighten the adjusting bolt to the specified torque (Figure 3-46). Then, back off the adjusting screw the specified number of turns and tighten the adjusting bolt locknut while holding the adjusting stem stationary. Reinstall the oil pan with a new gasket and refill the transmission with fluid. If the transmission problem still exists, an oil pressure test or transmission teardown must be done.

CAUTION: Do not excessively back off the adjusting stem as the anchor block may fall out of place, making it necessary to remove and disassemble the transmission to fit it back in place.

Parking Pawl

Any time you have the oil pan off, you should inspect the transmission parts that are exposed. This is especially true of the parking pawl assembly (Figure 3-47). This component is typically not hydraulically activated, rather the gearshift linkage moves the pawl into position to lock the output

An automatic transmission tune-up usually includes fluid and filter change, linkage checks, and adjustment of the bands.

Classroom Manual
Chapter 3, page 74

Special Tools

Large drain pan

Inch-pound torque wrench

Classroom Manual
Chapter 3, page 75

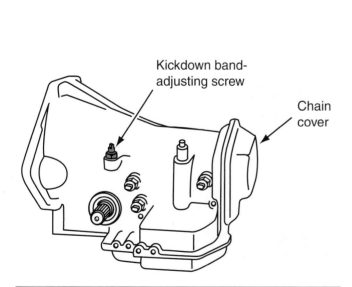

Figure 3-45 Location of external band-adjusting screw.

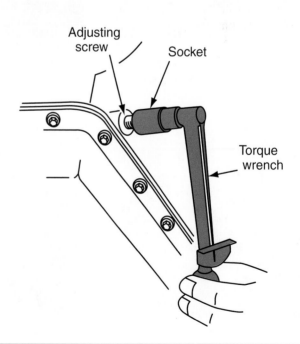

Figure 3-46 Bands are typically adjusted to a specific inch-pound torque setting.

Photo Sequence 4
Typical Procedure for Adjusting Transmission Bands

P4-1 Make sure you properly identify the transmission before making band adjustments.

P4-2 Locate band-adjusting screw.

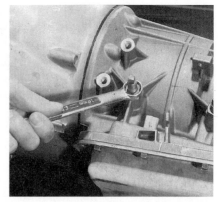

P4-3 Loosen adjusting screw locknut without allowing the screw to turn. If the locknut has a fluid seal, do not reuse the locknut. Install a new nut.

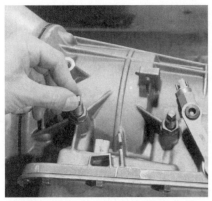

P4-4 Loosen the adjusting screw so that the band can relax around the drum and all tension is off of the adjusting screw.

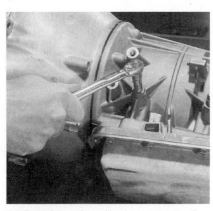

P4-5 Tighten the adjusting screw to the specified torque.

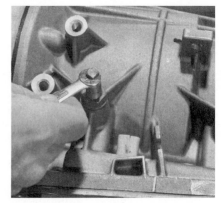

P4-6 Back off the adjusting screw the exact number of turns that is specified in the service manual.

P4-7 Position a wrench on the adjusting screw and over the locknut so that you can tighten the locknut without moving the adjusting screw.

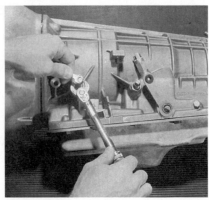

P4-8 Hold the adjusting screw in position and tighten the locknut to the specified torque.

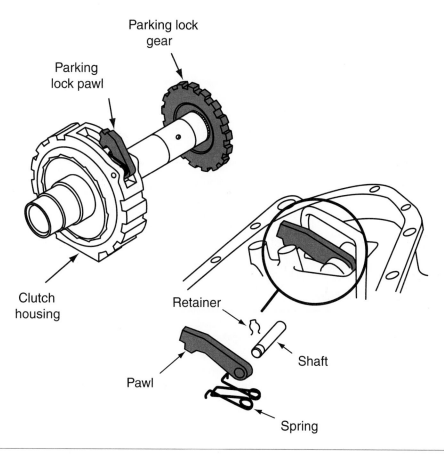

Figure 3-47 The entire parking pawl and gear assembly should be carefully inspected for wear or defects.

shaft of the transmission. Unless the customer's complaint indicates a problem with the parking mechanism, no test will detect a problem here.

Check the pawl assembly for excessive wear and other damage. Also check to see how firmly the pawl is in place when the gear selector is shifted into the PARK mode. If the pawl can be easily moved out, it should be repaired or replaced.

CASE STUDY

A customer with a late-model Ford pickup equipped with an E4OD transmission had complained that every time he hit a bump, the transmission would downshift. Then, when driven on smooth surfaces, the transmission would begin to cycle on its own between third and fourth gears.

The technician originally thought the problem was caused by a bad fourth gear clutch that couldn't hold the planetary gearset in overdrive or by a problem in the lockup torque converter circuit. Beginning the diagnosis with a visual inspection of the transmission and the many sensors that provide information to the E4OD's control module, the technician found faulty signals from the manual lever position (MLP) sensor. These signals explained the erratic shifting, because the computer uses the sensor to determine what gear the transmission should be in and how much modified line pressure to supply.

The technician used the required special tool to realign the sensor with the gear selector, but still found that the resistance readings were out of specifications. The sensor was replaced and the problem of erratic shifting solved.

Terms to Know

Aeration

Manifold absolute pressure (MAP) sensor

Oxidation

Throttle position sensor

Vehicle speed sensor

Variable force motor (VFM)

ASE-Style Review Questions

1. While diagnosing noises apparently coming from a transaxle assembly:
   Technician A says a knocking sound at low speeds is probably caused by worn CV joints.
   Technician B says a clicking noise heard when the vehicle is turning is probably caused by a worn or damaged outboard CV joint.
   Who is correct?
   A. A only   **C.** Both A and B
   B. B only   **D.** Neither A nor B

2. *Technician A* says if the shift for all forward gears is delayed, a slipping front or forward clutch is normally indicated.
   Technician B says a slipping rear clutch is indicated when there is a delay or slip when the transmission shifts into any forward gear.
   Who is correct?
   A. A only   **C.** Both A and B
   B. B only   **D.** Neither A nor B

3. While discussing the results of an oil pressure test:
   Technician A says when the fluid pressures are high, internal leaks, a clogged filter, low oil pump output, or a faulty pressure regulator valve are indicated.
   Technician B says if the fluid pressure increased at the wrong time, an internal leak at the servo or clutch seal is indicated.
   Who is correct?
   A. A only   **C.** Both A and B
   B. B only   **D.** Neither A nor B

4. *Technician A* says low engine vacuum will cause a vacuum modulator to sense a load condition when it actually is not present, causing delayed and harsh shifts.

Technician B says poor engine performance can cause delayed shifts through the action of the TV assembly.
Who is correct?
A. A only   **C.** Both A and B
B. B only   **D.** Neither A nor B

5. *Technician A* says delayed shifting can be caused by worn planetary gearset members.
   Technician B says delayed shifts or slippage may be caused by leaking hydraulic circuits or sticking spool valves in the valve body.
   Who is correct?
   A. A only   **C.** Both A and B
   B. B only   **D.** Neither A nor B

6. While discussing proper band adjustment procedures:
   Technician A says on some vehicles the bands can be adjusted externally with a torque wrench.
   Technician B says a calibrated inch-pound torque wrench is normally used to tighten the band-adjusting bolt to a specified torque.
   Who is correct?
   A. A only   **C.** Both A and B
   B. B only   **D.** Neither A nor B

7. While checking the condition of a car's ATF:
   Technician A says if the fluid has a dark brownish or blackish color and/or a burned odor, the fluid has been overheated.
   Technician B says if the fluid has a milky color, engine coolant has been leaking into the transmission's cooler.
   Who is correct?

A. A only   **C.** Both A and B
B. B only   **D.** Neither A nor B

8. While discussing the proper way to diagnose a kickdown switch:
Technician A says when the throttle pedal is fully depressed, a click should be heard just before the pedal reaches its travel stop. If the click is not heard, the switch should be replaced.
Technician B says if the transmission cannot be forced to automatically downshift, the kickdown switch is open and should be replaced.
Who is correct?
A. A only   **C.** Both A and B
B. B only   **D.** Neither A nor B

9. While discussing a pressure test:
Technician A says this test is the most valuable diagnostic check for slippage in one gear.

Technician B says the test can identify the cause of late or harsh shifting.
Who is correct?
A. A only   **C.** Both A and B
B. B only   **D.** Neither A nor B

10. While checking the engine and transmission mounts on a FWD car:
Technician A says any engine movement may change the effective length of the shift and throttle cables and therefore may affect the engagement of the gears.
Technician B says delayed or missed shifts are caused by hydraulic problems, not linkage problems.
Who is correct?
A. A only   **C.** Both A and B
B. B only   **D.** Neither A nor B

ASE Challenge Questions

1. *Technician A* says engine lugging during downshifts may be caused by a faulty lockup torque converter.
Technician B says poor engine performance may result in delay shifts on a TV cable controlled transmission or transaxle.
Who is correct?
A. A only   **C.** Both A and B
B. B only   **D.** Neither A nor B

2. *Technician A* says a leaking torque converter may be indicated by the presence of a red or reddish-orange fluid in the bell housing.
Technician B says transmission oil on the converter's rear shell indicates a leaking torque converter.
Who is correct?
A. A only   **C.** Both A and B
B. B only   **D.** Neither A nor B

3. Worn or broken engine/transaxle mounts may cause all of the following EXCEPT:
A. Delay shifts   **C.** No PARK gear
B. Change in the linkage effective length   **D.** Early or late kickdown shifts

4. *Technician A* says a ruptured vacuum modulator may be indicated by the color of the engine's exhaust.
Technician B says a stuck modulator actuating pin may be indicated by a no-reverse engagement condition.
Who is correct?
A. A only   **C.** Both A and B
B. B only   **D.** Neither A nor B

5. Pressure testing revealed low pressure in all gears.
Technician A says using a reverse stall test could isolate the faulty component.
Technician B says a reverse stall test will determine whether the low-reverse servo is the malfunctioning component.
Who is correct?
A. A only   **C.** Both A and B
B. B only   **D.** Neither A nor B

Job Sheet 5

Name _____ Date _____

Visual Inspection of an Automatic Transmission

Upon completion of this job sheet, you should be able to conduct a preliminary inspection of the transmission.

Tools and Materials

A vehicle   Safety glasses
A torque wrench   Service manual

Describe the vehicle being worked on:

Year _____ Make _____ VIN _____

Model _____

Model and type of transmission _____

Procedure

1. Service manual referred to:

_____

2. Check the transmission housing for damage, cracks, and signs of leaks. Comments:

_____

_____

_____

3. Check the slip joint area in the transmission extension housing for leaks. Comments:

_____

_____

_____

4. Check for leaks at everything attached to the transmission. Comments:

_____

_____

_____

5. Check the transmission's linkages for looseness, wear, and damage. Comments:

_____

_____

_____

Visual Inspection of an Automatic Transmission (continued)

6. Check any transmission cables for binding, wear, and damage. Comments:

_____

_____

7. Check the fluid condition and level. Comments:

_____

_____

8. What are your conclusions from these checks?

_____

_____

Instructor's Response _____

_____

_____

Job Sheet 6

Name _____ Date _____

Road Test a Vehicle to Check the Operation of the Automatic Transmission

Upon completion of this job sheet, you should be able to road test a vehicle with an automatic transmission to verify a customer's complaint and begin the diagnostic procedure.

Tools and Materials

A vehicle with an automatic transmission
Service manual
Clean shop rag
Pad of paper and a pencil

Describe the vehicle being worked on:

Year _____ Make _____ VIN _____

Model _____

Model and type of transmission _____

Procedure

	Task Completed
1. Park the vehicle on a level surface.	☐
2. Wipe all dirt off of the protective disc and the dipstick handle.	☐
3. Start the engine and allow it to reach operating temperature.	☐
4. Remove the dipstick and wipe it clean with a lint-free cloth or paper towel.	☐

5. Reinsert the dipstick, remove it again, and record the reading.

_____

6. Describe the condition of the fluid (color, condition, and smell):

_____

_____

7. What is indicated by the fluid's condition?

_____

_____

8. Find and duplicate the chart from a service manual that shows the band and clutch application for different gear selector positions. Using these charts will greatly simplify your diagnosis of automatic transmission problems. It is also wise to have a notebook or piece of paper to jot down notes about the operation of the transmission.

9. Inspect the transmission for signs of fluid leakage. Comments:

_____

_____

10. Drive the vehicle at normal speeds to warm up the engine and transmission. Describe the behavior of the transmission and torque converter.

_____

_____

11. Place the shift selector into the DRIVE position and allow the transmission to shift through all of its normal shifts. Describe the operation of the transmission and torque converter.

_____

_____

12. Check for proper operation in all forward ranges, especially the 1–2, 2–3, and 3–4 upshifts and converter lockup during light throttle operation. Describe the operation.

_____

_____

13. Force the transmission to kick down and record the quality of this shift and the speed at which it downshifts. Comments:

_____

_____

14. Manually cause the transmission to downshift. How did it react?

_____

_____

15. Record any vibrations or noises that occur during the test drive.

_____

_____

16. Park the vehicle and move the shifter into each gear range. Pay attention to shift quality as each range (including PARK) is selected. Record the results:

_____

_____

17. Compare your notes with the specifications and shifting chart. What are your conclusions about the transmission and torque converter? What transmission parts could be causing the problem? Use the clutch/band application chart to answer this.

_____

_____

_____

_____

Instructor's Response _____

_____

Job Sheet 7

Name _____ Date _____

Pressure Testing a Transmission

Upon completion of this job sheet, you should be able to conduct a pressure test on a transmission.

ASE Correlation

This job sheet is related to the ASE Automatic Transmission and Transaxle Test's Content Area: *General Transmission and Transaxle Diagnosis.*
Task: Perform pressure tests; determine necessary action.

Tools and Materials

A vehicle with an automatic transmission
Hoist
Tachometer
Pressure gauges
Vacuum gauge

T-fitting and vacuum hose
Lint-free shop towels
Service manual

Describe the vehicle that was assigned to you:

Year _____ Make _____ VIN _____

Model _____

Model and type of transmission _____

Procedure

Task Completed

1. Describe the pressure specifications and conditions that are listed in the service manual.

_____

_____

_____

_____

2. Start the engine and allow the engine and transmission to reach normal operating temperature. Then turn off the engine. ☐

3. Connect the tachometer to the engine if the vehicle is not equipped with one. ☐

4. Have a fellow student sit in the vehicle to operate the throttle, brakes, and transmission during this test. ☐

5. Raise the vehicle on the hoist to a comfortable working height. ☐

6. Use the service manual to locate the pressure test ports. Describe their locations.

_____

_____

_____

_____

☐   **7.** Connect the pressure gauges to the appropriate service ports.

☐   **8.** If the transmission has a vacuum modulator, use the T-fitting and vacuum hose to connect the vacuum gauge into the modulator circuit.

☐   **9.** Start the engine. Have your helper move the gearshift selector into the first test position.

☐   **10.** Run the engine at the specified speed. Move the gear selector as required by the specifications.

11. Observe and note the pressure and vacuum readings in the various gear ranges.

_____

_____

12. Move the pressure gauges to the appropriate test ports for the next range to be tested. Describe this location.

_____

_____

☐   **13.** Restart the engine and have your helper move the gear selector into the range to be tested and increase the engine's speed to the test point.

14. Observe and note the pressure and vacuum readings in the various gear ranges.

_____

_____

☐   **15.** Have your helper slowly apply the vehicle's brakes. Once the engine has returned to idle, turn it off.

☐   **16.** Repeat this sequence until the recommended test sequence has been completed.

17. Summarize the results of these tests.

_____

_____

_____

_____

_____

_____

Instructor's Response _____

_____

_____

Job Sheet 8

Name _____ Date _____

Air Pressure Testing

Upon completion of this job sheet, you should be able to conduct an air pressure test on a transmission.

ASE Correlation

This job sheet is related to the ASE Automatic Transmission and Transaxle Test's Content Area: *Off-Vehicle Transmission/Transaxle Repair Friction and Reaction Units.*
Tasks: Air test the operation of clutch and servo assemblies.

Tools and Materials

An automatic transmission or transaxle on a bench, or
   a vehicle with an automatic transmission
Compressed air and rubber tipped air nozzle
Inch-pound torque wrench

Eye protection
Air pressure testing plate
Service manual
Lint-free shop towels

Describe the transmission that was assigned to you:

Year _____ Make _____ VIN _____

Model _____

Model and type of transmission _____

Procedure

Task Completed

1. If this job sheet will be completed on a transmission installed in a vehicle: raise the vehicle on a hoist. Then drain the transmission fluid and remove the oil pan, oil filter, and valve body. ☐

2. If the job sheet will be completed with the transmission on a bench, remove the valve body. ☐

3. Refer to the service manual and identify the fluid passages required for testing. ☐

4. If you have an air pressure testing plate, install it with the appropriate gaskets and tighten the plates to the manufacturer's specifications. The specified torque is

_____ .

5. Apply air pressure to the servo and clutch passages and listen for 5–10 seconds. Describe the sounds you hear and tell what they indicate.

_____

_____

_____

6. Release the air pressure and reapply pressure. Listen and record the results.

_____

_____

Air Pressure Testing

☐

7. Remove the testing plate.

8. Summarize the results of this test.

_____

_____

_____

_____

Instructor's Response _____

_____

_____

Job Sheet 9

Name _____ Date _____

Transmission Cooler Inspection and Flushing

Upon completion of this job sheet, you should be able to inspect, test, and flush the cooler system for an automatic transmission.

ASE Correlation

This job sheet is related to the ASE Automatic Transmission and Transaxle Test's Content Area: *In-Vehicle Transmission and Transaxle Repair.*
Tasks: Check condition of engine cooling system: inspect, test, flush, and replace transmission cooler, lines, and fittings.

Tools and Materials

A vehicle with an automatic transmission   Various lengths of rubber hose
   or transaxle   Drain pan
Line wrenches   Compressed air and air nozzle
Tubing cutter   Supply of clean solvent or mineral spirits
Tubing flaring kit   Lint-free shop towels
ATF pressure tester   Service manual

Describe the vehicle that was assigned to you:

Year _____ Make _____ VIN _____

Model _____

Model and type of transmission _____

Procedure

Task Completed
_____

1. Remove the transmission dipstick and describe the condition, smell, and color of the fluid.

   _____

   _____

2. Open the radiator cap (after the engine is cooled down) and check the coolant for traces of ATF. Also check the cap and gasket for signs of ATF. Record your findings.

   _____

   _____

3. Replace the radiator cap.   ☐

4. Inspect the metal lines and fittings to and from the transmission cooler. Look for damage and signs of leakage. Describe your findings.

   _____

   _____

5. Summarize the results of your inspection.

_____

_____

☐ **6.** Place a drain pan under the fittings that connect the cooler lines to the radiator.

☐ **7.** Disconnect both cooler lines from the radiator. Plug or cap the lines from the transmission.

☐ **8.** Plug or cap one fitting at the radiator.

☐ **9.** Remove the radiator cap.

☐ **10.** Hold the compressed air nozzle tightly against the open fitting.

☐ **11.** Apply air pressure through the fitting (no more than 75 psi).

12. Check the coolant in the radiator for signs of air bubbles or air movement. Describe your findings.

_____

_____

☐ **13.** Unplug the other fitting and cooler lines.

☐ **14.** Reconnect the cooler lines to the radiator.

15. Summarize the results of the leak test.

_____

_____

_____

_____

☐ **16.** Set the parking brake and start the engine. Allow the engine and transmission to reach normal operating temperature before proceeding. Turn off the engine.

☐ **17.** Remove the transmission dipstick. Place a funnel in the dipstick tube.

☐ **18.** Raise the vehicle on a hoist.

☐ **19.** Disconnect the cooler return line at the point where it enters the transmission case.

☐ **20.** Attach a rubber hose to the disconnected cooler return line.

☐ **21.** Lower the vehicle and place the end of the hose into the funnel.

☐ **22.** Start the engine and run it about 1000 rpm.

23. Observe the flow of fluid into the funnel. Describe the flow and summarize what this indicates.

_____

_____

_____

_____

☐ **24.** Turn off the engine.

Transmission Cooler Inspection and Flushing (continued)

25. Raise the vehicle and remove the rubber hose from the return line. ☐

26. Reconnect the return line. ☐

27. Lower the vehicle and reinstall the dipstick. ☐

The remainder of this job sheet covers a procedure for flushing the transmission cooler. Check with your instructor before proceeding.

1. Raise the vehicle on a hoist. ☐

2. Place a drain pan under the engine's radiator (or place it under the external transmission cooler if the vehicle is so equipped). ☐

3. Disconnect both cooler lines at the ends of the lines and at the radiator or cooler. ☐

4. Clean the fittings at the ends of the lines and at the radiator or cooler. ☐

5. Connect a rubber hose to the inlet fitting at the radiator. ☐

6. Place the other end of the hose into the drain pan. ☐

7. Attach a hose funnel to the cooler return fitting. Install a funnel in the other end of this hose. ☐

8. Pour small amounts of mineral spirits into the funnel. Observe the flow and the condition of the fluid moving into the drain pan. What did it look like?

_____

_____

9. If there is little flow, apply some air pressure to the hose connected to the return line. ☐

10. Continue pouring mineral spirits into the funnel until the fluid entering the drain pan is clear. ☐

11. Then disconnect the cooler lines at the transmission and place the drain pan at one end of the lines. ☐

12. Install the rubber hose and funnel to the other end of the lines. ☐

13. Pour mineral spirits through the lines until the flow is clear. Describe your findings.

_____

_____

14. Now pour ATF through the cooler lines to remove the mineral spirits. ☐

15. Remove all hoses and reconnect the cooler lines to the transmission and radiator. ☐

16. Check the transmission's fluid level and correct it if necessary. ☐

Instructor's Response _____

_____

_____

_____

Electrical and Electronic System Diagnosis and Service

Upon completion and review of this chapter, you should be able to:

❏ Diagnose electronic control systems and determine needed repairs.

❏ Conduct a road test to determine if the fault is electrical or hydraulic.

❏ Use common electrical test instruments to locate problems.

❏ Conduct preliminary checks on EAT systems and determine needed repairs or service.

❏ Retrieve trouble codes from common electronically controlled automatic transmissions and determine needed repairs or service.

❏ Follow the prescribed diagnostic procedures according to DTC and/or

symptom and determine needed repairs or service.

❏ Perform converter clutch system tests and determine needed repairs or service.

❏ Inspect, test, and replace electrical/ electronic switches.

❏ Inspect, test, and replace electrical/ electronic sensors.

❏ Inspect, test, bypass, and replace electrical/ electronic solenoids.

❏ Inspect, test, adjust, and/or replace transmission-related electrical/electronic components.

One of the first tasks during diagnosis of an electronically controlled transmission is to determine if the problem is caused by the transmission or by electronics. To determine this, the transmission must be observed to see if it responds to commands given by the computer. Identifying whether the problem is related to the transmission or is electrical will determine what steps need to be followed to diagnose the cause of the problem.

An electronically controlled transmission (Figure 4-1) will work only as well as the commands it receives from the computer, even if the hydraulic and mechanical parts of the transmission are fine. All diagnostics should begin with a scan tool to check for trouble codes in the system's computer. After the received codes are addressed, you can begin a more detailed diagnosis of the system and transmission. Your next step may be to manually activate the shift solenoids by connecting a jumper wire to them or by using a transmission tester that allows you to manually activate the solenoids. Prior to doing this, the wiring to the solenoids should be studied to determine if the computer activates them by supplying voltage to them or by completing the ground circuit. Also you need to know what gear certain solenoids are activated in. This information can be found in the service manual.

Basic EAT Testing

The best way to diagnose an electronically controlled transmission is to approach solving the problem in a logical way. To do this, you should follow these seven steps, in order:

1. Verify the customer's complaint.
2. Conduct preliminary inspections and checks.
3. Check all service information, including service bulletins and recall notices, for information about the complaint and the system.

Basic Tools

Basic mechanic's tool set

Scan tool

DMM

Jumper wires

Appropriate service manual

Classroom Manual
Chapter 4, page 97

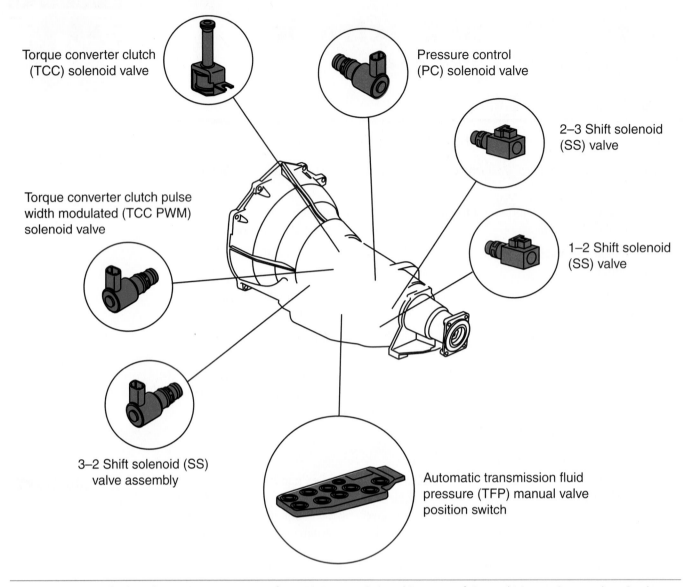

Figure 4-1 The key electronic components of a 4L60-E transmission. (Courtesy of General Motors Corporation, Service Operations Group)

Labels in figure:
- Torque converter clutch (TCC) solenoid valve
- Pressure control (PC) solenoid valve
- 2–3 Shift solenoid (SS) valve
- Torque converter clutch pulse width modulated (TCC PWM) solenoid valve
- 1–2 Shift solenoid (SS) valve
- 3–2 Shift solenoid (SS) valve assembly
- Automatic transmission fluid pressure (TFP) manual valve position switch

4. Follow the diagnostic procedures outlined in the service manual for the specific complaint.

5. Interpret and respond to all diagnostic codes.

6. Define and isolate the cause of the complaint or problem.

7. Fix the problem and verify the repair.

It is important that you totally understand what the complaint or problem is before you venture in and try to find the cause. This is the purpose of the first four steps. This may include an interview or road test with the customer to thoroughly define the complaint and to identify when and where the problem occurs.

Since many EAT problems are caused by the basics, it is wise to conduct all of the preliminary checks required for a non-electronically controlled transmission. In addition, you should conduct a thorough inspection of the electronic system. This inspection should include the retrieval of diagnostic codes. These codes will not only allow you to see what the PCM (Figure 4-2) sees as a transmission problem, but will also allow you to check for any engine problems. When-

EAT is a commonly used acronym for Electronic Automatic Transmission.

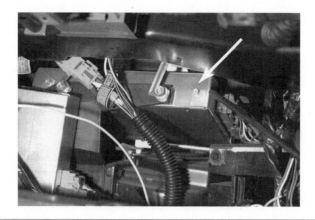

Figure 4-2 A typical PCM mounted under the dash.

ever diagnosing a transmission, remember that an engine problem can and will cause the transmission to act abnormally.

Often, accurately defining the problem and locating related information in TSBs and other materials can indicate the cause of the problem. No manufacturer makes a perfect vehicle or transmission and, as the manufacturer recognizes common occurrences of a problem, it will issue a statement regarding the way to fix the problem. Also, for many DTCs and symptoms, service manuals will give a simple diagnostic chart or path for identifying the cause of the problem (Figure 4-3). These are designed to be followed step-by-step and will lead to a conclusion if you follow the path matched exactly to the symptom. Check all available information before moving on in your diagnostics.

Sometimes the symptom will not match any of those described in the service manual. This doesn't mean it is time to guess. It means it is time to clearly identify what is working right. By eliminating those circuits and components that are working correctly from the list of possible causes of the problem, you can identify what may be causing the problem and what should be tested further.

Although the first steps in troubleshooting include retrieving diagnostic codes, there are problems that will not be made evident by a code. These problems may be solved with the diagnostic charts or pure logic. This logic must be based on a thorough understanding of the transmission and its controls.

Diagnostic codes should be used to recognize and locate problems. Codes that relate to transmission faults can be caused or detected by engine input or transmission input and/or output devices. Although codes may appear to be caused only by faulty input sensors or outputs, they may actually be caused by internal transmission problems. These problems may cause the inputs or outputs to appear out-of-normal range by the computer. To pinpoint the exact cause of a transmission problem, you will need to use basic electrical troubleshooting equipment, such as wiring diagrams, diagnostic charts, DMMs, lab scopes, and special transmission testers, as well as scan tools. There are many varieties of electronically controlled transmissions, but all of them work in similar ways. Therefore, troubleshooting should always be approached in the same way.

A TSB is a Technical Service Bulletin in which the latest fixes and updates for a component or system are found.

Guidelines for Diagnosing EATS

1. Make sure the battery has at least 12.6 volts before troubleshooting the transmission.
2. Check all fuses and identify the cause of any blown fuses.
3. Compare the wiring to all suspected components against the colors given in a service manual.
4. When testing electronic circuits, always use a high-impedance test light or DMM.

Step	Action	Value(s)	Yes	No
1	Did you perform the Powertrain Diagnostic System Check?	—	Go to Step 2	Go to *A Powertrain On Board Diagnostic (OBD) System Check* in Engine Controls
2	1. Install a *Scan Tool*. 2. Turn ON the ignition, with the engine OFF. **Important**: Before clearing the DTC, use the *Scan Tool* in order to record the Freeze Frame and Failure Records. Using the Clear Info function erases the Freeze Frame and Failure Records from the PCM. 3. Record the DTC Freeze Frame and Failure records. 4. Clear the DTC. 5. Drive the vehicle in the D4 drive range in second, third or fourth gear under steady acceleration, with a TP angle at 20%. While the *Scan Tool* TCC Enable status is No, does the *Scan Tool* display a TCC Slip Speed within the specified range?	−20 to +30 RPM	Go to Step 3	Go to Diagnostic Aids
3	The TCC is hydraulically stuck ON. Inspect for the following conditions: • A clogged exhaust orifice in the TCC solenoid valve. • The converter clutch apply valve is stuck in the apply position. • A misaligned or damaged valve body gasket. • A restricted release passage. • A restricted transmission cooler line. Did you find and correct the condition?	—	Go to Step 4	—
4	Perform the following procedure in order to verify the repair: 1. Select DTC. 2. Select Clear Info. 3. Drive the vehicle in D4 under the following conditions: Hold the throttle at 25 percent and accelerate to 88 km/h (55 mph). Ensure that the *Scan Tool* TCC Slip Speed is 130 to 2000 RPM for 4 seconds, with the TCC OFF. 4. Select Specific DTC. 5. Enter DTC P0742. Has the test run and passed?	—	System OK	Go to Step 1

Figure 4-3 An example of a manufacturer's diagnostic chart for a DTC. (Courtesy of General Motors Corporation, Service Operations Group)

5. If an output device is not working properly, check the power circuit to it.

6. If an input device is not sending the correct signal back to the computer, check the reference voltage it is receiving and the voltage it is sending back to the computer.

7. Compare the voltages in and out of a sensor with the voltages the computer is sending out and receiving.

8. Before replacing a computer, check the solenoid isolation diodes according to the procedures outlined in the service manual.

9. Make sure computer wiring harnesses do not run parallel with any high-current wires or harnesses. The magnetic field created by the high current may induce a voltage in

the computer harness. You should also be aware that antenna cables and CB radios can cause interference.

10. Take necessary precautions to prevent the possibility of static discharge while working with electronic systems.

11. While checking individual components, always check the voltage drop of the ground circuits. This becomes more and more important as cars are made of less material that conducts electricity well.

12. Make sure the ignition is off whenever you disconnect or connect an electronic device in a circuit.

13. All sensors should be checked in cold and hot conditions.

14. All wire terminals and connections should be checked for tightness and cleanliness.

15. Use TV-tuner cleaning spray to clean all connectors and terminals.

16. Use dielectric grease at all connections to prevent future corrosion.

17. If you must break through the insulation of a wire to take an electrical measurement, make sure you tightly tape over the area after you are finished testing.

Electrostatic Discharge

Some manufacturers mark certain components and circuits with a code or symbol to warn technicians that they are sensitive to electrostatic discharge (Figure 4-4). Static electricity can destroy or render a component useless.

When handling any electronic part, especially those that are static sensitive, follow the guidelines below to reduce the possibility of electrostatic build-up on your body and the inadvertent discharge to the electronic part. If you are not sure whether a part is sensitive to static, treat it as if it were.

Classroom Manual
Chapter 4, page 91

1. Always touch a known good ground before handling the part. This should be repeated while handling the part and more frequently after sliding across a seat, sitting down from a standing position, or walking a distance.

2. Avoid touching the electrical terminals of the part unless you are instructed to do so in the written service procedures. It is good practice to keep your fingers off all electrical terminals as the oil from your skin can cause corrosion. Use an anti-static strap, if one is available.

3. When you are using a voltmeter, always connect the negative meter lead first.

4. Do not remove a part from its protective package until it is time to install the part.

Figure 4-4 A manufacturer's warning symbol to show that a circuit or component is sensitive to electrostatic discharge (ESD). Courtesy of the Chevrolet Motor Division of General Motors)

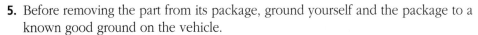

5. Before removing the part from its package, ground yourself and the package to a known good ground on the vehicle.

6. When replacing a programmable read only memory (PROM) unit, ground your body by putting a metal wire around your wrist and connect the wire to a known good ground.

Although diagnostic trouble codes (DTCs) are very helpful during diagnosis, they also can become a stumbling block. This is especially true if more than one DTC is present. If two or more codes are present, you should look at the relationship of the codes to identify if they could have a common cause. Reacting to the individual codes may move your troubleshooting efforts beyond the common cause.

It is also important to remember that codes can be set by out-of-range signals. This doesn't mean the sensor or sensor circuit is bad. It could mean the sensor is working properly but there is a mechanical or hydraulic problem causing the abnormal signals. Not only can internal transmission problems cause codes to be set, so can basic electrical problems. Problems such as loose connections, broken wires, corrosion, and poor grounds will affect the signals in that circuit.

Basic Electrical Testing

Although the emphasis of this chapter is electronically controlled transmissions, basic electrical checks must be covered first. Many of the preliminary checks involve the use of common electrical test equipment. The proper use of test equipment is also necessary to test and identify the exact cause of problems after diagnostic codes are retrieved.

Ammeters

Measuring current flow can provide a summary of the activity in a circuit. If the current flow is lower than expected, there is some extra resistance in the circuit. This resistance can be the result of corrosion, frayed wires, or a loose connection. When current is higher than expected, a short or low resistance is indicated. A normally operating circuit will have the expected amount of current flow.

An **ammeter** is used to measure current flow. Basically there are two types of ammeters and both require a different hook up into the circuit. Prior to using an ammeter in a circuit, make sure the meter is capable of handling the current in the circuit. Circuits like the starting motor circuit have very high amounts of current and only specially designed meters are able to withstand the current flow of these circuits. Often ammeters have ranges of measurement. Always select the proper range prior to connecting the meter.

One type of ammeter has two leads, a positive lead and a negative lead. In order for this type of ammeter to accurately read current flow, the meter must be connected in series to the circuit (Figure 4-5). The circuit should be disconnected from its power source, then opened by disconnecting a connector in the circuit and inserting the red lead in the connector half coming from the positive side of the battery and inserting the black lead into the other half. The circuit is now completed through the meter and, when activated, current will flow through the circuit and the meter, which will indicate the amount.

Inductive ammeters (Figure 4-6) are not connected into series with the circuit. This type of meter utilizes an inductive pick-up, which monitors the magnetic field formed by current flowing through a conductor. These meters normally have three leads: positive, negative, and inductive. The positive and negative leads are connected to their appropriate posts of the battery. The inductive lead is clamped around a conductor in the circuit. When the circuit is activated, current flow will be read on the meter. Meters used to measure current flow in the starting and charging systems normally are the inductive type.

The static electricity that can be generated by a vinyl seat is 20,000 volts. It only takes 35 volts to destroy any electronic automotive device.

Classroom Manual
Chapter 4, page 86

An **ammeter** is a test instrument used to measure electrical current.

Normally the positive lead is red and the negative lead is black.

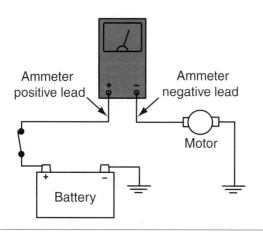

Figure 4-5 An ammeter is always connected in series with the circuit. This allows circuit current to flow through the meter.

Figure 4-6 An ammeter with an inductive pickup.

 SERVICE TIP: Always wipe the meter leads off with a damp cloth after checking a battery. This will prevent the battery acid from damaging the ends of the leads.

Voltmeters

Electrical circuits and components cannot operate properly if the proper amount of voltage is not available to them. Not only must the battery be able to deliver the proper amount of voltage but also the proper amount of voltage must be available to operate the intended component. As current flows through a resistance, voltage is lost or dropped. In circuits where undesired resistance is present, the voltage available for the circuit is decreased. Because of these conditions, voltage is measured at the source, at the electrical loads, and across the loads and circuits.

A **voltmeter** is used to measure voltage and has two leads, a positive lead and a negative lead. Connecting the positive lead to any point within a circuit and connecting the negative lead to a ground will measure the voltage at the point where the positive lead was connected. If battery voltage is being measured, the positive lead is connected to the positive terminal of the battery and the negative lead to the negative post or to a clean spot on the common ground. The voltmeter will read the potential difference between the two points and display that difference on its scale. In the case of a 12-volt battery, the potential at the positive post is 12 volts and the potential at the negative is zero. Therefore the meter will read 12 volts.

The positive lead can be connected anywhere within the circuit and it will measure the voltage present at that point. To measure the amount of available voltage, the negative lead can be connected to any point in the common ground circuit. To measure the amount of voltage drop across any wire or component of the circuit, the positive lead is connected to the battery side of the component and the negative lead is placed directly to the other side of the component. For example, to measure the voltage drop of a light bulb in a simple circuit, connect the positive lead to the battery or power lead at the bulb and the other lead to the groundside of the bulb (Figure 4-7). The meter will read the amount of voltage drop across the bulb when the circuit is activated. If no other resistances are present in the circuit, the amount of voltage drop will equal the amount of source voltage. However, if there is a resistance present in the ground connection, the bulb will have a voltage drop that is less than battery voltage. If the ground is bad, the voltage drop across the ground will equal the source voltage minus the amount of voltage dropped by the light. In order to measure voltage drop and available voltage at various points within a circuit, the circuit must be activated (Figure 4-8).

Classroom Manual
Chapter 4, page 85

A **voltmeter** is a test instrument used to measure voltage.

Battery voltage is typically called source voltage.

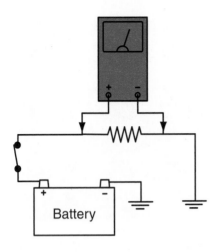

Figure 4-7 To measure voltage drop, the voltmeter is connected in parallel or across the component being tested.

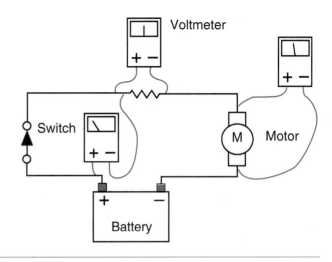

Figure 4-8 Voltmeter connections for measuring voltage drop across various parts of a circuit.

Classroom Manual
Chapter 4, page 86

SERVICE TIP: Some DMMs have an autoranging feature that will automatically change scales while you are measuring something. This is a convenient feature but may cause undue concern when testing some components, such as a throttle position (TP) sensor. The automatic change of range may look like a glitch. Therefore, always lock the DMM into its manual mode when testing TP sensors and other variable resistors.

Ohmmeters

An **ohmmeter** is a test meter used to measure electrical resistance and/or continuity.

Also recheck the calibration after changing to another range.

An Infinite reading simply means that the resistance is greater than what can be measured in that scale on the ohmmeter.

Digital multimeters may display "O.L" when there is an infinite measurement.

Ohmmeters measure resistance and should never be connected into an activated circuit. The meter uses its own power to determine the amount of resistance present between two points. If the circuit has its own power source and an ohmmeter is connected to it, the meter will easily burn out from the excessive power. An ohmmeter is frequently used to test for circuit completeness or continuity. If a circuit is complete, the meter will display a low resistance, whereas if the circuit is open, a large amount of resistance and an infinite reading will be displayed.

Most ohmmeters have a variety of scales to measure within. A technician chooses between scales of 1, 100, or 10,000 ohms. The anticipated resistance determines the scale to be selected. After the scale has been selected, the meter must be zeroed in that scale. To do this, the leads of the meter are held together and the needle or number display is adjusted until it reads zero. Typically, an ohmmeter is equipped with an adjustment control to zero the meter.

To measure the resistance of a circuit or component, disconnect the power from the circuit or remove the component from the circuit (Figure 4-9). With the meter zeroed in the appropriate scale, connect one of the meter's leads to the power side of the component and the other lead to the groundside. Make sure you are not holding the metal leads while taking measurements. If you do this, you will become a parallel circuit with the component or circuit you are checking and you will get inaccurate results. Always hold the leads with the plastic handle. The meter will display the amount of resistance present between the two points. If the resistance is greater than the selected scale, the meter will show a reading of "infinity" (Figure 4-10). When this reading results, the technician should move up a scale on the meter. Whenever the scales have been changed, the meter must be zeroed for that range prior to measuring the resistance. If subsequent scale changes result in continued readings of infinity, it can be assumed that there is no continuity between the two measured points.

Ohmmeters can also be used to compare the resistance of a component to the value it should have. Many electrical components have a specified resistance value, which is listed in the

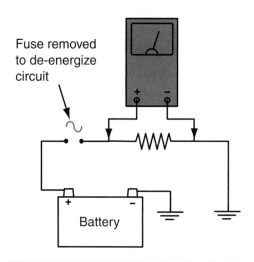

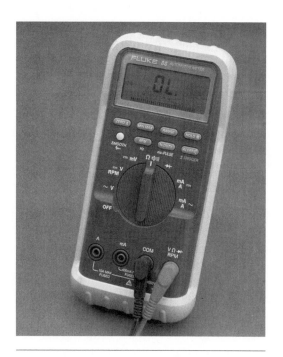

Figure 4-9 Using an ohmmeter to measure resistance. Note that the circuit's fuse has been removed.

Figure 4-10 A digital ohmmeter showing an infinite reading.

service manuals. This resistance value is important, as it controls the amount of voltage dropped and the amount of current that will flow in the circuit. If a component does not have the proper amount of resistance, the circuit will not operate properly.

Prior to testing a component or circuit with an ohmmeter, the service manual should be checked for precautions regarding the impedance of the meter. Ohmmeters supply their own power to measure resistance and often electronic components can be damaged by causing excessive current to flow in them. Using an ohmmeter with the proper impedance will avoid damage to the components. If an impedance specification is not given in the manual, it can be safely assumed that any ohmmeter is appropriate for that circuit.

Test Lights

Often a test light is used to verify the presence of voltage at a particular point. Although the test light cannot give a reading of the amount of voltage present, it can clearly display the presence of voltage. Normally a test light has one wire lead. This is connected to ground. The positive lead is actually a piercing probe that is slipped into the circuit at the desired point to verify voltage. If voltage is present, the bulb in the test light will shine.

A self-powered test light is often used to test for continuity instead of an ohmmeter. A self-powered test light does not rely on the power of the circuit to light its bulb. Rather it contains a battery for a power supply and when connected across a completed circuit the bulb will light. This type test light is connected to the circuit in the same manner as an ohmmeter.

Either type of test light should not be used to test electronic circuits or components. The use of test lights should be limited to checking for battery voltage at various points within an electrical circuit.

Multimeters

Ohmmeters, ammeters, and voltmeters can be separate meters or can be combined in one meter called a **multimeter.** Most multimeters can measure DC volts, DC current (amps), AC volts, and

A **multimeter** is a meter that combines the functions of a voltmeter, ammeter, and ohmmeter.

ohms. Multimeters have either a needle type analog display, or a numeric digital display. To use this type of meter, the desired function and the scale are selected prior to connecting the meter. Once the meter is set in a function, the **digital multimeter (DMM)** should be used as an individual function type meter.

A DMM with high **impedance** is normally required for testing electronic components and circuits. The high impedance prevents a surge of high current through the sensitive semiconductors when the tester is connected into a circuit. High current can destroy electronic circuits and components.

Using Meters

Meters can be used to test electrical components, such as switches, loads, and semiconductors. Switches can be tested for operation and for excessive resistance with a voltmeter, test light, or ohmmeter. To check the operation of a switch with a voltmeter or a test light, connect the meter's positive lead to the battery side of the switch. With the negative lead attached to a good ground, voltage should be measured at this point. Without closing the switch, move the positive lead to the other side of the switch. If the switch is open, no voltage will be present at that point. The amount of voltage present at this side of the switch should equal the amount on the other side when the switch is closed. If the voltage decreases, the switch is causing a voltage drop due to excessive resistance. If no voltage is present on the groundside of the switch with it closed, the switch is not functioning properly and should be replaced.

If a switch has been removed from the circuit, it can be tested with an ohmmeter or a self-powered test light. By connecting the leads across the switch connections, the action of the switch should open and close the circuit.

Ohmmeters are used to test semiconductors. Because semiconductors allow current flow only in certain directions, they can be tested by connecting an ohmmeter to their connections. Some digital meters have a feature that allows for accurate testing of diodes. This feature should be used. Analog meters are the preferred instrument for checking diodes because the results are easily observed. By placing the leads of the meter on the semiconductor connections, continuity should be observed. Reversing the meter leads to the same connections should result in different readings of continuity. For example, a diode should show good continuity (low resistance) when the leads are connected to it and poor continuity (high resistance) when the leads are reversed.

In addition to meters and test lights, troubleshooting electrical problems may involve the use of jumper wires, circuit breakers, and/or short detection devices. A jumper wire is simply a length of wire normally equipped with alligator clips on either end. These clips allow the wire to be easily and safely connected to various points within a circuit. Jumper wires are primarily used to bypass a particular point within a circuit and should contain an in-line fuse or circuit breaker to prevent damage to the circuit being tested and the wire itself.

Electrical Problems

Classroom Manual
Chapter 4, page 88

Normally, the circuits in an automobile have a circuit protection device placed in the common path from the battery. This protection device is usually a fuse or a circuit breaker. They are designed to protect the wires and components from excessive current. When a great amount of current flows through the fuse or breaker, an element will burn out, causing the circuit to be opened and stopping current flow. This action prevents the high current from burning up the wires or components in the circuit.

High current is caused by low resistance (Figure 4-11). A decrease in the amount of resistance is typically the result of a **short**. A short is best defined as an additional and unwanted path to ground. Most shorts, such as a bare wire contacting the frame of the car, create an ex-

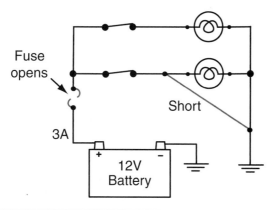

Figure 4-11 A short to ground.

tremely low resistance parallel branch. A slow-turning motor can also cause low resistance and high current.

A short to ground can be present before the load in the circuit or internally within the load or component. A short can also connect two or more circuits together, causing additional parallel legs and uncontrolled operation of components. An example of a possible result from a wire-to-wire short would be the horn blowing each time the brake pedal is depressed. This could be caused by a wire-to-wire short between the horn and brake light circuits. Shorts are one of the three common types of electrical problems.

A circuit breaker is a protection device that resets itself after it has been tripped by high current. Circuit breakers open due to the heat of high current. While open, the breaker cools and closes again to complete the circuit. When a fuse is blown or burned out by high current, it must be replaced to reactivate the circuit. To diagnose a circuit with a short, a completed circuit makes locating the fault easier. By inserting a circuit breaker in place of the fuse, the circuit can be activated and still protected.

A number of techniques and test instruments are used to locate a short. The most common of these devices is called a short detector, which utilizes the magnetic field formed by current flow to indicate the location of a short. The cycling of the circuit breaker will be indicated by the sweeps of the detector's needle as it is moved along the length of the circuit. When the needle no longer sweeps in response to the breaker, the location of the short can be assumed to be before that point in the circuit. Carefully moving the detector back through the circuit should locate the exact place of the short.

Another common electrical fault is the **open.** An open causes an incomplete circuit and can result from a broken or burned wire, loose connection, or a faulty component. If a circuit is open, there is no current flow and the component will not operate. If there is an open in one leg of a parallel circuit, the remaining part of that parallel circuit will operate normally.

To determine the location of an open, a test light or voltmeter is used. By probing along the circuit, the completed path can be traced. The point at which voltage is no longer present is the point at which the open is. If the circuit is open, there will be no current flow. This means that if the open is after the load, source voltage will be measured after the load because there is no voltage drop in the circuit. Often a jumper wire is used to verify or isolate the problem area.

Excessive resistance at a connector (Figure 4-12), internally in a component, or within a wire is also a common electrical problem. High resistance will cause low current flow and the component will not be able to operate normally, if at all.

The source of high resistance is easily found by measuring the voltage drop across the circuit. It is important to remember that excessive resistance can occur anywhere in the circuit. The best way to identify the source of high resistance is to divide the affected circuit into three parts: the component, the ground circuit, and the power circuit. If the voltage drop across the

An **open** is a break in the circuit that stops current flow.

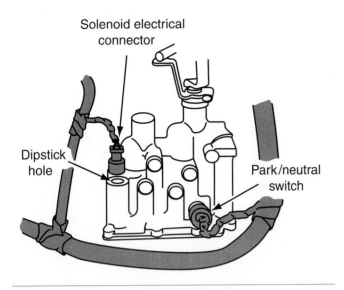

Figure 4-12 The various transmission connectors that need to be inspected and checked for high resistance problems.

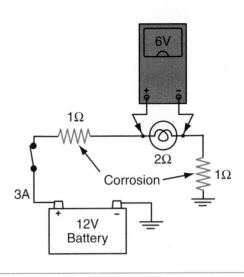

Figure 4-13 A simple light circuit with unwanted resistance. Notice the reduced voltage drop across the lamp and the reduced circuit current.

component is equal to source voltage (Figure 4-13), there are no other resistances in the circuit. If the voltage drop is less than source voltage, check the voltage drop across the ground circuit. If the voltage drop is more than 0.1 volts, there is high resistance in that part of the circuit. The source can be located by separating that circuit into small measurable parts. Then measure each part. The part that has the excessive voltage drop across it is the source for the high-resistance problem. The power side of the circuit is checked in the same way.

Lab Scopes

Special Tools

Lab scope

Assortment of jumper wires

Lab scope is the name given to a small (sometimes hand-held) oscilloscope. An oscilloscope is actually a visual voltmeter. Lab scopes have become the diagnostic tool of choice for many good technicians. A scope allows you to see voltage changes over time (Figure 4-14). Voltage is displayed across the screen of the scope as a waveform. By displaying the waveform, the scope

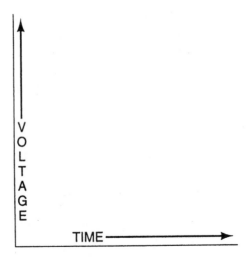

Figure 4-14 Voltage and time axes on a scope's screen.

shows the slightest changes in voltage. This is a valuable feature for a diagnostic tool. With a scope, precise measurement of voltage is possible. When measuring voltage with an analog voltmeter, the meter only displays the average values at the point being probed. Digital voltmeters simply sample the voltage several times each second and update the meter's reading at a particular rate. If the voltage is constant, good measurements can be made with both types of voltmeters. A scope will display any change in voltage as it occurs. This is especially important for diagnosing intermittent problems.

The screen of a lab scope is divided into small divisions of time and voltage (Figure 4-15). These divisions set up a grid pattern on the screen. The horizontal movement of the waveform represents time. Voltage is measured with the vertical position of the waveform. Since the scope displays voltage over time, the waveform moves from the left (the beginning of measured time) to the right (the end of measured time). The value of the divisions can be adjusted to improve the view of the voltage waveform. For example, the vertical scale can be adjusted so that each division represents 0.5 volts and the horizontal scale can be adjusted so that each division equals 0.005 seconds (5 milliseconds). This allows you to view small changes in voltage that occur in a very short period of time. The grid serves as a reference for measurements.

Since a scope displays actual voltage, it will display all electrical noises or disturbances that accompany the voltage signal (Figure 4-16). Noise is primarily caused by **radio frequency interference (RFI)**, which may come from the ignition system. RFI is an unwanted voltage signal that rides on a signal. This noise can cause intermittent problems with unpredictable results. The noise causes slight increases and decreases in the voltage. When a computer receives a voltage signal with noise, it will try to react to the minute changes. As a result, the computer responds to the noise rather than the voltage signal.

Electrical disturbances or **glitches** are momentary changes in the signal. These can be caused by intermittent shorts to ground, shorts to power, or opens in the circuit. These problems can occur for only a moment or may last for some time. A lab scope is handy for finding these and other causes of intermittent problems. By observing the voltage signal and wiggling or pulling a wiring harness, any looseness can be detected by a change in the voltage signal.

RFI is **radio frequency interference**.

A **glitch** is best defined as abnormal movement of the waveform.

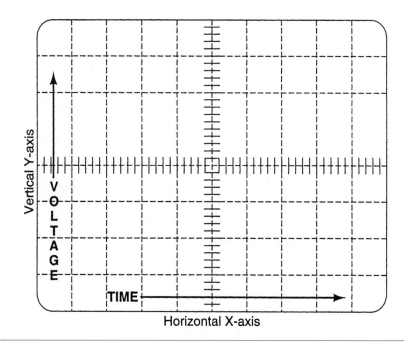

Figure 4-15 Grids on a scope screen serve as time and voltage references.

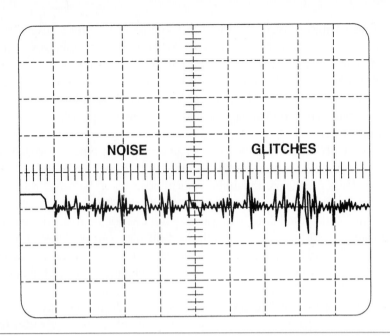

Figure 4-16 RFI noise and glitches may appear on the voltage signal.

Analog vs. Digital Scopes

Analog scopes show the actual activity of a circuit and are called real-time scopes. This simply means that what is taking place at the point being measured or probed is what you see on the screen. Analog scopes have a fast update rate, which allows for the display of activity without delay.

A digital scope, commonly called a **digital storage oscilloscope (DSO)**, converts the voltage signal into digital information and stores it in its memory. Some DSOs send the signal directly to a computer or a printer, or save it to a disk. To help in diagnostics, a technician can "freeze" the captured signal for close analysis. DSOs also have the ability to capture low-frequency signals. Low-frequency signals tend to flicker when displayed on an analog screen. To have a clean waveform on an analog scope, the signal must be repetitive and occur in real time. The signal on a DSO is not quite real time; rather, it displays the signal as it occurred a short time before.

This slight delay is actually very slight. Most DSOs have a sampling rate of one million samples per second. This is quick enough to serve as an excellent diagnostic tool. This fast sampling rate allows slight changes in voltage to be observed. Slight and quick voltage changes cannot be observed on an analog scope.

A DSO uses an analog-to-digital (A/D) converter to digitize the input signal. Since digital signals are based on binary numbers, the trace appears slightly choppy when compared to an analog trace. But the voltage signal is sampled more often, which results in a more accurate waveform. The waveform is constantly being refreshed as the signal is pulled from the scope's memory. Remember, the sampling rate of a DSO can be as high as a million times per second.

Both an analog and a digital scope can be dual trace scopes (Figure 4-17). This means they both have the capability of displaying two traces at one time. By watching two traces simultaneously, you can watch the cause and effect of a sensor, as well as compare a good or normal waveform to the one being displayed.

Waveforms

A waveform represents voltage over time. Any change in the **amplitude** of the trace indicates a change in the voltage. When the trace is a straight horizontal line, the voltage is constant (Figure

DSO is a common acronym for **digital storage oscilloscope**.

The update rate is the time it takes the trace from the end of the screen on the right to move back to the left of the screen.

A captured signal is a signal stored in a DSO's memory.

The sweep rate of a scope is simply how fast the electron beam moves across the screen.

Binary numbers are strings of zeros and ones, or ons and offs, which represent a numeric value.

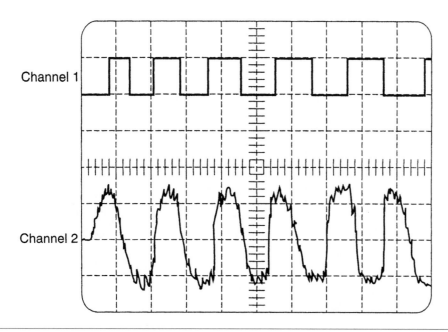

Figure 4-17 Some scopes can display two traces at one time; these are called dual trace scopes.

4-18). A diagonal line up or down represents a gradual increase or decrease in voltage. A sudden rise or fall in the trace indicates a sudden change in voltage.

Scopes can display AC and DC voltage either one at a time or both together, as in the case of noise caused by RFI. Noise results from AC voltage riding on a DC voltage signal. The consistent change of polarity and amplitude of the AC signal causes slight changes in the DC voltage signal. A normal AC signal changes its polarity and amplitude over a period of time. The waveform created by AC voltage is typically called a sine wave (Figure 4-19). One complete sine wave shows the voltage moving from zero to its positive peak, then moving down through zero to its

The height of a waveform is called its **amplitude**.

A sine wave is a waveform of a single frequency alternating current.

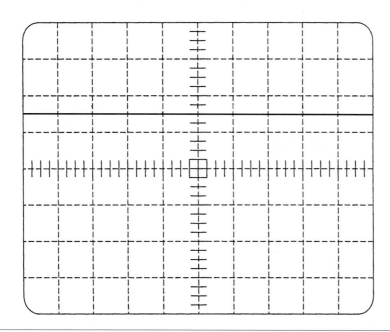

Figure 4-18 A constant voltage waveform.

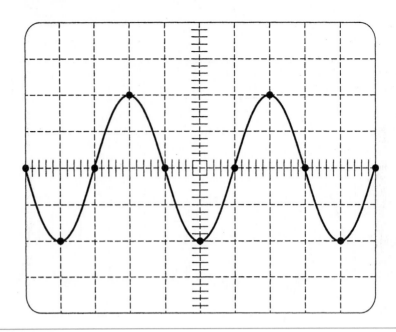

Figure 4-19 An AC voltage sine wave.

negative peak and returning to zero. If the rise and fall from positive and negative is the same, the wave is said to be **sinusoidal**. If the rise and fall are not the same, the wave is non-sinusoidal. Therefore, it can be said that all AC voltage waveforms are not sine waves.

One complete sine wave is a **cycle**. The number of cycles that occur per second is the frequency of the signal. Checking frequency or cycle time is one way of checking the operation of some electrical components. Input sensors are the most common components that produce AC voltage. Permanent magnet voltage generators produce an AC voltage that can be checked on a scope (Figure 4-20). AC voltage waveforms should also be checked for noise and glitches, which may send false information to the computer.

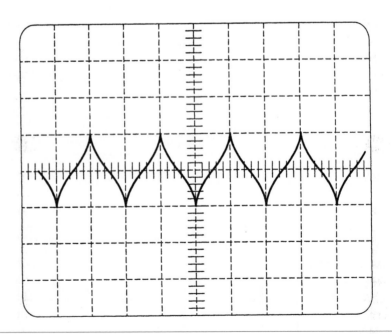

Figure 4-20 An AC voltage trace from a typical permanent magnet generator-type pickup or sensor.

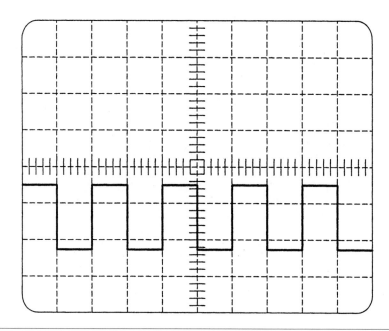

Figure 4-21 A typical square wave.

DC voltage waveforms may appear as a straight line or line showing a change in voltage. Sometimes a DC voltage waveform will appear as a square wave or digital signal, which shows voltage making an immediate change (Figure 4-21). Square waves are identified by having straight vertical sides and a flat top. This type of wave represents voltage being applied (circuit being turned on), voltage being maintained (circuit remaining on), and no voltage applied (circuit is turned off). Of course, a DC voltage waveform may also show gradual voltage changes.

Scope Controls

Depending on the manufacturer and model of the scope, the type and number of its controls will vary. However, nearly all scopes have these: intensity, vertical (Y-axis) adjustments, horizontal (X-axis) adjustments, and trigger adjustments. The intensity control is used to adjust the brightness of the trace. This allows for clear viewing regardless of the light around the scope screen.

The vertical adjustment actually controls the voltage displayed. The voltage setting of the scope is the voltage that will be shown per division (Figure 4-22). If the scope is set at 0.5 (500 milli) volts, this means a 5-volt signal will need 10 divisions. Likewise, if the scope is set to one volt, 5 volts will need only 5 divisions. While using a scope, it is important to set the vertical so that voltage can be accurately read. Setting the voltage too low may cause the waveform to move off the screen, while setting it too high may cause the trace to be flat and unreadable. The vertical position control allows the vertical position of the trace to be moved anywhere on the screen.

The horizontal position control allows the horizontal position of the trace to be set on the screen. The horizontal control is actually the time control of the trace (Figure 4-23). Setting the horizontal control is setting the time base of the scope's sweep rate. If the time per division is set too low, the complete trace may not show across the screen. Also, if the time per division is set too high, the trace may be too crowded for detailed observation. The time per division (TIME/DIV) can be set from very short periods of time (millionths of a second) to full seconds.

Trigger controls tell the scope when to begin a trace across the screen. Setting the trigger is important when trying to observe the timing of something. Proper triggering will allow the trace to repeatedly begin and end at the same points on the screen. There are typically numerous trigger controls on a scope. The trigger mode selector has a NORM and AUTO position. In the NORM

Examples of permanent magnet generators are some crankshaft and camshaft sensors, ignition pick-up units, vehicle speed sensors, and wheel speed sensors.

Some of the commonly found components powered by digital signals are fuel injectors, mixture-control solenoids, and digital EGR valves.

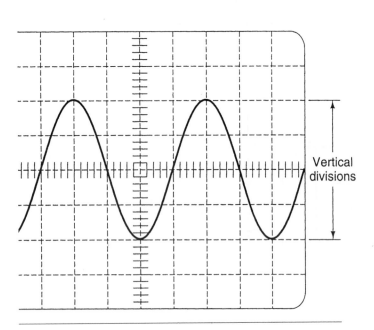

Figure 4-22 Vertical divisions represent voltage.

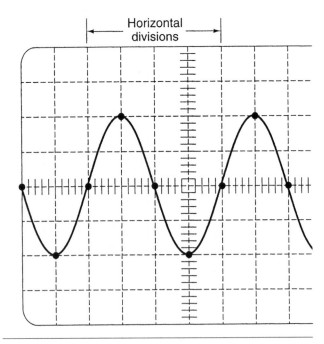

Figure 4-23 Horizontal divisions represent time.

setting, no trace will appear on the screen until a voltage signal occurs within the set time base. The AUTO setting will display a trace regardless of the time base.

Slope and level controls are used to define the actual trigger voltage. The slope switch determines whether the trace will begin on a rising or falling of the voltage signal (Figure 4-24). The level control determines where the time base will be triggered according to a certain point on the slope.

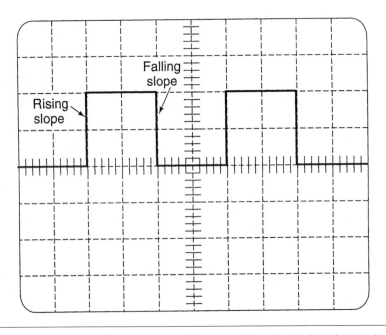

Figure 4-24 The trigger can be set to start the trace with a rise or fall of the voltage.

A trigger source switch tells the scope which input signal to trigger on. This can be Channel 1, Channel 2, line voltage, or an external signal. External signal triggering is very useful when desiring to observe a trace of a component that may be affected by the operation of another component. An example of this would be observing pressure control solenoid activity when changes in vehicle speed are made. The external trigger would be voltage change at the VSS. The displayed trace would be the cycling of the pressure control solenoid. Channel 1 and Channel 2 inputs are determined by the points of the circuit being probed. Some scopes have a switch that allows inputs from both channels to be observed at the same time or alternately.

Scan Tools

There are a variety of computer scan tools available today. A **scan tool** (Figure 4-25) is a microprocessor designed to communicate with the vehicle's computer. Connected to the computer through diagnostic connectors, a scan tool can access trouble codes, run tests to check system operations, activate outputs, and monitor the activity of the system. Trouble codes and test results are displayed on an LED screen, or printed out on the scanner printer.

Scan tools will retrieve fault codes from a computer's memory and digitally display these codes on the tool. A scan tool may also perform many other diagnostic functions depending on the year and make of the vehicle. Most aftermarket scan tools have removable modules that are updated each year. These modules are designed to test the computer systems on various makes of vehicles. For example, some scan testers have a 3-in-1 module that tests the computer systems on Chrysler, Ford, and General Motors vehicles. A 10-in-1 module is also available to diagnose computer systems on vehicles imported by 10 different manufacturers. These modules plug into the scan tool.

A **scan tool** is actually a computer designed to communicate with the vehicle's computer.

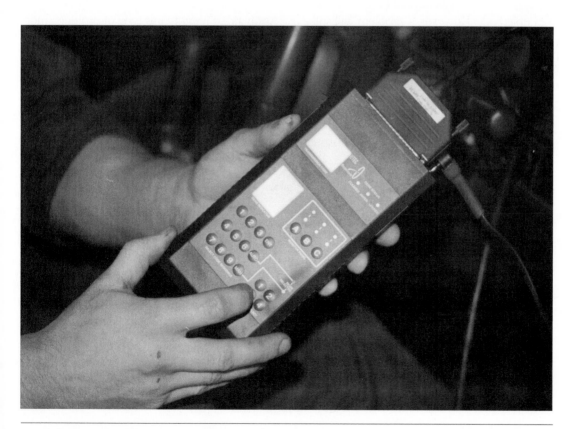

Figure 4-25 A typical scan tool.

Scan tools are capable of testing many onboard computer systems, such as transmission controls, engine computers, antilock brake computers, air bag computers, and suspension computers, depending on the year and make of the vehicle and the type of scan tester. In many cases, the technician must select the computer system to be tested with the scanner after it has been connected to the vehicle.

The scan tool is connected to specific diagnostic connectors on various vehicles. Most manufacturers have one diagnostic connector. This connects the data wire from each onboard computer to a specific terminal in this connector. Other vehicle manufacturers have several different diagnostic connectors on each vehicle, and each of these connectors may be connected to one or more onboard computers. A set of connectors is supplied with the scanner to allow tester connection to various diagnostic connectors on different vehicles.

The scanner must be programmed for the model year, make of vehicle, and type of engine. With some scan tools, this selection is made by pressing the appropriate buttons on the tester, as directed by the digital tester display. On other scan testers, the appropriate memory card must be installed in the tester for the vehicle being tested. Some scan testers have a built-in printer to print test results, while other scan testers may be connected to an external printer.

As automotive computer systems become more complex, the diagnostic capabilities of scan testers continue to expand. Many scan testers now have the capability to store, or "freeze," data into the tester during a road test, and then play back this data when the vehicle is returned to the shop.

Some scan testers now display diagnostic information based on the fault code in the computer memory. Service bulletins published by the manufacturer of the scan tester may be indexed in the tester after the vehicle information is entered in the tester. Other scan testers will display sensor specifications for the vehicle being tested.

The vehicle's computer only sets trouble codes when a voltage signal is entirely out of its normal range. The codes help technicians identify the cause of the problem when this is the case. If a signal is within its normal range but is still not correct, the vehicle's computer will not display a trouble code. However, a problem will still exist. As an aid to identify this type of problem, most manufacturers recommend that the signals to and from the computer be carefully looked at. This is accomplished with a scan tool by observing the **serial data**, or communications to and from the computer (Figure 4-26).

On systems fitted with OBD II systems, the diagnostic connectors are located in the same place on all vehicles. Also, a scan tool designed for an OBD-II system will work on all OBD-II systems; therefore, the need to have designated scan tools is eliminated. An OBD-II scan tool also has the ability to run diagnostic tests on all systems and has "freeze frame" capabilities.

The communications to and from the computer are commonly referred to as the system's **serial data**.

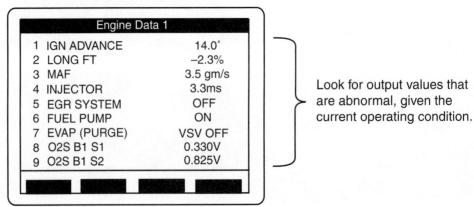

Figure 4-26 Use serial data to identify abnormal conditions.

General EAT Diagnosis

Some electronic transmissions are only partially controlled; that is, only the engagement of the converter clutch and third-to-fourth shifting is electronically controlled. Other models feature electronic shifting into all gears, plus electronic control of the TCC (Figure 4-27).

The controls of an EAT direct hydraulic flow by using solenoid-controlled valves. When used to control TCC operation, the solenoid opens a hydraulic circuit to the TCC spool valve, causing the spool valve to move and direct mainline pressure to apply the clutch. Electronically controlled shifting is accomplished in much the same way. Shifting occurs when a solenoid is either turned on or turned off (Figure 4-28). At least two shift solenoids are incorporated into the system and shifting takes place by controlling the solenoids. The desired gear is put into operation through a combination of on and off solenoids.

Classroom Manual
Chapter 4, page 104

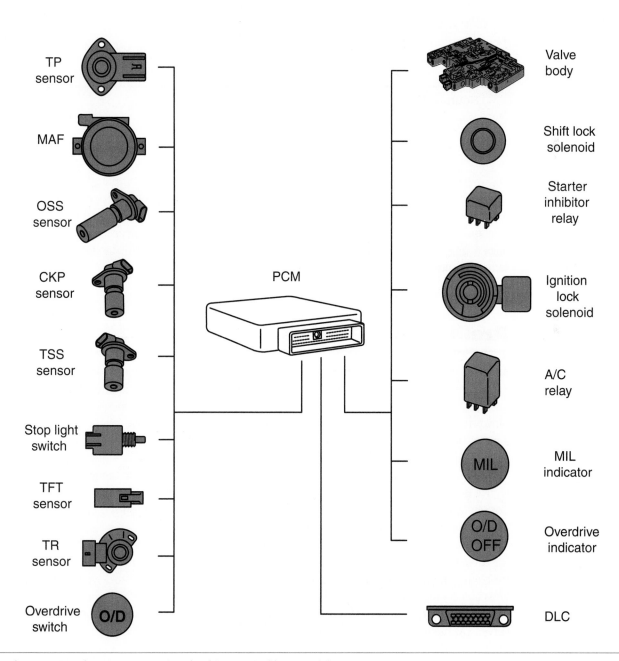

Figure 4-27 The components involved in a typical late-model EAT system.

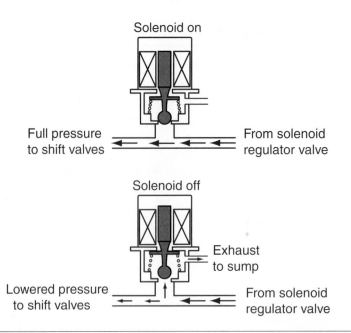

Figure 4-28 Shift solenoid action.

Several sensors and switches are used to inform the control computer of the current operating conditions. Most of these sensors are also used to calibrate engine performance. The computer then determines the appropriate shift time for maximum efficiency and best feel. The shift solenoids are controlled by the computer, which either supplies power to the solenoids or supplies a ground circuit. The techniques for diagnosing electronic transmissions are basically the same techniques used to diagnose TCC systems.

Preliminary Checks

Although EATs are rather reliable, they have introduced new problems for the automatic transmission technician to contend with. Some of the common problems that affect the shift timing and quality, as well as the timing and quality of TCC engagement are: incorrect battery voltage, a blown fuse, poor connections, a defective TP sensor or VSS, defective solenoids, wires to the solenoids or sensors crossed, corrosion at an electrical terminal, or faulty installation of some accessory, such as a cellular telephone.

Electrical circuit problems, faulty electrical components, or bad connectors, as well as a defective governor or governor drive gear assembly can cause improper shift points. Most EATs do not rely on the hydraulic signals from a governor; rather, they rely on the electrical signals from electrical sensors to determine shift timing.

Often computer-controlled transmissions will start off in the wrong gear. This can happen for several reasons, either because of internal transmission problems or external control system problems. Internal transmission problems can be faulty solenoids or stuck valves. External problems can be the result of a complete loss of power or ground to the control circuit or a fail-safe protection strategy initiated by the computer to protect itself or the transmission from an observed problem. Typically the default gear is simply the gear that is applied when the shift solenoids are off.

Basic System Checks

Troubleshooting a transmission's electronic control system is like troubleshooting any other electronic system; you need to make preliminary checks of the system before moving into specific checks. The first step in a preliminary check is verifying system voltage. With a DMM, check the open-circuit voltage of the battery. Minimum voltage at the battery should be 12.6 volts. If the battery is below this, recharge the battery. If the voltage is still low after recharging, replace it.

Continue by checking the condition of the battery cables. Conduct a voltage drop test across each cable. Make sure the system is activated when doing a voltage drop test. There should be no more than 0.1 volt dropped across the positive or the negative cable. If the voltage drop exceeds that amount, clean or replace the cables.

A visual inspection of the circuit should follow. Carefully look at the entire system and check for damaged or corroded wires, loose connections, and damaged connectors (Figure 4-29). Pay particular attention to the connectors. Make sure there are no bent or broken terminals. If any are present, replace the connector. Also look inside each connector for signs of corrosion. If any is found, clean the wires and terminals. If the corrosion cannot be cleaned, replace the wires and/or the connector.

Continue your basic inspection with checking the fuse or fuses to the control module. To accurately check a fuse, either test it for continuity with an ohmmeter or check each side of the fuse for power when the circuit is activated.

CAUTION: Remember to disconnect power to the component before checking it with an ohmmeter. Failure to do so can result in the meter being destroyed.

CUSTOMER CARE: Vehicles equipped with control computers may require a re-learn procedure after the battery is disconnected. Many computers memorize and store vehicle operation patterns for optimum driveability and performance. When the vehicle's battery is disconnected, this memory is lost. The computer will use default data until new

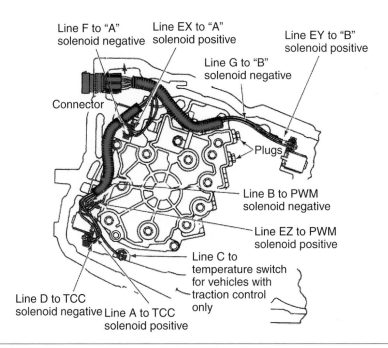

Figure 4-29 The position and condition of the transmission wiring harness and its connectors should be carefully checked.

data from each key start is stored. Customers often complain of driveability problems during the re-learn stage because the vehicle acts differently than it did before it was serviced. To reduce the possibility of complaints, the vehicle should be road tested and the correct re-learn procedure followed.

If the system has no apparent problems, continue testing the system. The diagnostic procedure for most EAT systems would now include checking the system with a scan tool. The scan tool will display any trouble codes in the system. More importantly to the technician, most scan tools will display serial data. The serial data stream allows you to monitor system sensor and actuator activity during operation. Comparing the test values to the manufacturer's specifications will greatly help in diagnostics.

It is possible that the data displayed by a scan tool is not the actual value. Most computer systems will disregard inputs that are well out of range and rely on a default value held in its memory. These default values are hard to recognize and do little for diagnostics; this is why detailed testing with a DMM or lab scope is preferred by many technicians. These test instruments are also used to further test the system after a scan tool has identified a problem.

Road Test

Critical to proper diagnosis of EAT and TCC control systems is a road test. The road test should be conducted in the same way as one for a nonelectronic transmission except that a scanner tool should also be connected to the circuit to monitor engine and transmission operation.

During the road test, the vehicle should be driven in the normal manner. All pressure and gear changes should be noted. Also the various computer inputs should be monitored and the readings recorded for future reference. Some scan tools have the capability of printing out a report of the test drive. Critical information from the inputs includes engine speed, vehicle speed, manifold vacuum, operating gear, and the time it took to shift gears. If the scanner does not have the ability to give a summary of the road test, you should record this same information after each gear or operating condition change.

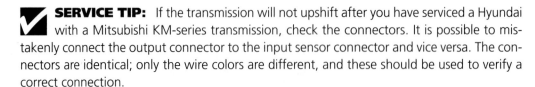 **SERVICE TIP:** If the transmission will not upshift after you have serviced a Hyundai with a Mitsubishi KM-series transmission, check the connectors. It is possible to mistakenly connect the output connector to the input sensor connector and vice versa. The connectors are identical; only the wire colors are different, and these should be used to verify a correct connection.

Basic Transmission Circuits

To summarize how the circuitry of an electronically controlled transmission interrelates to a computerized engine control system, a summary of the transmission controls for a Ford E4OD follows. This system is similar to the other systems used by Ford, as well as the other manufacturers. The transmission is controlled by the PCM of the engine control system. The computer receives inputs from both the engine and the transmission. Based on this information, the PCM can send signals to operate components in both the engine and transmission. There are many inputs to the PCM from the transmission. The transmission range (TR) sensor informs the PCM what gear has been selected. The vehicle speed sensor, located in the same place as the governor would be, inputs the mph at which the vehicle is traveling. The turbine speed sensor (not found on all models) sends signals to inform the PCM of the speed of the input to the transaxle. The transmission oil temperature sensor monitors the temperature of the ATF. This signal determines whether the PCM should go to a cold-start shift cycle or shift according to its schedule.

There are four nontransmission inputs: the TP sensor, MAF, PIP, and brake on switch. The system uses five solenoids, which are located in the transmission: the electronic pressure control solenoid that regulates transmission operating pressure, a modulated solenoid that controls the

Classroom Manual
Chapter 4, page 99

As a comparison of cold-start shifting and schedule shifting, if the engine is cold, the transmission will shift later and converter clutch engagement will also be later. When the engine warms up, the transmission will shift more quickly and have an earlier converter clutch engagement.

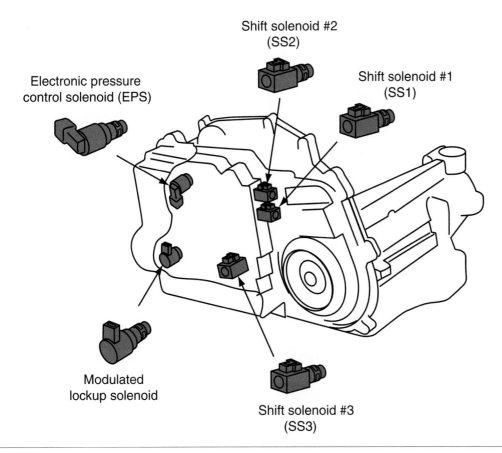

Figure 4-30 Location of output solenoids on an AXOD-E transaxle.

converter clutch, and three shift solenoids that control fluid flow to the various holding and clutching devices (Figure 4-30).

Because the engine and transmission control systems share information, a common component may affect both systems. Nearly all computer control systems have some form of self-diagnostics. During this mode of operation, the computer will display a trouble code that represents a circuit or component it determines is faulty. Part of the basic procedure for diagnosing an EAT is to retrieve the trouble codes from the computer. This information will determine what step you should take next (Figure 4-31).

As you can readily tell, in this system, as well as the others, engine performance problems can cause the transmission to perform unsatisfactorily. Since the computer controls the outputs to the engine and the transmission based on input sensors common to both, it is very important that you perform basic engine checks before spending time diagnosing the transmission. This was important when diagnosing non-electronically controlled transmissions and is even more important when diagnosing an EAT.

On-Board Diagnostics

Nearly all EATs have the ability to check input and output circuits during operation. When one of these circuits operates outside of the acceptable range, a diagnostic code is set. The setting of many of these codes will also cause the malfunction indicator lamp (MIL) to illuminate. This lamp informs the driver that a problem is present. It is important to remember that not all codes will cause the MIL to illuminate; therefore, an unlit MIL does not mean there are no DTCs in the computer's memory.

Classroom Manual
Chapter 4, page 104

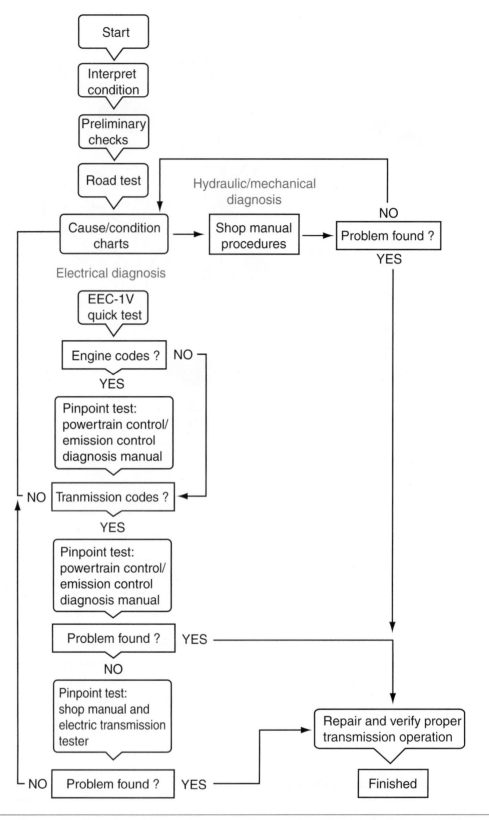

Figure 4-31 A flow chart showing the major steps that should be followed while diagnosing an EAT.

Diagnostic Trouble Codes

The name **diagnostic trouble codes** indicates what they are. They are designed to help technicians identify and locate problems in the transmission's system. DTCs from pre-OBD II systems can indicate a variety of things. Each manufacturer (and sometimes each vehicle model) used different codes to identify detected problems. It is very important that you refer to the service manual when interpreting DTCs.

On OBD II systems, there are two types of codes. Some codes are universal in that they are the same for all manufacturers and vehicle models. The other set of codes are manufacturer specific and must be interpreted through reference to the appropriate service manual

In a typical EAT system, DTCs are set by the computer when a signal from a component or circuit is not within the normal operating range for the operating conditions. A code will also be set when there is no signal from an input circuit or to an output device. EATs also have the ability to conduct a self-test. During this test, the sensor circuits are checked and, if a problem is found, a DTC is set.

There are basically two types of DTCs. A hard code is a DTC that represents a problem that is present at the time of retrieval. These are the codes that should be responded to first, during diagnostics. Soft codes are those DTCs that are not currently present. These codes can be retrieved and represent an intermittent problem or a problem that existed but is no longer present.

Basic Diagnostics of Non-OBD II Systems

Diagnosis on vehicles equipped with non-OBD II electronic control systems can be done by the self-diagnostics mode of the system's computer and/or by plugging in a manufacturer-provided or aftermarket-designed scan tool into the **DLC** or **ALDL**. Use the following discussions for learning purposes only. They are intended to expose you to the world of electronic transmission controls, not to guide you totally through a diagnostic procedure on a particular model transmission. Much of the information listed under one manufacturer will also be true for others.

Always refer to the correct service manual when diagnosing electronic control systems. The manufacturers list the procedures for self-diagnostics in their service manuals. Also listed are the interpretation tables for the trouble codes retrieved from the computer. Keep in mind, by law, all 1996 and newer vehicles must have an OBD II compliant system. Therefore it is fair to say the following procedures apply only to vehicles produced in 1995 or before.

Chrysler

Chrysler has basically two electronic control systems. One was first used in 1988 and has been used in a variety of models. The other system is a multiplexed system, which is used on later model mini-vans and cars. The original system is quite basic and similar to those used by other manufacturers. The control computer has self-diagnostic capabilities. The recommended way to retrieve trouble codes is through the use of Chrysler's scan tool, called a DRB-II.

The DRB-II is connected to the Serial Communication Interface, which is located by the left front shock tower. The red power supply lead is connected to the positive post of the battery. The scan tool has a variety of cartridges that contain the test parameters for the vehicle being tested. After the correct cartridge has been installed in the tester and the scan tool is connected to the vehicle, the tester's display will show a test pattern. After a few seconds, the display will change to read out the copyright date and revision level of the cartridge. After a few seconds, the display will ask for the vehicle's model year. The model year is selected by pressing either right or left arrow keys. Once the display shows the correct year, press the down arrow key.

The display will then ask for the system that will be tested. The system is selected by pressing either left or right arrow keys. Once the display indicates powertrain, press down arrow key. Other typical system choices are engine or EFI/Turbo.

Special Tools

Scan tool (non-OBD II)

Assortment of test cartridges

Jumper wires

Diagnostic key (for GM vehicles)

DLC stands for **data link connector**.

ALDL stands for **assembly line data link**.

Today's Chrysler transmissions require the use of the DRB-III.

The display will then ask for the desired test selection. The desired test is selected by pressing either the left or right arrow keys. The diagnostic test mode is used to see if there are any fault codes stored in the on-board diagnostic system memory of the computer. The circuit actuation test mode (ATM) is used to make the computer cycle a solenoid on and off. The sensor test mode is used to see output signals of certain sensors as the computer receives them when the engine is not running. The engine running test mode is used to see sensor output signals as received by the computer while the engine is running.

To enter into the diagnostic mode, set the DRB-II to the engine off test. Turn the ignition switch to "ON-OFF, ON-OFF, ON" within five seconds. Then record all codes displayed on the tester. By depressing the "READ/HOLD" key until "HOLD" is displayed, you can stop the display of codes. To continue the display, press the "READ/HOLD" key.

The following are typical codes and the condition that would cause the computer to display them:

> *Code 15*—Vehicle Speed Sensor: This vehicle speed sensor code will light the check engine lamp on California vehicles only. The code will be set if a speed sensor signal is not detected for over an 11–13 second period.
>
> *Code 22*—Engine Coolant Sensor: This code will cause the check engine lamp to light. The code will be set if the coolant sensor voltage is above 4.96 volts when engine is cold or below 0.51 volts when the engine is warm.
>
> *Code 24*—Throttle Position Sensor: This code will light the check engine lamp, and will be set if the signal from the TP sensor is below 0.16 or above 4.7 volts.
>
> *Code 37*—Transmission Lock-up Solenoid: This code will not illuminate the check engine lamp. The code will be set if the lock-up solenoid does not turn on and off when it should.
>
> *Code 55*—End of Diagnostic Mode: This code will not illuminate the check engine lamp and appears at the end of the diagnostic series.
>
> *Code 88*—Start of Diagnostic Mode: This code must appear at start of the diagnostic series. If it does not, the fault codes will be inaccurate.

To erase the fault codes, put the system into ATM test mode 10. Then press the "READ/HOLD" key until "HOLD" is displayed. The display will then flash alternating "0's." When the flashing "0's" stop and the display shows "00," all fault codes have been erased.

The Chrysler Collision Detection system is referred to as the CCD system.

The other transmission control system is the Chrysler Collision Detection system, which is a multiplex serial data bus or network. The network consists of a pair of twisted wires that interconnect the CCD modules. Each CCD module uses this network to communicate and exchange data with other modules on the network. The number of modules varies with the model of vehicle.

A bus is a common connector and is used as an information source for the vehicle's various control units.

Diagnosis of this transaxle system is relatively simple. Nearly all of the important information is available through the use of a scan tool and the transaxle is equipped with many pressure taps. Both of these features give you an accurate look at the operation of the computer.

When the transaxle controller detects a problem, it will automatically go into the "limp-in" mode of operation. This mode defaults the transaxle to second gear only with second gear starts, reverse, and park operations. The computer selects limp-in mode because it received a signal from an input device that it interpreted as an out-of-range signal. The computer will go into this mode if there is an electrical control system problem, a hydraulic apply circuit problem, an internal mechanical problem, or sticking valve problem.

To determine the reason why the computer has defaulted to this mode, connect a scan tool to the blue six-pin CCD connector under the dash. Retrieve and record the trouble codes that are displayed. Then clear the memory and drive the vehicle until the complaint occurs again. Retrieve and write down the codes again. These codes are the most current and should be compared to the original codes.

If the scan tool is unable to display a code or is unable to recognize the computer or its circuits, the problem may be a faulty or dead bus. A dead bus is commonly caused by a body computer, transaxle controller, or any other module on the CCD bus. The exact bad module can be identified by disconnecting each module, one at a time, from the bus.

After you have one module disconnected, rerun the start-up of the scan tool. If the bus becomes active, the disconnected control module was the cause of the previous dead bus. Repeat this process, by reconnecting the disconnected module and unplugging another. The module that is disconnected when the bus becomes active is the dead module.

If the cause of the dead bus was not identified, voltage checks should be made. Measure the voltage drop across the groundside of the diagnostic connector by inserting the positive voltmeter lead to the ground lead of the connector and the negative lead to the negative terminal of the battery. The maximum allowable voltage drop across the ground is 0.1 volts. The source voltage to the computer should also be checked. If battery voltage is not present at the computer, check the voltage of the battery. If it is okay, check the voltage drop in the supply circuit. If there is more than 0.1 volts dropped, identify the source of the resistance and repair the problem. Also check the bus bias voltage by connecting the voltmeter's positive lead to either bias terminal in the diagnostic connector and the negative lead to the ground terminal in the connector. The bias voltage should be between 2.1 to 2.6 volts. Any incorrect voltages could be the cause for a dead bus and the cause of the problems should be corrected. After the electrical problems are repaired, the scan tool test should be conducted.

SERVICE TIP: On some A-604-equipped vehicles built before 1/16/90 there was a problem of surging above 45 mph. This was caused by a headlamp/dash wiring harness that generated a radio frequency that the computer misinterpreted and started to cycle the torque converter clutch on and off, creating the surge. The repair procedure involves some alterations to the wiring harness. Check the appropriate service manual or technical service bulletin for the proper repair.

Using the scan tool, the trouble codes are retrieved and compared to the trouble code chart given in the service manual. These codes will identify problem circuits. The service manual also gives step-by-step procedures for pinpoint diagnostics of each trouble code. Always follow these; they are efficient and exact.

Ford

Most Ford EAT systems are based on their EEC IV or EEC V systems. Although the exact make-up of the systems varies according the transmission used and the model vehicle being tested, all follow similar procedures. The following equipment is recommended to diagnose and test an EEC system.

Self-Test Automatic Read-out (STAR) Diagnostic Tester—This tester is recommended but not required. The tester was designed for EEC systems and is used to display the service codes. There are also many aftermarket testers available for testing these systems. An analog volt/ohmmeter with a 0–20 V DC range can be used as an alternate to a diagnostic tester.

DMM—This multimeter must have a minimum impedance of 10-megohms.

Breakout Box—This is a jumper wire assembly that connects between the vehicle harness and the PCM. The breakout box is required to perform certain tests on the system. During individual circuit tests, the procedures will call for probing particular pin numbers. These pin numbers relate to the pins of the breakout box.

Vacuum Gauge and Vacuum Pump—This can be one assembly or separate units.

Tachometer—This must have a 0–6000 rpm range defined in 20-rpm increments.

Spark Tester—A modified side electrode spark plug with the electrode removed and an alligator clip attached may be used in place of the spark tester.

MAP/BP Tester—This tester plugs into the MAP/BP sensor circuit and the DMM. It is used to check input and output signals from the sensor, which produces a frequency signal. This signal may also be monitored on a scope set to a milli-volt scale.

Other equipment—Timing light, fuel injection pressure gauge, non-powered 12-volt test light, and a jumper wire, about 15 inches long.

CAUTION: Some of this test equipment is required to perform diagnostics on an EEC system. *Never* attempt to test this system without the proper equipment. Damage to vehicle components will result if improper test equipment is used.

The EEC system offers a variety of test sequences. These steps are called the Quicktest and should be carefully followed in sequence to get accurate readings. Failure to do so may result in misdiagnosis or the replacement of nonfaulty components. The steps are: test preparation and equipment hook-up, Key On Engine Off (KOEO) self-test, Key On Engine Running (KOER) self-test, and continuous self-test. Photo Sequence 5 shows the basic procedure for diagnosing Ford EAT problems.

After all tests, servicing, or repairs have been completed, the Quicktest should be repeated to ensure that all systems are working properly. The KOEO and KOER self-tests are designed to identify faults present at the time of the testing and not intermittent faults. Intermittent faults are detected by the Continuous self-test mode.

Before hooking up any equipment to diagnose the EEC system, the following checks should be made in preparation for further testing:

1. Verify the condition of the air cleaner and air ducting.

2. Check all vacuum hoses for leaks, restrictions, and proper routing.

3. Check the electrical connectors of the EEC system for corrosion, bent or broken pins, loose wires or terminals, and proper routing.

4. Check the PCM, sensors, and actuators for physical damage.

5. Set the parking brake and place shift lever in "P." *Do not* move the shift lever during the test unless the procedures call for it.

6. Turn off all lights and accessories. Make sure the vehicle's doors are closed when making voltage or resistance readings.

7. Check the level of the engine's coolant.

8. Start the engine and allow it to idle until the upper radiator hose is hot and pressurized.

9. Check for leaks around the exhaust manifold, exhaust gas oxygen sensor, and vacuum hose connections.

To retrieve codes with the analog VOM, turn the ignition key off. Then connect the jumper wire from Self-Test Input (STI) pigtail to pin #2 at the SELF-TEST connector. Set the VOM to the correct DC voltage scale. Connect the positive lead of VOM to the positive battery terminal. The negative lead of the meter is connected to the #4 pin of the SELF-TEST connector. Connect the timing light and proceed to the KOEO SELF-TEST. Codes are shown as voltage pulses (needle sweeps). Pay attention to the length of the pauses in order to read the codes correctly.

To retrieve codes with the STAR tester, turn the ignition off. Then connect the color-coded adapter cable leads to the diagnostic tester. Connect the adapter cable's two service connectors to the vehicle's SELF-TEST connectors. Then connect the timing light and proceed to the KOEO SELF-TEST.

Photo Sequence 5
Typical Procedure for Diagnosing EAT Problems

P5-1 Check fluid level and condition.

P5-2 Check all connections.

P5-3 Road test the vehicle to verify the complaint.

P5-4 Connect the STAR tester.

P5-5 Observe the codes.

P5-6 Connect the breakout box.

P5-7 Install the template over the breakout box.

P5-8 Perform pinpoint tests according to the service manual.

P5-9 After the fault has been corrected, rerun the Quicktest.

SERVICE TIP: The SELF-TEST connector is found in different locations, depending upon the model of the vehicle. On most FWD models, the SELF-TEST connector is located in the right rear of the engine compartment, near the firewall.

All service codes are 2- or 3-digit numbers that are generated one digit at a time. There will be a 2-second pause between each DIGIT in a code. There will be a 4-second pause between each CODE. The continuous memory codes are separated from the functional test service codes by a 6-second delay, a single 1/2-second sweep, and another 6-second delay. Always record the codes in the order received. If a diagnostic tester is used, it will count the pulses and display them as a digital code. If the "CHECK ENGINE" light is used, numeric service codes will be displayed at the light.

The following is the proper procedure for conducting the KOEO self-test:

1. Turn ignition switch to "OFF" position.

2. Connect tester.

3. Turn ignition on.

4. The PCM will run through a test cycle, opening and closing various solenoids and switches. After the completion of the test cycle, the codes will appear. Record all displayed codes.

The first set of codes that will be displayed are the KOEO self-test codes. These codes will be repeated twice. They are followed by the Separator Pulse signal and the continuous memory codes. These codes will also be repeated twice. A KOEO code 11 (or 111) indicates the system has passed the self-test. Whenever a repair is made, repeat the Quicktest. To clear codes, remove the jumper wire while the codes are being displayed and before turning off the ignition.

General Motors

The 4T60-E and other GM transaxles use two electric solenoids to control transaxle upshifts and downshifts. Each solenoid is turned on and off by the PCM. The 4L80-E transmission functions in much the same way; however, it is also fitted with a force motor, which controls hydraulic line pressure and a TCC solenoid. Some models have a transmission control module (TCM) in addition to the PCM. In all systems, the PCM has self-diagnostic capabilities, which help identify which parts or circuits may need further testing.

The PCM constantly monitors all electrical circuits. If the PCM detects a problem or an out-of-range sensor input, it records a trouble code in its memory. If the problem continues for some time, the CEL or SES light will glow. It is possible for the PCM to have detected a trouble and not light the CEL/SES lamp. However, the appropriate code may be stored in its memory.

Many methods are commonly used to retrieve the trouble codes from the PCM's memory. The simplest method is to use the CEL/SES light. Other methods include the use of special scan tools.

To display the trouble codes with the CEL/SES lamp, locate the ALDL connector. Then, with the ignition on and engine off, jump across the A and B terminals of the connector. The CEL/SES light should begin to flash codes. Each code will be repeated three times. The first series of flashes is the first digit of the code and the second series is the second digit. Trouble codes are displayed starting with the lowest-numbered code. Codes will continue until the jumper is disconnected from the ALDL.

A scan tool connected into the ALDL (Figure 4-32) can provide access to circuit voltage information, as well as trouble codes. By observing the various input voltages, you can identify out-of-specification input signals. Out-of-specification signals will not set a code unless they are out of the normal operating range. However, these signals will cause a driveability problem. To retrieve trouble codes with a scan tool, simply plug the tester lead into the ALDL. Then, program the scan tool for that particular vehicle according to the scan tools' instructions. Trouble codes will appear, digitally, on the tester.

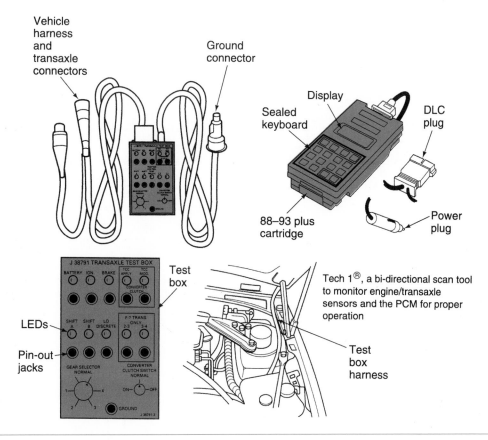

Figure 4-32 A transmission test box and Tech 1® scan tool used to diagnose early GM EATs. (Courtesy of General Motors Corporation, Service Operations Group)

To erase the codes after repairs have been made, turn the ignition off. With the jumper wire still connected, remove the control module fuse for 10 seconds. If the fuse cannot be located, disconnect the negative battery lead for 10 seconds. After the power has been disconnected from the PCM, the codes will be removed. So will the operating instructions on some models. For this reason it is important that you follow the re-learn procedures for the transmission.

Honda

Honda and Acura transaxle trouble codes are communicated by either flashing an LED on a side of the transaxle controller, or by flashing the S or D4 shifter status light on the instrument panel after connecting a service connector (Figure 4-33). Always refer to the correct service manual when attempting to retrieve codes.

> **✓ SERVICE TIP:** Honda/Acura systems flash a shift status light on the instrument panel when a problem is first noticed by the computer. This initial flashing is *not* a trouble code; rather, it is an indicator that a code is stored.

The connector is a two-pin, two-wire connector that is not connected to anything. Using a jumper wire, hook the two pins together, turn the ignition switch ON, and watch either the S or D4 light on the instrument panel. The three possible locations for the connector are: under the right side of the dash, behind the right side of the center console, or behind the left front edge of the passenger's carpet up against the firewall. If a trouble code is present, the light will either give

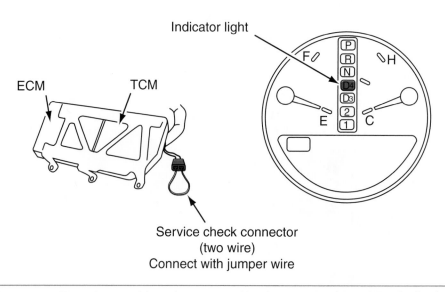

Figure 4-33 To retrieve trouble codes on some Hondas, connect the two terminals of the Service Check Connector and observe the D4 indicator.

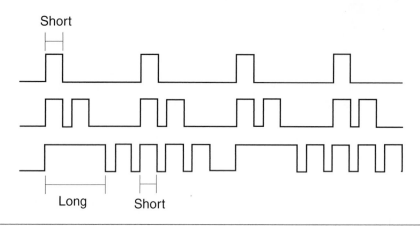

Figure 4-34 Honda trouble codes are displayed in a series of short and long flashes.

short or long and short flashes. Short flashes are either a one- or two-digit trouble code (Figure 4-34). Long flashes are the tens digit of a two-digit code. Short flashes are about a half second long and long flashes are about a second and a half long. There is about a one-second pause between each code. All codes will be shown in sequence, lowest number first, up to the highest code. The sequence will repeat as long as the key is on and the connector jumped across.

Some models have the diagnostic LED on the side of the transaxle controller. Read the flashes of the LED to determine the trouble code. The LED blinks all short flashes. Whenever the key is on and a problem exists, the LED will blink a series of short flashes, will pause, and then will either repeat the same number of flashes (if there is only one code) or advance to the next number of blinks. Just count each series of flashes, and you have got the code. The computer for Honda cars is located either under the front edge of the front passenger's carpet, behind the center console, or under the driver's seat. Acuras locate the computer behind the left side of the instrument panel, next to the inner fender panel, or under the right or left front seat with the LED facing the rear of the car.

The single- and double-digit codes should be compared to the appropriate trouble code chart provided in the service manual. Like other computer self-diagnostic systems, these systems do not determine the exact fault. Rather they identify the problem area. After repairs have been made, the codes should be erased.

To erase the codes, remove the appropriate fuse for at least one minute or pull the negative cable of the battery for a minute. Because Honda and Acura systems use different fuses for different models to keep the memory alive, make sure you check the service manual to identify the proper fuse.

● **CUSTOMER CARE:** Before disconnecting the battery, record the preset stations on the radio. Then reset the stations and the clock before delivering the vehicle to the customer.

Mazda

Trouble codes in Mazda EAT systems are displayed by either the HOLD light or the CHECK ENGINE light. If no problem exists, the HOLD or CHECK ENGINE light will come on for three seconds after turning the key on, then it will go out. If a trouble does exist, the light will flash in a regular pattern until the trouble is no longer there. This is not a trouble code; rather, it is an indication that a problem exists.

To retrieve the codes, locate the Diagnostic Request lead. This lead may be a single pin and wire blue connector, a single pin and wire green connector, or one of the pins in an integrated Diagnostic connector. Connect the lead to a good ground with a jumper wire. When the diagnostic pin is grounded, the HOLD or CHECK ENGINE light will display the code in long and short flashes. Long flashes, 1.2 seconds long, are the tens digit of a two-digit trouble code; short flashes, 0.4 seconds long, are the ones digit of a one- or two-digit code. The lowest number codes will be shown first, then the higher number codes. There is a four-second pause between codes. The series of codes will repeat as long as the Diagnostic Request pin is grounded (Figure 4-35). If no codes are present when you ground this pin, the indicator light will remain off.

Nissan

Transmission trouble code retrieval on Nissan computer-shifted transmissions and transaxles are easy and uniform throughout the different models of vehicles (Figure 4-36). Simply turn the ignition switch to the ON position and move the shifter and the O/D OFF switch. Then an instrument panel light will flash codes. The codes are displayed by the O/D OFF light, POWER, or A/T CHECK light.

To begin the procedure, make sure the MODE switch is in the AUTO position. Then run the engine until it has reached normal operating temperature. Turn the key off, set the parking brake, and do not touch the brake or throttle pedals until the procedure calls for you to do so. Move the shifter to D and turn the O/D switch OFF. Turn the ignition switch back to the ON position. After waiting at least two seconds, move the shifter to "2" and turn the O/D switch ON. Then, move the shifter to "1" and turn the O/D switch OFF. Now, slowly push the throttle pedal to the floor and slowly release it. The codes will now appear on the instrument panel.

The codes will appear in a sequence of eleven flashes. This sequence always begins with a long, 2-second, starter flash, followed by ten shorter flashes. If a problem does not exist, the ten following flashes will be very short flashes, 0.2 seconds. If a problem exists, one of the ten short flashes will be a little longer, 0.8 seconds, than the rest in that sequence. To identify the code, count the short flashes and determine which one was the longer one. Begin counting after the longer initial flash. If the first flash was the longest short flash, the code is 1. After each sequence of 11 flashes, there is a 2.5-second pause, then either the code or codes will repeat. All codes will automatically clear if the problem is fixed and the engine has been started two times after the system has been repaired.

Code #	Buzzer Pattern	Diagnosed Circuit	Condition	Point	Memorized
01	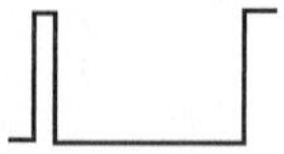	Engine rpm signal (Ne1 signal)	No input signal from distributor Ne1 signal while driving at drum speed above 600 rpm in D, S, or L ranges	-Distributor connector -Wiring from distributor to powertrain control module (transmission)(PCMT)	Yes
06	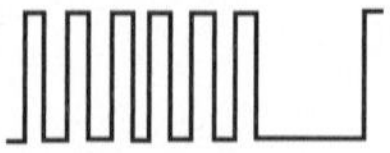	Vehicle speed sensor	No input signal from vehicle speed sensor while driving at drum speed above 600 rpm in D, S, or L ranges	-Vehicle speed sensor connector -Wiring from vehicle speed sensor to instrument cluster -Wiring from instrument cluster to powertrain control module (transmission) (PCMT) -Vehicle speed sensor resistance	Yes
12		Throttle position sensor	Open or short circuit	-Throttle position sensor connector -Wiring from throttle position sensor to powertrain module (transmission)(PCMT) -Throttle position sensor resistance	Yes
14	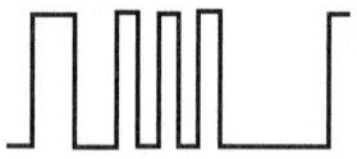	Barometric absolute pressure sensor	Open or short circuit	-Barometric absolute pressure sensor connector -Wiring from atmospheric pressure sensor to powertrain control module (PCME)	Yes
55		Vehicle speed pulse generator	No input signal from vehicle speed pulse generator while driving at vehicle speed 40 km/h (25 mph) or higher in D, S, or L range	-Vehicle speed pulse generator connector -Wiring from vehicle speed pulse generator to powertrain control module (transmission) (PCMT) -Vehicle speed pulse generator resistance	Yes
56	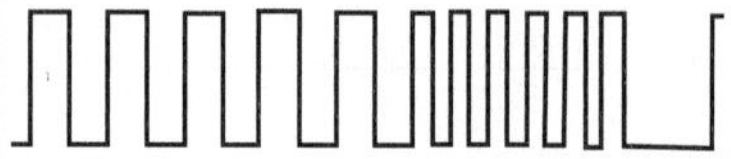	AFT thermo-sensor	Open or short circuit	-ATF thermo sensor connector -Wiring from ATF thermo sensor to powertrain control module (transmission)(PCMT)	Yes
57		Reduce torque signal 1	Open or short circuit of reduce torque signal 1 wire harness	-Wiring form powertrain control module (PCME) to powertrain control module (transmission)(PCMT)	Yes
58		Reduce torque signal 2	Open or short circuit of reduce torque signal 2 wire harness	-Wiring form powertrain control module (PCME) to powertrain control module (transmission)(PCMT)	Yes

(A)

Figure 4-35 Trouble code chart for Mazda vehicles.

Code #	Buzzer Pattern	Diagnosed Circuit	Condition	Point	Memorized
59		Torque reduced signal/ engine coolant temperature signal	Open or short circuit of torque reduced signal/ engine coolant temperature signal wire harness	-Wiring form powertrain control module (PCME) to powertrain control module (transmission)(PCMT)	Yes
60		Solenoid valve (1-2 shift)	Open or short circuit of solenoid valve and/or wiring	-Solenoid valve connector -Wiring from solenoid valve to powertrain control module (transmission) (PCMT) -Solenoid valve resistance	Yes
61		Solenoid valve (2-3 shift)	Open or short circuit of solenoid valve and/or wiring	-Solenoid valve connector -Wiring from solenoid valve to powertrain control module (transmission) (PCMT) -Solenoid valve resistance	Yes
62		Solenoid valve (3-4 shift)	Open or short circuit of solenoid valve and/or wiring	-Solenoid valve connector -Wiring from solenoid valve to powertrain control module (transmission) (PCMT) -Solenoid valve resistance	Yes
63		Solenoid valve (lockup control)	Open or short circuit of solenoid valve and/or wiring	-Solenoid valve connector -Wiring from solenoid valve to powertrain control module (transmission) (PCMT) -Solenoid valve resistance	Yes
64	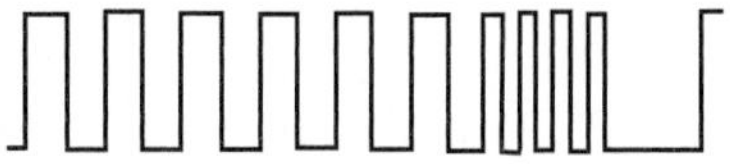	Solenoid valve (3-2 timing)	Open or short circuit of solenoid valve and/or wiring	-Solenoid valve connector -Wiring from solenoid valve to powertrain control module (transmission) (PCMT) -Solenoid valve resistance	Yes
65		Solenoid valve (lockup)	Open or short circuit of solenoid valve and/or wiring	-Solenoid valve connector -Wiring from solenoid valve to powertrain control module (transmission) (PCMT) -Solenoid valve resistance	Yes
66	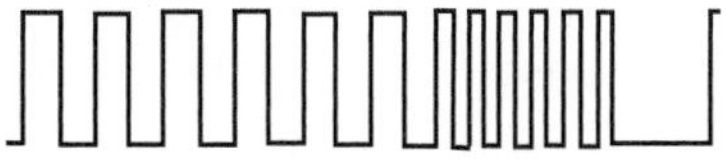	Solenoid valve (line pressure)	Open or short circuit of solenoid valve and/or wiring	-Solenoid valve connector -Wiring from solenoid valve to powertrain control module (transmission) (PCMT) -Solenoid valve resistance	Yes

(B)

Figure 4-35 Trouble code chart for Mazda vehicles (continued).

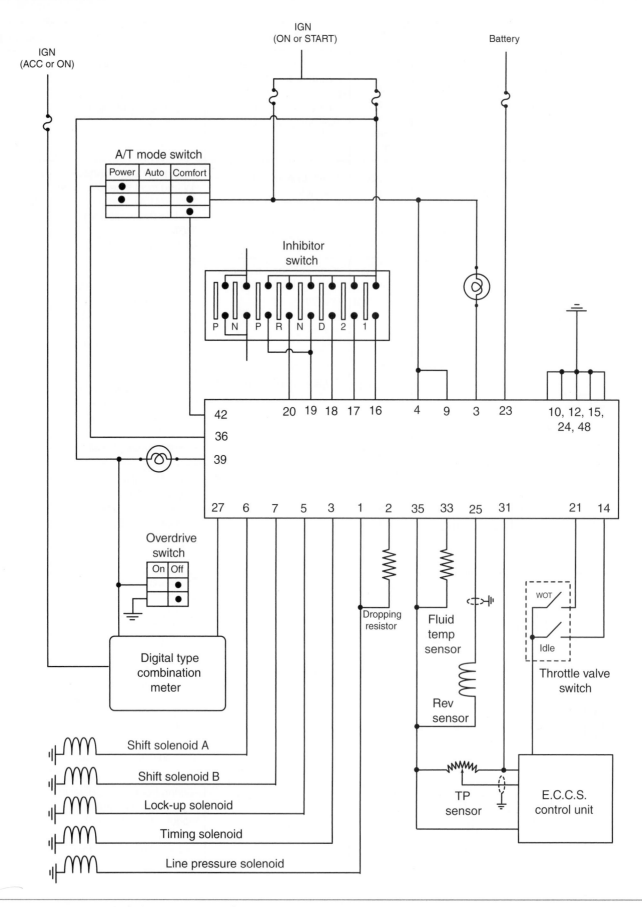

Figure 4-36 A typical Nissan EAT wiring diagram.

Saturn

Saturn EATs use five electronic valves that are controlled by the PCM to control shift feel and timing. The PCM has a self-diagnostic function. If an electrical problem occurs, the SES or "Shift to D2" lamp will come on or will start flashing. The PCM stores trouble codes whenever it has detected a fault in the engine or transaxle circuit. Problems are stored as hard or intermittent problems. The PCM will also store other problems but may not turn on the SES light in response to the problem. These codes are stored in the memory to aid in diagnostics.

The PCM recognizes three different types of faults: hard, intermittent, and malfunction history/information flags. Hard codes cause the SES light to glow and remain on until the problem is repaired. These codes can be interpreted by looking up the code in the appropriate trouble code chart. Intermittent codes cause the SES to flicker or glow and go out approximately 10 seconds after the fault disappears. However, the corresponding trouble code will remain in memory to be retrieved by a technician. If the intermittent fault does not occur for 50 engine starts, the code will be erased from the computer's memory. Engine information flags will not cause the SES light to glow. Unlike the other types of codes, information flags and malfunction history will not be erased from the PCM memory. These codes can only be retrieved and erased from memory by using a Saturn Portable Diagnostic Tool (PDT).

Trouble codes are read by counting the flashes of the SES lamp. The first series of flashes represents the first digit of the trouble code and the second series represents the second digit. The first code will always be code "12," followed by all other stored codes. Each code is repeated three times. When code "11" appears, that signals the end of the engine's self-diagnostic test and the beginning of the transaxle self-test. The "Shift to D2" lamp will start flashing any transaxle-related codes stored in the transaxle controller. If the SES doesn't flash a code "11," there are no transaxle codes stored in memory.

To set the PCM into the self-diagnostic mode, turn the ignition on with the engine off. Connect a jumper wire from terminal B to terminal A at the PCM diagnostic connector. The SES should begin to flash codes. To end the self-diagnostic mode, turn the ignition off and remove the jumper wire.

To erase the trouble codes, turn the ignition on and make contact three times within five seconds with a jumper wire between terminals A and B at the diagnostic connector.

The PCM has a re-learn function and the proper learn process should be followed anytime the battery has been disconnected.

Toyota

A self-diagnostic feature is built into all Toyota EATs. A warning of a fault is indicated by the overdrive OFF indicator lamp. If a malfunction occurs in the speed sensor or solenoid circuits, the overdrive OFF lamp will blink to warn the driver of the fault. On some models, the diagnostic codes can be read by the number of blinks on the overdrive OFF lamp when terminals ECT and E2, in the diagnostic connector, are shorted together. Other models require that terminals TE and E be shorted together (Figure 4-37). Make sure you refer to the appropriate service manual to determine which terminals should be shorted. Once the computer is set into its diagnostic mode, codes will begin to be displayed.

If a malfunction is in memory, the overdrive OFF lamp will blink once every 0.5 seconds. The first number of blinks equals the first digit of a two-digit code. After a 1.5-second pause, the second number of blinks will equal the second digit. If there are two or more codes, there will be a 2.5-second pause between each code. All codes should be interpreted by using the trouble code chart provided in the service manual.

The diagnostic codes can be erased from the computer's memory by removing the DOME, ECU +B, or EFI fuse for at least 10 seconds with the key off.

 SERVICE TIP: Warning and diagnostic codes can only be read when the overdrive switch is on. When the switch is off, the lamp is on continuously and will not flash.

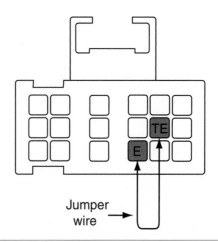

Jumper
wire

Figure 4-37 A typical Toyota diagnostic connector with a jumper wire inserted to activate the self-test mode.

Basic Diagnostics of OBD II Systems

Special Tools

Scan tool (OBD II compliant)

Classroom Manual
Chapter 4, page 99

On-Board Diagnostics II (OBD II) systems were developed in response to the federal government's and the state of California's emission control system monitoring standards for all automotive manufacturers. The main goal of OBD II was to detect when engine or system wear or when component failure caused exhaust emissions to increase by 50% or more. OBD II also called for standard service procedures without the use of dedicated special tools. To accomplish these goals, manufacturers needed to change many aspects of their electronic engine control systems. According to the guidelines of OBD II, all vehicles have:

❏ A universal diagnostic test connector, known as the data link connector (DLC), with dedicated pin assignments.
❏ A standard location for the DLC. It must be under the dash on the driver's side of the vehicle and easily accessible (SAE standard J1962).
❏ A standard list of diagnostic trouble codes (DTCs); this is SAE's standard J2012.
❏ A standard communication protocol (SAE standard J1850).
❏ The use of common scan tools on all vehicle makes and models (SAE standard J1979).
❏ Common diagnostic test modes (SAE standard J2190).
❏ Vehicle identification must be automatically transmitted to the scan tool.
❏ Stored trouble codes must be able to be cleared from the computer's memory with the scan tool.
❏ The ability to record, and store in memory, a snapshot of the operating conditions that existed when a fault occurred.
❏ The ability to store a code whenever something goes wrong and affects exhaust quality.
❏ A standard glossary of terms, acronyms, and definitions must be used for all emissions-related components in the electronic control systems (SAE standard J1930).

OBD II systems must illuminate the MIL (Figure 4-38) if the vehicle conditions would allow emissions to exceed 1.5 times the allowable standard for that model year based on a Federal Test Procedure (FTP). When a component or strategy failure allows emissions to exceed this level, the MIL is illuminated to inform the driver of a problem and a diagnostic trouble code is stored in the PCM.

Besides enhancements to the computer's capacities, some additional hardware is required to monitor the emissions performance closely enough to fulfill the tighter constraints and beyond merely keeping track of component failures. In most cases, this hardware consists of additional

EAT indicator

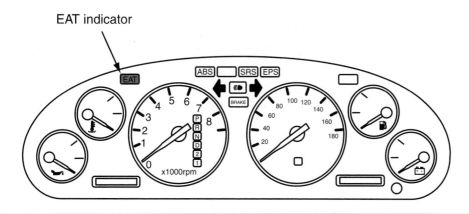

Figure 4-38 An example of a MIL that illuminates to warn the driver that transmission control has been switched to the default mode.

and improved sensors, upgrading specific connectors, and components to last the mandated 100,000 miles or 10 years, and a new standardized 16-pin DLC.

Instead of a fixed, unalterable PROM, the PCMs in most OBD II systems have an **electronically erasable PROM (EEPROM)** to store a large amount of information. The EEPROM is soldered into the PCM and is not replaceable. The EEPROM stores data without the need for a continuing source of electrical power.

The EEPROM is an integrated circuit that contains the program used by the PCM to provide powertrain control. It is possible to erase and reprogram the EEPROM without removing this chip from the computer. When a modification to the PCM operating strategy is required, it is no longer necessary to replace the PCM. The EEPROM may be reprogrammed through the DLC using the special equipment.

For example, if the vehicle calibrations are updated for a specific car model, the recalibrating or reprogramming equipment can be used to erase the EEPROM. After the erasing procedure, the EEPROM is reprogrammed with the updated information. The reprogramming information is provided by the manufacturer and is typically directed by a service bulletin or recall letter.

Computer systems without OBD II have the ability to detect component and system failure. Computer systems with OBD II are also capable of monitoring the ability of systems and components to maintain low emission levels.

Data Link Connector

OBD II and the Society of Automotive Engineers (SAE) standards require that the DLC for OBD II systems be mounted in the passenger compartment out of sight of vehicle passengers. The standard DLC (Figure 4-39) is a 16-pin connector. The same seven pins are used for the same information, regardless of the vehicle's make, model, and year. The connector is "D"-shaped and has guide keys that allow the scan tool to only be installed one way. Using a standard connector design and by designating the pins allows data retrieval with any scan tool designed for OBD II. Some European vehicles meet OBD II standards by providing the designated DLC along with their own connector for their own scan tool.

 SERVICE TIP: If a vehicle has a 16-pin DLC, it does not necessarily mean that the vehicle is equipped with OBD II.

The DLC is designed only for scan tool use. You cannot jump across any of the terminals to display codes on an instrument panel or other indicator lamp. The MIL is only used to inform the driver that the vehicle should be serviced soon. It also informs a technician that the computer has set a trouble code.

Classroom Manual
Chapter 4, page 97

An **EEPROM** is an electrically erasable programmable read only memory chip.

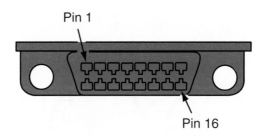

Pin 1

Pin 16

Pin 1   Secondary UART 8192 baud serial data
          Class B 160 baud serial data
Pin 2   J1850 Bus + line on 2 wire systems, or single wire (class 2)
Pin 3   Ride control diagnostic enable
Pin 4   Chassis ground pin
Pin 5   Signal ground pin
Pin 6   PCM/VCM Diagnostic enable
Pin 7   K line for International Standards Organization (ISO) application
Pin 8   Keyless entry enable, or MRD theft diagnostic enable
Pin 9   Primary UART
Pin 10 J1850 Bus-line for J1850-2 wire applications
Pin 11 Electronic Variable Orifice (EVO) steering
Pin 12 ABS diagnostic, or CCM diagnostic enable
Pin 13 SIR diagnostic enable
Pin 14 E&C bus
Pin 15 L line for International Standards Organization (ISO) application
Pin 16 Battery power from vehicle unswitched (4 AMP MAX)

Figure 4-39 The purpose of each of the sixteen pins in a typical DLC for an OBD II system.

The DLC must be easily accessible while sitting in the driver's seat (Figure 4-40). All DLCs must be located somewhere between the left end of the instrument panel and a position 300 millimeters to the right of the center. The DLC cannot be hidden behind panels and must be accessible without tools. Any generic scan tool can be connected to the DLC and can access the diagnostic data stream.

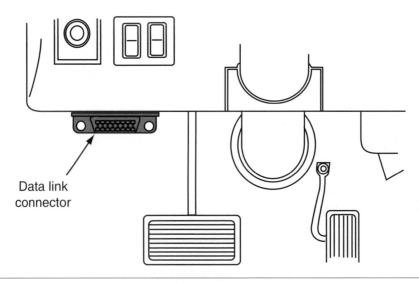

Data link
connector

Figure 4-40 Standard location for the DLC in an OBD II system.

The connector pins are arranged in two rows and are numbered consecutively. Seven of the sixteen pins have been assigned by the OBD II standard. The remaining nine pins can be used by the individual manufacturers to meet their needs and desires.

Malfunction Indicator Lamp Operation

An OBD II system continuously monitors the entire system, switches on a MIL if something goes wrong, and stores a fault code in the PCM when it detects a problem. The codes are well defined and can lead a technician to the problem. A scan tool must be used to access and interpret emission-related DTCs regardless of the make and model of the vehicle.

Many of the trouble codes from an OBD II system will mean the same thing regardless of manufacturer. However, some of the codes will pertain only to a particular system or will mean something different with each system. The DTC is a five-character code with both letters and numbers (Figure 4-41). This is called the alphanumeric system. The first character of the code is a letter, which defines the system where the code was set. Currently there are four possible first character codes: "B" for body, "C" for chassis, "P" for powertrain, and "U" for undefined. The U-codes are designated for future use.

The second character is a number. This defines the code as being a mandated code or a special manufacturer code. This number will either be a "0," "1," "2," or "3." A "0" code means that the fault is defined or mandated by OBD II. A "1" code means that the code is manufacturer specific. Codes of "2" or "3" are designated for future use.

The third through fifth characters are numbers. These describe the fault. The third character of a powertrain code tells you where the fault occurred. A "7" or "8" indicates the problem is

---

The SAE J2012 standards specify that all DTCs will have a five-digit alphanumeric numbering and lettering system. The following prefixes indicate the general area to which the DTC belongs:

1. P — powertrain
2. B — body
3. C — chassis

The first number in the DTC indicates who is responsible for the DTC definition.

1. 0 — SAE
2. 1 — manufacturer

The third digit in the DTC indicates the subgroup to which the DTC belongs. The possible subgroups are:

0 — Total system
1 — Fuel-air control
2 — Fuel-air control
3 — Ignition system misfire
4 — Auxiliary emission controls
5 — Idle speed control
6 — PCM and I/O
7 — Transmission
8 — Non-EEC power train

The fourth and fifth digits indicate the specific area where the trouble exists. Code P1711 has this interpretation:

P — Powertrain DTC
1 — Manufacturer-defined code
7 — Transmission subgroup
11 — Transmission oil temperature (TOT) sensor and related circuit

---

Figure 4-41 An explanation of the five digits in an OBD II trouble code.

directly related to the transmission. The remaining two characters describe the exact condition that set the code. OBD II codes are quite definitive and can lead you to the cause of the problem. For example, there are five codes that relate to the transmission fluid temperature circuit. Code P0710 indicates there is a fault in the sensor's circuit, P0711 indicates the circuit is operating out of the acceptable range, P0712 indicates the circuit has lower than normal input signals, P0713 indicates the input signals are higher than normal, and P0714 indicates the circuit appears to have an intermittent problem.

Test Modes

All OBD II systems have the same basic test modes. These test modes must be accessible with an OBD II scan tool. Mode 1 is the Parameter Identification (PID) mode. It allows access to certain data values, analog and digital inputs and outputs, calculated values, and system status information. Some of the PID values will be manufacturer specific; others are common to all vehicles.

Mode 2 is the Freeze Frame Data Access mode. This mode permits access to emission-related data values from specific generic PIDs. These values represent the operating conditions at the time the fault was recognized and logged into memory as a DTC. Once a DTC and a set of freeze frame data are stored in memory, they will stay in memory even if other emission-related DTCs are stored. The number of these sets of freeze frames that can be stored are limited.

There is one type of failure that is an exception to this rule—misfire. Fuel system misfires will overwrite any other type of data except for other fuel system misfire data. This data can only be removed with a scan tool. When a scan tool is used to erase a DTC, it automatically erases all freeze frame data associated with the events that led to that DTC.

The basic advantage of the snapshot feature is the ability to look at the existing conditions when a code was set. This will be especially valuable for diagnosing intermittent problems. Whenever a code is set, a record of all related activities will be stored in memory. This allows you to look at the action of sensors and actuators when the code was set, which helps identify the cause of the problem.

Mode 3 permits scan tools to obtain stored DTCs. The information is transmitted from the PCM to the scan tool following an OBD II Mode 3 request. Either the DTC, its descriptive text, or both will be displayed on the scan tool.

The PCM reset mode (Mode 4) allows the scan tool to clear all emission-related diagnostic information from its memory. Once the PCM has been reset, the PCM stores an inspection maintenance readiness code until all OBD II system monitors or components have been tested to satisfy an OBD trip cycle without any other faults occurring. Specific conditions must be met before the requirements for a trip are satisfied.

Mode 5 is the oxygen sensor monitoring test. This mode gives the oxygen sensor fault limits and the actual oxygen sensor outputs during the test cycle. The test cycle includes specific operating conditions that must be met to complete the test. This information helps determine the effectiveness of the catalytic converter.

Mode 6 is the output state mode (OTM), which allows a technician to activate or deactivate the system's actuators through the scan tool. It is important to note that all scan tools do not have this capability. When the OTM is engaged, the actuators can be controlled without affecting the radiator fans. The fans are controlled separately; this gives a pure look at the effectiveness and action of the outputs.

OBD II Diagnostics

Diagnosing and servicing OBD II systems is much the same as diagnosing and servicing any other electronic control system, except that OBD II systems give more useful information than the previous designs and, with OBD II, each make and model vehicle does not have its own basic testing and servicing procedures. General testing of the engine and its major systems are done in the same way as is done on non-OBD systems. Testing of individual components is important in diagnosing OBD II systems. Although the DTCs in these systems give much more detail on the problems, the PCM does not know the exact cause of the problem. That is the technician's job.

OBD II systems note the deterioration of certain components before they fail, which allows owners to bring in their vehicles at their convenience and before it is too late. OBD II diagnosis is best done with a strategy-based procedure based on the flow charts and other information given in the service manuals. Check for any published TSBs relating to the exhibited symptoms. This should include videos, newsletters, and any electronically transmitted media. Do not depend solely on the diagnostic tests run by the PCM. A particular system may not be supported by one or more DTCs.

When the ignition is initially turned ON, the MIL will momentarily flash ON then OFF and remain ON until the engine is running or there are no DTCs stored in memory. Now connect the scan tool. Enter the vehicle identification information, and then retrieve the DTCs with the scan tool. The scan tool can be used to do more than simply retrieve codes. The information that is available, whether or not a DTC is present, can reduce diagnostic time. Some of the common uses of the scan tool are: identifying stored DTCs, clearing DTCs, performing output control tests, and reading serial data (Figure 4-42).

The scan tool must have the appropriate connector to fit the DLC on an OBD II system and the proper software for the vehicle being tested. Cable adapters are used to connect a standard scan tool to an OBD II system. The software inserts make the scan tool compatible with the PCM in the vehicle. Make sure the recommended insert for that make and model of vehicle is

SCAN TOOL DATA LIST (CONT'D)

Engine Idling/Radiator Hose Hot/Closed Throttle/Park or Neutral/Closed Loop/Accessories Off			
Scan Tool Parameter	**Data List**	**Parameter Range/Units**	**Typical Data Values**
TP Sensor 2	Eng 1, TAC	Volts	4.0–1.4V
TP Sensors Disagree	TAC	Yes/No	No
TP Sensor 1 Out of Range	TAC	Yes/No	No
TP Sensor 2 Out of Range	TAC	Yes/No	No
Traction Control	Eng 1, 3, TAC, CC, FF, FR	Active/Inactive	Inactive
Transmission Fluid Temp. (Export-manual transmission only)	FF, FR	C°/F°	Varies
Transmission OSS	FF, FR	RPM	Varies
Transmission Range	Eng 1, 2, 3, CC, FF, FR	Transaxle Gear Position	Park/Neutral
Vehicle Speed	ENG 1, 2, 3, EE, TAC, FT, CC, FF, FR	km/h/mph	0 km/h (0 mph)
VTD Auto Learn Timer	Eng 3	Active/Inactive	Inactive
VTD Fuel Disable	Eng 3	Active/Inactive	Inactive
VTD Fuel Disable Until Ign. Off	Eng 3	Yes/No	No
Warm-Ups w/o Emis. Faults	Eng 2	Counts	0–255
Warm-Ups w/o Non-Emis. Faults	Eng 2	Counts	0–255

Figure 4-42 A small sample of the data available from an OBD II compliant scan tool. (Courtesy of General Motors Corporation, Service Operations Group)

used in the scan tool. Always follow the instructions given in the scan tool's manual when testing the system.

If DTCs are stored in memory, the scan tool will display them. Record them. If there are multiple DTCs, they should be interpreted in the following order:

1. PCM error DTCs
2. System voltage DTCs
3. Component level DTCs
4. System level DTCs

CAUTION: Do not clear DTCs unless directed by a diagnostic procedure. Clearing DTCs will also clear valuable freeze frame and failure records data.

After retrieving the codes, use the scan tool to command the MIL to turn OFF. If it turns OFF, turn it back ON and continue. If it stays ON, follow the manufacturer's diagnostic procedure for this problem.

The next step in diagnostics depends on the outcome of the DTC display. If there were DTCs stored in the PCM's memory, proceed by following the designated DTC diagnostic table to make an effective diagnosis and repair. If no DTCs were displayed, match the symptom to the symptoms listed in the manufacturer's symptom tables and follow the diagnostic paths or suggestions to complete the repair, or refer to the applicable component/system check. If there is not a matching symptom, analyze the complaint. Then develop a plan for diagnostics. Utilize the wiring diagrams and theory of component and system operation. Call technical assistance for similar cases where repair history may be available. It is possible that the customer's complaint is a normal operating condition for that vehicle. If this is suspected, compare the complaint to the driveability of a known good vehicle.

After the DTCs and the MIL have been checked, the OBD II data stream can be selected. In this mode, the scan tool displays all of the data from the inputs. The actual available data will depend on the vehicle and the scan tool; however, OBD II codes dictate a minimum amount of data. Always refer to the appropriate service manual and the scan tool's operating instructions for proper identification of normal or expected data.

One of the selections on a scan tool is DTC data. This selection contains two important categories of information: freeze frame, which displays the conditions that existed when a DTC was set, and failure records. The latter is valuable when the MIL was not turned on but a fault was detected. The failure records display the conditions present when the system failed the diagnostic test.

With the scan tool, record the freeze frame and failure records information. By storing the freeze frame and failure records on the scan tool, an electronic copy of the conditions that were present when the fault occurred is made and stored in the scan tool. Using the scan tool with a lab scope is extremely handy for watching the activity of sensors and actuators.

Electronic Defaults

A transmission's **default mode** is a mode of operation that allows for limited use in the case of electronic failure.

Some electronically controlled transmissions will exhibit strange characteristics if the control computer senses a problem within its system. These transmissions will go into a **default mode** of operation in which the computer will disregard all input signals. During this period of time, the transmission may lock itself into a single forward gear or will bypass all electronic controls. While diagnosing a problem in an electronically controlled transmission, always refer to the appropriate service manual to identify the normal "default" operation of the transmission. By not recognizing that the transmission is operating in default, you could spend time tracing the wrong problem.

When the computer senses something wrong, the transmission will go into a default mode of operation. While in this mode, the transmission may only operate in second gear with second gear starts and Reverse and Park. The transmission goes into the default mode because a device, usually an input device, has sent faulty or erroneous signals to the transaxle's control computer.

Whenever the computer sees a potential problem that may increase wear and/or damage to the transmission, the system also defaults to limp-in. Minor slipping can be sensed by the computer through its input and output sensors. This slipping will cause premature wear and may cause the computer to move into its default mode. Some systems may increase fluid pressure to compensate for the problem. A totally burned clutch assembly will cause limp-in operation, as will some internal pressure leaks that may not be apparent until pressure tests are run.

Detailed Testing of System

At times it may be difficult to determine if the problem is caused by something inside the transmission or if the cause is electronic. If you face this dilemma, take the vehicle on a road test and note all of the things that do not seem right. Then, remove the fuse for the transmission control unit. Doing this disables electronic control of the transmission and will cause it to operate in its default mode. Road test the vehicle again and observe the operation of the transmission. Keep in mind that, during default operation, there will be no forward upshifts. If the problem or problems observed in the first road test still exist, the cause must be in the transmission.

Computer Checks

Although there are DTCs that relate directly to the PCM or TCM, it may be difficult to identify a faulty computer unit. Since the computer reacts to the inputs it receives, abnormal input signals may result in abnormal computer operation. Also, most computers are designed to react to out-of-range inputs by ignoring the signals and using a fixed value held in its memory. This results in less-than-efficient operation.

Basically, if the computer system's inputs and outputs and the engine and hydraulic and mechanical functions of the transmission are all good, but a problem exists, the computer or its circuit is bad. A check of the computer circuit must always follow the suspicion that the computer is faulty.

To check the circuit, check for battery voltage to the computer. Use the wiring diagram in the service manual to identify the power feed for the computer. If less than battery voltage is available, conduct a voltage drop test to determine the location of the unwanted resistance.

If the computer is receiving battery voltage, check the computer's ground. The best way of doing this is to place one probe of a voltmeter on the case or ground connection of the computer and the other at a known good ground. With the system powered, the meter should read less than 0.1 volt. If the reading is higher than that, conduct voltage drop tests on the rest of the ground circuit.

Most scan tools give you the capability of controlling shift points by manually turning on shift solenoids. If your scan tool does not have this feature, special testers are available to do this. This check operates the solenoids on command, thereby bypassing the computer. Doing this, you can observe the operation of the solenoids and the transmission. If the customer's complaint is still evident when the computer is bypassed, you know the problem is not the computer. If the problem is not evident, you know the problem is the computer and/or computer circuit.

TCC Functional Test

On most vehicles, the engagement of the torque converter clutch can be checked during a road test. Drive the vehicle at a steady speed of 40 to 45 mph. Note the engine's speed on the tachometer. Then lightly tap the brake pedal, while maintaining road speed. Note the tachometer reading. If the brake switch caused the converter clutch to disengage, the engine's speed should

Figure 4-43 A torque converter clutch solenoid.

Classroom Manual
Chapter 4, page 100

increase. Shortly after disengagement, the clutch should re-engage and engine speed should drop. If you observed no evidence of clutch engagement or disengagement, the TCC system (Figure 4-43) needs further testing. Photo Sequence 6 covers a typical procedure for checking non-OBD II TCC systems with a scan tool.

Detailed Testing of Inputs

There are many different designs of sensors, depending on the operation of the system they monitor. Some sensors are nothing more than a switch that completes a circuit. Others are complex devices that react to chemical reactions and generate their own voltage during certain conditions. If the self-test sequence pointed to a problem in an input circuit, the input circuit should be tested to determine the exact malfunction. Often the manufacturers list specific procedures for specific sensors; always follow them.

Testing Switches

Many different switches are used as inputs or control devices for EATs. Most of the switches are either mechanically or hydraulically controlled. The operation of these switches can be easily checked with an ohmmeter (Figure 4-44). With the meter connected across the switch's leads, there should be continuity or low resistance when the switch is closed and there should be infinite resistance across the switch when it is open. A test light can also be used. When the switch is closed, power should be present at both sides of the switch. When the switch is open, power should be present at only one side.

Classroom Manual
Chapter 4, page 95

The neutral safety switch and the kickdown switch are two examples of the switches used on some transmissions. There are many more possible switches connected to a transmission and they can serve many different purposes. However, all of these switches either complete or open an electrical circuit. By completing a circuit these switches either provide a ground for the circuit or connect two wires together. To open the circuit, switches either disconnect the ground from the circuit or disconnect two wires.

Prior to the use of electronic controls and their associated sensors, grounding switches were often used to control the engagement of torque converter clutches. These switches were controlled by governor pressure and supply the ground for the TCC solenoid when a particular speed is reached. Different TCC engagement speeds are possible through the use of different spring tensions on the switch. To engage the clutch earlier, a weaker spring is used and when the clutch engagement should occur later, a stronger spring is used. Grounding switches are also used to indicate or sense what gear the transmission is in. Other grounding type switches are used to control the back-up lamp circuit.

Photo Sequence 6
Typical Procedure for Diagnosing a Computer-Controlled TCC System

P6-1 Turn the ignition on with the engine off. Verify that the "Check Engine" or "Service Engine Soon" lamp is lit.

P6-2 Locate the DLC.

P6-3 Ground the diagnostic test terminal by connecting a jumper wire from terminal A to terminal B at the DLC connector.

P6-4 The MIL should flash a code 12. The code 12 is indicated by a flash, a pause, and two flashes.

P6-5 Connect a scan tool to the DLC.

P6-6 Connect the scan tool power lead to the cigarette lighter receptacle.

P6-7 The scan tool must be programmed with the correct identification of the vehicle.

P6-8 The scan tool will display the trouble codes numerically.

P6-9 A scan tool not only will display the trouble codes held in the memory of the PCM, but also can be used to display the data or signals that the PCM is receiving from inputs or is sending to outputs.

(continued)

P6-10 Compare the trouble codes that are received with the appropriate trouble code chart.

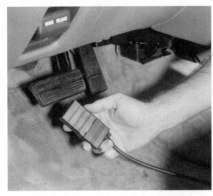

P6-11 After the codes have been retrieved, remove the jumper wire or scan tool.

P6-12 Then remove the PCM fuse to erase the codes from memory.

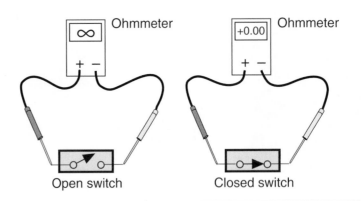

Figure 4-44 Measuring resistance with an ohmmeter. The meter is placed in parallel with the component after the power is removed from the circuit.

The manual lever position (MLP) or transmission range (TR) sensor (Figure 4-45) is a switch that provides information to the computer as to what operating range has been selected by the driver. Based on that information, the computer determines the proper shift strategy for the transmission. The following is a list of problems that may result from a faulty TR sensor:

❑ No upshifting
❑ Slipping out of gear
❑ High line pressure in transmissions equipped with a pressure control solenoid
❑ Delayed gear engagements
❑ Engine starts in other lever positions besides park and neutral

Because this switch is open or closed, depending on position, it can be checked with an ohmmeter. By referring to a wiring diagram, you should be able to determine when the switch should be open. Doing this for some switch designs may be a little difficult, as the switch assembly may be made up of more than one switch. Then connect the meter across the input and out of the switch. Move the lever into the desired position and measure the resistance. An infinite

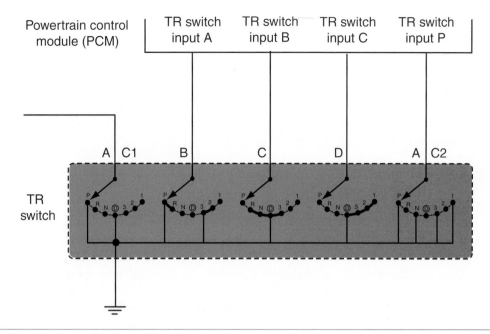

Figure 4-45 A transmission range (TR) switch. (Courtesy of General Motors Corporation, Service Operations Group)

reading is expected when the switch is open. If there is any resistance, the switch should be replaced. The switch should also be replaced if there is some resistance when the switch is closed.

The pressure switches (Figure 4-46) used in today's transmissions either complete or open an electrical circuit. These switches are either normally open or normally closed. Normally open switches will have no continuity across the terminals until oil pressure is applied to it. Normally closed switches will have continuity across the terminals until oil pressure is applied to it. Refer to the wiring diagram to determine the type of switch and test the switch with an ohmmeter. An

Classroom Manual
Chapter 4, page 103

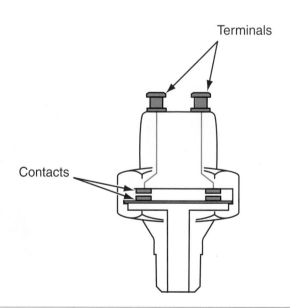

Figure 4-46 To check a normally open pressure switch, apply air to the bottom oil passage and check for continuity across the terminals. (Courtesy of General Motors Corporation, Service Operations Group)

ohmmeter can be used to identify the type of switch being used and can be used to test the operation of the switch. There should be continuity when the switch is closed and no continuity when the switch is open. Base your expected results on the type of switch you are testing.

Pressure switches can be checked by applying air pressure to the part that would normally be exposed to oil pressure. When applying air pressure to these switches, check them for leaks. Although a malfunctioning electrical switch will probably not cause a shifting problem, it will if it leaks. If the switch leaks off the applied pressure in a hydraulic circuit to a holding device, the holding member may not be able to function properly.

CAUTION: Always disconnect an electrical switch or component from the circuit or from a power source before testing it with an ohmmeter. Failure to do so will result in damage to the meter.

When possible, you should check pressure switches when they are installed and controlled by the vehicle. By watching an ohmmeter connected to a governor switch you can check its spring rating. Connect the ohmmeter to the "A" terminal and ground. Bring the engine's speed up. When the ohmmeter reads about 25 ohms, the governor switch is closed. The speed at which this occurs is the speed required to close the grounding switch. Other grounding type switches can be checked in the same way. However, it is important that you identify whether it is a normally open or closed switch before testing.

Throttle Position Sensor

Classroom Manual

Chapter 4, page 100

Another type of switch is a potentiometer (Figure 4-47). Instead of opening and closing a circuit, a potentiometer controls the circuit by varying its resistance in response to something. A TP sensor is a potentiometer (Figure 4-48). It sends very low voltage back to the computer when the throttle plates are closed and increases the voltage as the throttle is opened.

A TP sensor sends information to the control computer as to what position the throttle is in. If this sensor fails, the following problems can result:

❏ No upshifts
❏ Quick upshifts
❏ Delayed shifts
❏ Line-pressure problems with transmissions that have a line pressure control solenoid
❏ Erratic converter clutch engagement

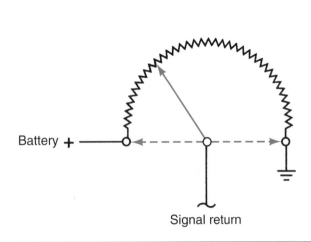

Figure 4-47 A potentiometer is used to send a signal voltage from the wiper of the switch.

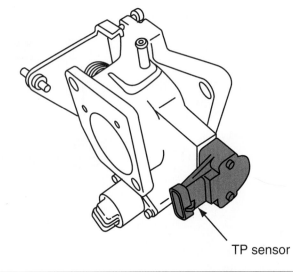

Figure 4-48 A TP sensor. (Courtesy of General Motors Corporation, Service Operations Group)

Most TP sensors receive a reference voltage of 5 volts. What the TP sensor sends back to the computer is determined by the position of the throttle. When the throttle is closed, it sends approximately 0.5 volts back to the computer. When the throttle begins to open, the voltage begins to increase. When the throttle reaches the wide-open position, approximately 4.5 volts are sent back to the computer. This increase in voltage should rise and fall smoothly with a change in throttle position. All changes in throttle position should result in a change in voltage.

A TP sensor can be checked with an ohmmeter or a voltmeter. If checked with an ohmmeter, you should be able to watch the resistance across the TP sensor change as the throttle is opened and closed. Often there will be a resistance specification given in the service manual. Compare your reading to this.

With a voltmeter, you will be able to measure the reference voltage and the output voltage. Both of these should be within specified amounts. If the reference voltage is lower than normal, check the voltage drop across the reference voltage circuit from the computer to the TP sensor. If the TP sensor is found to be defective, it should be replaced.

Potentiometer testing with a lab scope is a good way to watch the sweep of the resistor. The waveform (Figure 4-49) is a DC signal that moves up as the voltage increases. Most potentiometers in computer systems are fed a reference voltage of 5 volts. Therefore, the voltage output of these sensors will range from 0.5 to 4.5 volts. The change in voltage should be smooth. Look for glitches in the signal. These can be caused by changes in resistance or an intermittent open.

Throttle position (TP), EGR valve position, and vane air flow sensors are commonly used sensors that are actually potentiometers.

Mass-Airflow Sensor

A mass-airflow sensor (Figure 4-50) is used to determine engine load by measuring the mass of the air being taken into the throttle body. The computer uses this information for fuel and air control. When this sensor fails or sends faulty signals, the engine runs roughly and tends to stall as soon as you put the transmission into gear. The mass airflow sensor is a wire, located in the intake air stream that receives a fixed voltage. The wire is designed so that it changes resistance in response to temperature changes. When the wire is hot, the resistance is high and less current flows through the wire. When the wire is cold, the resistance is low and larger amounts of current pass through it. The amount of air passing over the wire determines the amount of resistance the wire has. The computer monitors the amount of current flow and interprets the flow as the mass or volume of the air.

Classroom Manual
Chapter 4, page 101

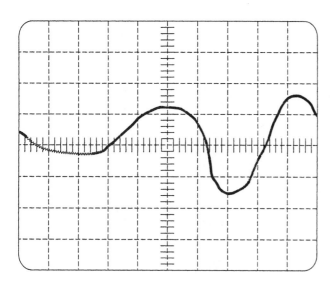

Figure 4-49 A DC signal for a good TP sensor, measured from closed throttle to wide-open throttle.

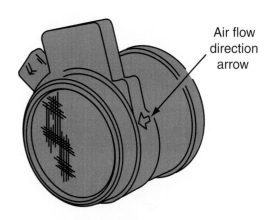

Figure 4-50 A mass air flow sensor. (Courtesy of General Motors Corporation, Service Operations Group)

This sensor can typically be measured with a multimeter set to the Hz frequency range. Check the service manual for specific values. Normally at idle, 30 Hz is measured with the frequency increasing as the throttle opens. A scan tool can also be used to test this sensor; most scanners have a test mode that monitors mass-airflow sensors. The output of some MAFs can be observed with a DMM, as their output is variable DC voltage. While diagnosing these systems, keep in mind that cold air is denser than warm air.

Temperature Sensors

Classroom Manual
Chapter 4, page 95

Temperature sensors are designed to change resistance with changes in temperature. A temperature sensor is based on a thermistor. Some thermistors increase resistance with an increase of temperature. Others decrease the resistance as temperature increases. Obviously these sensors can be checked with an ohmmeter. To do so, remove the sensor. Then determine the temperature of the sensor and measure the resistance across it. Compare your reading to the chart of normal resistances given in the service manual (Figure 4-51)

A thermistor's activity can be monitored with a lab scope. Connect the scope across to the output of the thermistor or temperature sensor. Run the engine and watch the waveform. As the temperature increases, there should be a smooth increase or decrease in voltage. Look for glitches in the signal. These can be caused by changes in resistance or an intermittent open.

DIAGNOSTIC AID
TRANSMISSION SENSOR—TEMP TO RESISTANCE (APPROXIMATE)

°C	°F	Ω MINIMUM RESISTANCE	Ω NORMAL RESISTANCE	Ω MAXIMUM RESISTANCE
−40°	−40°	80965	100544	120123
−30°	−20°	42701	52426	62151
−20°	−04°	23458	28491	33524
−10°	14°	13366	16068	18770
0°	32°	7871	9379	10869
10°	50°	4771	5640	6508
20°	68°	2981	3500	4018
30°	86°	1915	2232	2550
40°	104°	1260	1460	1660
50°	122°	848.8	977.1	1105
60°	140°	584.1	668.1	753.4
70°	158°	410.3	467.2	524.2
80°	176°	293.7	332.7	371.7
90°	194°	213.9	241.0	268.2
100°	212°	158.1	177.4	196.8
110°	230°	118.8	132.6	146.5
120°	248°	90.40	100.6	110.8
130°	266°	69.48	77.29	85.11
140°	284°	53.96	60.13	66.29
150°	302°	42.43	47.31	52.20
160°	320°	32.51	36.13	39.73
170°	338°	25.13	27.92	30.71
180°	356°	19.42	21.58	23.74
190°	374°	15.01	16.68	18.35
200°	392°	11.60	12.89	14.18

Figure 4-51 A chart showing the relationship between temperature and resistance for a temperature sensor. (Courtesy of General Motors Corporation, Service Operations Group)

Speed Sensors

Electronically controlled transmissions rely on electrical signals from a speed sensor to control shift timing (Figure 4-52). The use of this type sensor negates the need for hydraulic signals from a governor. When this sensor fails or sends faulty readings, it can cause complaints that are similar to those caused by a bad TP sensor. The most common complaints are no overdrive, no converter-clutch engagement, and no upshifts. Vehicle speed sensors provide road speed information to the computer.

There are basically two types of speed sensors: permanent magnetic (PM) generator sensors and reed style sensors. Both of these rely on magnetic principles. The reed style is simply a switch that closes every time a magnet passes by. The magnet is attached to a rotating shaft, typically the output shaft. The activity of the sensor can be checked with an ohmmeter. If the switch is disconnected and the output shaft rotated one complete revolution, the ohmmeter should show an open and closed circuit within the one revolution of the shaft.

PM generators rely on a stationary magnet and a rotating shaft fitted with iron teeth (Figure 4-53). Each time a tooth passes through the magnetic field an electrical pulse is present. By counting the number of teeth on the output shaft, you can determine how many pulses per one revolution you will measure with a voltmeter set to AC volts.

The operation of a PM generator can be tested with a DMM set to measure AC voltage. Raise the vehicle on a lift. Allow the wheels to be suspended and allowed to rotate freely. Connect the meter to the speed sensor. Start the engine and put the transmission in gear. Slowly increase the engine's speed until the vehicle is at approximately 20 mph, and then measure the voltage at the speed sensor. Slowly increase the engine's speed and observe the voltmeter. The voltage should increase smoothly and precisely with an increase in speed.

Classroom Manual
Chapter 4, page 103

CAUTION: While conducting this procedure, it is possible to damage the CV joints if the drive wheels are allowed to dangle in the air. Place safety stands under the front suspension arms to maintain proper drive axle operating angles.

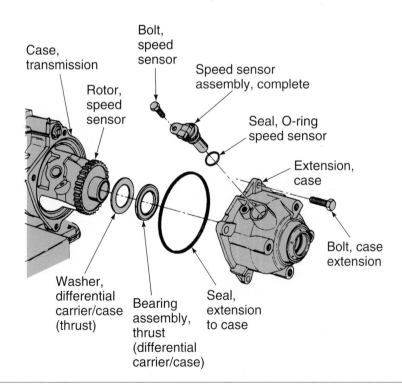

Figure 4-52 AC generators rely on a stationary magnet, the sensor assembly, and a rotor fitted with teeth. (Courtesy of General Motors Corporation, Service Operations Group)

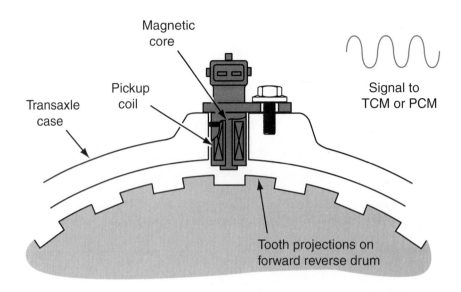

Magnetic
core

Pickup
coil

Transaxle
case

Signal to
TCM or PCM

Tooth projections on
forward reverse drum

Figure 4-53 The vehicle speed sensor will display a pulse each time a tooth of the rotor passes by the sensor.

⚠️ **WARNING:** When running the vehicle on a lift, make sure you stay clear of the wheels and other rotating parts. If you get caught by something powered by the engine, serious injury can result.

Magnetic pulse generators can be tested with a lab scope. Instead of connecting a voltmeter across the sensor's terminals, connect the lab scope leads. The expected pattern is an AC signal that should be a perfect sine wave when the speed is constant. When the speed is changing, the AC signal should change in amplitude and frequency. If the readings are not steady and do not smoothly change with a change in speed, suspect a faulty connector, wiring harness, or sensor.

A speed sensor can also be tested with it out of the vehicle. Connect an ohmmeter across the sensor's terminals. The desired resistance readings across the sensor will vary with each and every individual sensor; however, you should expect to have continuity across the leads. If there is no continuity, the sensor is open and should be replaced. Reposition the leads of the meter so that one lead is on the sensor's case and other touches a terminal. There should be no continuity in this position. If there is any measurable amount of resistance, the sensor is shorted.

Detailed Testing of Actuators

If you were unable to identify the cause of a transmission problem through the previous checks, you should continue your diagnostics with testing the solenoids. This will allow you to determine if the shifting problem is related to the solenoids or their control circuit, or if it is a hydraulic or mechanical problem in the transmission.

Before continuing, however, you must first determine if the solenoids are case grounded and fed voltage by the computer or if they always have power applied to them and the computer merely supplies the ground. While looking in the service manual to find this information, also find the section that tells you which solenoids are on and which are off for each of the different gears.

To begin this test you should secure the tools and/or equipment necessary to manually activate the solenoids. Switch panels are available that connect into the solenoid assembly and allow the technician to switch gears by depressing or flicking a switch. Photo Sequence 7 guides you through this procedure using a commonly available solenoid tester.

Many vehicle speed sensors (VSS), wheel speed sensors, and ignition pickup units are magnetic pulse generators.

Classroom Manual
Chapter 4, page 105

Special Tools

Solenoid tester

Jumper wire

20A fuse or circuit breaker

Photo Sequence 7
Checking the Condition of Solenoids with a Solenoid Tester

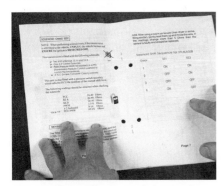

P7-1 In the service manual, locate the wiring diagram for the solenoids and determine if the solenoids are power controlled or ground controlled. Also while looking in the service manual, locate the section that tells you which solenoids are on and which are off for each of the different gears.

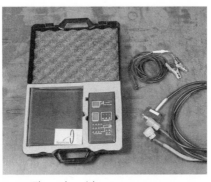

P7-2 The solenoid tester.

P7-3 Make sure the ignition switch is off and disconnect the car's wiring harness that leads to the solenoids.

P7-4 If the solenoids are controlled by their ground, connect a jumper wire with an in-line 20-amp fuse to the tester and a known good ground.

P7-5 Complete the connection of the tester to the transmission and then program it according to the tester's instructions.

P7-6 Start the engine and move the shift lever into Drive.

P7-7 Now turn on solenoid #1; the transmission should be in first gear. Pay attention to how it shifted.

P7-8 Increase your speed, and then turn on solenoid #2. The transmission should have immediately shifted into second gear. Pay attention to how it shifted.

P7-9 When third gear is selected, solenoid #1 will be turned off. Pay attention to how it shifted.

(continued)

Photo Sequence 7 (cont'd.)
Checking the Condition of Solenoids with a Solenoid Tester

P7-10 Likewise when you want fourth gear, solenoid #2 will be turned off. Pay attention to how it shifted.

P7-11 After checking all gears under light, half, and full throttle, return to the shop and summarize your findings.

P7-12 Disconnect the tester from the system and proceed to check solenoid circuits that seemed not to work correctly.

SERVICE TIP: This type of tester is easily made. Simply secure a harness for the transmission you want to test. Connect the leads from the harness to simple switches. Follow the solenoid/gear pattern when doing this. To change gears, all you will need to do is turn off one switch and turn on the next.

With the tester, the solenoids will be energized in the correct pattern and observe the action of the solenoids. A simple electronically controlled transmission has two shift solenoids and the computer sends voltage to the solenoids to activate them. In first gear the #1 solenoid is on. In second gear, both solenoids are on. In third gear, only solenoid #2 is on and neither of the solenoids is on in fourth gear. One wire in the transmission harness goes to solenoid #1 and another connects to solenoid #2.

To begin this test sequence, disconnect the wiring harness that leads to the solenoids. Then connect 12 volts to the switch panel. Start the engine and move the shift lever into Drive. Now turn on solenoid #1; the transmission should be in first gear. Increase your speed, and then turn on solenoid #2. The transmission should have immediately shifted into second gear. When you want to shift into third, turn solenoid #1 off. Likewise, when you want fourth gear, turn solenoid #2 off.

WARNING: When jumping solenoids with a hot wire, be sure to have a 20-amp fuse in line to prevent damage to the solenoids and to your hand. A shorted solenoid will draw very high current, which will melt the wire and seriously burn your hand.

To totally test the transmission, you should shift gears under light, half, and full throttle. If the transmission shifts fine with the movement of the switches, you know that the transmission is fine. Any shifting problem must be caused by something electrical. If the transmission didn't respond to the switch movements, the problem is probably in the transmission.

This same technique can be used to check solenoids with a controlled ground. Instead of running a 12-volt hot wire, you run a good ground wire. This check can be conducted on all transmission solenoids except those that have a duty cycle. Check the service manual to identify this type of solenoid.

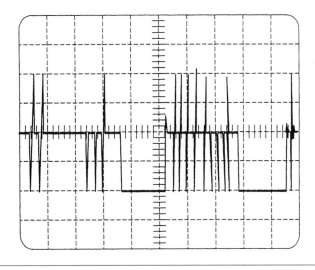

Figure 4-54 A typical control signal for a pulse width modulated solenoid.

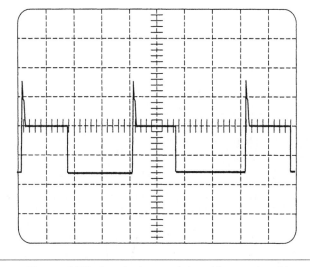

Figure 4-55 A typical control signal for a solenoid.

At times, a solenoid valve will work fine during light throttle operation but may not exhaust enough fluid when pressure increases. To verify that the valve is not exhausting, activate the solenoid, then increase engine speed while pulling on the throttle cable. If the solenoid valve cannot exhaust, the transmission will downshift. Restricted solenoids should be suspect whenever the transmission shifts roughly early under heavy loads or full throttle but shifts fine under light throttle.

Testing Actuators with a Lab Scope

Actuators are electro-mechanical devices, meaning that they are electrical devices that cause some mechanical action. When actuators are faulty, it is because they are electrically faulty or mechanically faulty. By observing the action of an actuator on a lab scope, you will be able to watch its electrical activity. Normally, if there is a mechanical fault, this will affect its electrical activity as well. Therefore, you get a good sense of the actuator's condition by watching it on a lab scope.

Some actuators are controlled pulse-width modulated signals (Figure 4-54). These signals show a changing pulse width. These devices are controlled by varying the pulse width, signal frequency, and voltage levels. By watching the control signal, you can see the turning on and off of the solenoid (Figure 4-55). The voltage spikes are caused by the discharge of the coil in the solenoid.

Both waveforms should be checked for amplitude, time, and shape. You should also observe changes to the pulse width as operating conditions change. A bad waveform will have noise, glitches, or rounded corners. You should be able to see evidence that the actuator immediately turns off and on according to the commands of the computer.

Testing Actuators with an Ohmmeter

Solenoids can be checked for circuit resistance and shorts to ground. This can typically be done without removing the oil pan. The test can be conducted at the transmission case connector. By identifying the proper pins in the connector, individual solenoids can be checked with an ohmmeter. Remember that lower than normal resistance indicates a short, whereas higher than normal resistance indicates a problem of high resistance. If you get an infinite reading across the solenoid, the solenoid windings are open. The ohmmeter can also be used to check for shorts to ground. Simply connect one lead of the ohmmeter to one end of the solenoid windings and the other lead

Component	Pass Thru Pins	Resistance at 20°C	Resistance at 100°C	Resistance to Ground (Case)
1–2 Shift Solenoid Valve	A, E	19–24 Ω	24–31 Ω	Greater than 250 M Ω
2–3 Shift Solenoid Valve	B, E	19–24 Ω	24–31 Ω	Greater than 250 M Ω
TCC Solenoid Valve	T, E	21–26 Ω	26–33 Ω	Greater than 250 M Ω
TCC PWM Solenoid Valve	U, E	10–11 Ω	13–15 Ω	Greater than 250 M Ω
3–2 Shift Solenoid Valve Assembly	S, E	20–24 Ω	29–32 Ω	Greater than 250 M Ω
Pressure Control Solenoid Valve	C, D	3–5 Ω	4–7 Ω	Greater than 250 M Ω
*Transmission Fluid Temperature (TFT) Sensor	M, L	3088–3942 Ω	159.3–198.0 Ω	Greater than 10 M Ω
Vehicle Speed Sensor	A, B, VSS CONN	1420 Ω @ 25°C	2140 Ω @ 150°C	Greater than 10 M Ω

Important: The resistance of this device is necessarily temperature dependent and will therefore vary far more than any other device. Refer to *Transmission Fluid Temperature (TFT) Sensor Specifications.*

Figure 4-56 Service manuals list the resistance values and test points for transmission solenoids and other components. (Courtesy of General Motors Corporation, Service Operations Group)

to a good ground. The reading should be infinite. If there is any measurable resistance, the winding is shorted to ground.

Solenoids can also be tested on a bench. Resistance values are typically given in service manuals for each application (Figure 4-56). A solenoid may be electrically fine but still may fail mechanically or hydraulically. A solenoid's check valve may fail to seat or the porting may be plugged. This is not an electrical problem; rather, it could be caused by the magnetic field collecting metal particles in the ATF and clogging the port or check valve. These would cause erratic shifting, no shift conditions, wrong gear starts, no or limited passing (kickdown) gear, or binding shifts. When a solenoid affected in this way is activated, it will make a slow dull thud. A good solenoid tends to snap when activated.

Chrysler Solenoids

The typical Chrysler solenoid assembly consists of four solenoids, four ball valves, three pressure switches, and three resistors. The ball valves are operated by the solenoids and together they cause the transaxle to shift. The three pressure switches and resistors are used to signal the computer that a shift has occurred. The solenoids should be checked for proper operation. This can be done through the eight-pin connector at the assembly.

Using an ohmmeter with the harness disconnected from the assembly, check the resistance across pins 1 and 4, 2 and 4, and 3 and 4. The resistance should be between 270 and 330 ohms.

With this test you will have checked the resistor for each of the switches. To check the action of these switches, apply air pressure to the feed holes for the overdrive, low-reverse, and overdrive pressure switches. With air applied, the resistance reading across each of the pin combinations should now be zero ohms.

Each of the four solenoids can also be checked with an ohmmeter. Connect the meter across pins 4 and 5, 4 and 6, 4 and 7, and 4 and 8. The resistance across any of these combinations should be 1.5 ohms.

Ford Solenoids

Ford's E4OD can be shifted without the computer by energizing the #1 and #2 shift solenoids in the correct sequence. Refer to Figure 4-57 for the location of the pins in the E4OD harness connector. Connect a 12-volt power source to pin #1, then connect a ground lead to the following pins to shift the transmission: first gear—pin #3, second gear—pins #3 and #2, third gear—pin #2, and fourth gear—no pins grounded. If the transmission doesn't shift, check the solenoids.

Using an ohmmeter, check the resistance of the #1 solenoid by measuring the resistance across pins #1 and #3. To check solenoid #2, measure across pins #1 and 2. And to check solenoid #3, measure the resistance across pins #1 and #4. The resistance across any of these should be 20 to 30 ohms.

To check the solenoids on an AX4S with an ohmmeter, disconnect the harness to the solenoid assembly. Measure the resistance across the appropriate pins and compare your readings to specifications. The shift solenoids should have a reading of 12–30 ohms; the EPC solenoid should have a reading of 2.5–6.5 ohms; the modulated converter clutch solenoid should have a reading of 0.75–22.0 ohms; and the regular converter clutch solenoid should have 16–40 ohms across its terminals.

GM Solenoids

The solenoids of 4T60-E transaxles can be checked with an ohmmeter (Figure 4-58). The terminals for each of these solenoids are accessible at the transmission harness (Figure 4-59).

As are most solenoids, these solenoids are simple on/off devices. Because of this they can be checked through normal diagnostic methods. However, some transmissions, such as the GM 4L80-E, control line pressure according to engine load signals. A simple on/off shift solenoid cannot regulate the flow of oil in a metered manner. Therefore, the 4L80-E is equipped with a solenoid that controls line pressure. This solenoid is a three-port, spool valve, electronic pressure

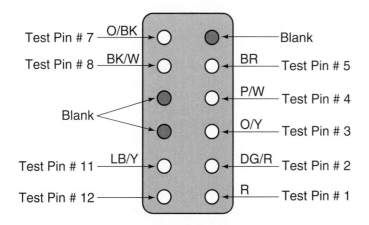

Figure 4-57 The transmission harness connector for an E4OD. The pin numbers are needed for testing of the solenoids.

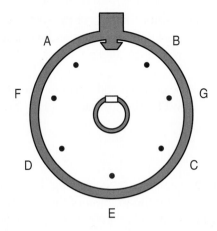

SOLENOID	ACROSS	the RESISTANCE should be:
SOLENOID A	E to F	20 – 40 ohms
SOLENOID B	E to G	20 – 40 ohms
PWM SOLENOID	E to B	10 – 15 ohms
ALL	A to D	20 – 40 ohms
ALL	A to D	2 – 15 ohms (reversed leads)

Figure 4-58 The transmission harness connector for a 4T60-E. The pin numbers are needed for testing of the solenoids.

regulator that controls pressure based on current flow through its coil winding. The amount of current flow through the solenoid is controlled by the PCM. As current through the pressure control solenoid increases, line pressure decreases.

Solenoids that are controlled by duty cycle or a constantly changing signal are more active than the simple on/off solenoids. Therefore, they are more likely to wear. Many transmission rebuilders recommend that this type solenoid be automatically replaced during a transmission overhaul. Duty cycle solenoids are less prone to clogging because they are cycled full on every 10 seconds. This full on time helps to keep the regulator valve clean.

The action of a pressure control solenoid can be monitored on a DSO. Connect the DSO directly to the solenoid's circuit. The pattern should be compared to those given in the service manual. A PWM solenoid is used to apply and release the TCC in such a manner that the TCC will apply and release smoothly under all conditions. The PWM solenoid in the 4L80-E uses a negative duty cycle and its activity can also be monitored on a scope (Figure 4-60).

Honda Solenoids

Honda's shift solenoids are normally closed. When the computer sends battery voltage to them, they open and allow fluid to flow through them. When the fluid flows through the solenoid, it releases pressure that was preventing a shift valve from moving. When the pressure is exhausted, the valve is able to move and the transaxle shifts.

A Honda transmission can be shifted, without the computer, by jumping 12 volts to the shift solenoids. Apply voltage to solenoid B to operate in first gear, apply voltage to both shift solenoids for second gear, to solenoid A for third, and to none of them for fourth gear. If the transaxle shifts normally when the solenoids are energized, the solenoid's sensors or the controller are at fault and the mechanical and hydraulics of the transaxle are normal.

To observe the action of the solenoids, connect a pressure gauge to the transaxle taps for the solenoids. Observe the pressure readings when the transaxle upshifts. If the pressure is greater than 3 psi when the solenoid valve is open, the valve has a restriction or the solenoid is not open-

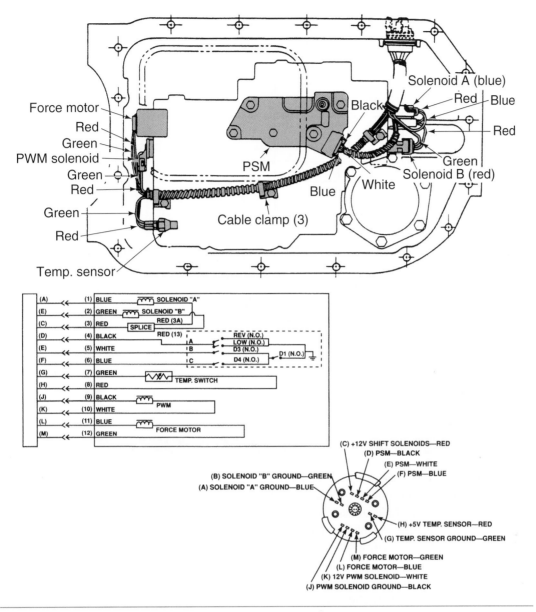

Figure 4-59 Testing of the electrical components in a 4L80-E can be done through the transmission harness connector. (Courtesy of General Motors Corporation, Service Operations Group)

ing fully and not allowing a full release of pressure. If the valving is restricted, the solenoids should be replaced.

✓ **SERVICE TIP:** There are small filters located in the valve body to protect the solenoids. If these filters are plugged, they will cause a restriction to oil flow. The filters are mounted in a preformed rubber gasket and can be cleaned; however, the rubber gaskets should be replaced on every rebuild.

Use an ohmmeter to electrically check the solenoids. The resistance across a shift solenoid should be 12–24 ohms and 14–30 ohms across the lock-up solenoid. If the resistance of the solenoids is not within these figures, they should be replaced.

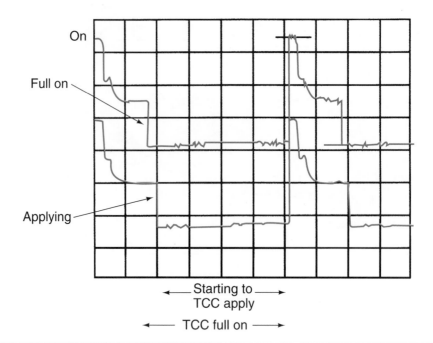

Figure 4-60 A normal scope pattern for a pulse width modulated solenoid. (Courtesy of General Motors Corporation, Service Operations Group)

C A S E   S T U D Y

A customer came into the shop with a high-mileage Hyundai Excel with a Mitsubishi KM-175 transmission. A tow truck delivered the car because the transmission wouldn't shift gears. The technician assigned to the car pulled the transmission and overhauled it. After installing an overhaul kit and inspecting all other transmission parts, the transmission was reassembled and installed back into the car. The transmission still didn't shift.

Then the technician got serious and began to check the electronics involved with the transmission. After a little research in the service manual, he found that the transmission was in the fail-safe mode. With a multimeter, the technician then checked the solenoids, connectors, and wiring, only to find nothing wrong. However, after becoming frustrated, he began to check complete circuits and found that the computer was doing nothing.

He located the computer under the passenger's seat and found heavy corrosion on the computer and the wiring harness terminals. The technician then cleaned the area up and replaced the computer. The transmission then shifted well.

Terms To Know

Ammeter	Data link connector (DLC)	Electronically erasable PROM (EEPROM)
Amplitude	Default mode	Glitch
Assembly line data link (ALDL)	Digital multimeter (DMM)	Impedance
Cycle	Digital storage oscilloscope (DSO)	Multimeter

Ohmmeter   Scan tool   Sinusoidal

Open   Serial data   Voltmeter

Radio frequency interference (RFI)   Short

ASE-Style Review Questions

1. *Technician A* says a faulty TP sensor can cause delayed shifts.
   Technician B says delayed shifts can be caused by an open shift solenoid.
   Who is correct?
   A. A only   **C.** Both A and B
   B. B only   **D.** Neither A nor B

2. While checking a transmission range sensor:
   Technician A says typically the switch should be open in all positions except Park and Neutral
   Technician B says a faulty TR sensor may allow the engine to start in other gear positions besides Park and Neutral.
   Who is correct?
   A. A only   **C.** Both A and B
   B. B only   **D.** Neither A nor B

3. *Technician A* says scan tools are needed to retrieve codes on all models of cars.
   Technician B says all scan tools provide a historical report of the computer system.
   Who is correct?
   A. A only   **C.** Both A and B
   B. B only   **D.** Neither A nor B

4. *Technician A* says an ammeter without an inductive pick-up is always connected in series with the circuit it is testing.
   Technician B says an ammeter is only used when the power for the circuit is disconnected.
   Who is correct?
   A. A only   **C.** Both A and B
   B. B only   **D.** Neither A nor B

5. While checking voltage drops:
   Technician A connects the voltmeter across the load with the circuit activated.
   Technician B connects the negative meter lead to ground and the positive lead to the component being tested.
   Who is correct?
   A. A only   **C.** Both A and B
   B. B only   **D.** Neither A nor B

6. *Technician A* says some shift solenoids can be activated by providing a ground for the solenoid.
   Technician B says some shift solenoids can be activated by applying hydraulic pressure to their valve.
   Who is correct?
   A. A only   **C.** Both A and B
   B. B only   **D.** Neither A nor B

7. While checking the trouble codes on a Honda without OBD II:
   Technician A says the short flashes should be counted, as this is the trouble code.
   Technician B says there is about a one-second delay in the flashing between codes.
   Who is correct?
   A. A only   **C.** Both A and B
   B. B only   **D.** Neither A nor B

8. While checking a switch:
   Technician A uses a voltmeter at the output of the switch in the various switch positions.
   Technician B uses a test light at the output of the switch in the various switch positions.
   Who is correct?
   A. A only   **C.** Both A and B
   B. B only   **D.** Neither A nor B

9. *Technician A* says an ohmmeter can be used to test semiconductors.
   Technician B says a low impedance ohmmeter should be used on electronic circuits.
   Who is correct?
   A. A only   **C.** Both A and B
   B. B only   **D.** Neither A nor B

10. While checking the codes on a Ford AX4S with an OBD II compliant computer system:
    Technician A says during the KOEO self-test, the codes are repeated only twice.
    Technician B says the continuous memory codes appear after the separator pulse signal.
    Who is correct?
    A. A only   **C.** Both A and B
    B. B only   **D.** Neither A nor B

ASE Challenge Questions

1. An electronically controlled transmission has erratic shifting.
   Technician A says a poor PCM ground will cause this problem.
   Technician B says a bad AC generator-to-battery circuit will cause erratic performance of a transmission or transaxle.
   Who is correct?
   - **A.** A only
   - **B.** B only
   - **C.** Both A and B
   - **D.** Neither A nor B

2. A scope is displaying a nonsinusoidal wave at a rate of 10,000 times per second.
   Technician A says the device being checked may not be functioning correctly because the wave is nonsinusoidal.
   Technician B says the frequency cannot be calculated because the wave is nonsinusoidal.
   Who is correct?
   - **A.** A only
   - **B.** B only
   - **C.** Both A and B
   - **D.** Neither A nor B

3. A glitch appears in the waveform of a vehicle speed sensor. Which of the following is *not* a probable cause of the problem?
   - **A.** A loose connector
   - **B.** A damaged wire
   - **C.** A poorly mounted sensor
   - **D.** A damaged magnet in the sensor

4. A hot-wire MAF has low resistance across it when the wire is hot.
   Technician A says this condition will cause the engine to stall as soon as you put the transmission into gear.
   Technician B says this condition will cause the transmission to shift late.
   Who is correct?
   - **A.** A only
   - **B.** B only
   - **C.** Both A and B
   - **D.** Neither A nor B

5. *Technician A* says the second character code in an OBD II DTC is a designated code for future use and has little value to a technician.
   Technician B says that if the third character of a powertrain code is a 7 or 8, the problem is directly related to the transmission.
   Who is correct?
   - **A.** A only
   - **B.** B only
   - **C.** Both A and B
   - **D.** Neither A nor B

Job Sheet 10

Name _____ Date _____

Using a DSO on Sensors and Switches

Upon completion of this job sheet, you should be able to connect a DSO and observe the activity of various sensors and switches.

ASE Correlation

This job sheet is related to the ASE Automatic Transmission and Transaxle Test's Content Area: General Transmission and Transaxle Diagnosis.
Task: Diagnose mechanical and vacuum control systems; determine necessary action.

Tools and Materials

A vehicle with accessible sensors and switches   A DSO
Service manual for the above vehicle   A DMM
Component locator manual for the above vehicle

Describe the vehicle being worked on:

Year _____ Make _____ VIN _____

Model _____

Procedure

1. Connect the DSO across the battery. Make sure the scope is properly set. Observe the trace on the scope. Is there evidence of noise? Explain.

   _____

   _____

   _____

2. Locate the A/C compressor clutch control wires. Start and run the engine. Connect the DMM to read available voltage. Observe the meter, then turn the compressor on. What happened on the meter?

   _____

   _____

   Now connect the DSO to the same point with the compressor turned off. Observe the waveform, then turn the compressor on. What happened to the trace?

   _____

   _____

3. Turn off the engine but keep the ignition on. Locate the TP sensor. List each wire and describe the purpose of each.

   _____

   _____

   _____

   _____

4. Connect the DMM to read reference voltage at the TP sensor. What do you read?

_____

_____

Now move the leads to read the output of the sensor. Starting with the throttle closed, slowly open the throttle until it is wide open. Watch the voltmeter while doing this. Describe your readings.

_____

_____

5. Now connect the DSO to read reference voltage at the TP sensor. What do you see on the trace?

_____

_____

Now move the leads to read the output of the sensor. Starting with the throttle closed, slowly open the throttle until it is wide open. Watch the trace while doing this. Describe your readings.

_____

_____

6. Now run the engine. Locate the oxygen sensor and identify the purpose of each wire to it. Connect the DMM to read voltage generated by the sensor. (To do this, you may use an electrical connector for the O_2 sensor that is positioned away from the hot exhaust manifold). Watch the meter and describe what happened.

_____

_____

_____

_____

7. Now connect the DSO to read voltage output from the sensor. Watch the trace and describe what happened.

_____

_____

_____

_____

8. Explain what you observed as the differences between testing with a DMM and a DSO.

_____

_____

_____

_____

Instructor's Response _____

_____

Job Sheet 11

11

Name _____ Date _____

Test an ECT Sensor

Upon completion of this job sheet, you should be able to check the operation of an engine coolant temperature sensor.

ASE Correlation

This job sheet is related to the ASE Automatic Transmission and Transaxle Test's Content Area: *General Transmission and Transaxle Diagnosis*.
Task: Diagnose mechanical and vacuum control systems; determine necessary action.

Tools and Materials

DMM

Describe the vehicle being worked on:

Year _____ Make _____ VIN _____

Model _____

Procedure

1. Describe the location of the ECT sensor.

_____

_____

2. What color wires are connected to the sensor?

_____

_____

3. Record the resistance specifications for a normal ECT sensor for this vehicle.

_____

_____

_____

_____

_____

_____

_____

4. Disconnect the electrical connector to the sensor.

5. Measure the resistance of the sensor. It was _____ ohms at approximately _____ degrees F.

6. Conclusions from this test.

_____

_____

_____

_____

Instructor's Response _____

_____

_____

Job Sheet 12

12

Name _____ Date _____

Check the Operation of a TP Sensor

Upon completion of this job sheet, you should be able to test the operation of a throttle position sensor with a variety of test instruments.

ASE Correlation

This job sheet is related to the ASE Automatic Transmission and Transaxle Test's Content Area: *General Transmission and Transaxle Diagnosis*.
Task: Diagnose mechanical and vacuum control systems; determine necessary action.

Tools and Materials

DMM   Lab scope

Describe the vehicle being worked on:

Year _____ Make _____ VIN _____

Model _____

Procedure

Task Completed

1. Connect the lab scope across the TP sensor.   ☐

2. With the ignition on, move the throttle from closed to fully open and then allow it to   ☐
   close slowly.

3. Observe the trace on the scope while moving the throttle. Describe what the trace
   looked like.

   _____

   _____

   _____

4. Based on the waveform of the TP sensor, what can you tell about the sensor?

   _____

   _____

5. With a voltmeter, measure the reference voltage to the TP sensor. The reading should
   be _____ volts. It was _____ volts.

6. What is the output voltage from the sensor when the throttle is closed? _____ volts

7. What is the output voltage from the sensor when the throttle is open? _____ volts

8. Move the throttle from closed to fully open and then allow it to close slowly. Describe
   the action of the voltmeter.

   _____

   _____

Check the Operation of a TP Sensor (continued)

9. Conclusions from these tests.

_____

_____

_____

Instructor's Response _____

_____

_____

Job Sheet 13

13

Name _____ Date _____

Testing a MAP Sensor

Upon completion of this job sheet, you should be able to test a manifold absolute pressure sensor in a variety of ways.

ASE Correlation

This job sheet is related to the ASE Automatic Transmission and Transaxle Test's Content Area: *General Transmission and Transaxle Diagnosis.*
Task: Diagnose mechanical and vacuum control systems; determine necessary action.

Tools and Materials

Hand-operated vacuum pump   DMM
Lab scope

Describe the vehicle being worked on:

Year _____ Make _____ VIN _____

Model _____

Procedure

Task Completed

1. If the MAP sensor produces an analog voltage signal, follow this procedure.

2. With the ignition switch on, backprobe the 5-volt reference wire. ☐

3. Connect a voltmeter from the reference wire to ground. The reading is _____ volts.

4. If the reference wire is not supplying the specified voltage, what should be checked next?

   _____

5. With the ignition switch on, connect the voltmeter from the sensor ground wire to the battery ground. ☐

6. What is the measured voltage drop? _____ volts

7. What does this indicate?

   _____

   _____

8. Backprobe the MAP sensor signal wire and connect a voltmeter from this wire to ground with the ignition switch on. ☐

9. What is the measured voltage? _____ volts

10. What does this indicate?

   _____

   _____

11. How do you determine the barometric pressure based on these voltage readings?

_____

_____

_____

_____

☐

12. Turn the ignition switch on and connect a voltmeter to the MAP sensor signal wire.

13. Connect a vacuum hand pump to the MAP sensor vacuum connection and apply 5 inches of vacuum to the sensor. Record the voltage reading: _____ volts.

Note: On some MAP sensors, the sensor voltage signal should change 0.7–1.0 volt for every 5 inches of vacuum change applied to the sensor. Always use the vehicle manufacturer's specifications. If the barometric pressure voltage signal was 4.5 volts with 5 inches of vacuum applied to the MAP sensor, the voltage should be 3.5 V–3.8 V. When 10 inches of vacuum is applied to the sensor, the voltage signal should be 2.5 V–3.1 V. Check the MAP sensor voltage at 5-inch intervals from 0–25 inches.

If the MAP sensor voltage is not within specifications at any vacuum, replace the sensor.

14. Record the results of all vacuum checks.

_____

_____

15. What did these tests indicate?

_____

_____

☐

16. Connect the scope to the MAP output and a good ground.

17. Accelerate the engine and allow it to return to idle. Observe and describe the trace.

_____

_____

18. What did the trace show about the sensor?

_____

_____

_____

_____

Note: If the MAP sensor produces a digital voltage signal of varying frequency, check the 5-volt reference wire and the ground wire with the same procedure used on other MAP sensors. This sensor diagnosis is based on the use of a MAP sensor tester that changes the MAP sensor varying frequency voltage to an analog voltage. Follow these steps to test the MAP sensor voltage signal:

☐

1. Turn off the ignition switch, and disconnect the wiring connector from the MAP sensor.

☐

2. Connect the connector on the MAP sensor tester to the MAP sensor.

☐

3. Connect the MAP sensor tester battery leads to a 12-volt battery.

☐

4. Connect a pair of digital voltmeter leads to the MAP tester signal wire and ground.

5. Turn on the ignition switch and observe the barometric pressure voltage signal on the meter. Observe and record the voltmeter readings.

_____

_____

_____

_____

6. Supply the specified vacuum to the MAP sensor with a hand vacuum pump.　□

7. Observe the voltmeter reading at each specified vacuum. Record the readings.

_____

_____

_____

_____

8. What do these readings indicate?

_____

_____

Instructor's Response _____

_____

_____

Job Sheet 14

Name _____ Date _____

Conduct a Diagnostic Check of an Engine Equipped with OBD II

Upon completion of this job sheet, you should be able to conduct a system inspection and retrieve codes from the PCM of an OBD II equipped engine.

ASE Correlation

This job sheet is related to the ASE Automatic Transmission and Transaxle Test's Content Area: *General Transmission and Transaxle Diagnosis.*
Task: Diagnose mechanical and vacuum control systems; determine necessary action.

Tools and Materials

A vehicle equipped with OBD II   Service manual
Scan tool

Describe the vehicle being worked on:

Year _____ Make _____ VIN _____

Model _____

Engine size and type _____

Procedure

1. Check all vehicle grounds, including the battery and computer ground, for clean and tight connections. Comments:

   _____

   _____

2. Perform a voltage drop test across all related ground circuits. State where you tested and what your findings were.

   _____

   _____

   _____

3. Check all vacuum lines and hoses, as well as the tightness of all attaching and mounting bolts in the induction system. Comments:

   _____

   _____

4. Check for damaged air ducts. Comments:

   _____

   _____

Conduct a Diagnostic Check of an Engine Equipped with OBD-II (continued)

5. Check the ignition circuit, especially the secondary cables, for signs of deterioration, insulation cracks, corrosion, and looseness. Comments:

_____

_____

6. Are there any unusual noises or odors? If there are, describe them and tell what may be causing them.

_____

_____

7. Inspect all related wiring and connections at the PCM. Comments:

_____

_____

☐ 8. Gather all pertinent information about the vehicle and the customer's complaint. This should include detailed information about the symptom from the customer, a review of the vehicle's service history, published TSBs, and the information in the service manual.

9. Are there any vacuum leaks? _____ Yes _____ No

10. Is the engine's compression normal? _____ Yes_____ No

11. Is the ignition system operating normally? _____ Yes _____ No

12. Are there any obvious problems with the air/fuel system? _____ Yes _____ No

13. Your conclusions from the preceding questions and answers.

_____

_____

14. Check the operation of the MIL by turning the ignition on. Describe what happened and what this means.

_____

_____

☐ 15. Connect the scan tool to the DLC.

☐ 16. Enter the vehicle identification information into the scan tool.

☐ 17. Then retrieve the DTCs with the scan tool.

18. List all codes retrieved by the scan tool.

_____

_____

_____

19. Conclusions from these tests and checks.

_____

_____

_____

Instructor's Response _____

_____

Rebuilding Transmissions and Transaxles

Upon completion and review of this chapter, you should be able to:

❏ Diagnose noise and vibration problems and determine needed repairs.

❏ Remove and install a transmission/transaxle assembly from a car or light truck.

❏ Disassemble, clean, and inspect a transmission/transaxle.

❏ Inspect, repair, and replace transmission case(s).

❏ Inspect and repair the bores, passages, bushings, vents, and mating surfaces of a transmission case.

❏ Inspect, repair, and replace extension housing and extension housing bushings and seals.

❏ Inspect and replace speedometer drive gear, driven gear, and retainers.

❏ Inspect and replace external seals and gaskets.

❏ Reassemble a transmission/transaxle after servicing it.

Basic Tools

Basic mechanic's tool set

Clean lint-free rags

Appropriate service manual

Diagnosis of Noise and Vibration Problems

Abnormal noises and vibrations can be caused by faulty bearings, damaged gears, worn or damaged clutches and bands, or a bad oil pump, as well as by contaminated fluid or an improper fluid level. Torque converter and cracked flexplate problems can also be the cause of vibrations (Figure 5-1).

A customer will often complain of a transmission noise that in reality is caused by something else in the driveline and not the transmission or torque converter. Bad CV or U-joints, wheel bearings, brake calipers, and dragging brake pads can generate noises that customers, and unfortunately some technicians, mistakenly blame on the transmission or torque converter. The entire driveline should be checked before assuming the noise is transmission related.

Most vibration problems are caused by an unbalanced torque converter assembly, a poorly mounted torque converter, or a faulty engine or transmission mount, or a defective output shaft. The key to determining the cause of the vibration is to pay particular attention to the vibration in relation to engine and vehicle speed. If the vibration changes with a change in engine speed, the cause of the problem is most probably the torque converter. If the vibration changes with vehicle speed, the cause is probably the output shaft or the driveline connected to it. The latter type of problem can be a bad extension housing bushing or universal joint, which would become worse at higher speeds.

Begin your diagnosis by determining if the cause of the problem is the driveline or the transmission. To do this, put the transmission in gear and apply the foot brakes. If the noise is no longer evident, the problem must be in the driveline or the output of the transmission. If the noise is still present, the problem must be in the transmission or torque converter.

Noise problems are also best diagnosed by paying a great deal of attention to the speed and the conditions at which the noise occurs. The conditions to pay most attention to are the operating gear and the load on the driveline. If the noise is engine speed related and is present in all gears, including PARK and NEUTRAL, the most probable source of the noise is the oil pump because it rotates whenever the engine is running. However, if the noise is engine related and is present in all gears except PARK and NEUTRAL, the most probable sources of the noise are those parts that rotate in all gears, such as the drive chain, the input shaft, and the torque converter.

This chapter covers the mechanical aspects of diagnosis, disassembly, inspection, service, and assembly of an automatic transmission. The hydraulic components are detailed in Chapter 8.

Refer to a good manual transmission textbook for procedures on diagnosing the entire driveline.

Classroom Manual
Chapter 5, page 119

Problem	Probable Cause(s)
Ratcheting noise	The return spring for the parking pawl is damaged, weak, or misassembled
Engine speed sensitive whine	Torque converter is faulty
	A faulty pump
Popping noise	Pump cavitation—bubbles in the ATF
	Damaged fluid filter or filter seal
Buzz or high-frequency rattle Whine or growl	Cooling system problem
	Stretched drive chain
	Broken teeth on drive and/or driven sprockets
	Nicked or scored drive and/or driven sprocket bearing surfaces
	Pitted or damaged bearing surfaces
Final drive hum	Worn final drive gear assembly
	Worn or pitted differential gears
	Damaged or worn differential gear thrust washers
Noise in forward gears	Worn or damaged final drive gears
Noise in specific gears	Worn or damaged components pertaining to that gear
Vibration	Torque converter is out of balance
	Torque converter is faulty
	Misaligned transmission or engine
	The output shaft bushing is worn or damaged
	Input shaft is out of balance
	The input shaft bushing is worn or damaged

Figure 5-1 A basic noise and vibration diagnostic chart.

Noises that occur only when a particular gear is operating must be related to those components responsible for providing that gear, such as a band or clutch. On some RWD vehicles, a defective rear transmission mount can cause noise only in first gear. If the noise is vehicle related, the most probable causes are the output shaft and final drive assembly. Often, the exact cause of noise and vibration problems can be identified only through a careful inspection of a disassembled transmission.

Prior to tearing down a transmission, it is a good idea to inspect the material trapped in the transmission's filter. Use a magnet to determine if the particles are steel or aluminum. Generally, steel deposits are caused by something in the transmission, whereas aluminum deposits are from the stator of the torque converter. Some transmissions use aluminum clutch drums and supports. If you are working on a transmission so equipped, the aluminum deposits may be from the transmission as well as the torque converter. Plastic deposits may also be from the torque converter. Some converters, including most Ford converters, are fitted with stators made of phenolic plastic.

Regardless of the exact fault in the transmission that is causing a vibration or noise, the transmission will need to come out and the entire unit will have to be checked and carefully inspected. Proper diagnosis prior to disassembling the transmission will identify the specific areas that should be carefully looked at and can prevent unnecessary transmission removal and teardown.

Transmission/Transaxle Removal

> ⚠️ **WARNING:** Be sure to wear safety glasses or goggles when working under the vehicle and when handling ATF. ATF, as well as rust and dirt, can cause serious damage to your eyes.

RWD Vehicles

Removing the transmission from a rear-wheel-drive car is generally more straightforward than removing one from a front-wheel-drive model, as there is typically one crossmember, one driveshaft, and easy access to cables, wiring, cooler lines, and bell housing bolts. Transmissions in FWD cars, because of their limited space, can be more difficult to remove because you may need to disassemble or remove large assemblies such as engine cradles, suspension components, brake components, splash shields, or other pieces that would not usually affect RWD transmission removal.

The following is a list of components typically removed or disconnected while removing an automatic transmission from a RWD vehicle. This list is arranged in a suggested order of events. Some vehicles require more than this, others require less.

Battery ground cable

Transmission oil pan

Torque converter access plate

Torque converter drain plug

Transmission dipstick tube

Transmission cooler lines

Speedometer cable

Vacuum hose to modulator

Electrical connectors to solenoids

Electrical connectors to sensors

Gear selector linkage

Throttle pressure linkage

Kickdown linkage or electrical connector to switch

Neutral safety switch

Reverse lamp switch

Starter motor

Exhaust heat shields

Electrical connectors to oxygen sensors

Exhaust pipes and catalytic converters

Drive shaft

Torque converter to flexplate bolts or nuts

Transmission mounts

Crossmember

Bell housing to engine bolts

> ✔️ **SERVICE TIP:** If you plug the cooler line fittings on the housing and the lines themselves, you will prevent the frustration of having ATF drip down your neck or into your eyes while removing the transmission.

Safe removal of transmissions requires the purchase or fabrication of tools to help support the engine in the chassis and to lift and carry the transmission away from the vehicle.

Special Tools

Drain pan

Transmission jack

Engine support and/or heavy-duty chain

The exact procedure for removing a transmission will vary with each year, make, and model of vehicle. Always refer to the service manual for the procedure you should follow. Normally the procedure begins with placing the vehicle on a hoist so that you can easily work under the vehicle and under the hood.

Once the vehicle is in position, disconnect the negative battery cable and place it away from the battery. Disconnect and remove any transmission linkages connected to the engine. Also remove the ATF dipstick.

Then raise the vehicle and disconnect the parts of the exhaust system that may interfere with transmission removal. Disconnect all electrical connections at the transmission (Figure 5-2). Make sure you place them away from the transmission so they are not damaged during transmission removal or installation.

Remove the torque converter inspection plate or dust cover (Figure 5-3). Place an index mark on the converter and the flexplate to ensure that the two will be properly mated during installation. Using a flywheel turning tool, rotate the flywheel until some of the converter-to-flexplate bolts are exposed. Loosen and remove the bolts. Then rotate the flywheel until more bolts are accessible. Remove them and continue the process until all the mounting bolts are removed. Once the bolts are removed, slide the converter back into the transmission.

Place a drain pan under the transmission and drain the fluid from the transmission. Once the fluid is out, place the oil pan back onto the transmission and keep it in place with three or four bolts. Move the drain pan under the rear of the transmission. Then remove the drive shaft.

Disconnect the shift and other linkages from the transmission. Then remove the speedometer cable (if there is one). Disconnect the cooler lines from the transmission. Cap the ends of the lines to prevent leakage and dirt from entering the cooling system. Disconnect the brackets for the transmission vent and dipstick tube. Then remove the vent and dipstick tubes (Figure 5-4).

Place a transmission jack under the transmission (Figure 5-5) and secure the transmission to it. The use of a transmission jack allows for easier access to parts hidden by crossmembers or hidden in the space between the transmission and the vehicle's floor pan. Using a chain or other holding fixture, secure the transmission to the jack's pad. Using two chains in an "X" pattern around the transmission works well to secure the transmission to the jack. If the transmission begins to slip on the jack while you are removing it, never try to catch it. Let it fall.

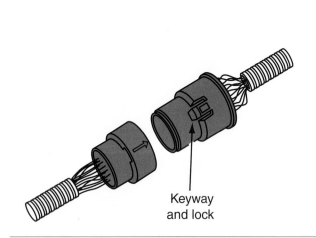

Figure 5-2 Be sure to examine the electrical connectors before attempting to separate them. Most have locks that must be unlocked. (Courtesy of General Motors Corporation, Service Operations Group)

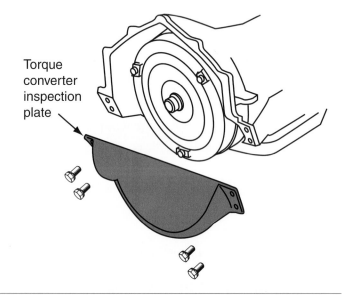

Figure 5-3 Location of a typical torque converter (and flexplate) inspection cover and bolts. (Courtesy of General Motors Corporation, Service Operations Group)

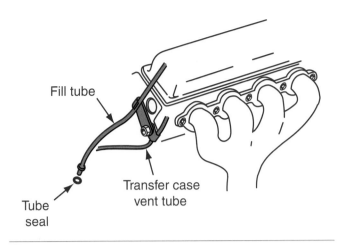

Fill tube

Transfer case
vent tube

Tube
seal

Figure 5-4 Disconnect the brackets for the transmission vent and dipstick tube and then remove the vent and dipstick tubes.

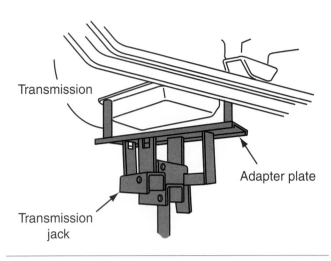

Transmission

Adapter plate

Transmission
jack

Figure 5-5 Place a transmission jack under the transmission and secure the transmission to it.

Remove the transmission mounting bolts. Then remove the crossmember at the transmission. After the mounts are free from the transmission, lower the transmission slightly so you can easily access the top transmission-to-engine bolts. Loosen and remove these bolts. Then remove the remaining transmission-to-engine bolts.

Move the transmission away from the engine until the unit is clear of the alignment dowels at the rear of the engine. Then slowly lower the transmission (Figure 5-6). Make sure the converter hub and any associated shafts have a clear path while you are lowering the transmission. Once the transmission is out of the vehicle, carefully move it to the work area and mount it to a stand.

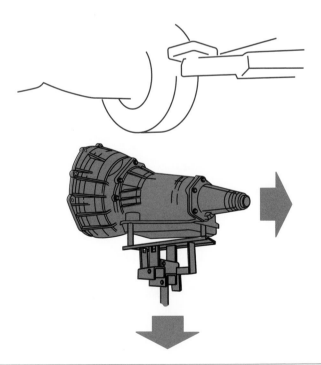

Figure 5-6 Move the transmission away from the engine until the unit is clear of the alignment dowels at the rear of the engine and the converter hub has a clear path while lowering the transmission.

FWD Vehicles

Obviously the procedure for removing a transmission from a FWD vehicle is different. Because all the components are confined to a very small area, the exact items that need to be removed or disconnected vary from vehicle to vehicle. The following is a list of components that are typically removed or disconnected. This list is arranged in a suggested order of events.

Battery ground cable

Underhood electrical connectors to transaxle

Front wheels

Electrical connectors at the wheel brake units

Brake calipers

Steering knuckles

Drive axles

Transmission oil pan

Transaxle to engine brackets

Torque converter access plate

Torque converter drain plug

Transmission dipstick tube

Transmission cooler lines

Speedometer cable

Electrical connectors to solenoids

Electrical connectors to sensors

Gear selector linkage

Throttle pressure linkage

Kickdown linkage or electrical connector switch

Neutral safety switch

Reverse lamp switch

Starter motor

Exhaust heat shields

Electrical connectors to oxygen sensors

Exhaust pipes and catalytic converters

Torque converter to flexplate bolts or nuts

Transaxle to engine mount

Crossmember

Bell housing to engine bolts

On some vehicles, the recommended procedure may include removing the engine with the transaxle. Always refer to the appropriate service information before proceeding to remove the transaxle. You may waste much time and energy if you don't check the manual first.

● **CUSTOMER CARE:** Because much of your time will be spent working under the hood while you are removing an engine, make sure you use fender covers and take extra care not to damage the sheet metal or finish of the vehicle.

Begin removal by placing the vehicle on a lift. However, before raising the vehicle, take a look around the engine bay to see if any interference will occur between the firewall and engine

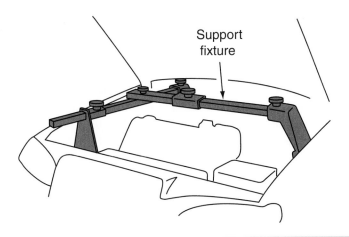

Support
fixture

Figure 5-7 A typical engine support fixture for a FWD vehicle.

components, such as distributors, fans and fan shrouds, fuel lines, exhaust systems or electrical components, when the transaxle is removed. If any causes for interference are found, these problems should be corrected before continuing. Also, any bell housing bolts, wiring, or T.V. cables accessible from above should be removed. Before raising a FWD vehicle to begin transaxle removal, a support fixture should be attached to the engine (Figure 5-7).

SERVICE TIP: Some technicians use instant cameras or video recording cameras to help recall the locations of underhood items by taking pictures before work is started. This technique can be valuable considering how complex the underhood systems of cars have become.

CAUTION: Always begin the transmission removal procedure by disconnecting the battery ground cable. This is a safety precaution to help avoid any electrical surprises when removing starters or wiring harnesses. It is also possible to send voltage spikes, which may kill the PCM if wiring is disconnected when the battery is still connected.

Raise the vehicle to a comfortable height and remove all but the three or four corner bolts of the transmission pan, depending on the shape of the pan. Place a drain pan large enough to catch all the oil under the transaxle. Carefully remove bolts from one side of the pan. Back off the bolt or bolts on the other side just enough to allow the pan to drop slightly as you pry it loose. Be careful, as some pans will come loose without being pried and if you loosen the pan too much, you may have a large mess to clean up! When the fluid stops draining, replace the pan with a minimum of bolts, taking care not to lose the remaining bolts.

SERVICE TIP: To control the mess, some technicians disconnect a cooler line at the radiator and place the end of the line in a drain pan. The engine is then cranked and fluid pumped into the drain pan. This can be done to move most of the fluid out of the transmission. Do not crank the engine for more than 30 seconds at one time.

To remove FWD driveshafts, you must first loosen the large nut that retains the outer CV joint, which is splined on the shaft to the hub.

CAUTION: It is recommended that this nut be loosened with the vehicle on the floor and the brakes applied to reduce possible damage to the CV joints and wheel bearings.

Now raise the vehicle and remove the front wheels. Tap the splined CV joint shaft with a soft-faced hammer to see if it is loose. Most will come loose with a few taps. Many Ford FWD cars use an interference fit spline at the hub. You will need a special puller for this type CV joint. The tool pushes the shaft out and on installation pulls the shaft back into the hub.

The lower ball joint must now be separated from the steering knuckle. The ball joint (Figure 5-8) will be either bolted to the lower control arm or held into the knuckle with a pinch bolt. Once the ball joint is loose, the control arm can be pulled down and the knuckle can be pushed outward to allow the splined CV joint shaft to slide out of the hub (Figure 5-9). The inboard joint

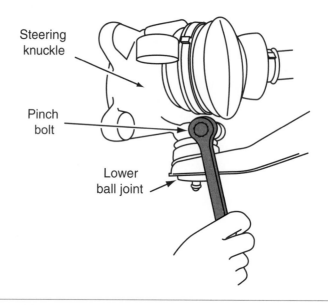

Figure 5-8 Typical ball joint to steering knuckle attachment.

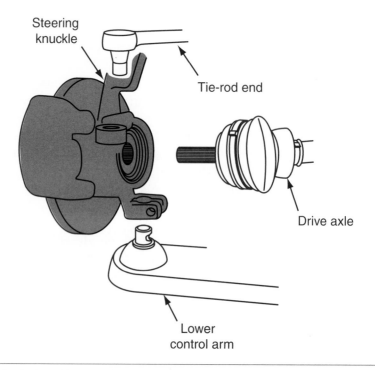

Figure 5-9 The knuckle assembly is moved out of the way to allow the stub shaft to be pulled out of the hub.

can then be pried out or it will slide out. Some transaxles have retaining clips that must be removed before the inner joint can be removed.

The speedometer drive gears may need to be removed before pulling out the driveshafts, as they may be damaged when the shafts are removed. On some cars, the inner CV joints have flange-type mountings. These must be unbolted for removal of the shafts. In some cases, the flange-mounted driveshafts may be left attached to the wheel and hub assembly and unbolted only at the transmission flange. The free end of the shafts should be supported and placed out of the way. Doing this will greatly decrease the amount of time needed to remove and install the transmission.

At this time, you should study the underbody layout to aid reinstallation. This is another time when use of an instant camera will help your memory when it comes time to reinstall the transmission. In many shops, the technician repairing the transmission is not the same person who removes and reinstalls it. If the installer does not remember the exact location of all the underbody and engine bay components, or the correct adjustments of control cables, the transmission may not operate properly when it is reinstalled. Improper installation can ruin a transmission or cause driveability, noise, and vibration problems.

Now the shift linkages, vacuum hoses, electrical connections, speedometer drives, and control cables should be disconnected (Figure 5-10). The exhaust system may also need to be lowered or partially removed. The inspection cover between the transaxle and engine should be removed to allow access to the torque converter bolts. There will be three to six bolts or nuts securing the converter to the flexplate, depending on the application.

CAUTION: To prevent accidental damage, pay attention to the proper location of all mounting bolts and hardware when the transmission is removed. For example, on some torque converters, using longer than normal flywheel-to-converter bolts can dimple the torque converter clutch friction surface, causing the converter clutch to chatter.

Mark the position of the converter to the flexplate to help maintain balance or runout. It will be necessary to rotate the crankshaft to remove the converter bolts. This can be done by using a long ratchet and socket on the front crankshaft bolt or by using a flywheel turning tool, if space permits.

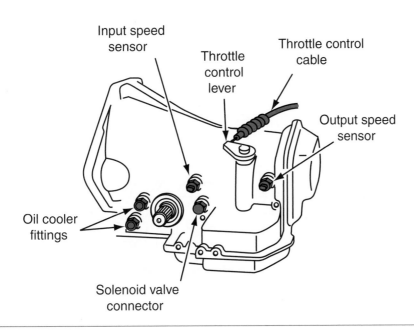

Figure 5-10 Location of the various switches, connectors, and levers on a typical transaxle.

Carefully remove the cooler lines by holding the case fitting with one wrench and loosening the line nut with a line wrench. Doing this assures that you will not twist the steel lines, which would damage them and restrict their flow.

With the transmission jack supporting the transmission, remove the transmission mounts (FWD). If the car is equipped with an engine cradle that will separate, remove the half of the cradle that allows for transaxle removal.

Now remove the starter. The starter wiring may be left connected but you will need to hang the starter with heavy mechanic's wire to avoid damage to the cables. You can also remove the starter from the vehicle to get it totally out of the way.

 **CAUTION:** Never allow the starter to hang by the wires attached to it. The weight of the starter can damage the wires or, worse, break the wire and allow the motor to fall, possibly on you or someone else. Always securely support the starter and position it out of the way after you have unbolted it from the engine.

Now pull the transaxle away from the engine. It may be necessary to use a pry bar between the transaxle and engine block to separate the two units. Make sure the converter comes out with the transaxle. This prevents bending the input shaft, damaging the oil pump, or distorting the drive hub. After separating the transaxle from the engine, retain the torque converter in the bell housing. This can be done simply by bolting a small combination wrench to a bell housing bolt hole across the outer edge of the converter.

CAUTION: Never force the torque converter back into the oil pump if it has slipped out.

Installation

Transmission installation is generally a reverse of the removal procedure. Care must be taken to avoid destroying the new or rebuilt transmission during installation. A quick check of the following list will greatly simplify your installation and reduce the chances of destroying something during installation.

❏ Make sure the block alignment pins (**dowels**) are in the appropriate bores and are in good shape and that the alignment holes in the bell housing are not damaged (Figure 5-11).

Alignment pins are commonly referred to as **dowels**.

CAUTION: Dowels or locating pins are used to keep the engine and transmission in perfect alignment. Torque converter housing bolts alone cannot maintain perfect alignment. Perfect alignment is necessary to prevent flexplate breakage.

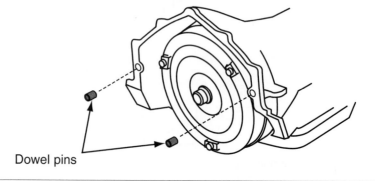

Dowel pins

Figure 5-11 The dowel pins may remain in the engine or the transmission case when the transmission is removed. In either case, they should be inspected.

❑ Make sure the pilot hole in the crankshaft is smooth and not out-of-round. This will allow the converter to move in and out on the flexplate.

❑ Make sure the pilot hub of the converter is smooth and cover it with a light coating of chassis lubricant to prevent chafing or rust.

❑ Make sure the converter's drive hub is smooth and coat it with trans gel or petroleum jelly.

❑ Secure all wiring harnesses out of the way to prevent their being pinched between the bell housing and engine block. If the wires get pinched, not only will there be a large electrical short, but also you may destroy the car's computer.

❑ Flush out the oil cooler lines and the cooler itself to remove any material that could damage the transmission. Likewise, the converter should be flushed. It is recommended that clutch-type converters be replaced, as it is not possible to tell how much the unit has been damaged.

❑ Always perform an endplay check and check the overall height before reinstalling a torque converter or installing a fresh unit out of the box.

❑ Pour one quart of the recommended fluid into the converter before mounting the converter to the transmission. This will assure that all parts in the converter have some lubrication before start-up.

Classroom Manual
Chapter 5, page 124

Slide the converter into the transmission, making sure that all drives are engaged. Double-check this by using the height dimension you measured during transmission removal. In older transmissions, the torque converter must engage into the input shaft stator splines and the oil pump, whereas later models have those and possibly a direct drive shaft or an oil pump drive shaft.

Secure the converter in the transmission as you did during removal. Then transfer the transmission to the jack and move it under the car. Raise the transmission to get close alignment to the engine block. If the transmission has a full-circle bell housing, you will need to align the converter drive studs or bolt pads before you push the transmission up against the block, as this is difficult once the transmission is against the engine block. Make sure you set the torque converter onto the flexplate in alignment with the marks made during removal. These will provide proper balance for the unit. Once the converter is aligned, be sure the block dowel pins line up with the bell housing, then push the transmission against the engine block. Check to be sure that nothing has been caught between the block and the bell housing. Start two bell housing bolts across from each other and slowly tighten them. Check for free rotation of the torque converter while doing this. Never use the bolts to pull the transmission against the engine block. Then, install the rest of the bolts and torque them to specifications.

On FWD cars, if the engine is held in place with a support bar, the transmission jack may be removed at this time. If the car has a split cradle-type subframe, it should be installed now.

On RWD vehicles, it may be necessary to leave the transmission jack in place while components such as exhaust crossovers or frame crossmembers are installed. Do not connect the gearshift linkage until the transmission is mounted to the cross member and the transmission jack is removed. Once the rear transmission mount is attached to the crossmember, the jack may be removed.

Install the converter drive bolts (or nuts). You should notice while installing these fasteners that the converter has some noticeable fore and aft movement. This amount varies with different transmissions, but it is generally between 1/8 and 1/4 inch. This is normal and necessary as it allows the converter to move on the flexplate and also allows for a noninterference fit at the oil pump drive gear. If there were no movement, premature pump failure would result.

> **CAUTION:** Never use an impact wrench on torque converter bolts. Impact wrenches can drive the bolts through the cover, which will warp the inside surface and prevent proper clutch apply or may damage the clutch pressure plate.

The cooler lines should now be installed. Make sure the lines are connected to the bores they were originally in. Then, tighten the line fittings, making sure you do not twist or distort the line. The remaining components—starter, throttle cables, electrical connections, dipstick tube, and so on—can now be installed.

> **CAUTION:** After the transmission has been installed, make sure everything is properly realigned. Repositioning the transmission an inch or less can have a big effect on the manual shift linkage adjustment.

While connecting the manual shift linkage, pay attention to the condition of the plastic bushings. If these are worn or missing, hard or inaccurate linkage movement will result. Replace the bushings if needed.

Some other areas that require special attention during installation are the grounding straps and any rubber tubes. These are often overlooked during transmission removal and installation and can cause problems if not checked. All rubber tubes are suspect, as they are exposed to heat, which causes them to crack. New parts should be used if necessary. The ground straps must be in good condition and free of corrosion, as these provide an electrical ground path to the body of the car during operation. Failure to clean or attach these straps or cables can also result in poor signals to the PCM, voltage spikes that can damage the PCM, electrolysis through the fluid that welds internal transmission components, or even fused manual linkage cables as the current looks for a path to flow through.

On RWD vehicles, the driveshaft must be installed using the marks you made during removal (Figure 5-12). Be sure to coat the slip yoke with ATF before sliding it into the extension housing. The driveshafts of FWD cars are installed in the reverse manner in which they were removed and the related components (speedometer gears, strut lower ball joint bolts, and so on) should be installed just as they were before they were removed. Use a new nut on the outer CV joint stud, if the nut is the self-locking type. The torque of this nut is critical. Torquing the nut should be done with the car's tires on the ground and with the brakes held. Air-type impact tools should not be used, as they can damage the wheel bearing or hub.

Connect the battery cables. Then add about half of the total quantity of the proper ATF to the transmission. This amount varies, but an average amount is four quarts. Some technicians will also connect a pressure gauge to the transmission and take pressure readings during the initial operation of the fresh transmission. Some imported-transmission manufacturers do require that this be done because line pressures must be set after start-up. Connecting the pressure gauges will also give you an indication of whether there is sufficient oil pressure to road test the car.

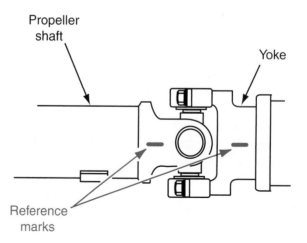

Figure 5-12 The index marks on the drive shaft flange yoke and the pinion flange ensure that the drive shaft is installed at the correct phase.

SERVICE TIP: On governor-equipped transmissions, put the transmission into reverse as soon as the engine starts. Many transmissions do not send fluid to the governor in reverse; putting the transmission in reverse prevents the possibility of moving debris from the rebuild into the governor, which would cause it to stick or seize.

Put the vehicle on a lift, then start the engine and apply the brakes. Move the shift selector through all of the gear ranges. Then place the selector into PARK. Check the fluid level and add ATF until you reach the ADD or COLD mark. Then, inspect the transmission for any signs of leakage and loose bolts. If a pressure gauge was attached to the transmission and the pressure reading was fine, disconnect the gauge after you shut down the engine.

Road test the vehicle to check the operation of the transmission and anything else that may have been affected by your work. Make any necessary adjustments. The road test should cover at least 20 miles in order to completely warm up the transmission. Recheck and adjust the fluid level. It should be between the ADD and FULL marks on the dipstick. DO NOT OVERFILL! After the road test, again visually inspect the transmission for signs of leakage. Also carefully look around the engine and transmission to see if any wires, hoses, or cables are disconnected or positioned in an undesirable spot.

SERVICE TIP: If the transmission is computer-controlled, check the service manual before taking it on its initial road test. Some computer-controlled automatic transmissions require that a "Learning Procedure" be followed, which includes various driving conditions. Since you need to road test the transmission and need to teach the transmission, you may as well do both at the same time.

Computer Relearn Procedures

Vehicles equipped with engine or transmission computers may require a relearn procedure after the battery is disconnected. Many vehicle computers memorize and store vehicle operation patterns for optimal driveability and performance. When the vehicle battery is disconnected, this memory is lost. The computer will use default data until new data from each key start are stored. As the computer memorizes vehicle operation for each new key start, driveability is restored. A vehicle computer may memorize a vehicle's operation patterns for 40 or more key starts. Always refer to the service manual, as some transmissions have the ability to learn quickly when certain conditions are met.

Customers often complain of driveability problems during the relearn stage, because the vehicle acts differently than before it was serviced. Depending on the type and make of the vehicle and how it is equipped, the following complaints may exist: harsh or poor shift quality, rough or unstable idle, hesitation or stumble, rich or lean running, and poor fuel economy. These symptoms and complaints should disappear after a number of drive cycles have been memorized. To reduce the possibility of complaints, after any service that requires that the battery be disconnected, the vehicle should be road tested. If a specific relearn procedure is not available, the following procedure may be used on vehicles equipped with an automatic transmission:

1. Set the parking brake and start the engine in P or N. Allow the engine to warm up to normal operating temperature or until the electric cooling fan cycles on.
2. Allow the vehicle to idle for about a minute in the N position, then move the gear selector into the D position and allow it to idle in gear for one minute.
3. Road test the vehicle. Accelerate at normal throttle openings (20%–50%) until the vehicle shifts into top gear.
4. Then maintain a cruising speed with a light-to-medium throttle.
5. Decelerate to a stop, make sure you allow the transmission to downshift, and use the brakes to bring the vehicle to a stop.
6. If a driveability problem still exists, repeat the sequence.

Some manufacturers recommend a specific relearn procedure, which is designed to establish good driveability during the relearn process. These procedures are especially important for all vehicles equipped with an electronically controlled converter or transmission. Always complete the procedure before returning the vehicle to the customer.

Chrysler recommends the following procedure for all vehicles equipped with 41TE and 42LE transaxles:

1. Warm the transaxle to normal operating temperatures by allowing the vehicle to idle.

2. Operate the vehicle and maintain a constant throttle opening during upshifts. This sets the transaxle into the upshift relearn process.

3. Accelerate the vehicle with a throttle opening of 10–50 degrees. Accelerating the vehicle from a stop to 45 mph with a moderate throttle is sufficient for this part of the procedure.

4. Operate the vehicle until the transaxle completes 1–2, 2–3, and 3–4 shifts at least 20 times.

5. Operate the vehicle with a vehicle speed of less than 25 mph and force downshifts with a wide-open throttle. Repeat this at least eight times.

6. Operate the vehicle at speeds above 25 mph and in fourth gear and force downshifts with a wide-open throttle. Repeat this at least eight times. The forced downshifts allow the computer to relearn kickdown operation.

Ford Motor Company also specifies a relearn procedure for some of their transmissions and transaxles. All the specific procedures begin with a fluid check and warming the ATF to normal temperatures. The procedure has two segments: an idle relearn and a drive relearn. The idle relearn procedure begins with starting the vehicle in P, with all accessories off and the parking brake set. Then move the gear selector to N and allow the engine to idle for one minute. Move the gear selector to D and again allow the engine to idle for one minute. After the idle relearn procedure is complete, the drive relearn procedure can begin. The drive relearn procedures are specific for the different transmissions.

If the vehicle is equipped with a 4R7OW or AX4S transmission, road test the vehicle. With the gear selector in overdrive, moderately accelerate the vehicle to 50 mph and hold that speed for a minimum of 15 seconds. The transmission should be in fourth gear at the end of that time. Then hold the speed with a steady throttle and lightly apply and release the brake for about 5 seconds. Then stop and park the vehicle with the gear selector in the D position for a minimum of 20 seconds. Repeat this procedure five times.

If the vehicle is equipped with an E4OD transmission, the drive procedure is as follows:

1. Put the gear selector in the D position and depress the O/D cancel button. (The LED should light.)

2. Moderately accelerate to 40 mph for a minimum of 15 seconds. The transmission should be in third gear at this time.

3. Hold the throttle steady and depress the O/D cancel button. (The LED should go out.) Accelerate to 50 mph. The transmission should be in fourth gear at this time.

4. Hold that speed for 15 seconds, then lightly apply and release the brake. Maintain 50 mph while applying the brake.

5. Then stop and park the vehicle with the gear selector in the D position for a minimum of 20 seconds.

6. Repeat this procedure five times.

The disassembly of both a transmission and transaxle are similar; therefore, disassembly, inspection, and reassembly guidelines for both transaxles and transmissions can be safely grouped together.

Disassembly

Once the transmission has been removed from the vehicle, certain things should be checked before beginning to disassemble the transmission. Measure the depth that the converter fits into the transmission case (Figure 5-13). To do this, hold a straightedge across the bell housing and measure into the pilot hub or a drive pad on the converter. Record this dimension and save it for use when installing the converter.

Also inspect the flexplate for warpage and cracks (Figure 5-14). Pay attention to the condition of the teeth on the starter ring gear. Also check the flexplate for excessive runout and elongated bolt holes. Replace the flexplate if there is evidence of damage. Also inspect the converter attaching bolts. Replace any worn bolts with suitable equivalents. While inspecting the flexplate, check the starter ring gear for looseness and damage. Pay special attention to the welds that secure the ring gear to the plate. It is common for these and the areas around them to crack.

Check the flexplate for cracks at the crankshaft mounting bolt holes. The best way to do this is to remove the plate and hold it up toward a bright light. If there are cracks, the light will shine through the cracks. If there is any flexplate damage, the flexplate should be replaced. Before reinstalling the flexplate, check the service manual to make sure the attaching bolts are either reusable or should be replaced. Also make sure the flexplate is installed in the correct direction and that all spacers are in place.

Check the drive hub of the torque converter. It should be smooth and show no signs of wear. Pay particular attention to the area the seal rides on. If the hub is worn, the torque converter should be replaced and the oil pump drive should be carefully inspected for scoring or other damage.

Before disassembling the transmission, check out the causes of any leakage. If there is evidence of leakage or leakage is the reason for the **teardown**, determine the path of the leakage before cleaning the area around the seals. At times, the leakage may be from sources other than the seal. Leakage could be from worn gaskets, loose bolts, cracked housings, or loose line connections.

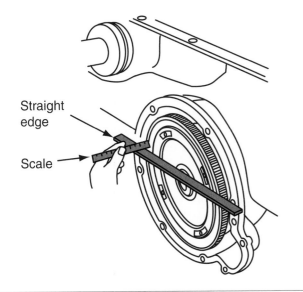

Figure 5-13 After the transmission and engine have been separated, measure and record the depth of the converter into the housing. This measurement will be used during reassembly to ensure that the torque converter and oil pump are properly assembled.

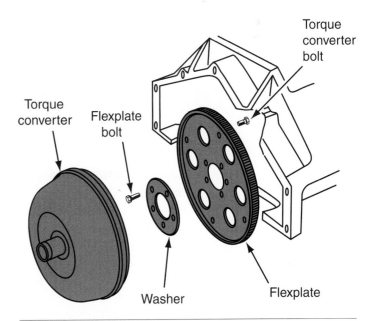

Figure 5-14 Carefully inspect the flexplate and torque converter mounting hardware.

Inspect the outside sealing area of the seal to see if it is wet or dry. If it is wet, see whether the oil is running out or if it is merely a lubricating film. Check both the inner and outer parts of the seals for wet oil, which means leakage.

When removing a seal, inspect the sealing surface, or lips, before cleaning it. Look for signs of unusual wear, warping, cuts and gouges, or particles embedded in the seal. On spring-loaded lip seals, make sure the spring is seated around the lip, and that the lip was not damaged when first installed. If the seal's lip is hardened, this was probably caused by heat from either the shaft or the fluid.

If the seal is damaged, check all shafts for roughness, especially at seal contact areas. Look for deep scratches or nicks that could have damaged the seal. Determine if a shaft spline, keyway, or burred end could have caused a nick or cut in the seal lip during installation. Inspect the bore into which the seal was fitted. Look for nicks and gouges that could create a path of oil leakage. A coarsely machined bore can allow oil to seep out through a spiral path. Sharp corners at the bore edges could have scored the metal case of the seal when it was installed. These scores can make a path for oil leakage.

Cleaning and Inspection

 SERVICE TIP: Record the model and serial number of the unit you are working on. You will need this information to purchase the correct parts during rebuild.

Classroom Manual
Chapter 5, page 137

Before disassembling the automatic transmission, care should be taken to clean away any dirt, undercoating, grease, or road grime on the outside of the case. This ensures that dirt will not enter the transmission during disassembly. Once the transmission is clean outside, you may begin the disassembly.

When cleaning automatic transmission parts, avoid the use of solvents, degreasers, and detergents that can decompose the friction composites used in a transmission. It is best not to attempt to clean the friction members, as this will damage the parts. Use compressed air to dry components; don't wipe down parts with a rag. The lint from a rag can easily destroy a transmission after it has been rebuilt.

There are many different ways to clean automatic transmission parts. Some rebuilding shops use a parts washing machine, which does an excellent job in a small amount of time, to thoroughly clean a transmission case, converter housing, and extension housing. These parts washers use hot water and a special detergent that are sprayed onto the parts as they rotate inside the cleaner. The key to good cleaning with these machines is to use a very small amount of soap with very hot water. Many rebuilders simply clean the parts in a mineral spirits tank, where the parts are brushed and hand cleaned. No matter what type of cleaning procedure is followed, the transmission and parts should be rinsed with water, then thoroughly dried with compressed air before reassembly.

Special Tools

Solvent

Scrub brush

Compressed air
   nozzle

⚠ **WARNING:** Always wear safety goggles when using compressed air to dry something. The air pressure can easily move dirt, metal, or other debris around the work area. If these get into your eye, they can cause permanent damage.

After the case is clean, remove the torque converter and carefully inspect it for damage. Check the converter hub for grooves caused by hardened seals. Also check the bushing contact area. To remove the converter, slowly rotate it as you pull it from the transmission, and have a drain pan handy to catch the fluid. It should come right out without binding. This is a good time to check the input shaft splines, stator support splines, and the converter's pump drive hub for any wear or damage. Converters with direct drive shafts should be checked to be sure that no excessive play is present at the drive splines of the shaft or the converter. If any play is found in the converter, the converter or the shaft must be replaced.

SERVICE TIP: It is best to mount the transmission on a bench mount made especially for working on transmissions, like the one shown (Figure 5-15). These mounts are available from many tool suppliers.

Position the transmission to perform an endplay check. The transmission endplay checks can provide the technician with information about the condition of internal bearings and seals, as well as clues to possible causes of improper operation found during the road test. These measurements will also determine the thickness of the thrust washer(s) during reassembly. Thrust-washer thickness sets the endplay of various components. Excessive endplay allows clutch drums to move back and forth too much, causing the transmission case to wear. If there is insufficient clearance, rapid wear may occur due to inadequate oil clearances. Assembled endplay measurements should be between minimum and maximum specifications, but preferably at the low end of the specifications. Photo Sequence 8 covers a typical procedure for checking endplay.

 SERVICE TIP: Record endplay measurements before disassembly and during reassembly.

Position the transmission so that the shaft centerline is vertical. This allows the weight of the internal components to load the shafts toward the rear of the transmission. Most GM transmissions can be checked by mounting a dial indicator to read the input shaft movement. Zero the dial indicator and lift upward on the input shaft. Most Chrysler and some Ford transmissions require removing the oil sump pan and valve body and prying the input shell upward. Chrysler and Ford transmissions may be measured horizontally (Figure 5-16), but you may want to center the output shaft with a slip yoke for more accurate readings.

Input shaft endplay is measured with a dial indicator. The dial indicator should be solidly clamped to the bell housing and the plunger positioned so that it is centered on the end of the input shaft. Move the input shaft in and out of the case and note the reading on the indicator. Compare this reading with the specifications for that transmission. If the endplay is incorrect, it should be corrected during assembly by installing a thrust washer, snap ring, or spacer of the correct thickness between the input shaft and the output shaft. Some transmissions require additional endplay checks during disassembly. The manufacturer's recommended procedure for checking endplay should always be followed.

Some transaxles may require more than one check. As most transaxles have their shaft centerlines to one side, the transaxle may need to be partially disassembled in order to gain access to the shafts.

It is possible that an output or final drive endplay check should be made on either a transmission or transaxle (Figure 5-17). The purpose of these settings is to provide long gear life and

Classroom Manual
Chapter 5, page 133

The 4R7OW and 4T60 are examples of transmissions that use a direct drive shaft.

Special Tools

Dial indicator mounting fixture
Dial indicator
Transmission stand

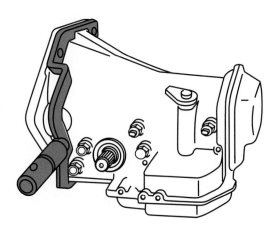

Figure 5-15 A typical transaxle holding fixture.

Photo Sequence 8
Measuring Input Shaft Thrust Play (Endplay)

P8-1 Position the transmission so the input shaft is facing up.

P8-2 Mount a dial indicator to the transmission case so the indicator's plunger can move with the input shaft as the shaft is pulled up and pushed down.

P8-3 Lightly pull up on the shaft, then push it down until it stops. Then zero the indicator.

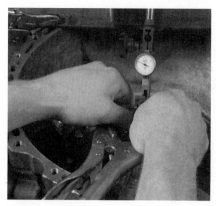

P8-4 Pull the shaft up until it stops and then read the indicator. This is the total amount of endplay.

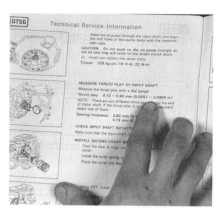

P8-5 Locate the endplay specifications in the service manual.

P8-6 Compare your measurements to the specifications.

to eliminate engagement "clunk." All measurements taken during this phase of teardown should be kept during overhaul and used as a guide during the rebuild. Having this information gives the technician places to look for worn parts during the teardown and helps obtain the necessary shims to correct undesired endplay. If the endplay is excessive, this may indicate that the thrust washers or bearings are worn or the sealing rings and grooves are worn. During rebuild, endplay settings should be kept to the minimum allowable amount. Once the endplay checks have been made and recorded, disassembly can begin.

> ✓ **SERVICE TIP:** When working on automatic transmissions, there is no such thing as being too clean. All dirt, grease, and other materials should be cleaned off any parts that are going to be reused.

A clean work area, clean tools, and an orderly sequence during teardown will help you avoid lost time and frustration during reassembly. As you remove the various assemblies from the transmission, place them on the workbench in the order that they were removed. The correct po-

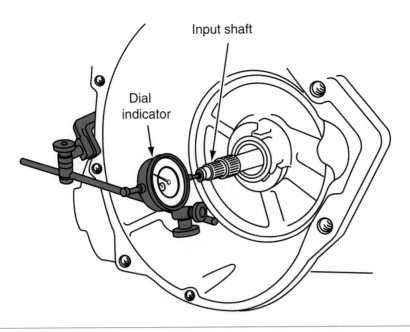

Figure 5-16 The endplay of the input shaft should be checked during disassembly and reassembly.

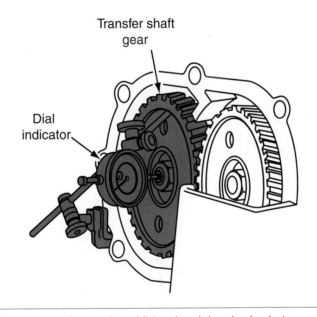

Figure 5-17 Some transaxles require additional endplay checks during assembly and disassembly.

sition of thrust washers, snap rings, and even bolts and nuts is of great importance and should be noted. Unless the transmission suffered a major breakdown or was overheated, the internal components of the transmission will be fairly clean due to the nature of the fluid. A light cleaning with mineral spirits or a simple blowing off with low pressure air is all that is necessary for cleaning most parts for inspection.

Each subassembly should be completely disassembled and the parts thoroughly cleaned and checked for signs of excessive wear, scoring, overheating, chipping, or cracking. If there is any question as to the condition of a part, replace it.

Basic Disassembly

Special Tools

Transmission mount

Mineral spirits

Snapring pliers

Dial indicator

Assortment of special tools as needed for particular transmission

The following is a general procedure for disassembling a transmission. Always refer to the manufacturer's recommended procedure for the specific transmission.

With the transmission mounted on a fixture, remove the oil pan. Carefully inspect the oil pan for types of foreign matter. An analysis of the type of matter may help you determine what parts need to be carefully examined.

Remove the bell housing from the transmission case, if it is not part of the case. Unscrew and remove all externally mounted solenoids and switches, such as the electrical kickdown solenoid. Remove the O-rings and seals for each after the part is removed.

If the transmission has a vacuum modulator, unscrew and remove it. Remove the speedometer drive assembly, with its gear and O-ring.

If the transmission has a solenoid assembly, unbolt and remove it from the valve body or case (Figure 5-18). Pay attention to the wiring harness and connectors. Check them for fraying, corrosion, or other damage. During disassembly, it is a good idea to set aside and save all the old gaskets. This will allow you to compare the new with the old, making sure the gaskets are correct.

✓ SERVICE TIP: The magnets inside electronic shift control solenoids will attract any ferrous metal that is floating around the inside of the transmission. Thoroughly clean or replace these solenoids during a transmission rebuild.

Unbolt and remove the valve body from the case. Then remove the manual valve from the valve body to prevent the valve from dropping out. Back off the servo piston stem locknut and

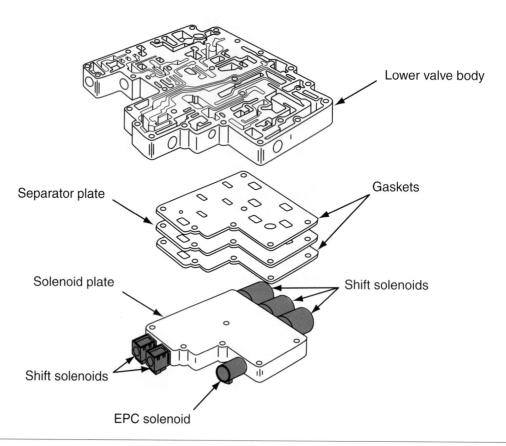

Figure 5-18 If the transmission has a solenoid assembly, unbolt it and remove it from the valve body or case.

snugly tighten the piston stem to prevent the front clutch drum from dropping out when removing the front pump.

 CAUTION: When removing the valve body, the steel check balls may fall out. Be prepared for this to happen, and take steps so that you don't lose them. Better yet, try to remove the valve body so the balls will remain in place while the valve body is separated from the case.

Using the correct puller, remove the front pump from the case (Figure 5-19). Lay this assembly to the side for further inspection.

Remove the front clutch thrust washer and bearing race. Now back off the front brake band servo piston stem to release the band. Remove the brake band strut and front brake band. The drum and band may be removed together.

Remove the front and rear clutch assemblies (Figure 5-20). Note the positions of the front pump thrust washers and rear clutch thrust washer, if the transmission has them.

Remove the rear clutch hub, front planetary carrier, and connecting shell. Note the positions of the thrust bearings and the front planetary carrier and thrust washer.

Remove the output shaft snap ring. It will often be easier to remove the snap ring if the carrier is removed first. Remove the carrier snap ring and remove the carrier. Now remove the output shaft snap ring.

Remove the rear connecting drum from the housing. Then using a screwdriver, remove the large retaining snap ring of the rear brake assembly. Tilt the extension housing upward and remove the rear brake assembly.

Then unbolt and remove the extension housing. Be careful not to lose the parking pawl, spring, and retaining washer. Pull out the output shaft, without the governor. Now remove the governor with its attachments (such as the oil distributor, thrust washer, and needle bearing assembly).

Remove the inner race of the one-way clutch, the thrust washer, piston return spring, and thrust ring. Using compressed air, remove the rear brake piston and front servo.

WARNING: Cover the servo with a rag to prevent ATF from blowing into your face and the servo piston from popping into your face.

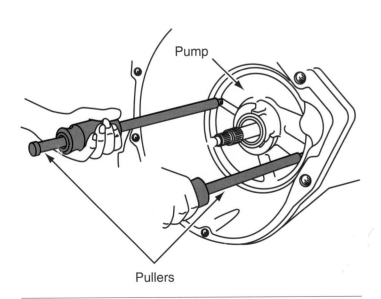

Figure 5-19 The proper tool should be used to pull the oil pump from the transmission housing.

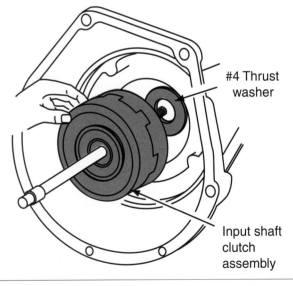

Figure 5-20 When removing clutch assemblies, be careful not to lose track of the placement of the thrust washers.

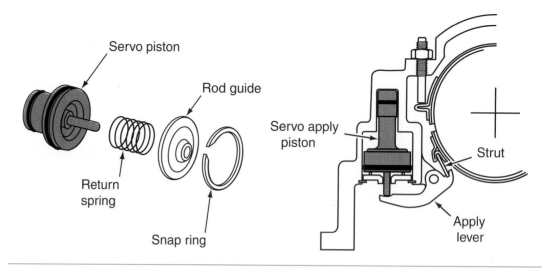

Figure 5-21 All servos and accumulators should be removed from their bores in the case and the bores carefully inspected.

The inspection, service, and testing of all hydraulic components are detailed in Chapter 7 of this manual.

Band servos and accumulators are pistons with seals in a bore held in position by springs and retaining rings. Remove the retaining rings and pull the assembly from its bore for cleaning (Figure 5-21). Check the condition of the piston and springs. Cast iron seal rings may not need replacement, but elastomer seals should always be replaced.

Once the transmission has been disassembled into its various subassemblies, each subassembly should be disassembled, cleaned, inspected, and reassembled.

Speedometer Drive Service

Speedometer gear problems normally result in an inoperative speedometer. However, this problem can also be caused by a faulty cable, drive gear, driven gear, or speedometer. A damaged drive gear can cause the driven gear to fail; therefore, both should be carefully inspected during a transmission overhaul (Figure 5-22). On some transmissions, the speedometer drive gear is a set of gear teeth machined into the output shaft. Inspect this gear. If the teeth are slightly rough, they can be cleaned up and smoothened with a file. If the gear is severely damaged, the entire output shaft must be replaced. Other transmissions have a drive gear that is splined to the output shaft, held in place by a clip (Figure 5-23), or driven and retained by a ball that fits into a depression in the shaft. If a clip is used, it should be carefully inspected for cracks or other damage.

The drive gear can be removed and replaced if necessary. The driven gear is normally attached to the transmission end of the speedometer cable (Figure 5-24). If the drive gear is dam-

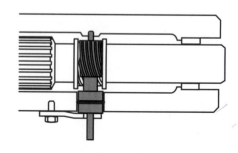

Figure 5-22 The relationship of the speedometer drive and driven gears.

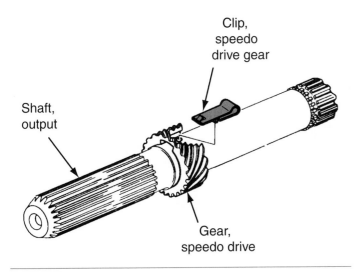

Figure 5-23 The speedometer drive gear is often retained to the output shaft by a clip. (Courtesy of General Motors Corporation, Service Operations Group)

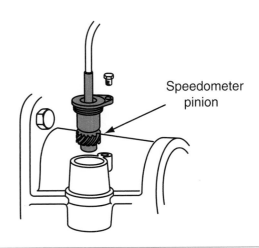

Figure 5-24 The speedometer cable and gear normally fit into a bore in the transmission case or extension housing.

aged, it is very likely that the driven gear will also be damaged. Most driven gears are made of plastic to reduce speedometer noise; therefore, they are weaker than most drive gears. Also check the retainer of the driven gear on the speedometer cable.

Transmission Case Service

The transmission case should be thoroughly cleaned and all passages blown out. Make sure all electrical components have been removed from the case before cleaning it. After the case has been cleaned, all bushings, fluid passages, bolt threads, clutch plate splines, and the governor bore should be checked. The passages can be checked for restrictions and leaks by applying compressed air to each one. If the air comes out the other end, there is no restriction. To check for leaks, plug off one end of the passage and apply air to the other. If pressure builds in that passage, there are probably no leaks in it.

Modern transmission cases are made of aluminum, primarily to save weight. Aluminum is a soft material that is susceptible to corrosion and can be deformed, scratched, cracked, or scored much more easily than cast iron. Special attention should be given to the following areas: clutch, oil pump, servo, and accumulator bores. All bores should be smooth to avoid scratching or tearing the seals. The servo piston could also hang up in a bore that is deeply scored. Check the fit of the servo piston in the bore without the seal, if possible, to be sure it has free travel. There should be no tight spots or binding over the whole range of travel. Any deep scratches or gouges that cause binding of the piston will require case replacement.

Case-mounted accumulator bores are checked the same as servo bores. The oil pump bore at the front of the case should be free of any scratches that would keep the O-ring from sealing the outer diameter of the pump to the front of the case. Case-mounted hydraulic clutch bores are prone to the same problems as servo bores. Look for any scratches or gouges in the sealing area that would affect the rubber seals. It is possible to damage these areas during disassembly, so be careful with tools used during overhaul.

Sealing surfaces of the case should be inspected for surface roughness, nicks, or scratches where the seals ride (Figure 5-25). Any problems found in servo bores, clutch drum bores, or governor support bores can cause pressure leakage in the affected circuit. Imperfections in steel or cast iron parts can usually be polished out with **crocus cloth**. Care should be taken not to disturb the

Classroom Manual
Chapter 5, page 124

Crocus cloth is a very fine polishing paper. It is designed to remove very little metal, so it is safe to use on critical surfaces. Never use regular sandpaper, as it will remove too much metal.

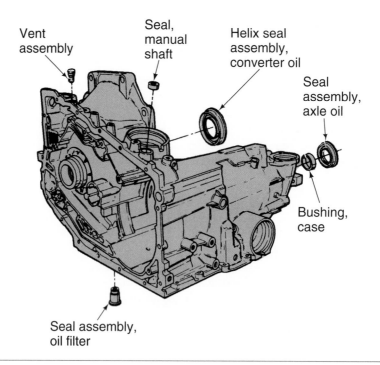

Vent assembly

Seal, manual shaft

Helix seal assembly, converter oil

Seal assembly, axle oil

Bushing, case

Seal assembly, oil filter

Figure 5-25 All sealing surfaces and bores of the transmission case should be carefully inspected for cracks, grooves, and scratches. (Courtesy of General Motors Corporation, Service Operations Group)

original shape of the bore. Under no circumstances should sandpaper be used. Sandpaper will leave too deep a scratch in the surface. Use the crocus cloth inside clutch drums to remove the polished marks left by the cast iron sealing rings. This will help the new rings rotate with the drum as designed. As a rule, all sealing rings, either cast iron or Teflon, are replaced during overhaul, as this gives the desired sealing surface required for proper operation.

Passages in the case guide the flow of fluid through the case. Although not common, porosity in this area can cause cross-tracking of one circuit to another. This can cause bind-up (two gears at once) or a slow bleed-off of pressure in the affected circuit, which can lead to slow burnout of a clutch or band. If this is suspected, try filling the circuit with solvent and watching to see if the solvent disappears or leaks away. If the solvent goes down, you will have to check each part of the circuit to find where the leak is. Be sure to check that all necessary check balls were in position during disassembly.

Check the valve body mounting area for warpage with a straightedge and feeler gauge. This should be done in several locations so that any crossover from one worm track to another is evident. If there is a slight burr or high spot, it can be removed by flat filing the surface.

A long straightedge should be laid across the lower flange of the case to check for distortion. Any warpage found here may result in circuit leakage, causing any number of hydraulically related problems. Case warpage should be less than 0.002 inch. Cases with bolted-in center supports, such as 3L80, 4L80, and E4OD models, should be checked with the support bolted in place. Cases are being made lighter and often will distort during service. This causes a pressure loss to some circuits, causing band or clutch burnouts.

SERVICE TIP: The Ford E4OD has shown this to be a problem with as much as 0.0625-inch warpage across the lower surface. Since the E4OD case costs approximately $625, replacing the case is rather expensive. It is possible to push the case back into the correct shape and then, with the center support bolted in, flat file the case to obtain the desired flatness. This type of repair is used in the aftermarket to save an otherwise good case and to reduce costs. This is not a Ford-authorized repair.

The oil passages in the case are commonly called worm tracks.

☑ **SERVICE TIP:** The oil pump gears may seize to the pump plate on Ford A4LD transmissions if the bell housing is warped around the bushing bore. Be extra careful when inspecting this bell housing as this fault can destroy a transmission rebuild.

Be sure to check all bell housing bolt holes and dowel pins. Cracks around the bolt holes indicate that the case bolts were tightened with the case out of alignment with the engine block. The case should be replaced if the following problems are present: broken worm tracking, cracked case at the oil pump to case flange, cracked case at clutch housing pressure cavity, ears broken off the bell housing, or oil pan flange broken off the case. Although it is possible to weld the aluminum case, it is not possible to determine if the repair will hold. A transmission case is very thin and welding may distort the case.

If any of the bolts that were removed during disassembly have aluminum on the threads, the thread bore is damaged and should be repaired. Thread repair entails installing a thread insert, which serves as new threads for the bolt, or retapping the bore. After the threads have been repaired, make sure you thoroughly clean the case to remove all metal filings.

The small screens found during teardown should be inspected for foreign material. These screens are used to prevent valve hang-up at the pressure regulator and governor and they must be in place. Most screens can be removed easily. Care should be taken when cleaning, because some cleaning solvents will destroy the plastic screens. Low air pressure (approximately 30 psi) can be used to blow the screens out in a reverse direction.

Bushings in a transmission case are normally found in the rear of the case and require the same inspection and replacement techniques as other bushings in the transmission. Always be sure that the oil passage to a pressure-fed bushing or bearing is open and free of dirt and foreign material. It does no good to replace a bushing without checking to be sure it has good oil flow.

Classroom Manual
Chapter 5, page 134

Vents are located in the pump body or transmission case and provide for equalization of pressures in the transmission. These vents can be checked by blowing low-pressure air through them, squirting solvent or brake cleaning spray through them, or by pushing a small-diameter wire through the vent passage. A clean, open passage is all you need to verify proper operation.

Extension Housing

Check the extension housing for cracks, especially around the case mounting surface and the pad that attaches to the transmission mount. Using a straightedge and feeler gauge set, check the flatness of the mating surface. Lay the straightedge across the surface and attempt to insert feeler gauges of various sizes between the bottom of the straightedge and the surface. If a 0.002-inch gauge fits under the straightedge, the surface may need to be resurfaced or the housing replaced. Minor problems may be corrected by filing down the surface. However, filing should only be done when it is necessary. To correctly file the surface, select the largest single-cut file available. Place the file across one end of the surface and pull or draw the file across the surface to the opposite end. Lift the file off the surface and place it back at its original position, then draw the file to the other end. Repeat this process until the surface is corrected. Never move the file from side to side.

Carefully inspect all threaded and non-threaded bores in the housing. All damaged threaded bores should be repaired by running a tap through the bore or by installing threaded inserts. If any condition exists that cannot be adequately repaired, the case should be replaced.

At the rear of the extension housing is the slip yoke bushing. This bushing will normally wear to one side due to loads imparted on it during operation. Oil-feed holes at this bushing must be checked to make sure oil can get to this bushing. The speedometer drive gear is often responsible for throwing oil back to the rear bushing. A sheared or otherwise inoperative speedometer gear could cause the extension housing bushing to fail. Always make sure this bushing is aligned correctly during replacement or premature failure can result.

Before installing the rear extension housing, assemble the parking pawl pin, washer, spring, and pawl, and any other assembly that is enclosed by the extension housing. Be sure they are assembled properly. Install the housing and tighten the bolts to specifications.

Gaskets

Classroom Manual
Chapter 5, page 137

Gaskets are used to create a seal between two flat surfaces. This seal must be able to prevent fluid leaks while undergoing changes in pressure and temperature. Whenever a transmission is overhauled, new gaskets should be used throughout the transmission. The only exception to this is the transmission oil pan gasket, which can be reused if it is in good condition. If the oil pan gasket is damaged or deformed in any way, do not reuse it. Most rebuilders don't reuse oil pan gaskets; they always replace them. The risk of the gasket not sealing is not worth the money saved.

While disassembling a transmission, keep all of the old gaskets. These will not be reinstalled, but will be used to select the correct gaskets for reassembly. Often transmission rebuilding kits come with a few different gaskets for the same part. To help select the correct gasket for the transmission and the part, compare the old gasket with the new ones. Obviously, the new gasket that matches the old one is the gasket that should be used.

When installing a gasket, make sure both surfaces are clean and flat. Any imperfections in a sealing surface should be corrected before installing a new gasket. These imperfections will prevent the gasket from providing a good seal. To remove minor scratches from a sealing surface, use a fine flat file. Make sure you don't remove more metal than is necessary. Also make sure the surface is flat after you have filed away the scratches.

To ensure flatness, spread a piece of medium grade (approximately 300 grit) sandpaper on a flat surface. Then set the sealing surface of the object on the paper. Move the object, in a figure 8 pattern, around the surface of the sandpaper. Apply an even amount of pressure on the object while doing this. When the sandpaper leaves a mark on all parts of the surface, the sealing surface is flat. Stop sanding it and clean it off.

Oil pans are typically made of stamped steel (Figure 5-26). The thin steel tends to become distorted around the attaching bolt holes. These distortions can prevent the pan from fitting tightly against the transmission case. To flatten the pan, place the mounting flange of the pan on a block of wood and flatten one area at a time with a ball-peen hammer.

Transmission gaskets should not be installed with any type of liquid adhesive or sealant, unless specifically noted by the manufacturer. If any sealer gets into the valve body, severe damage can result. Also, sealant can clog the oil filter. If a gasket is difficult to install, a thin coating of transmission assembly lube or petroleum jelly can be used to hold the gasket in place while assembling the parts.

One type of gasket that presents unique installation problems is the cork-type gasket. These gaskets tend to change shape and size with changes in humidity. If a cork gasket is slightly larger than it should be, soak it in water and lay it flat on a warm surface. Allow it to dry before installing it. If the gasket is slightly smaller than required, soak it in warm water prior to installation.

SERVICE TIP: Another way to make a cork gasket grow is to lay it on a flat, clean, and hard surface, and then, with a hammer, strike it all the way around until it is the correct size.

Whenever you are using a cork gasket to create a seal between two parts, make sure you properly tighten the two surfaces together. Tighten the attaching bolts or nuts, in a staggered pattern, to the specified torque so that the gasket material is evenly squeezed between the two surfaces. If too much torque is applied, the gasket may split.

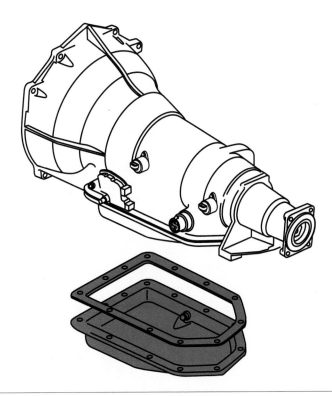

Figure 5-26 A transmission oil pan shown with its gasket. (Courtesy of General Motors Corporation, Service Operations Group)

Seals

Four types of seals are used in automatic transmissions: O-ring, square-cut, lip, and sealing rings (Figure 5-27). These seals are designed to stop fluid from leaking out of the transmission and from moving into another circuit of the hydraulic circuit. The latter ensures proper shifting of the transmission.

O-ring and square-cut seals are used to seal nonrotating parts (Figure 5-28). These seals can provide a good seal at high pressure points within the transmission. Common points that are sealed with these seals are oil pumps, servos, clutch pistons, speedometer drives, and vacuum modulators.

When installing a new O-ring or square-cut seal, coat the entire surface of the seal with assembly lube or petroleum jelly. Make sure you don't stretch or distort the seal while you work it into its holding groove. Some stretching may be necessary to work the seal over a shaft or fitting, but do not stretch it more than is needed. After a square-cut seal is installed, double-check to make sure it is not twisted. The flat surface of the seal should be parallel with the bore. If it is not, fluid will easily leak past the seal.

Lip seals are commonly used around rotating shafts and apply pistons. Lip seals that are used to seal a shaft typically have a metal flange around their outside diameter. The shaft rides on the lip seal at the inside diameter of the seal assembly. The rigid outer diameter provides a mounting point for the lip seal and is pressed into a bore. Once pressed into the bore, the outer diameter of the seal prevents fluid from leaking into the bore while the inner lip seal prevents leakage past the shaft.

Piston lip seals are set into a machined groove on the piston. These lip seals are not housed in a rigid metal flange. They are designed to be flexible and provide a seal while the piston moves up and down. While the piston moves, the lip flexes up and down. Any distortion or

The square-cut seal is commonly called a lathe-cut seal.

Most seal problems after a rebuild are caused by incorrect installation. Never rush through the installation process and always use the correct tool for the job.

Classroom Manual
Chapter 5, page 137

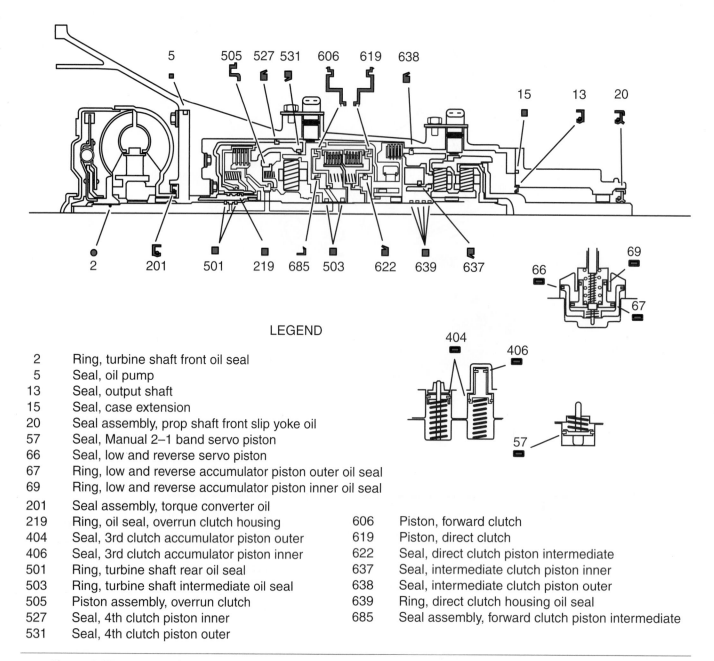

LEGEND

2	Ring, turbine shaft front oil seal
5	Seal, oil pump
13	Seal, output shaft
15	Seal, case extension
20	Seal assembly, prop shaft front slip yoke oil
57	Seal, Manual 2–1 band servo piston
66	Seal, low and reverse servo piston
67	Ring, low and reverse accumulator piston outer oil seal
69	Ring, low and reverse accumulator piston inner oil seal
201	Seal assembly, torque converter oil
219	Ring, oil seal, overrun clutch housing
404	Seal, 3rd clutch accumulator piston outer
406	Seal, 3rd clutch accumulator piston inner
501	Ring, turbine shaft rear oil seal
503	Ring, turbine shaft intermediate oil seal
505	Piston assembly, overrun clutch
527	Seal, 4th clutch piston inner
531	Seal, 4th clutch piston outer

606	Piston, forward clutch
619	Piston, direct clutch
622	Seal, direct clutch piston intermediate
637	Seal, intermediate clutch piston inner
638	Seal, intermediate clutch piston outer
639	Ring, direct clutch housing oil seal
685	Seal assembly, forward clutch piston intermediate

Figure 5-27 Location of the various seals used in today's transmissions. (Courtesy of General Motors Corporation, Service Operations Group)

damage to the seal will allow fluid to escape. If fluid escapes from around a piston, the piston will not move with the force and speed that it should.

The most important thing when installing a lip seal is to make sure the lip is facing in the correct direction. The lip should always be aimed toward the source of pressurized fluid. If installed backward, fluid under pressure will easily leak past the seal. Also remember to make sure the surfaces to be sealed are clean and not damaged.

Teflon or metal sealing rings are commonly used to seal servo pistons, oil pump covers, and shafts. These rings may be designed to provide for a seal, but they may also be designed to allow a controlled amount of fluid leakage. Leakage may be allowed to lubricate some shaft bushings in the transmission. Sealing rings are either solid or cut rings. Cut sealing rings are one of three designs: open-end, butt-end, or locking-end.

Locking-end sealing rings may be called locking rings if they are made of metal, or scarf-cut rings if they are made of Teflon.

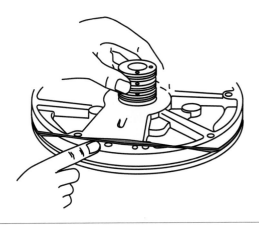

Figure 5-28 O-ring and square-cut seals are used to seal non-rotating parts, such as the front pump.

Solid sealing rings are commonly used in late-model transmissions. These rings are made of a Teflon-based material and are never reused. To remove them, simply (but carefully) cut the seal after it has been pried out of its groove. Installing a new solid sealing ring requires special tools. These tools allow you to stretch the seal while pushing it into position. Once in position, the tool holds the seal in place until the transmission components are assembled together. Never attempt to install a solid seal without the proper tools. Because these seals are soft, they are easily distorted and damaged.

Open-end sealing rings fit loosely into a machined groove. The ends of the rings do not touch when they are installed. This type of ring is typically removed and installed with a pair of snap ring pliers. The ring should be expanded just enough to move it off or onto the shaft.

Butt-end sealing rings are designed so that their ends butt or touch each other once the seal is in place. This type of seal can be removed with a small screwdriver. The blade of the screwdriver is used to work the ring out of its groove. To install this type of ring, use a pair of snap ring pliers and expand the ring to move it into position.

Locking-end rings may have hooked ends that connect or ends that are cut at an angle to hold the ends together. These seals are removed and installed in the same way as butt-end rings. After these rings are installed, make sure the ends are properly positioned and touching.

Reassembly and Testing

Before proceeding with the final assembly of all components, it is important to verify that the case, housing, and parts are clean and free from dust, dirt, and foreign matter (use an air gun). Have a tray available with clean ATF for lubricating parts. Also have ready a jar or tube of Vaseline for securing washers during installation.

> ☑ **SERVICE TIP:** When tightening any fastener that directly or indirectly involves a rotating shaft or other part, rotate that part during and after tightening to ensure that the part does not bind.

Coat all parts with the proper type of ATF. Soak bands and clutches in the fluid for at least 15 minutes before installing them. All new seals and rings should have been installed before beginning final assembly.

Automatic transmission parts distributors package part kits to fit the level of service to be performed. Two common kits are the resealing and reconditioning kits. The resealing kit includes all the necessary parts, such as gaskets and cast iron, Teflon, rubber, and metal rings and seals to reseal a transaxle or transmission. A reconditioning kit includes all the necessary seals, gaskets, filters, bushings, clutch friction and steel discs, and brake bands.

Appropriate set of
seal removers and
installers

Fresh ATF

Feeler gauge

Torque wrench

Classroom Manual
Chapter 5, page 141

Many Teflon seals
may look different
than the ones shown
in these figures and
they may require the
application of heat
for installation.
Always refer to the
service manual when
installing seals and
other transmission
components.

Seal Installation

All seals should be checked in their own bores. They should be slightly smaller or larger (± 3 percent) than their groove or bore. If a seal is not the proper size, find one that is. Do not assume that because a particular seal came with the overhaul kit, it is the correct one.

Never install a seal when it is dry. The seal should slide into position and allow the part it seals to slide into it. A dry seal is easily damaged during installation. Coat all seals with transmission assembly lube. Following are some guidelines for installing a seal in an automatic transmission:

1. Install only genuine seals recommended by the manufacturer of the transmission.

2. Use only the proper fluids as stated in the appropriate service manual.

3. Keep the seals and fluids clean and free of dirt.

4. Before installing seals, clean the shaft or bore area. Carefully inspect these areas for damage. File or stone away any burrs or bad nicks and polish the surfaces with a fine crocus cloth. Then clean the area to remove the metal particles. In dynamic applications, the sliding surface for the seal should have a mirror finish for best operation.

5. Lubricate the seal, especially any lips, with transmission assembly lube to ease installation.

6. All metal sealing rings should be checked for proper fit. Since these rings seal on their outer diameter, the seal should be inserted in its bore and should feel tight there. If the seal has some form of locking ends, these should be interlocked prior to trying the seal in its bore (Figures 5-29 and 5-30). The fit of the sealing rings in their shaft groove should also be checked. If the ring can move laterally in the groove, the groove is worn and this will cause internal fluid leaks. To check the side clearance of the ring, place the ring into its groove and measure the clearance between the ring and the groove with a feeler gauge. Typically, the clearance should not exceed 0.003 inch.

7. While checking the clearance, look for nicks in the grooves and for evidence of groove taper or stepping. If the grooves are tapered or stepped, the shaft will need to be replaced. If there are burrs or nicks in the grooves, they can be filed away.

8. Any distorted or undersized sealing rings should be replaced.

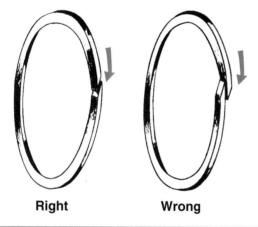

Figure 5-29 Installation of Teflon oil seal rings. (Courtesy of General Motors Corporation, Service Operations Group)

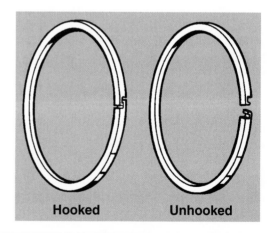

Figure 5-30 Installation of cast iron oil seal rings. (Courtesy of General Motors Corporation, Service Operations Group)

9. Always use the correct driver when installing a seal (Figure 5-31) and be careful not to damage the seal during installation. When possible, press the seal into position. This usually prevents the garter spring from moving out of position during installation.

Transmission Reassembly

Prior to reassembly, check the service manual for the order and procedure for installing the different components and units. Also be sure to examine all thrust washers carefully and coat them with petroleum jelly before placing them in the housing. Install the thrust ring, piston return spring, thrust washer, and one-way clutch inner race into the case. Align and start the bolts into the inner race from the rear of the case. Torque the bolts to specifications.

CAUTION: Check and verify that the return spring is centered onto the race before tightening.

Lubricate and install the rear piston into the case. After determining the correct number of friction and steel plates, install the steel dished plate first. Make sure the steel disc is facing in the correct direction, then install the steel and friction plates, the retaining plate, and snap ring (Figure 5-32).

SERVICE TIP: The ends of snap rings are slightly tapered. When installing snap-rings, make sure the taper faces up or toward the outside of the shaft. This will provide a good gripping surface for the snap ring pliers during installation and for removal the next time the transmission is taken apart.

Using a suitable blow gun with a rubber tapered tip, air check the rear brake operation. After the rear brake has been completely assembled, measure the clearance between the snap ring and the retainer plate. Select the proper thickness of retaining plate that will give the correct ring-to-plate clearance if the measurement does not meet the specified limits.

Slide the governor distributor assembly onto the output shaft from the front of the shaft. Install the shaft and governor distributor into the case, using care not to damage the distributor rings.

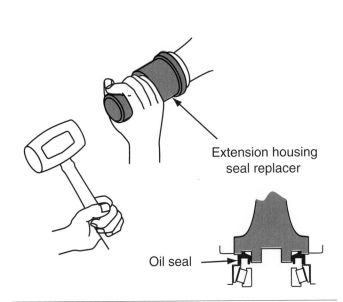

Extension housing
seal replacer

Oil seal

Figure 5-31 Use a special driver to install an oil seal. Note how the tool prevents damage to the seal.

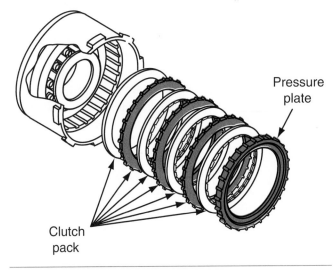

Pressure
plate

Clutch
pack

Figure 5-32 Be sure the steel disc is facing in the correct direction, then install the steel and friction plates, the retaining plate, and snap ring when assembling a clutch pack.

On some models, the output shaft, bearing, and appropriate gauging shims are placed into the transmission housing. The output shaft washer and bolt are then installed. While holding the output shaft and gear assembly, torque the output shaft nut to specifications. Then install a dial indicator and check the travel of the output shaft as it is pushed and pulled. Remove the gauging shims and install the correct sizes of service shims, output shaft gear, washer, and nut. Torque the output shaft nut to specifications. Using an inch-pound torque wrench, check the turning torque of the output shaft and compare this reading to specifications.

Place the small thrust washer on the pilot end of the transaxle output shaft. Then place the rear clutch assembly, front clutch drum, turbine shaft, and thrust washer into the housing. Locate and align the rear clutch over its hub. Gently move the rear clutch and turbine shaft around, rotating the assembly to engage the teeth of the friction discs with the rear clutch hub. Align the direct clutch assembly over the front clutch hub. Move the input shaft back and forth, rotating it so the front clutch friction disks engage with the front clutch hub.

Position the thrust washer to the back of the rear planetary carrier. Install the rear planetary carrier and thrust washer into the housing to engage the rear planetary ring gear. Install the front thrust washer and the drive shell assembly, engaging the common sun gear with the planetary pinions in the rear planetary carrier. Assemble the front planetary gear assembly into the front planetary ring gear. Make sure the planetary pinion gear shafts are securely locked to the planetary carrier.

Install the one-way sprag into the one-way clutch outer race with the arrow on the sprag facing the front of the transmission. Some overrunning clutches do not have an arrow to use for guidance. For these, the manufacturers recommend that an index mark be made on the clutch and the transmission case (Figure 5-33). This index mark should be made at the non-threaded bore in the clutch. The bores for the mounting of the cam are slightly countersunk on one side. The clutch needs to be installed with this side facing rearward or toward the piston retainer. Install the connecting drum with sprag by rotating the drum clockwise, using a slight pressure and wobbling to align the plates with hub and sprag assembly. The connecting drum should now be free to rotate clockwise only. This check will verify that the sprag is correctly installed and operative. Now install the rear internal gear and the shaft's snap ring.

Secure the thrust bearing with petroleum jelly and install the rear planet carrier and the snap ring.

SERVICE TIP: This snap ring may be thinner than the clutch drum snap ring, so be sure you are using the correct size. Should you have insufficient space to install the snap ring into the drum groove, pull the connecting drum forward as far as possible. This will give you sufficient groove clearance to install the drum snap ring.

Assemble the front and rear clutch drum assemblies together and lay them flat on the bench. Be sure the rear hub thrust bearing is properly seated, then measure from the face of the front clutch drum to the top of the thrust bearing. Install the thrust washer and pump front bearing race to the pump. Then measure from the pump shaft (bearing race included) to the race of the thrust washer. Normally, the difference in measurements should be about 0.02 inch. If the thrust washer is not within the limits, replace it with one of the correct thickness.

Total endplay should now be checked. Set the transmission case on end, front end up. Be sure the thrust bearings are secure with petroleum jelly. Pick up the complete front clutch assembly and install it into the case. Be sure all parts are seated before proceeding with the measurement. Using a dial indicator or caliper, measure the distance from the rear hub thrust bearing to the case. Next measure the pump with the front bearing race and gasket installed. Tolerance should fall within specifications. If the difference between the measurements is not within tolerance, select the proper size front bearing race. If it is necessary to change the front bearing race, be sure to change the front clutch thrust washers the same amount.

Install the brake band servo. Use extreme care not to damage the O-rings. Lubricate around the seals. Then install and torque the retainer bolts to specifications. Loosen the piston stem. In-

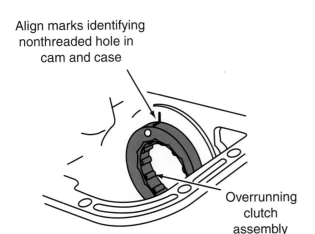

Align marks identifying
nonthreaded hole in
cam and case

Overrunning
clutch
assembly

Figure 5-33 For proper installation, an overrunning clutch should be indexed to the case prior to removal.

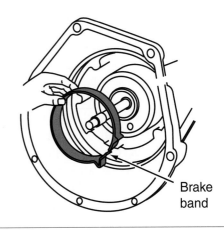

Brake
band

Figure 5-34 Install the brake band and strut and finger tighten the band servo piston stem just enough to keep the band and strut snug or from falling out.

stall the brake band and strut (Figure 5-34) and finger tighten the band servo piston stem just enough to prevent the band and strut from falling out. Do not adjust the band at this time. Air check for proper performance.

Place some petroleum jelly in two or three spots around the oil pump gasket and position it on the transaxle housing. Next align the pump and install it with care (Figure 5-35). Tighten the pump attaching bolts to specifications in the specified order. After the bolts are tight, check the rotation of the input shaft. If the shaft does not rotate, disassemble the transmission to locate the misplaced thrust washer.

Before installing the bell housing, check the bolt hole alignment. Install the bell housing and torque the retaining bolts to specifications.

Now adjust the band after you check to make sure that the brake band strut is correctly installed. Torque the piston stem to specifications. Back off two (or the number specified by the manufacturer) full turns and secure with a locknut. Tighten the locknut to specifications.

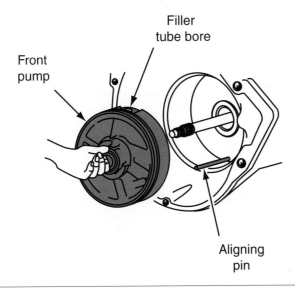

Filler
tube bore

Front
pump

Aligning
pin

Figure 5-35 When installing the oil pump, make sure it is in the proper position. Some manufacturers recommend the use of an alignment pin to aid in this process.

Before proceeding with the installation of the valve body assembly, it is good practice to perform a final air check of all assembled components. This will ensure that you have not overlooked the tightening of any bolts or damaged any seals during assembly.

Assemble the parking pawl assembly. Place the assembly into its position and install the extension housing with a new gasket, then tighten the attaching bolts to the proper specifications.

On transaxles, the differential assembly should be disassembled, cleaned, inspected, and reassembled. After it has been reassembled, measure its endplay with gauging shims. Then select a shim thick enough to correct the endplay. After installing the proper shims, measure the differential turning torque with an inch-pound torque wrench. Increase or decrease the shim thickness to provide the correct turning torque.

Install the valve body. Be sure the manual valve is in alignment with the selector pin. Tighten the valve body attaching bolts to the specified torque.

CAUTION: Be certain that the proper length bolts are installed in the related depth holes. The lengths of the bolts vary.

Before installing the vacuum modulator valve, it is good practice to measure the depth of the hole in which it is inserted. This measurement determines the correct rod length to ensure proper performance. Refer to the service manual to determine the correct rod length based on your measurements. You should note that the actual rod size is slightly longer than the measurement taken. If you do not have the correct chart, it is fairly safe to simply add 0.07 inch to the measurement taken.

Before installing the solenoids, check to verify that they are operating properly. Connect a solenoid checker to each solenoid. When the solenoids are activated, you should hear the solenoid click on. You can also check the integrity of a solenoid with an ohmmeter. If the solenoid is good, install it. If the solenoid does not check out, replace it.

Install the kickdown switch. Again check the operation of the switch. This is best done by connecting an ohmmeter across the switch and moving the switch through its different settings.

Before installing the oil pan, check the alignment and operation of the control lever and parking pawl engagement. Make a final check to be sure all bolts are installed in the valve body. Install the oil pan with a new gasket. Torque the bolts to specifications.

Lubricate the oil pump's lip seal and the converter neck before installing the converter. Also put a coat of petroleum jelly on the seal's garter spring. This will prevent the spring from being knocked loose while installing the torque converter. Install the converter, making sure that the converter is properly meshed with the oil pump drive gear.

SERVICE TIP: It is a good idea to partially prefill all torque converters before installing them in the vehicle.i

The transmission is now ready for installation into the vehicle. Use the reverse of the removal procedures. Never bolt the converter onto the flexplate and then try to install the transmission onto the converter. Remember to follow the correct fluid filling procedure and check the service manual for transmission relearn instructions.

CASE STUDY

A customer brought a late-model GM vehicle with a 2.8 L engine into a transmission rebuilding shop to have the transmission rebuilt. This was the second time in three years that the transmission went out.

The technician diagnosed the problem as low pump pressure in all gears. A faulty pump was suspected. While pulling out the transmission, she noticed that the flexplate was cracked. This somewhat verified the diagnosis, since a faulty or excessively worn oil pump body bushing would cause this problem. She assumed that by replacing the oil pump and the flexplate, the customer would be happy and have many years of troublefree transmission service.

After completing the repairs and during the installation of the transmission, the technician noticed the dowel pins. There was nothing wrong with them, but they made her remember something she had read in a technical bulletin. She researched the TSBs, and found the one she was looking for. The two 15.0-mm dowel pins at the rear of the engine needed to be replaced with two 19.0-mm dowels. This extra length gave more rigidity to the mating of the transmission and the engine.

The slight shifting of the transmission against the engine caused both the oil pump bushings and the flexplate to go bad. Had the technician not read the TSB before, she would have replaced the flexplate and the oil pump only to have the customer come back again for the same problem.

Terms to Know

Crocus cloth   Dowels   Teardown

ASE Style Review Questions

1. *Technician A* says before tearing down a transmission, you should inspect the material trapped in the fluid filter.
   Technician B says installing an overhaul kit will take care of all the problems with a transmission.
   Who is correct?
   A. A only   **C.** Both A and B
   B. B only   **D.** Neither A nor B

2. While discussing endplay checks:
   Technician A says these checks should be taken before the transmission is disassembled.
   Technician B says these checks should be taken after the transmission is reassembled.
   Who is correct?
   A. A only   **C.** Both A and B
   B. B only   **D.** Neither A nor B

3. *Technician A* says abnormal noises from a transmission will never be caused by faulty clutches or bands.
   Technician B says abnormal noises from a transmission are typically caused by a faulty torque converter.
   Who is correct?
   A. A only   **C.** Both A and B
   B. B only   **D.** Neither A nor B

4. An output shaft or final drive endplay check should be made on either a transmission or transaxle.
   Technician A says this is done to provide long gear life and to eliminate engagement "clunk."
   Technician B says endplay should be measured during teardown which helps you obtain the correct shims necessary to correct undesired endplay.
   Who is correct?
   A. A only   **C.** Both A and B
   B. B only   **D.** Neither A nor B

5. *Technician A* says abnormal noises and vibrations can be caused by damaged or worn gears, and by damaged clutches and/or bands.
   Technician B says abnormal noises can be caused by a bad oil pump or contaminated fluid.
   Who is correct?
   A. A only   **C.** Both A and B
   B. B only   **D.** Neither A nor B

6. *Technician A* says vibration problems can be caused by an unbalanced torque converter.
   Technician B says vibration problems can be caused by a faulty output shaft.
   Who is correct?
   A. A only   **C.** Both A and B
   B. B only   **D.** Neither A nor B

7. *Technician A* says once the transmission is removed and before it is disassembled, it should be cleaned in a solvent tank.
   Technician B says no matter what type of cleaning procedure is followed, the transmission and parts should be rinsed with water and then thoroughly dried with compressed air before reassembly.
   Who is correct?
   A. A only
   B. B only
   C. Both A and B
   D. Neither A nor B

8. While checking a transmission's vent:
   Technician A applies a vacuum to it and says that if the vent leaks, it should be replaced.
   Technician B runs ATF through the vent and says that if the vent cannot hold fluid, it must be replaced.
   Who is correct?

   A. A only
   B. B only
   C. Both A and B
   D. Neither A nor B

9. *Technician A* says the seals should be kept clean and free of dirt, before and during installation.
   Technician B says most seals should be installed dry.
   Who is correct?
   A. A only
   B. B only
   C. Both A and B
   D. Neither A nor B

10. While discussing computer relearn procedures:
    Technician A says this type of computer strategy is designed to optimize driveability.
    Technician B says the computer on some cars needs to learn about the vehicle before it can control the systems properly.
    Who is correct?
    A. A only
    B. B only
    C. Both A and B
    D. Neither A nor B

ASE Challenge Questions

1. The customer complains of a buzzing noise that increases with an increase in engine speed.
   Technician A says cavitation in the oil pump may be the noise source.
   Technician B says a leaking band apply servo may be the source.
   Who is correct?
   A. A only
   B. B only
   C. Both A and B
   D. Neither A nor B

2. *Technician A* says all shafts that extend out of the transmission should have the endplay measured before transmission disassembly.
   Technician B says to use a straightedge and ruler to measure shaft endplay during reassembly.
   Who is correct?
   A. A only
   B. B only
   C. Both A and B
   D. Neither A nor B

3. *Technician A* says compressed air may be used to remove some servos.
   Technician B says compressed air is used to check clutch and servo action during transmission reassembly.
   Who is correct?
   A. A only
   B. B only
   C. Both A and B
   D. Neither A nor B

4. *Technician A* says the best way to diagnose noise problems is to take a road test and pay attention to the operating gear, speed, and the conditions at which the noise occurs.
   Technician B says noise diagnosis should begin with putting the vehicle in gear and apply the brake. If the noise or vibration is not evident, the problem is undoubtedly in the driveline or output of the transmission.
   Who is correct?
   A. A only
   B. B only
   C. Both A and B
   D. Neither A nor B

5. After completing a transmission overhaul:
   Technician A checks the service manual for the relearn procedures before taking the vehicle out on a road test.
   Technician B connects a scan tool to the electronic control system and checks for DTCs before operating the transmission.
   Who is correct?
   A. A only
   B. B only
   C. Both A and B
   D. Neither A nor B

Job Sheet 15

15

Name _____ Date _____

Prepare to Remove a Transaxle or Transmission

Upon completion of this job sheet, you should be able to describe the procedures that must be followed in order to remove a transaxle or transmission from a vehicle.

ASE Correlation

This job sheet is related to the ASE Automatic Transmission and Transaxle Test's Content Area: *Off-Vehicle Transmission and Transaxle Repair, Removal, Disassembly, and Assembly*.
Task: Remove and replace transmission/transaxle; inspect engine core plugs, transmission dowel pins, and dowel pin holes.

Tools and Materials

A vehicle with an automatic transmission   Drain pan
Hoist   Droplight or good flashlight
Transmission jack   Service manual
Engine support fixture

Describe the vehicle being worked on:

Year _____ Make _____ VIN _____

Model _____

Model and type of transmission _____

Procedure

This job sheet is designed to allow you to take a good look at what would be involved in removing a transmission from an assigned vehicle.

1. Refer to the service manual and look for any precautions or special instructions that relate to the removal of a transmission from this vehicle. Describe them below:

_____

_____

2. Before beginning to remove the transmission from this vehicle, what is the first thing that should be disconnected? Why?

_____

_____

3. Look under the hood and identify everything that should be removed or disconnected from above before raising the vehicle on the hoist. List those items below:

_____

_____

_____

Prepare to Remove a Transaxle or Transmission (continued)

_____

_____

_____

_____

☐

4. Raise the vehicle to a comfortable working height.

5. Carefully examine the area around the transmission and identify everything that should be disconnected or removed before unbolting the transmission from the engine. List those items below:

_____

_____

_____

_____

_____

_____

_____

_____

_____

_____

Instructor's Response _____

_____

_____

Job Sheet 16

16

Name _____ Date _____

Disassemble and Inspect a Transmission

Upon completion of this job sheet, you should be able to disassemble and inspect the major components of a transmission.

ASE Correlation

This job sheet is related to the ASE Automatic Transmission and Transaxle Test's Content Area: *Off-Vehicle Transmission and Transaxle Repair, Removal, Disassembly, and Assembly.* *Task*: Disassemble, clean, and inspect.

Tools and Materials

An automatic transmission or transaxle on a bench   Lint-free shop towels
Compressed air and air nozzle   Service Manual
Supply of clean solvent

Describe the transmission being worked on:

Model and type of transmission _____

Year _____ Make _____ VIN _____

Model _____

Procedure

Task Completed

1. Disassemble the transmission into major units. Set each unit aside until this job sheet refers to it. Describe any problems you encountered while disassembling the transmission. Be sure to follow the procedures given in the appropriate service manual while taking the transmission apart.   ☐

_____

_____

_____

_____

2. Clean the planetary gearset in clean solvent and allow it to air dry. If compressed air is used to help the drying process, firmly hold the pinion gears to prevent the bearings from moving.   ☐

3. Clean all thrust washers, thrust bearings, and bushings in clean solvent and allow them to air dry.   ☐

4. Place one member of the planetary gearset on the sun gear. Rotate the gearset slowly. Do the same for the other parts of the gearset. Describe the feel of the gears' rotation:

_____

_____

_____

Disassemble and Inspect a Transmission (continued)

5. Inspect each member of the gearset for damaged or worn gear teeth. Describe their condition:

_____

_____

6. Inspect the splines of each gearset member and describe their condition:

_____

_____

_____

☐

7. Remove any buildup of material or dirt that may be present between the teeth of the gears.

8. Look at the carrier assemblies and check for cracks or other damage. Describe your findings.

_____

_____

_____

9. Inspect the entire gearset for signs of discoloration. Describe your findings and explain what is indicated by them.

_____

_____

_____

10. Check the output shaft and its splines for wear, cracking, or other damage. Describe their condition.

_____

_____

_____

11. Carefully inspect the driving shells and drive lugs for wear and damage. Describe their condition.

_____

_____

12. Summarize the condition of the planetary gearsets and shaft.

_____

_____

_____

_____

Disassemble and Inspect a Transmission (continued)

13. Check the thrust washers and bearings for damage, excessive wear, and distortion. Describe their condition.

_____

_____

_____

14. Check all of the bushings for signs of wear, scoring, and other damage. Describe their condition.

_____

_____

_____

_____

15. Check the shafts that ride in the bushings. Describe their condition.

_____

_____

_____

16. Summarize the condition of the thrust washers, bearings, and their associated shafts.

_____

_____

_____

Instructor's Response _____

_____

_____

Job Sheet 17

Name _____ Date _____

Reassembly of a Transmission/Transaxle

Upon completion of this job sheet, you should be able to properly measure and set endplay and preload during the assembly of a transmission or transaxle.

ASE Correlation

This job sheet is related to the ASE Automatic and Transaxle Test's Content Areas: *Off-Vehicle Transmission Repair; Removal, Disassembly, and Assembly and Gear Train, Shafts, Bearings, and Case.*

Tasks: Assemble transmission/transaxle and measure endplay or preload; determine necessary action.

Tools and materials

Basic hand tools   Air nozzle
Special tools for the transmission   Micrometer
Clean ATF in a tray   Dial indicator
Petroleum jelly

Describe the transmission that was assigned to you and the vehicle it came out of:

Year _____ Make _____ Model_____

VIN _____ Engine size and type _____

Model and type of transmission _____

Procedure

Task Completed

1. Before proceeding with the final assembly of all components, it is important to verify that the case, housing, and parts are clean and free from dust, dirt, and foreign matter. ☐

2. Coat all parts with the proper type of ATF. Soak bands and clutches in the fluid for at least 15 minutes before installing them. All new seals and rings should have been installed before beginning final assembly. ☐

3. Examine all thrust washers carefully and coat them with petroleum jelly before placing them in the housing. ☐

4. Install the thrust ring, piston return spring, thrust washer, and one-way clutch inner race into the case. Align and start the bolts into the inner race from the rear of the case. Torque the bolts to specifications. What are the specifications? ☐

   _____

5. Lubricate and install the rear piston into the case. ☐

6. After determining the correct number of friction and steel plates, install the steel dished plate first, then the steel and friction plates, and finally the retaining plate and snapring. How many steels did you have?

   _____

Reassembly of a Transmission/Transaxle (continued)

7. Using a suitable blowgun with a rubber tapered tip, air check the rear brake operation. What were the results?

_____

_____

8. After the rear brake has been completely assembled, measure the clearance between the snap ring and the retainer plate. Select the proper thickness of retaining plate that will give the correct ring to plate clearance if the measurement does not meet the specified limits. What were the results?

_____

_____

☐ 9. Slide the governor distributor assembly onto the output shaft from the front of the shaft; install the shaft and governor distributor into the case using care not to damage the distributor rings.

10. On some models, the output shaft, bearing, and appropriate gauging shims are placed into the transmission housing. The output shaft washer and bolt is then installed. While holding the output shaft and gear assembly, torque the output shaft nut to specifications. What are the specifications?

_____

11. Install a dial indicator and check the travel of the output shaft as it is pushed and pulled. What were the results?

_____

_____

☐ 12. Remove the gauging shims and install the correct sizes of service shims, output shaft gear, washer, and nut.

13. Torque the output shaft nut to specifications. Using an inch-pound torque wrench, check the turning torque of the output shaft and compare this reading to specifications. What were the results?

_____

_____

☐ 14. Place the small thrust washer on the pilot end of the transaxle output shaft.

☐ 15. Place the rear clutch assembly, front clutch drum, turbine shaft, and thrust washer into the housing.

☐ 16. Locate and align the rear clutch over its hub. Gently move the rear clutch and turbine shaft around, rotating the assembly to engage the teeth of the friction disks with the rear clutch hub. Align the direct clutch assembly over the front clutch hub. Move the input shaft back and forth, rotating it so the front clutch friction disks engage with the front clutch hub.

☐ 17. Position the thrust washer to the back of the rear planetary carrier.

Reassembly of a Transmission/Transaxle (continued)

18. Install the rear planetary carrier and thrust washer into the housing to engage the rear planetary ring gear. ☐

19. Install the front thrust washer and the drive shell assembly, engaging the common sun gear with the planetary pinions in the rear planetary carrier. ☐

20. Assemble the front planetary gear assembly into the front planetary ring gear. Make sure the planetary pinion gear shafts are securely locked to the planetary carrier. ☐

21. Install the one-way sprag into the one-way clutch outer race with the arrow on the sprag facing the front of the transmission. ☐

22. Install the connecting drum with sprag by rotating the drum clockwise using a slight pressure and wobbling to align the plates with hub and sprag assembly. The connecting drum should now be free to rotate clockwise only. This check will verify that the sprag is correctly installed and operative. What were the results? ☐

   _____

   _____

23. Install the rear internal gear and the shaft's snap ring. ☐

24. Secure the thrust bearing with petroleum jelly and install the rear planet carrier and the snap ring. ☐

25. Assemble the front and rear clutch drum assemblies together and lay them flat on the bench. ☐

26. Make sure the rear hub thrust bearing is properly seated, then measure from the face of the front clutch drum to the top of the thrust bearing. What did you measure and how does it compare to the specifications?

   _____

   _____

   _____

27. Install the thrust washer and pump front bearing race to the pump. ☐

28. Measure from the pump shaft (bearing race included) to the race of the thrust washer. If the thrust washer is not within the limits, replace it with one of the correct thickness. What were the results?

   _____

   _____

29. Total endplay should now be checked. Set the transmission case on end, front end up. ☐

30. Make sure the thrust bearings are secure with petroleum jelly. Pick up the complete front clutch assembly and install it into the case. Be sure all parts are seated before proceeding with the measurement. Using a dial indicator or caliper, measure the distance from the rear hub thrust bearing to the case. What were the results?

   _____

   _____

31. Measure the pump with the front bearing race and gasket installed. Tolerance should fall within specifications. If the difference between the measurements is not within tolerance, select the proper size front bearing race. If it is necessary to change the front bearing race, be sure to change the front clutch thrust washers the same amount. What were the results?

_____

_____

_____

32. Install the brake band servo. Use extreme care so as not to damage the O-rings. Lubricate around the seals.

33. Install and torque the retainer bolts to specifications. What are the specifications?

_____

34. Loosen the piston stem.

35. Install the brake band strut and finger tighten the band servo piston stem just enough to keep the band and strut snug or from falling out. Do not adjust the band at this time.

36. Air check for proper performance. What were the results?

37. Place some petroleum jelly in two or three spots around the oil pump gasket and position it on the transaxle housing.

38. Align the pump and install the pump with care.

39. Tighten the pump attaching bolts to specifications in the specified order. What are the specifications and the specified order?

_____

40. Check the rotation of the input shaft. If the shaft does not rotate, disassemble the transmission to locate the misplaced thrust washer. What were the results?

_____

_____

41. Install the bell housing and torque the retaining bolts to specifications. What are the specifications?

_____

42. Adjust the band, after you check to make sure that the brake band strut is correctly installed.

43. Torque the piston stem to specifications. What are the specifications?

_____

44. Back off two (or the number specified by the manufacturer) full turns and secure with the locknut. What are the specifications?

_____

45. Tighten the locknut to specifications. What are the specifications?

_____

Reassembly of a Transmission/Transaxle (continued)

46. Before proceeding with the installation of the valve body assembly, it is good practice to perform a final air check of all assembled components. This will ensure that you have not overlooked the tightening of any bolts or damaged any seals during assembly. What are the specifications?

_____

47. Assemble the parking pawl assembly. Place the assembly into its position and install the extension housing with a new gasket, then tighten the attaching bolts to the proper specifications. What are the specifications?

_____

48. On transaxles, the differential assembly should be disassembled, cleaned, inspected, and reassembled. After it has been reassembled, measure its endplay with gauging shims. Then select a shim thick enough to correct the endplay. What are the specifications?

_____

49. After installing the proper shims, measure the differential turning torque with an inch-pound torque wrench. What were your results?

_____

_____

50. Install the valve body. Be sure the manual valve is in alignment with the selector pin. Tighten the valve body attaching bolts to the specified torque. What are the specifications?

_____

51. Before installing the vacuum modulator valve, it is good practice to measure the depth of the hole in which it is inserted. This measurement determines the correct rod length to ensure proper performance. Refer to the service manual to determine the correct rod length based on your measurements. ☐

52. Before installing the kickdown solenoid or other solenoids, check to verify that they are operating properly. Connect the solenoid to a 12-volt source and ground the other terminal. What happened?

_____

_____

_____

53. Install the kickdown switch. ☐

54. Before installing the oil pan, check the alignment and operation of the control lever and parking pawl engagement. Make a final check to be sure all bolts are installed in the valve body. ☐

55. Install the oil pan with a new gasket. Torque the bolts to specifications. What are the specifications?

_____

56. Lubricate the oil pump's lip seal and the converter neck before installing the converter. ☐

Reassembly of a Transmission/Transaxle (continued)

☐ **57.** Install the converter, making sure that the converter is properly meshed with the oil pump drive gear.

☐ **58.** The transmission is now ready for installation into the vehicle. Use the reverse of the removal procedures. Remember to follow proper fluid filling procedures.

Instructor's Response _____

_____

_____

Torque Converter and Oil Pump Service

Upon completion and review of this chapter, you should be able to:

❏ Diagnose torque converter problems and determine needed repairs.

❏ Perform a stall test and determine needed repairs.

❏ Perform converter clutch system tests and determine needed repairs.

❏ Diagnose hydraulically and electrically controlled torque converter clutches.

❏ Inspect converter flexplate, converter attaching bolts, converter pilot, and converter pump drive surfaces.

❏ Measure torque converter endplay and check for interference.

❏ Check the stator clutch.

❏ Check a torque converter and transmission cooling system for contamination.

❏ Inspect, leak test, flush, and replace cooler, lines, and fittings.

❏ Inspect, measure, and replace an oil pump assembly and related components.

Basic Tools

Basic mechanic's tool set

DMM

Appropriate service manual

General Diagnostics

Many transmission problems are related to the operation of the torque converter. Normally torque converter problems will cause abnormal noises, poor acceleration in all gears, normal acceleration but poor high-speed performance, or transmission overheating.

To test the operation of the torque converter, many technicians perform a stall test. The stall test checks the holding capacity of the converter's stator overrunning clutch assembly, as well as the clutches and bands in the transmission.

However, torque converter problems can often be identified by the symptoms, and therefore the need for conducting a stall test is minimized. If the vehicle lacks power when it is pulling away from a stop or when passing, it has a restricted exhaust or the torque converter's one-way stator clutch is slipping. To determine which of these problems is causing the power loss, test for a restricted exhaust first.

The easiest way to test for an exhaust restriction is to use a vacuum gauge. Connect the vacuum gauge to a source of engine manifold vacuum. Observe the vacuum reading with the engine at idle. Quickly open the throttle plates and observe the vacuum reading. Then quickly release the throttle plates to allow them to close. The vacuum reading should show an increase of about 5 in. Hg. upon initial closing of the throttle plates (Figure 6-1). If the vacuum did not increase with the closing of the throttle plates, a restricted exhaust is indicated. Another way to check for a restricted exhaust is to insert a pressure gauge in the exhaust manifold's bore for the oxygen sensor. With the gauge in place, start the engine. Bring the engine to 2000 rpm and keep it there while observing the gauge. If the exhaust is not restricted, the pressure reading will be less than 1.25 psi. An extremely restricted exhaust will give a reading of over 2.75 psi. If there is no evidence of a restricted exhaust, it can be assumed that the torque converter's stator clutch is slipping and not allowing any torque multiplication to take place in the converter. To repair this problem, the torque converter should be replaced.

If the engine's speed flares up when it is accelerated in DRIVE and does not have normal acceleration, the clutches or bands in the transmission are slipping. This symptom is similar to the slipping of a clutch in a manual transmission. Often this problem is mistakenly blamed on the torque converter.

Many engine problems can cause a vehicle to behave as if the torque converter is malfunctioning, or actually cause the converter clutch to either lock up early or not at all. Clogged fuel injectors or bad spark plug wires can be misinterpreted as torque converter complaints. Bad vacuum lines, EGR valves, or engine speed sensors can prevent the clutch from locking up at the proper time.

Special Tools

Vacuum gauge

Various tee fittings

Various lengths of hoses

Common exhaust restrictions are a plugged catalytic converter or a collapsed exhaust pipe.

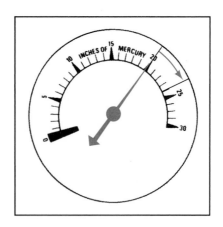

Figure 6-1 To check for an exhaust restriction, connect a vacuum gauge to the engine and observe the readings while the engine speed is raised and quickly lowered. (Courtesy of General Motors Corporation, Service Operations)

Sources for engine manifold vacuum are found in the intake manifold, below the throttle plate.

Technicians often blame the torque converter for problems simply based on the customer's complaint. Complaints of thumping or grinding noises are often blamed on the converter when they are really caused by bad thrust washers or damaged gears and bearings in the transmission. This type of noise can also be caused by non-transmission components, such as bad CV joints and wheel bearings.

Also, many engine problems can cause a vehicle to act as if it has a torque converter problem. This is especially true of converter clutches, which may engage early or not at all. Ignition and fuel injection problems can behave the same as a malfunctioning converter. Vacuum leaks and bad electrical sensors can prevent the converter from engaging at the correct time.

Testing Converter Clutches

Classroom Manual
Chapter 6, page 162

Late-model transmissions are equipped with a torque converter clutch. Most converter clutches are controlled by the powertrain control module (Figure 6-2). The computer turns on the converter clutch solenoid, which opens a valve and allows fluid pressure to engage the clutch. When the computer turns the solenoid off, the clutch disengages.

One of the trickiest parts of diagnosing a converter clutch problem is recognizing a normal-acting converter clutch, as well as an abnormal-acting clutch. You should pay attention to the ac-

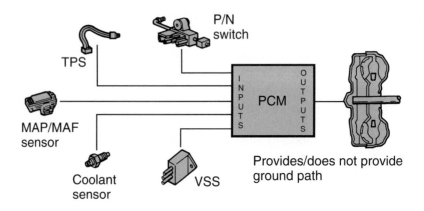

Figure 6-2 Typical electrical control circuitry for a lock-up torque converter. (Courtesy of General Motors Corporation, Service Operations)

tion of all converter clutches, whether or not they are suspected of having a problem. By knowing what a normal clutch feels like, it is easier to feel abnormal clutch activity. A malfunctioning converter clutch can cause a wide variety of driveability problems. Normally, the application of the clutch should feel like a smooth engagement into another gear. It should not feel harsh, nor should there be any noises related to the application of the clutch.

To properly diagnose converter clutch problems, you must know when they should engage and disengage and understand the function of the various controls involved with the system. Although the actual controls for a converter clutch vary with the different manufacturers and models of transmissions, they all will have certain operating conditions that must be met before the clutch can be engaged.

Care should be taken during diagnostics because abnormal clutch action can be caused by engine, electrical, clutch, or torque converter problems.

Before the converter clutch is applied, the vehicle must be traveling at or above a certain speed. The vehicle speed sensor sends this speed information to the computer. Also, the converter clutch will not engage when the engine is cold; therefore, a coolant temperature sensor provides the computer with information regarding engine temperature. During sudden deceleration or acceleration, the clutch should be disengaged. One of the sensors used to tell the computer when these driving modes are present is the TP sensor. Some transmissions use a third or fourth gear switch to inform the computer when the transmission is in those gears. The computer will then allow clutch engagement if other operating conditions are satisfactory. A brake switch is also used in some clutch circuits to disengage the clutch when the brakes are applied (Figure 6-3). These key sensors—the VSS, ECT, TP, third/fourth gear switch, and brake switch—should be visually checked as part of your diagnosis of converter problems.

Diagnosis of a converter clutch circuit should be conducted in the same way as any other computer system. The computer will recognize problems within the system and store trouble codes that reflect the problem area of the circuit. The codes can be retrieved and displayed by an instrument panel light or a hand-held scanner tool.

Classroom Manual
Chapter 6, page 162

Special Tools

Jumper wire

Paper clip

Diagnostic key

Scan tool

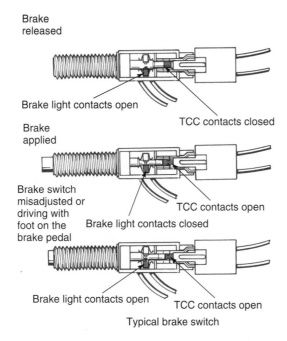

Figure 6-3 Proper adjustment of the brake light switch is essential for proper operation of a torque converter clutch. (Courtesy of General Motors Corporation, Service Operations)

Engagement Quality

Torque converter clutch shudder is sometimes called chatter or a stick-slip condition.

The ability of a converter clutch to hold torque is its torque capacity.

Classroom Manual
Chapter 6, page 164

Nearly all torque converters are welded-together units and therefore cannot be disassembled for inspection or repair. The primary objective when servicing a torque converter should be to check the serviceability of the unit.

The engagement of the clutch should be smooth. If the clutch prematurely engages or is not being applied by full pressure, a shudder or vibration results from the rapid grabbing and slipping of the clutch. The clutch begins to engage, then slips, because it cannot hold the engine's torque and become fully engaged. The torque capacity of the clutch is determined by the oil pressure applied to the clutch and the condition of the frictional surfaces of the clutch assembly.

If the shudder is only noticeable during the engagement of the clutch, the problem is typically in the converter. When the shudder is only evident after the engagement of the clutch, the cause of the shudder is the engine, transmission, or another component of the driveline. Often the problems will not only affect TCC engagement, but will also affect the operation of the transmission. Examples of these problems are worn valves or valve bores in the valve body, worn shafts, or bearings, and worn sealing rings. If the shudder is caused by the clutch, the converter must be replaced to correct the problem.

When clutch apply pressure is low and the clutch cannot firmly engage fully, shudder will occur. A faulty clutch solenoid or its return spring may cause this. The valve controlled by the solenoid is normally held in position by a coil-type return spring. If the spring loses tension, the clutch will be able to prematurely engage. Because insufficient pressure is available to hold the clutch, shudder occurs as the clutch begins to grab and then slips. If the solenoid valve and/or return spring is faulty, they should be replaced, as should the torque converter.

An out-of-round torque converter prevents full clutch engagement, which will also cause shudder, as will contaminated clutch frictional material. The frictional material can become contaminated by metal particles circulating through the torque converter and collecting on the clutch.

Broken or worn clutch dampener springs will also cause shudder.

Besides replacing the torque converter and/or replacing other components of the system, one possible correction for shudder is the use of an ATF with friction modifiers. Some rebuilders may recommend that an oil additive be added to the ATF. The additive is designed to improve or alter the friction capabilities of regular ATF.

⬤ **CUSTOMER CARE:** After replacing a converter because of clutch problems, be sure to explain to the customer that other engine or transmission problems may become more evident with the new converter clutch. Because the old converter didn't fully engage, the fluid in the converter dampened many of the vibrations from the engine and transmission. The new converter will now connect the engine to the transmission and engine and transmission vibrations will now be mechanically transmitted through the drivetrain.

TC-Related Cooler Problems

Special Tools
Short piece of hose
Graduated container
OSHA-approved air nozzle

Classroom Manual
Chapter 6, page 160

Vehicles equipped with a converter clutch may stall when the transmission is shifted into reverse gear. The cause of this problem may be plugged transmission cooler lines or the cooler itself may be plugged. Fluid normally flows from the torque converter through the transmission cooler. If the cooler passages are blocked, fluid is unable to exhaust from the torque converter and the converter clutch piston remains engaged. When the clutch is engaged, there is no vortex flow in the converter and therefore little torque multiplication is taking place in the converter.

To verify that the transmission cooler is plugged, disconnect the cooler return line from the radiator or cooler (Figure 6-4). Connect a short piece of hose to the outlet of the cooler and allow the other end of the hose to rest inside an empty container. Start the engine and measure the amount of fluid that flows into the container after 20 seconds. Normally one quart of fluid should flow into the container. If less than that filled the container, a plugged cooler is indicated.

To correct a plugged transmission cooler, disconnect the cooler lines at the transmission and the radiator. Blow air through the cooler, one end at a time, then through the cooler lines. The air will clear large pieces of debris from the transmission cooler. Always use low air pressure, no more

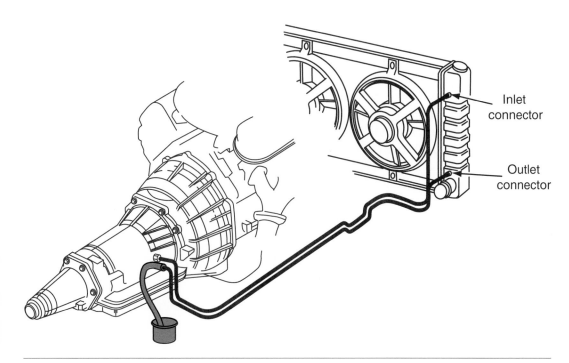

Figure 6-4 Location of cooler lines.

than 50 psi. Higher pressures may damage the cooler. If there is little air flow through the cooler, the radiator or external cooler must be removed and flushed or replaced.

General Converter Control Diagnostics

All testing of TCC controls should begin with a basic inspection of the engine and transmission. Too often technicians skip this basic inspection and become frustrated during diagnostics because of conflicting test results. Apparent transmission and torque converter problems are often caused by engine mechanical problems, broken or incorrectly connected vacuum hoses, incorrect engine timing, or incorrect idle speed adjustments. The basic inspection should include the following:

1. A road test to verify the complaint and further define the problem.
2. Careful inspection of the engine and transmission.
3. A check of the PCM for codes. Then do a check of the mechanical condition of the engine, the output of the ignition system, and the efficiency of the fuel system.
4. An idle speed and ignition timing check. If the timing is non-adjustable, check the operation of the electronic spark control system.
5. A check of the entire intake system for vacuum leaks.

When inspecting wires and hoses, look for burned spots, bare wires, and damaged or pinched wires. Make sure the harness to the electronic control unit has a tight and clean connection. Also check the source voltage at the battery before beginning any detailed tests on an electronic control system. If the voltage is too low or too high, the electronic system cannot function properly.

 SERVICE TIP: Nearly all electronic converter and transmission controls have a self-diagnostic mode and are capable of displaying trouble codes. However, the basic inspection is important because the computer will not display codes for low compression, open spark plug wires, or problems in the idle air control system.

TCC is a commonly used acronym for torque converter clutch.

Classroom Manual
Chapter 6, page 167

Special Tools

Hydraulic pressure
  gauge

Various tee fittings

Various lengths of
  hoses

On early TCC-equipped vehicles, clutch engagement was controlled hydraulically. A switch valve was controlled by two other valves, the lock-up and fail-safe valves, in the clutch control assembly (Figure 6-5). The lockup valve responds to governor pressure and prevents lock up at speeds below 40 mph. The fail-safe valve responds to throttle pressure and permits clutch engagement in high gear only. Problems with this system are diagnosed in the same way as other hydraulic circuits.

The clutch in most hydraulic TCC systems is applied when oil flow through the torque converter is reversed. This change can be observed with a pressure gauge. Using the pressure gauge is also a good way to diagnose the clutch's hydraulic system. Connect a pressure gauge to the hydraulic line, with a "tee" fitting, from the transmission to the cooler. Position the gauge so that it is easily seen from the driver's seat. Then raise the vehicle on a hoist with the drive wheels off the ground and able to spin freely. Operate the vehicle until the transmission shifts into high gear.

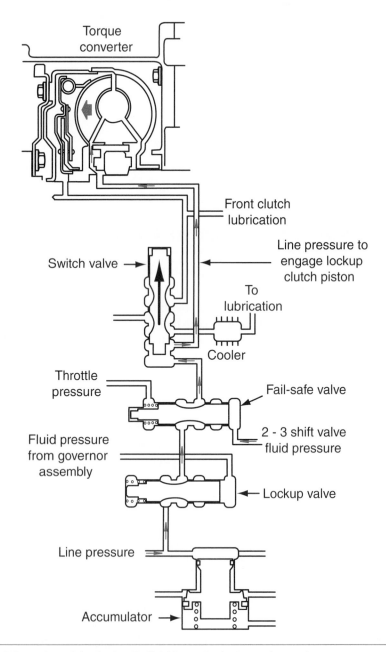

Figure 6-5 Action of the hydraulic fluid from the lockup valve to the torque converter.

Then maintain a speed of approximately 55 mph. Once the speed is maintained, watch the pressure gauge.

If the pressure decreases 5 to 10 psi, the converter clutch was applied. With this action you should feel the engagement of the clutch, as well as a drop in engine speed. If the pressure changed but the clutch did not engage, the problem may be inside the converter or at the end of the input shaft. If the input shaft end is worn or the O-ring at the end is cut or worn, there will be a pressure loss at the converter clutch. This loss in pressure will prevent full engagement of the clutch. If the pressure did not change and the clutch did not engage, suspect a faulty clutch valve or control solenoid, or a fault in the solenoid control circuit.

Diagnosis of electrical TCC controls isn't as hard as some would lead you to believe. In fact, it becomes easier as you understand how electricity works and how the component or system you are diagnosing works. Typically, TCC electrical problems are diagnosed through a three-step process: visual inspection, computer code retrieval, and electrical checks of specific circuits. Computer trouble code retrieval procedures are rather simple but vary with each manufacturer. Proper inspection procedures and the checking of electrical circuits require an understanding of basic electricity theory, electrical components, and electrical test instruments. These apply to all electrical circuits, no matter who manufactured the vehicle.

Electronic Converter Clutch Control Diagnostics

Each automobile manufacturer has specific test sequences that should be followed to test the system. However, there are certain checks that can be made on all systems. The first check is with the scan tool. After entering the appropriate information, retrieve the DTCs. If there are no trouble codes, check to make sure the converter clutch circuit has power. Check the circuit's fuse; make sure it is not blown. Then visually inspect all of the wires and connectors in the circuit. Make sure they are not loose, disconnected, or corroded.

Classroom Manual
Chapter 6, page 167

If the clutch does not engage, check for power to the solenoid. If power is available, make sure the ground of the circuit is good. If there is power available and the ground is good, check the voltage drop across the solenoid. The solenoids should drop very close to source voltage. If less than that is measured, check the voltage drop across the power and ground sides of the circuit. If the voltage drop testing produces good results, remove the solenoid and test it with an ohmmeter. If the solenoid checks out fine with the ohmmeter, suspect clutch material, dirt, or other material plugging up the solenoid valve passages. If a blockage is found, attempt to flush the valve with clean ATF. If the solenoid has a filter assembly (Figure 6-6), replace the filter after cleaning the fluid passages. If the blockage cannot be removed, replace the solenoid.

If the clutch engages at the wrong time, a sensor or switch in the circuit is probably the cause. If clutch engagement occurs at the wrong speed, check all speed-related sensors. A faulty temperature sensor may cause the clutch not to engage. If the sensor is not reading the correct temperature, the PCM may never realize that the temperature is suitable for engagement. Checking the appropriate sensors can be done with a scan tool, DMM, and/or lab scope. A check of the sensors is normally part of the manufacturer's system check.

The method used to diagnose computer-controlled TCC systems varies with each manufacturer (Figure 6-7). Although many of the procedural steps are quite different, all basically follow the same scheme. To illustrate a typical TCC diagnostic procedure, the following is a detailed look at diagnosing a TCC system fitted to a GM transmission and a GM computerized engine control system.

General Motors TCC Diagnostic Routine

GM's non-OBD II PCMs will detect failures in many of its systems or circuits and the failed system can be identified through displayed trouble codes.

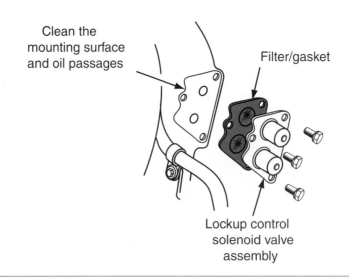

Clean the mounting surface and oil passages

Filter/gasket

Lockup control solenoid valve assembly

Figure 6-6 A torque converter clutch solenoid with a replaceable filter.

To retrieve trouble codes, turn the ignition ON with the engine OFF. Verify that the "Check Engine" or "Service Engine Soon" lamp is lit. This lamp should go off when the ignition switch is off. If the lamp fails to light, check that circuit before proceeding.

Then ground the diagnostic test terminal by connecting a jumper wire from terminal A to terminal B at the DLC connector located below the instrument panel on most GM vehicles. The malfunction indicator light (MIL) light should flash a code 12. The code 12 is indicated by a flash, a pause, and two flashes. A code 12 indicates that the system is now in its self-diagnostic mode. Code 12 will flash three times and will be followed by any trouble codes stored in the memory of the PCM. Each of these trouble codes will flash three times, starting with the lowest-numbered code. When all of the trouble codes have been displayed, a code 12 will be repeated. If no codes are present, the system will continue to flash a code 12 until the jumper wire at the DLC is removed or the engine is started.

SERVICE TIP: Instead of using a jumper wire to jump across the terminals of the DLC, some technicians use a GM diagnostic key. Both of these make the process easier and quicker than using a jumper wire.

The displayed codes should be recorded. By referring to the appropriate Trouble Code Chart, you will be able to identify the circuit or system that has the fault.

After the codes are retrieved, erase the codes from the memory of the PCM. To do this, locate and remove the PCM fuse. Keep the fuse out for at least ten seconds. This simple act stops power to the memory of the computer and erases the codes.

The codes are erased before diagnostics continue because you need to know if the new codes recorded during your road test were soft or hard codes. A hard code is a code that reappears after you clear the codes and perform a retest of the system. A soft code usually does not reappear after you clear the codes and perform a routine retest. Trouble code charts are designed to repair hard codes, not soft codes.

If no hard codes are present after you have started the engine, use the Field Service Mode or a scan tool to verify that the PCM has control over the fuel delivery and other systems. With the engine running, connect the jumper wire across terminals A and B. The MIL should respond to inputs from the oxygen sensor. If the lamp flashes two times per second, the system is in open loop. Open loop indicates that the voltage from the oxygen sensor is not being recognized by the PCM. The MIL will display open loop for 30 seconds to 2 minutes after the engine has been started or until the oxygen sensor reaches normal operating temperature. If the system never goes into closed loop, you should check the oxygen sensor circuit.

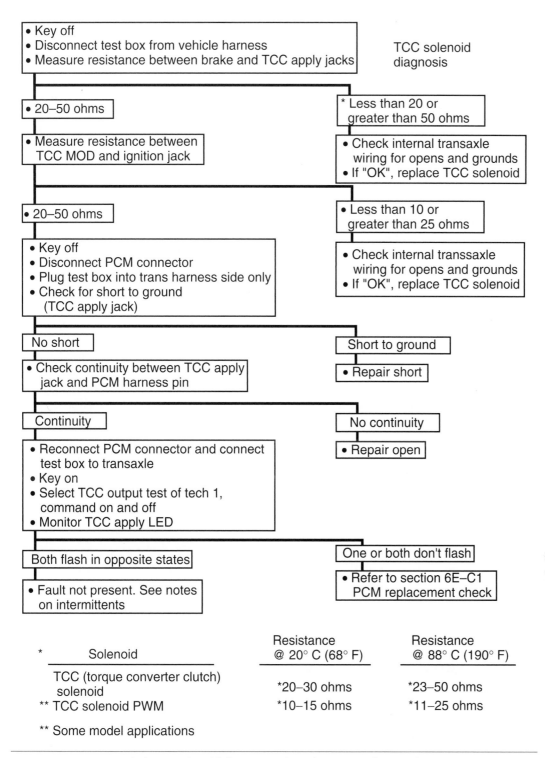

Figure 6-7 A typical TCC solenoid diagnostic chart. (Courtesy of General Motors Corporation, Service Operations)

Closed loop is indicated by the MIL flashing one time per second. Closed loop confirms that the oxygen sensor signal is recognized by the PCM to control fuel delivery and that the system is working normally.

When you have an intermittent problem, use any soft codes from previous self-diagnostics as a troubleshooting aid. Soft codes may allow you to isolate the faulty circuit. The customer's explanation of the problem is also of great help during diagnostics of these problems. Finding out when and where the problem occurs can lead you to properly identifying the problem area. This information will also give you the operating conditions in which the problem exists, allowing you to duplicate the conditions to verify the complaint.

The PCM controls the TCC solenoid, which in turn controls the flow of ATF to the torque converter clutch. The clutch mechanically connects the engine to the transmission; this action increases fuel economy and reduces heat in the transmission. The TCC solenoid receives power from the ignition switch through the brake light switch. The PCM provides the ground for the solenoid. The PCM completes the TCC solenoid circuit based on inputs from the ECT, VSS, MAF, MAP, and TP sensors. The presence of trouble codes 14, 15, 21, 24, 32, 34 can be clues as to the cause of faulty TCC operation.

The DLC contains a test terminal for the TCC circuit. This lead, terminal F, is to be used during pinpoint testing of the TCC circuit. By activating the TCC and watching the engine's speed, you can quickly determine if the TCC is actually being fully engaged. Firm and total engagement of the clutch should cause the engine's rpm to drop approximately 300 rpm. If the engine speed doesn't change, you know that the TCC has not been applied. If the change in engine speed is slight, the clutch may not be fully applied. Some transmissions have sensors that inform the PCM of what gear the transmission is operating in. They allow the PCM to apply the TCC in second, third, and fourth gears, depending on engine load and system design.

If the clutch does not engage when the DLC converter terminal is jumped to ground, check the clutch control solenoid and its circuit. Check for battery power to the solenoid. To identify the power feed wire, look at the appropriate wiring diagram in the service manual. If power is being supplied to the solenoid, ground the other end of the solenoid. This should activate the solenoid. If the solenoid does activate, check the ground circuit. If the solenoid did not respond, check the solenoid. Most solenoids have a resistance value specification. Measure the resistance across the solenoid and compare the readings to specifications.

All faulty wires and connectors should be repaired or replaced. TCC solenoids are not rebuildable and are replaced when they are faulty.

Stall Testing

Classroom Manual
Chapter 6, page 160

Two methods are commonly used to check the operation of the stator's one-way clutch: the stall test and the bench testing. Bench testing a converter will be covered later in this chapter, since we are still trying to determine if the converter is the cause of the customer's complaint. To bench test a converter, it must be removed from the vehicle and there is no need to do that unless we know the converter is at fault.

✔ **SERVICE TIP:** Stall testing is not recommended on many late-model transmissions. This test places extreme stress on the transmission and should only be conducted if recommended by the manufacturer.

CAUTION: If a stall test is not correctly conducted, the converter and/or transmission can be damaged.

⚠ **WARNING:** Make sure no one is around the engine or the front of the vehicle while a stall test is being conducted. A lot of stress is put on the engine, transmission, and brakes during the test. If something lets go, somebody can be seriously hurt.

Special Tools

Tachometer
Large fan
Wheel chocks

To conduct a stall test, connect a tachometer to the engine and position it so that it can be easily read from the driver's seat. Set the parking brake, raise the hood, and place blocks in front of the vehicle's non-driving tires (Figure 6-8). Conduct the test outdoors if possible, especially if it is a cold day. If the test is conducted indoors, place a large fan in front of the vehicle to keep

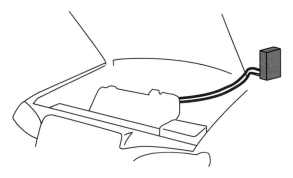

Trouble	Probable cause
Stall rpm high in $\boxed{D_4}$, $\boxed{2}$, $\boxed{1}$ & $\boxed{R}$	Low fluid level or oil pump output Clogged oil strainer Pressure regulator valve stuck closed Slipping clutch
Stall rpm high in $\boxed{R}$	Slippage of 4th clutch
Stall rpm high in $\boxed{2}$	Slippage of 2nd clutch
Stall rpm high in $\boxed{D_4}$	Slippage of 1st clutch or 1st gear one-way clutch
Stall rpm low in $\boxed{D_4}$, $\boxed{2}$, $\boxed{1}$ & $\boxed{R}$	Engine output low Torque converter one-way clutch slipping

Figure 6-8 Before conducting a stall test, chock the wheels and place the tachometer in a position where it can be easily seen from the driver's seat.

the engine cool. With the engine running, press and hold the brake pedal. Then move the gear selector to the DRIVE position and press the throttle pedal to the floor. Hold the throttle down for two seconds, then note the tachometer reading and immediately let off the throttle pedal and allow the engine to idle. Compare the measured stall speed to specifications (Figure 6-9).

If the torque converter and transmission are functioning properly, the engine will reach a specific speed. If the tachometer indicates a speed above or below specifications, a possible problem exists in the transmission or torque converter. If a torque converter is suspected of being faulty, it should be removed and the one-way clutch checked on the bench.

 CAUTION: To prevent serious damage to the transmission, follow these guidelines while conducting a stall test:

1. Never conduct a stall test if there is an engine problem.
2. Check the fluid levels in the engine and transmission before conducting the test.

TORQUEFLITE TRANSMISSION STALL SPEED CHART

Engine liter	Transaxle type	Converter diameter	Stall rpm
1.7	A-404	9-1/2 inches (241 millimeters)	2300–2500
2.2	A-413	9-1/2 inches (241 millimeters)	2200–2410
2.6	A-470	9-1/2 inches (241 millimeters)	2400–2630

Figure 6-9 A typical torque converter stall speed chart.

3. The engine should be at normal operating temperature during the test.
4. Never hold the throttle wide open for more than 5 seconds during the test.
5. Do not perform the test in more than two gear ranges without driving the vehicle a few miles to allow the engine and transmission to cool down.
6. After the test, allow the engine to idle for a few minutes to cool the transmission fluid before shutting off the ignition.

If the stall speed is below the specifications, a restricted exhaust or slipping stator clutch is indicated. If the stator's one-way clutch is not holding, ATF leaving the turbine of the converter works against the rotation of the impeller and slows down the engine. With both of these problems, the vehicle would exhibit poor acceleration. This is caused by either a lack of power from the engine or no torque multiplication occurring in the converter. If the stall speed is only slightly below normal, the engine is probably not producing enough power and should be diagnosed and repaired.

If the stall speed is above specifications, the bands or clutches in the transmission may be slipping and not holding properly.

If the vehicle has poor acceleration but had good results from the stall test, suspect a seized one-way clutch. Excessively hot ATF in the transmission is a good indication that the clutch is seized. However, other problems can cause these same symptoms; therefore, be careful during your diagnosis.

A normal stall test will generate a lot of noise, most of which is normal. However, if you hear any metallic noises during the test, diagnose the source of those noises. Operate the vehicle, at low speeds, on a hoist with the drive wheels free to rotate. If the noises are still present, the source of the noise is probably the torque converter.

Visual Inspection

If there was a noise coming from the torque converter during the stall test, visually inspect the converter before pulling the transmission out and testing or replacing the converter. Remove the torque converter access cover on the transmission (Figures 6-10 and 6-11) and rotate the engine with a remote starter button. While the torque converter is rotating with the engine, check to make sure the torque converter bolts are not loose and are not contacting the bell housing. Also, observe the action of the converter as it is spinning. If the converter wobbles, it may be due to a damaged flexplate or converter.

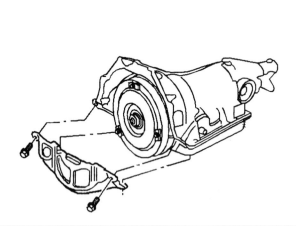

Figure 6-10 To inspect a torque converter while it is still in the vehicle, remove the access cover. (Courtesy of General Motors Corporation, Service Operations)

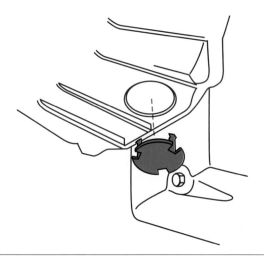

Figure 6-11 A torque converter access plug on a typical GM transmission. (Courtesy of General Motors Corporation, Service Operations)

> ⚠ **WARNING:** Before rotating the engine with a remote starter button, make sure the engine's ignition or fuel injection system is disabled. Also keep all parts of your body away from the rotating flexplate. Serious injury can result from being careless during this check.

Check the converter's balance weights to make sure they are still firmly attached to the unit. Carefully inspect the flexplate for evidence of cracking or other damage. Also check the condition of the starter ring gear, the teeth of which should not be damaged. The gear should be firmly attached to the flexplate.

Check the torque converter for **ballooning**. If excessive pressure was able to build up inside the converter, the converter will expand, or balloon. A stuck converter check valve typically causes this. If the converter is ballooned, it should be replaced and the cause of the problem also repaired.

A ballooned torque converter will also damage other driveline components. If the converter is ballooned toward the rear of the unit, the transmission's oil pump is most likely bad. If it is ballooned toward the front, the crankshaft's thrust bearings are undoubtedly worn or damaged.

Ballooning looks like the torque converter has been blown up like a balloon; it is caused by excessive pressure in the converter.

Bench Testing

Further inspection of the torque converter requires that it be removed from the vehicle. These inspections (Figure 6-12) should also be done anytime you have the torque converter out of a vehicle, especially if the oil pump is damaged or if the customer's complaint appears to be related to the torque converter.

When removing the transmission, pay particular attention to the condition of the flexplate and converter mounting bolts. Problems that appear to be torque converter problems may be caused by a damaged flexplate or a loose torque converter.

Once the converter is separated from the transmission, inspect the drive studs or lugs used to attach the converter to the flexplate. These are designed to hold the converter firmly to the flexplate and to keep the converter in line with the engine. Damaged studs or lugs will cause runout, misalignment, and vibration problems, and can result in damage to the bushings of the oil pump. If the threads are damaged slightly, they can be cleaned up with a thread file, tap, or die set. However, if they are badly damaged, the converter should be replaced. The converter should also be

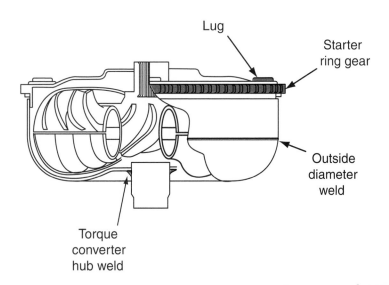

Figure 6-12 Some of the areas of a torque converter that need to be carefully checked.

replaced if the studs or lugs are loose or damaged. Also check the shoulder area around the lugs and studs for cracked welds or other damage. If any damage is found, the converter should be replaced. An exception to this is when the internal threads of a drive lug are damaged. These can often be repaired by tapping the threads or by installing a threaded insert. Also inspect the converter attaching bolts or nuts and replace them if they are damaged.

Check the pilot of the converter for wear and other damage. Also check the area around the pilot for cracks. In either case, the converter should be replaced. If the area around the pilot has dimples and looks like it has contacted the flywheel bolts, the converter has ballooned. If the converter has ballooned, it should be replaced and the cause of the ballooning corrected.

Inspect the flexplate for signs of damage, warpage, and cracks. Check the condition of the teeth on the starter ring gear. Replace the flexplate if there is any evidence of damage.

Check the drive hub of the torque converter; it should be smooth and not show any signs of wear. The easiest way to check the smoothness of the hub is to run a fingernail across the surface. If any irregularities are felt, the hub should be smoothed with crocus cloth or the converter replaced. If the hub is worn, carefully inspect the oil pump drive and replace the torque converter. Light scratches, burrs, nicks, or scoring marks on the hub surface can be polished with fine crocus cloth. Be careful not to allow dirt to enter into the converter while polishing the hub. Use a rag to cover the opening in the hub. Dirt and dust that enter the converter through this opening can cause the converter to wear rapidly. After polishing, clean the hub with solvent and a clean, lint-free rag. If the hub has deep scratches or other major imperfections, the converter should be replaced.

Since misalignment can cause many different problems, it is important that you check for excessive runout of the flexplate and the converter's hub. Check flexplate runout with a dial indicator. Mount the indicator onto the engine block and set its plunger on the flexplate (Figure 6-13). Rotate the flexplate and observe the readings on the indicator. Typically, the maximum allowable runout is 0.03 inches.

To check the runout of the converter's hub, mount the converter onto the flexplate. Tighten the retaining bolts to the specified torque. Position a dial indicator on the engine block. Set the plunger of the indicator on the hub (Figure 6-14). Rotate the torque converter and observe the readings on the indicator. Again, the typical maximum allowable runout is about 0.03 inches. If the runout is excessive and there is a normal amount of flexplate runout, remove the torque converter and set it at a new position on the flexplate. Check the runout again. If the runout is now within specifications, mark its position on the flexplate. If repositioning the converter on the flexplate does not correct hub runout, the converter should be replaced.

Because of the weight of torque converters, the above check is not always reliable. To get an accurate check of torque converter runout, mount the converter on two vee-blocks. Set the dial indicator so it can read lateral runout of the converter housing. To do this, place the indicator's

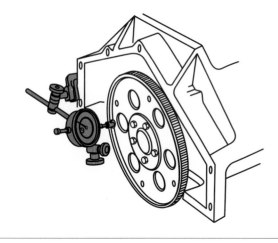

Figure 6-13 Setup for checking flexplate runout.

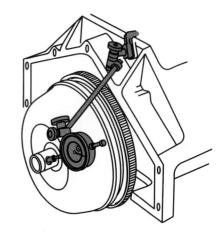

Figure 6-14 Setup for checking converter hub runout.

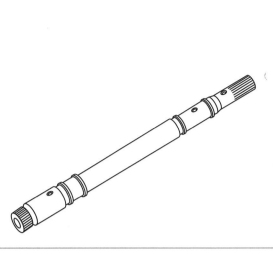

Figure 6-15 The pump drive shaft should be inspected in the areas that ride on the bushings and bearings. The shaft seals should always be replaced.

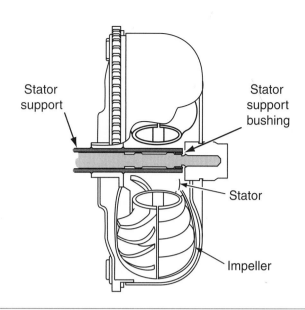

Figure 6-16 Typical stator support and oil flow in a torque converter.

plunger against the top of the converter housing's flywheel mounting surface. Rotate the housing and observe the runout. To check radial runout, place the dial indicator on the outside end of the pump hub. Rotate the housing and observe the runout. Then move the indicator on the hub and toward the housing. Again rotate and measure the runout.

In general, a torque converter should also be replaced if it has fluid leakage from its seams or welds, loose drive studs, worn drive stud shoulders, stripped drive stud threads, a heavily grooved hub, or excessive hub runout.

Transaxles that don't have their oil pump driven directly by the torque converter use a drive shaft (Figure 6-15) that fits into a support bushing inside the converter's hub (Figure 6-16). This bushing should be checked for wear. To do this, measure the inside diameter of the bushing and the outside diameter of the drive shaft. The difference between the two is the amount of clearance. This measurement should be compared to factory specifications. Normally the maximum allowable clearance is 0.004 inches. Excessive clearance can cause oil leaks. If the clearance is excessive, the bushing should be replaced.

Stator One-Way Clutch Diagnosis

The operation of the stator one-way clutch inside the torque converter is critical to overall effectiveness of the torque converter. If a problem occurs in this clutch assembly, the clutch will either fail to lock when rotated in either direction or it will fail to unlock when rotated in either direction. Although these problems are similar, they affect efficiency at opposite ends of the engine's operating speeds. However, in either case, fuel economy will be affected.

When the stator clutch does not lock, there is a disruption in vortex flow and a loss of torque multiplication in the torque converter. A vehicle with this problem will have sluggish low-speed performance, but will perform normally at higher speeds when the stator is supposed to free wheel.

A vehicle with a constantly locked stator will have good low-speed and poor high-speed performance. Torque multiplication will always occur, as will speed reduction. A vehicle with a constantly locked stator will show signs of overheating. If you suspect a locked stator, check for a bluish tint on the hub of the converter. This discoloration typically results from overheating. It is normal for some blue to be evident at the spot where the hub was welded to the housing. But, if the most of the hub is blue, the converter has overheated.

Classroom Manual
Chapter 6, page 153

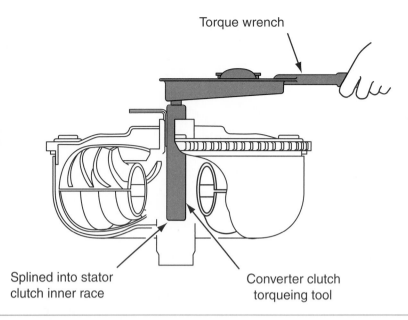

Torque wrench

Splined into stator
clutch inner race

Converter clutch
torqueing tool

Figure 6-17 Checking a one-way clutch with a special tool fixture.

To check the stator's one-way clutch with the converter on a bench, insert a finger into the splined inner race of the clutch. Snap ring pliers that have long and thin jaws work better than your finger; use them if they are available. Attempt to turn the inner race in both directions. You should be able to turn the race freely in one direction and feel lock-up in the opposite direction. If the clutch rotates freely in both directions or if the clutch is locked in both directions, the converter should be replaced.

Because this check does not put a load on the clutch assembly, it does not totally check the unit. Therefore, some manufacturers, such as Ford, recommend the use of a special tool set that holds the inner race and exerts a measurable amount of torque on the outer race, thereby allowing the technician to observe the action of the clutch while under load (Figure 6-17).

Internal Interference Checks

Internal converter parts hitting each other or hitting the housing may also cause noises. To check for any interference between the stator and turbine, place the converter, face down, on a bench (Figure 6-18). Then install the oil pump assembly. Make sure the oil pump drive engages with the

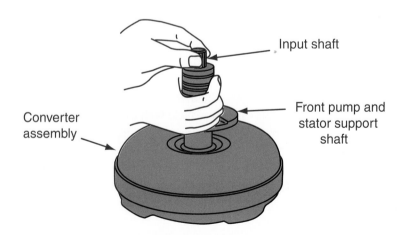

Input shaft

Front pump and
stator support
shaft

Converter
assembly

Figure 6-18 Checking stator-to-turbine interference.

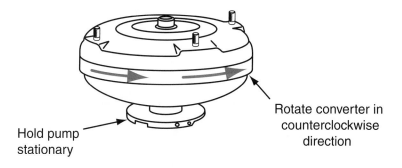

Hold pump stationary

Rotate converter in counterclockwise direction

Figure 6-19 Checking stator-to-impeller interference.

oil pump. Insert the input shaft into the hub of the turbine. Hold the oil pump and converter stationary, and then rotate the turbine shaft in both directions. If the shaft does not move freely and/or makes noise, the converter must be replaced.

To check for any interference between the stator and the impeller, place the transmission's oil pump on a bench and fit the converter over the stator support splines (Figure 6-19). Rotate the converter until the hub engages with the oil pump drive. Then hold the pump stationary and rotate the converter in a counterclockwise direction. If the converter does not freely rotate or makes a scraping noise during rotation, the converter must be replaced. Photo Sequence 9 covers the procedure for checking internal interference and endplay.

The input shaft may be referred to as the turbine shaft on some transaxles.

Endplay Check

The special tools required to check the internal endplay of a torque converter are typically part of the essential tool kit recommended by each manufacturer. However, these specialty tools can be individually purchased through specialty tool companies. Basically, the special tools are a holding tool and a dial indicator with a holding fixture. The holding tool is inserted into the hub of the converter and, once bottomed, it is tightened in place (Figure 6-20). This locks the tool into the splines of the turbine. The dial indicator is fixed onto the hub. The amount indicated on the dial indicator, as the tool is lifted up, is the amount of endplay inside the converter. If this amount exceeds specifications, replace the converter.

Special Tools

Manufacturer-specified holding tool

Dial indicator and mounting bracket

Converter Leakage Tests

If the initial visual inspection suggested that the converter has a leak, special test equipment can be used to determine if the converter is leaking. This equipment (Figure 6-21) uses compressed air to pressurize the converter. Leaks are found in much the same way as tire leaks are; that is, the converter is submerged in water and the trail of air bubbles lead the technician to the source of leakage.

Oil Pump Seal

If the backside of the torque converter was wet, it is very likely that the transmission pump seal is bad. There are many possible causes for leaks at the seal. These must also be looked at and corrected when the new seal is installed (Figure 6-22). Listed below are many of the causes of pump seal leakage:

- ❏ The seal has a missing or damaged spring.
- ❏ The lip of the seal is cut or damaged.
- ❏ The pump bore is worn, scratched, or damaged.
- ❏ The pump seal bore is off the centerline of the crankshaft.
- ❏ The pump drain hole is too small or is restricted.

Photo Sequence 9
Internal Interference and Endplay Checks

P9-1 To check for any interference between the stator and turbine, place the converter, face down, on a bench.

P9-2 Then install the oil pump assembly. Make sure the oil pump drive engages with the oil pump.

P9-3 Insert the input shaft into the hub of the turbine.

P9-4 Hold the oil pump and converter stationary, and then rotate the turbine shaft in both directions. If the shaft does not move freely and/or makes noise, the converter must be replaced.

P9-5 To check for interference between the stator and the impeller, place the transmission's oil pump on a bench and fit the converter over the stator support splines.

P9-6 Rotate the converter until the hub engages with the oil pump drive. Then hold the pump stationary and rotate the converter in a counterclockwise direction. If the converter does not rotate freely or makes a scraping noise during rotation, the converter must be replaced.

P9-7 Special tools are required to check the internal endplay of a torque converter. These include a holding tool and a dial indicator with a holding fixture.

P9-8 Insert the holding tool into the hub of the converter and once bottomed, tighten it in place. This locks the tool into the splines of the turbine.

P9-9 Fasten the dial indicator to the hub.

P9-10 Set the plunger of the dial indicator so it can read the movement of the tool.

P9-11 Zero the dial on the indicator.

P9-12 Lift up on the tool and observe the reading on the indicator. This is the amount of endplay inside the converter. Your reading should be compared to specifications.

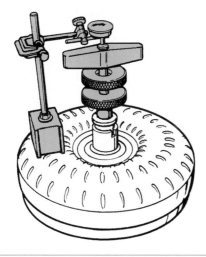

Figure 6-20 Checking converter endplay. (Courtesy of General Motors Corporation, Service Operations)

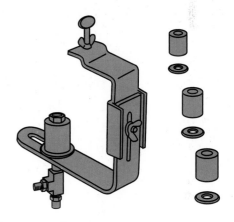

Figure 6-21 Converter leakage test kit.

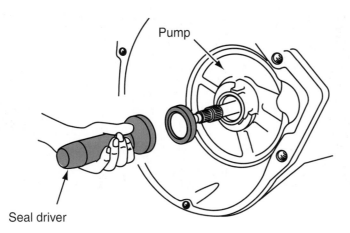

Figure 6-22 The converter/pump seal must be installed with the proper driver in order to seal.

- ❏ The pump bushing is worn, loose, or missing.
- ❏ The hub seal area is worn, rusted, or pitted.
- ❏ The hub has excessive runout.
- ❏ The hub is cracked or machined rough.
- ❏ The hub is the incorrect length.
- ❏ The hub or converter is out of balance.
- ❏ Excessive crankshaft end clearance (should be no more than 0.006-inch).
- ❏ Crankshaft pilot bore is worn.
- ❏ Crankshaft pilot sleeve is missing.
- ❏ The flexplate spacer is broken or missing.
- ❏ The flexplate is broken or has excessive runout.
- ❏ The converter housing is loose.
- ❏ The dowels in the converter housing are worn or missing.
- ❏ Replacement seal has incorrect elastomer or is missing an auxiliary lip.
- ❏ The inside or outside diameter of the seal is incorrect.
- ❏ The elastomer of the seal is too thick and is restricting the drain hole.
- ❏ The helix direction of the seal is incorrect.
- ❏ The seal was installed upside down.
- ❏ The seal was damaged during installation by the use of a defective or wrong driver.
- ❏ Seal was damaged by careless installation of the torque converter.

Summary of Checks

If certain conditions are found while inspecting a torque converter, the converter must be replaced. These conditions suggest that the converter should be replaced:

1. If there is a stator clutch failure.
2. If there is internal interference.
3. If the transmission's front pump is badly damaged.
4. If the converter hub is severely damaged or scored.
5. If there are signs of external fluid leaks.
6. If the drive studs or lugs are damaged or loose.
7. If there are signs of overheating.
8. If heavy amounts of metal were found in the fluid.
9. If any damage is evidence that the converter is no longer balanced.

Although some specialty shops will rebuild a converter, nearly all technicians replace the converter when any of the above exists. This is especially true of converters with a clutch. Rebuilding a converter is for the automatic transmission specialist and is not a normal task for an automobile technician. Because this procedure requires special equipment and knowledge, do not attempt to repair a faulty converter.

Starter Ring Gear Replacement

Torque converters fitted with a starter ring gear are typically found in Torqueflite and C-4 transmissions.

The starter ring gear is most often part of the flexplate. Therefore, whenever the teeth of the ring gear are damaged, the entire flexplate is replaced. There are some transmissions that are equipped with a torque converter fitted with a ring gear around the outer circumference of the torque converter cover. The ring gear is welded to the front of the converter cover and gear replacement involves breaking or cutting the welds to remove the old gear, then replacing the gear and welding it back onto the converter. This procedure is typically not recommended for torque converters fitted with a clutch, as the heat from welding can destroy the frictional surfaces inside the converter. In these cases, if the ring gear is damaged, the entire converter should be replaced.

Cleaning

SERVICE TIP: A torque converter can collect a lot of debris after a major transmission problem, and no amount of soaking and flushing will remove all of the dirt from the converter. A new converter should be installed.

A torque converter without a clutch should be flushed anytime it will be reused and there was contaminated fluid in the transmission or if the transmission overheated. There are many small corners inside a converter for dirt to become trapped in. Left over debris in the converter can lead to converter and transmission damage. Remember, the fluid that circulates through the converter circulates through the transmission next. Converters are typically cleaned by flushing the inside of the housing with solvent.

Flushing with solvent is not recommended for most converters with a clutch. It is recommended that a clutch-type converter be replaced, rather than flushed or cleaned.

Flushing removes all debris and sludge from the inside of the converter. Flushing is either done by a machine or by hand. Some transmission shops use a torque converter flushing machine that pumps a solvent through the converter as it is rotated by the machine. This method keeps the fluid moving inside the converter. The moving fluid is quite efficient at picking up any dirt present. The dirty solvent is pumped out of the converter and new solvent added during the flushing procedure.

In shops without a flushing machine, you can clean the inside of converters by hand. But this method is risky and not very effective, so it is not generally recommended. Basically, the procedure involves pouring about two quarts of clean solvent into the converter. Then, forcefully rotate and vibrate the converter. This action should dislodge any trapped debris. It is also helpful to use the input shaft and spin the turbine while the fluid is inside. The solvent is drained and the process repeated until the drained solvent is clean.

Since any amount of dirt can destroy a transmission, most rebuilders replace the torque converter during a transmission overhaul. There is no accurate way of knowing if dirt remains in the converter after cleaning and flushing it. The cleaning process may loosen up the debris, which will break down and contaminate the fluid once the torque converter starts spinning and gets hot.

Some transmission shops may cut the torque converter shell in half, then clean the parts, examine them for wear, and replace any that are worn or broken. The shell is then welded back together. This is a job only for shops that are equipped to do this job right.

After the converter has been flushed, disconnect it from the machine. If the converter has a drain plug, invert and drain the complete assembly. Converters without drain holes or plugs can be drilled with a 1/8-inch drill bit between the top end of the impeller fin dimples. This hole will act as an air bleed to maximize flushing. After flushing and draining the converter, the bleeder hole is sealed with a closed-end pop rivet covered with sealant.

Special Tools

Torque converter
 flushing machine

Torque Converter Replacement

Extra care should always be taken when replacing a torque converter. Size and fit are not the only important variables. Nor does size alone determine the stall speed of the converter. Even if the converter has exactly the same stall speed, it may have a different torque ratio and should not be used. Always check and double check the part or model number of the torque converter you are removing and compare it to the one you are going to install. Converters are typically identified by a sticker or a number code stamped into the converter housing.

When replacing a torque converter, never use an impact wrench on the torque converter bolts. Impact wrenches can drive the bolts through the cover, which will warp the inside surface and prevent proper clutch engagement, or it will damage the clutch's pressure plate.

Classroom Manual
Chapter 6, page 160

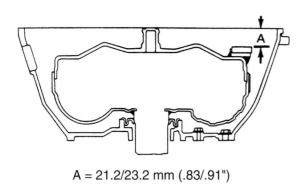

A = 21.2/23.2 mm (.83/.91")

Figure 6-23 Before removing a torque converter, check its installed depth in the bell housing.

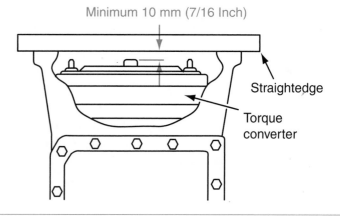

Minimum 10 mm (7/16 Inch)

Straightedge

Torque converter

Figure 6-24 Checking proper installation of a torque converter in a transmission.

Always perform an endplay check and check the depth of the torque converter in the bell housing (Figure 6-23) before reinstalling a torque converter or installing a new unit.

Correct installation of a torque converter requires that the converter's hub be fully engaged with the transmission. To help in seating the torque converter into the transmission, push on the converter while rotating it. Care should be taken not to damage the pump seal while installing the converter. To verify full engagement, place a straightedge across the bell housing or transaxle flange. Then measure from the straightedge to the pilot (Figure 6-24). Compare your readings to those specified by the manufacturer. If your reading is less than the specifications, the torque converter is not properly engaged in the transmission.

☑ **SERVICE TIP:** Many OBD-II systems have an operational mode that allows for converter clutch break-in. The purpose of this mode is to allow the system to readjust to a new clutch in the torque converter. Whenever a torque converter is replaced in a vehicle equipped with OBD-II, set the system into the clutch break-in mode. Doing this will allow the system to realize things have changed and will allow the converter clutch to work more efficiently.

Converter Balance

When replacing a torque converter, make sure the manufacturer or rebuilder balanced it. If the converter is not balanced, noise and vibrations can result, as well as damage to the transmission.

When reusing a torque converter, make sure it is positioned in its original position on the flexplate and make sure the mounting bolts or nuts are tightened to specifications. Sometimes manufacturers install a slightly unbalanced torque converter in a precise position on the flexplate. This is done to offset or counter engine vibrations, and you have no way of knowing if this is what they did.

When installing a known balanced converter, it can be positioned anywhere on the flexplate as long as it is fully seated against it and the mounting bolts or nuts are properly torqued.

Oil Pump Service

Classroom Manual
Chapter 6, page 183

The oil pump (Figure 6-25) of an automatic transmission should be carefully inspected during any transmission overhaul and especially when low line pressure was measured during a pressure test. Carefully remove the oil pump assembly (Figure 6-26). Some transmissions require the use of a

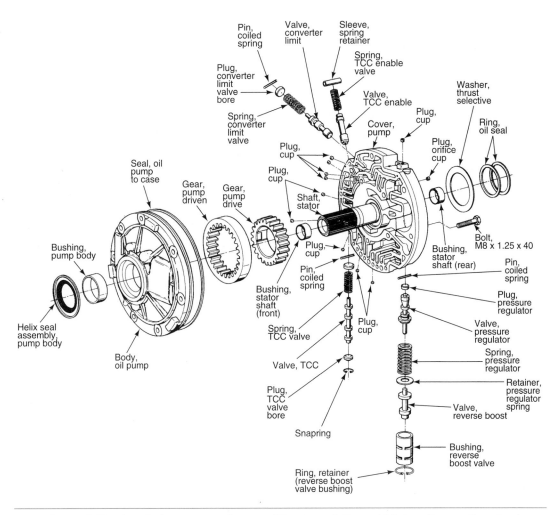

Figure 6-25 A typical gear-type oil pump. (Courtesy of General Motors Corporation, Service Operations)

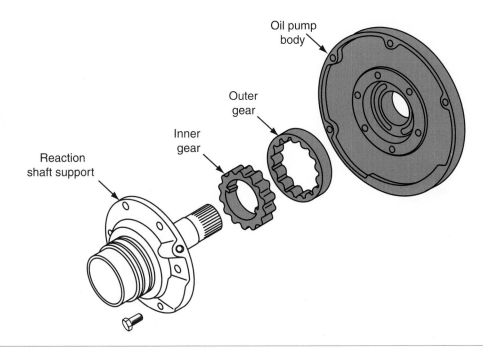

Figure 6-26 A pump assembly.

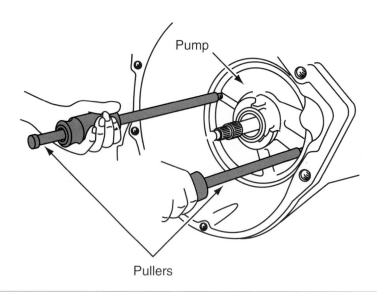

Pump

Pullers

Figure 6-27 Using a puller to remove an oil pump from a transmission case.

special puller to remove the oil pump from the transmission case (Figure 6-27). Never pry the pump out; it is easy to damage the case if this is done.

Disassembly

Before disassembling the pump, mark the alignment of the gears. If acceptably worn gears are re-installed in a position other than their wear pattern, excessive noise will result. When the pump halves have been separated, look at the relationship between the inner and outer gears. Most gears will have a mark on them indicating the top side of the gear (Figure 6-28).

If no mark is present, you should use a nondestructive type of marker to be sure you install the gears in the same position they were originally in (Figure 6-29). This ensures that the converter drive hub will mate correctly with the inner gear.

Another way of ensuring the proper position of the gears is to observe the existing wear pattern on the gears as they are being removed and make a note of this for use during reassembly.

Optional identification marks

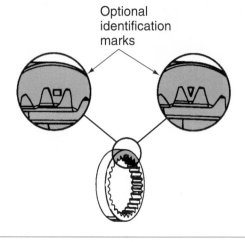

Figure 6-28 Identification marks on oil pump gearsets. (Courtesy of General Motors Corporation, Service Operations)

Figure 6-29 Marking the location of the outer and inner gears so that they can be properly meshed during reassembly. (Courtesy of General Motors Corporation, Service Operations)

If the inner gear is installed upside down, there will not be enough free movement, which can result in one or more of the following problems: broken inner gear, broken flexplate, pump cover scoring, or broken transmission case ears. The proper procedure for disassembling an oil pump is shown in Photo Sequence 10. This procedure is given as an example; always refer to the procedures for the specific transmission you are working on.

Inspection

Inspect the pump bore for scoring on both its bottom and sides. A converter that has a tight fit at the pilot hub could hold the converter drive hub and inner gear too far into the pump, causing cover scoring. A front pump bushing that has too much clearance may allow the gears to run off center, causing them to wear into the crescent and/or the sides of the pump body.

 SERVICE TIP: Some service manuals will call for a maximum front bushing wear of 0.004 inches; however, a dimension of less than that will avoid wear, as well as keep the pressure losses at the bushing to a minimum.

The stator shaft should be inspected for looseness in the pump cover (Figure 6-30). The shaft's splines and bushing should also be carefully looked at. If the splines are distorted, the shaft and the pump cover should be replaced. Because the bushings control oil flow through the converter and cooler, their fit must be checked (Figure 6-31). Bushings must be tight inside the shaft and provide the input shaft with a 0.0005 to 0.003 inch clearance.

Inspect the gears and pump parts for deep nicks, burrs, or scratches. Examine the pump housing for abnormal wear patterns. The fit of each gear into the pump body, as well as the centering effect of the front bushing, controls oil pressure loss from the high-pressure side of the pump to the low-pressure input side. Scoring or body wear will greatly reduce this sealing capability.

 SERVICE TIP: All pumps, valve bodies, and cases should be checked for warpage and should be flat filed to take off any high spots and burrs prior to reassembly.

On positive displacement pumps, use a feeler gauge to measure the clearance between the outer gear and the pump pocket in the pump housing (Figure 6-32). If possible, also check the clearance between the outer pump gear teeth and the crescent (Figure 6-33), and between the inner gear teeth and the crescent. Compare these measurements to the specifications. Use a

Gear-type and rotor-type pumps are positive displacement pumps.

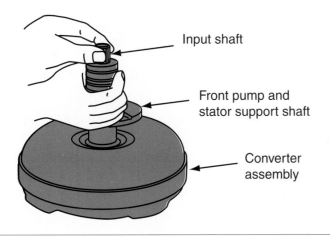

Figure 6-30 Checking the play between the splines of the oil pump and the input shaft.

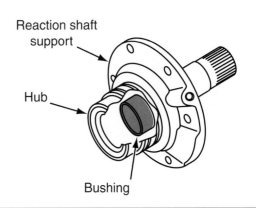

Figure 6-31 Location of a stator shaft bushing.

Photo Sequence 10
Proper Procedure for Disassembling an Oil Pump

P10-1 Remove the front pump bearing race, front clutch thrust washer, gasket, and O-ring. Inspect the pump bodies, pump shaft, and ring groove areas.

P10-2 Unbolt and separate the pump bodies.

P10-3 Mark the gears with machinist bluing ink or paint before removing them, so that the gears will remain in the same relationship during reassembly.

P10-4 Inspect the gears and all internal surfaces for defects and visible wear.

P10-5 Measure between the outer gear and the pump housing 0.003–0.006 in. or 0.08–0.15 mm. Replace the pump if the measurements exceed specifications.

P10-6 Measure between the outer gear teeth and the crescent. Clearance should be 0.006–0.007 in. or 0.015–0.18 mm. Replace the pump if clearance exceeds 0.008 in.

P10-7 Place the pump flat on the bench and, with a feeler gauge and straight-edge, measure between the gears and the pump cover. The clearance should be no more than 0.002 in. (0.05 mm). Replace the pump if the measurement exceeds specifications.

P10-8 Put the halves back together.

P10-9 Torque the securing bolts to specifications. Replace all O-rings and gaskets.

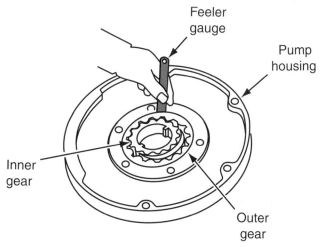

Figure 6-32 Measuring oil pump gear-to-pocket clearance.

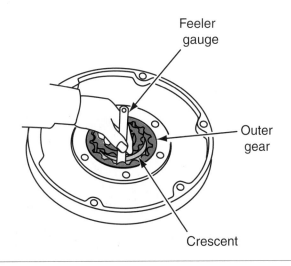

Figure 6-33 Measuring the clearance between the outer gear and the crescent.

straightedge and feeler gauge to check gear side clearance (Figure 6-34) and compare the clearance to the specifications. If the clearance is excessive, replace the pump.

Variable displacement vane-type pumps require different measuring procedures. However, the inner pump rotor to converter drive hub fit is checked in the same way as described for the other pumps. The pump rotor, vanes, and slide are originally selected for size during assembly at the factory (Figure 6-35). Changing any of these parts during overhaul can destroy this sizing and possibly the body of the pump. You must maintain the original sizing if any parts are found to need replacement. These parts are available in select sizes for just this reason.

☑ **SERVICE TIP:** Some early pump vanes cracked as a result of excessive load. These pumps did not have enough vanes to handle the hydraulic loads encountered during operation. Updated pumps with more vanes cured the problem.

The vanes are subject to edge wear, as well as cracking and subsequent breakage. The outer edge of the vanes should be rounded with no flattening (Figure 6-36). These pumps have an aluminum body and cover halves; therefore, any scoring indicates that they should be replaced.

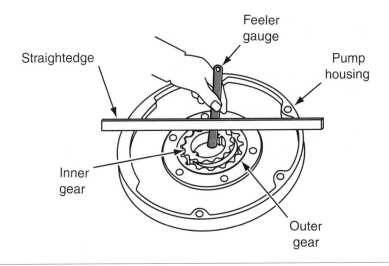

Figure 6-34 Measuring the side clearance of the pump's gears.

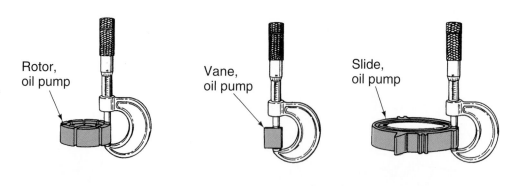

ROTOR SELECTION		VANE SELECTION		SLIDE SELECTION	
THICKNESS (mm)	THICKNESS (in.)	THICKNESS (mm)	THICKNESS (in.)	THICKNESS (mm)	THICKNESS (in.)
17.593 - 17.963	.7068 - .7072	17.943 - 17.961	.7064 - .7071	17.983 - 17.993	.7080 - .7084
17.963 - 17.973	.7072 - .7076	17.961 - 17.979	.7071 - .7078	17.993 - 18.003	.7084 - .7088
17.973 - 17.983	.7076 - .7080	17.979 - 17.997	.7078 - .7085	18.003 - 18.013	.7088 - .7092
17.983 - 17.993	.7080 - .7084			18.013 - 18.023	.7092 - .7096
17.993 - 18.003	.7084 - .7088				

Figure 6-35 Vane-type oil pump measurements and selection chart. (Courtesy of General Motors Corporation, Service Operations)

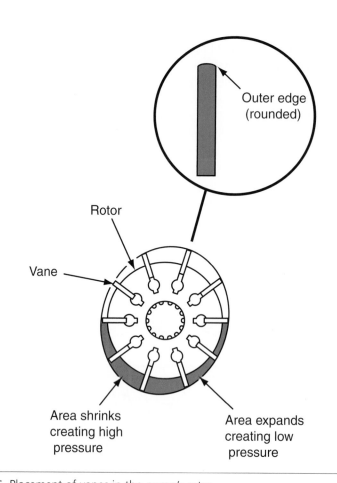

Figure 6-36 Placement of vanes in the pump's rotor.

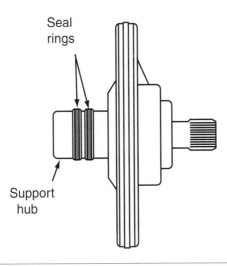

Figure 6-37 Location of oil seal rings on a typical oil pump assembly.

Inspect the reaction shaft's seal rings. If the rings are made of cast iron, check them for nicks, burrs, scuffing, or uneven wear patterns; replace them if they are damaged (Figure 6-37). Make sure the rings are able to rotate in their grooves. Check the clearance between the reaction shaft support ring groove and the seal ring. If the seal rings are the Teflon full-circle type, cut them out, and use the required special tool to replace them.

The outer area of most pumps utilizes a rubber seal (Figure 6-38). Check the fit of the new seal by making sure the seal sticks out a bit from the groove in the pump. If it does not, it will leak. The seal at the front of the pump is always replaced during overhaul. Most of these seals are the metal-clad lip seal type. Care must be taken to avoid damage to the seating area when removing the old seal.

Check the area behind the seal to be sure the drain back hole is open to the sump. If this hole is clogged, the new seal will possibly blow out. The drain back hole relieves pressure behind the seal. A loose-fitting converter drive hub bushing can also cause the front pump seal to blow out.

Reassembly

The pump seal can be installed with a hammer and seal driver. Apply a very thin coating of RTV sealant around the outside surface of the seal case, when installing the seal. Place some transmission

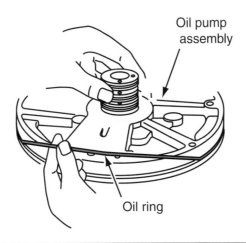

Figure 6-38 Outer oil pump seal.

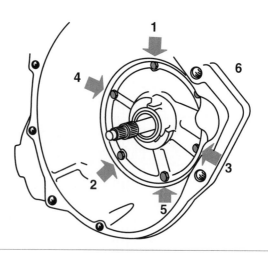

Figure 6-39 Oil pump bolts must be torqued in the sequence recommended by the manufacturer.

fluid in the pocket of the pump housing and install the gears into the housing according to their alignment marks. Align and install the reaction shaft support and tighten the bolts to the specified torque. Make sure the pump is not binding after you have tightened it by using the torque converter to rotate the pump.

Place a layer of clean ATF in two or three spots around the oil pump gasket and position it onto the transmission case. Tighten the pump attaching bolts to specifications in the specified order (Figure 6-39). After the bolts are tight, check the rotation of the input shaft. If the shaft does not rotate, disassemble the transmission to locate the misplaced thrust washer or misaligned friction plate.

Transmission Cooler Service

A **heat exchanger** may also be called an intercooler.

Heat is transferred because of a law of nature in that the heat of an object will always attempt to heat a cooler object. Heat is exchanged.

The process of cooling is actually the process of removing heat.

Vehicles equipped with an automatic transmission can have an internal or external transmission fluid cooler, or both (Figure 6-40). The basic operation of either type of cooler is that of a **heat exchanger**. Heat from the fluid is transferred to something else, such as a liquid or air. Hot ATF is sent from the transmission to the cooler, where it has some of its heat removed, then the cooled ATF returns to the transmission.

Internal type coolers are located inside the engine's radiator. Heated ATF travels from the torque converter to a connection at the radiator. Inside the radiator is a small internal cooler, which is sealed from the liquid in the radiator. ATF flows through this cooler and its heat is transferred to the liquid in the radiator. The ATF then flows out of a radiator connection, back to the transmission.

● **CUSTOMER CARE:** This type of cooler is less efficient as engine temperatures increase; therefore, you should recommend that an external cooler be installed in addition to the internal one, when the vehicle is used in ways that the engine will be operating at high temperatures, such as carrying heavy loads or pulling trailers.

External coolers are mounted outside the engine's radiator, normally just in front of it. Air flowing through the cooler removes heat from the fluid before it is returned to the transmission.

The engine's cooling system is the key to efficient transmission fluid cooling. If anything affects engine cooling, it will also affect ATF cooling. The engine's cooling system should be carefully inspected whenever there is evidence of ATF overheating or a transmission cooling problem.

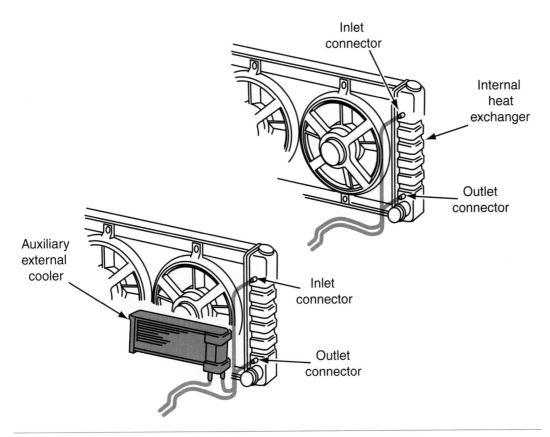

Figure 6-40 Typical transmission coolers.

If the problem is the transmission cooler, examine it for signs of leakage. Check the dipstick for evidence of coolant mixing with the ATF. Milky fluid indicates that engine coolant is leaking into and mixing with the ATF because of a leak in the cooler. At times, the presence of ATF in the radiator will be noticeable when the radiator cap is removed, as ATF will tend to float to the top of the coolant. A leaking transmission cooler core can be verified with a leak test.

External cooler leaks are evident by the traces or film buildup of ATF around the source of the leak and typically requires little time to determine the source of the leakage. However, internal coolers present a little bit more difficulty and a leak test should be conducted. To do this, place a catch pan under the fittings that connect the cooler lines to the radiator (Figure 6-41). Then, disconnect (Figure 6-42) and plug both cooler lines from the transmission (Figure 6-43). Tightly plug one end of the cooler and apply compressed air (50–75 psi) into the open end of the cooler.

 WARNING: Be sure the radiator is cool to the touch. Hot coolant may spray on you if the system is hot. The hot coolant can cause serious injuries. Wearing safety goggles is a good precaution when working around hot liquids.

 WARNING: Wear safety goggles when working with compressed air.

CAUTION: Do not apply more than 50 psi to the cooler. Extreme pressures can cause the cooler to explode.

Carefully check the coolant through the top of the radiator; if there is a leak in the transmission cooler, bubbles will be apparent in the coolant. If it leaks, the cooler must be replaced.

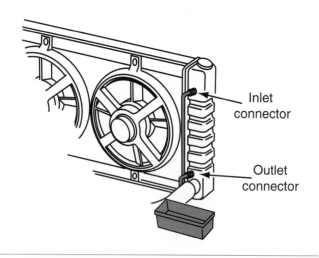

Figure 6-41 Cooler lines at an external cooler. (Courtesy of General Motors Corporation, Service Operations)

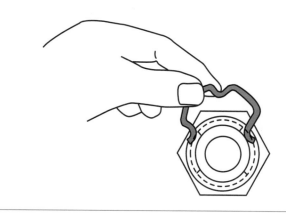

Figure 6-42 Cooler line fittings may use a retaining ring to lock the fitting to the line. These rings should never be reused and must be properly installed. (Courtesy of General Motors Corporation, Service Operations)

The inability of the transmission fluid to cool can also be caused by a plugged or restricted fluid cooling system. If the fluid cannot circulate, it cannot cool. To verify that the transmission cooler is plugged, disconnect the cooler return line from the radiator or cooler (Figure 6-44). Connect a short piece of hose to the outlet of the cooler and allow the other end of the hose to rest inside an empty container. Start the engine and measure the amount of fluid that flows into the container after 20 seconds. Normally, one quart of fluid should flow into the container. If less than that filled the container, a plugged cooler is indicated.

CAUTION: A clogged cooler will result in a subsequent failure of a new or rebuilt transmission.

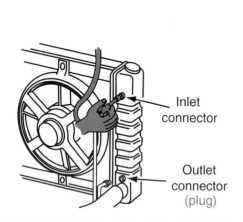

Figure 6-43 Before testing the cooling system, identify which line is the in line and which is the out line.

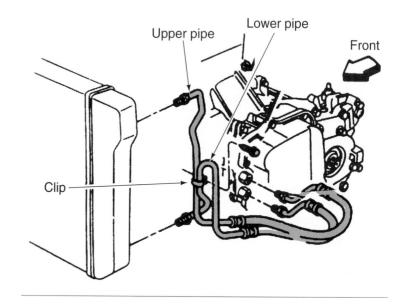

Figure 6-44 Typical routing of transmission oil cooler lines for a transaxle.

A tube-type transmission cooler can be cleaned by using cleaning solvent or mineral spirits and compressed air. A fin-type cooler, however, cannot be cleaned in this way. Therefore, the normal procedure includes replacing the radiator (which includes the cooler). In both cases the cooler lines should be flushed to remove any debris.

If the cooler is plugged, disconnect the cooler lines at the transmission. Apply compressed air through one port of the cooler, then to the other port. The air should blow any debris out of the cooler. Always use low air pressure, no more than 50 psi. Higher pressures may damage the cooler. If little air passes through the cooler, the cooler is severely plugged and may need to be replaced.

If the vehicle has two coolers, pull the inlet and outlet lines from the side of the transmission and, using a hand pump, pump **mineral spirits** into the inlet hose until clear fluid pours out of the outlet hose. To remove the mineral spirits from the cooler, release some compressed air into the inlet port, then pump some ATF into the cooler.

> ✔ **SERVICE TIP:** A transmission cooler is a good hiding place for debris when there has been a major transmission failure. Blowing out the cooler and lines will not always remove all of the dirt. Install a temporary auxiliary oil filter in the cooler lines to catch any left-over debris immediately after rebuilding a transmission.

The transmission's cooling system can be flushed. This will remove nearly all of the debris, leaving a clean, dry system. To do this, a transmission cooler flusher (Figure 6-45) is required. The flusher's container is filled with biodegradable flushing fluid. Compressed air pushes the fluid through the entire cooling system of the transmission. The system is then cleaned with forced hot water. When the discharge from the system is cleared, compressed air is expelled through the system until no moisture is evident in the discharge.

Check the condition of the cooler lines from their beginning to their ends. A line that has been accidentally damaged while the transmission has been serviced will reduce oil flow through the cooler and shorten the life of the transmission. If the steel cooler lines need to be replaced, use only double-wrapped and brazed steel tubing. Never use copper or aluminum tubing to replace steel tubing. The steel tubing should be double-flared and installed with the correct fittings.

Mineral spirits is an alcohol-based liquid that leaves no surface residue after it dries.

Mineral spirits can be purchased at most hardware stores.

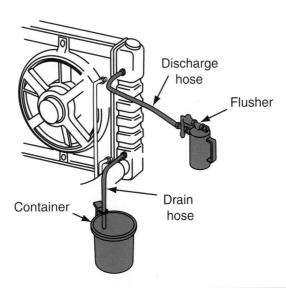

Figure 6-45 Setup for flushing a transmission's cooling system.

A customer brought her early model Ford Escort into the shop with a complaint that the transmission not shifting into the higher gears. The car had over 100,000 miles on it. The technician was assumed that after this many miles the transaxle needed to be rebuilt and assured the customer that this would take care of the problem. This was not an easy decision for the customer, as the car was worth only slightly more than the cost of the repair. She decided, however, to go ahead with the repair since she really liked the car.

While the technician was rebuilding the transaxle, he found nothing major wrong with it. There was just normal wear, especially considering the mileage on the car. When things wear, oil pressure goes down. That was probably the cause of the no-shift problem. After rebuilding the transaxle and installing it into the car, he road tested it. The thing still wouldn't shift.

He was baffled and frustrated. He knew he had the transmission together right and had used good pieces. Therefore, he called a friend, who also did transmission work, for advice. The first question his friend asked him was "Did you check the converter?" He had not!

What his friend had told him was that early style ATXs used the splitter gear-type torque converter and, if the planetary gear inside the converter was damaged, there would be no high gears. With this information, he knew what the problem was and proceeded to repair the transaxle.

Terms To Know

Ballooning   Heat exchanger   Mineral spirits

ASE-Style Review Questions

1. *Technician A* says a ballooned torque converter can cause damage to the oil pump.
   Technician B says a ballooned torque converter is caused by excessive pressure in the torque converter.
   Who is correct?
   A. A only   **C.** Both A and B
   B. B only   **D.** Neither A nor B

2. While checking torque converter endplay:
   Technician A says the torque converter must be installed in the transmission.
   Technician B says the torque converter endplay is corrected by installing a thrust washer between the oil pump and the direct or front clutch.

   Who is correct?
   A. A only   **C.** Both A and B
   B. B only   **D.** Neither A nor B

3. While servicing a variable displacement vane-type pump:
   Technician A says the pump rotor, vanes, and slide have selective sizes and may destroy the pump if the correct ones are not used.
   Technician B says the outer edge of the vanes should be flat.
   Who is correct?
   A. A only   **C.** Both A and B
   B. B only   **D.** Neither A nor B

4. *Technician A* says the gears in a gear-type oil pump should be replaced if the outer edges of the teeth are worn flat.
   Technician B says all parts of gear-type pumps are selectively sized.
   Who is correct?
   A. A only
   B. B only
   C. Both A and B
   D. Neither A nor B

5. *Technician A* says a fin-type cooler can be cleaned by using cleaning solvent or mineral spirits and compressed air.
   Technician B says a tube-type transmission cooler can be cleaned by using cleaning solvent or mineral spirits and compressed air.
   Who is correct?
   A. A only
   B. B only
   C. Both A and B
   D. Neither A nor B

6. *Technician A* says that if the stall speed of a torque converter is below specifications, a restricted exhaust or slipping stator clutch is indicated.
   Technician B says that if the stall speed is above specifications, the bands or clutches in the transmission may not be holding properly.
   Who is correct?
   A. A only
   B. B only
   C. Both A and B
   D. Neither A nor B

7. *Technician A* says torque converter clutch control problems are always caused by electrical malfunctions.
   Technician B says nearly all converter clutches are engaged through the application of hydraulic pressure on the clutch.
   Who is correct?
   A. A only
   B. B only
   C. Both A and B
   D. Neither A nor B

8. While checking the stall speed of a torque converter:
   Technician A maintains full-throttle power until the engine stalls.
   Technician B manually engages the TCC before running the test.
   Who is correct?
   A. A only
   B. B only
   C. Both A and B
   D. Neither A nor B

9. *Technician A* says a converter that has a tight fit at the pilot hub could hold the converter drive hub and inner gear too far into the pump, causing cover scoring.
   Technician B says a front pump bushing that has too much clearance may allow the gears to run off center, causing them to wear into the crescent and/or the sides of the pump body.
   Who is correct?
   A. A only
   B. B only
   C. Both A and B
   D. Neither A nor B

10. *Technician A* says a seized one-way stator clutch will cause the vehicle to have good low-speed operation but poor high-speed performance.
    Technician B says a freewheeling or non-locking one-way stator clutch will cause the vehicle to have poor acceleration.
    Who is correct?
    A. A only
    B. B only
    C. Both A and B
    D. Neither A nor B

ASE Challenge Questions

1. *Technician A* says a lower than specified stall speed may be caused by a faulty stator one-way clutch.
   Technician B says a higher than specified stall speed may be caused by faulty clutch packs.
   Who is correct?
   A. A only
   B. B only
   C. Both A and B
   D. Neither A nor B

2. *Technician A* says a shudder after the converter clutch engages could be caused by a damaged or missing clutch check ball.
   Technician B says driveline or converter shudder can be isolated by disconnecting the converter's clutch solenoid.
   Who is correct?
   A. A only
   B. B only
   C. Both A and B
   D. Neither A nor B

3. The backside of the torque converter was wet.
   Technician A says that may be caused by excessive torque converter hub runout.
   Technician B says that can be caused by insufficient input shaft endplay.
   Who is correct?
   A. A only
   B. B only
   C. Both A and B
   D. Neither A nor B

4. A customer complains that the engine seems to surge when driving at about 48 mph.

Technician A says there may be a problem with the torque converter clutch.

Technician B says a faulty vehicle speed sensor may cause this problem.

Who is correct?

A. A only

B. B only

C. Both A and B

D. Neither A nor B

5. *Technician A* says a plugged transmission cooler or cooler lines may cause the vehicle to stall when the transmission is shifted into a forward gear.

Technician B says a plugged transmission cooler or cooler lines may overheat the converter.

Who is correct?

A. A only

B. B only

C. Both A and B

D. Neither A nor B

Job Sheet 18

18

Name _____ Date _____

Road Test a Vehicle to Check the Operation of a Torque Converter Clutch

Upon completion of this job sheet, you should be able to road test a vehicle with an automatic transmission to verify and analyze the operation of a torque converter clutch.

ASE Correlation

This job sheet is related to the ASE Automatic Transmission and Transaxle Test's Content Area: *General Transmission and Transaxle Diagnosis.*
Task: Perform lockup converter mechanical/hydraulic system tests; determine necessary action.

Tools and Materials

A vehicle with an automatic transmission and lockup converter
Service manual
Pad of paper and a pencil

Describe the vehicle being worked on:

Year _____ Make _____ VIN _____

Model _____

Model and type of transmission _____

Procedure

Task Completed

1. Park the vehicle on a level surface.

☐

2. Describe the condition of the fluid (color, condition, smell).

_____

_____

3. What is indicated by the fluid's condition?

_____

_____

4. In the appropriate service manual, find the chart that shows the conditions that must be present in order for the converter clutch to be engaged. Describe these conditions below:

_____

_____

_____

_____

5. Allow the engine and transmission to cool by letting the vehicle sit.

☐

6. Then take the vehicle for a test drive. Bring the vehicle to normal lockup speed in the appropriate gear and observe the action of the clutch. Record the speed and feel of the clutch engagement:

_____

_____

_____

7. Was the engagement of the clutch normal for the engine temperature and vehicle speed? Why or why not?

_____

_____

8. Now drive the car through the following conditions and note the action of the torque converter clutch during each of these:

Manual upshifting of the transmission during acceleration

Clutch action: _____

Part-throttle downshifting

Clutch action: _____

Wide-open throttle downshifting

Clutch action: _____

Cruising at highway speed then lightly applying the brakes

Clutch action: _____

Cruising at highway speed then letting completely off the throttle but not applying the brakes

Clutch action: _____

9. Summary of clutch operation:

_____

_____

_____

_____

Instructor's Response _____

_____

_____

Job Sheet 19

19

Name _____ Date _____

Visual Inspection of a Torque Converter

Upon completion of this job sheet, you should be able to conduct a visual inspection of a torque converter.

ASE Correlation

This job sheet is related to the ASE Automatic Transmission and Transaxle Test's Content Area: *General Transmission and Transaxle Diagnosis.*
Task: Listen to driver's concern and road test vehicle to verify mechanical/hydraulic system problems; determine necessary action.

Tools and Materials

A vehicle with an automatic transmission
Clean white paper towels
Hoist

Remote starter switch
Droplight or good flashlight
Service manual

Describe the vehicle being worked on:

Year _____ Make _____ VIN _____

Model _____

Model and type of transmission _____

Procedure

Task Completed

1. Park the vehicle on a level surface.

☐

2. Describe the condition of the fluid (color, condition, smell).

_____

_____

3. What is indicated by the fluid's condition?

_____

_____

4. Connect a remote starter switch to the vehicle and disable the ignition. Explain how you did this:

_____

_____

5. Raise the vehicle on a hoist to a comfortable working height.

☐

6. Put on eye protection.

☐

7. Remove the torque converter access cover or shield.

☐

Visual Inspection of a Torque Converter (continued)

8. Inspect the converter through the access hole. Use the remote starter switch to observe all of the converter. Describe your findings below:

_____

_____

_____

_____

Instructor's Response _____

_____

_____

Job Sheet 20

20

Name _____ Date _____

Conduct a Stall Test

Upon completion of this job sheet, you should be able to conduct a stall test.

ASE Correlation

This job sheet is related to the ASE Automatic Transmission and Transaxle Test's Content Area: *General Transmission and Transaxle Diagnosis.*
Task: Perform stall tests; determine necessary action.

Tools and Materials

A vehicle with an automatic transmission   Stethoscope
Tachometer   Droplight or good flashlight
Hoist   Service manual

Describe the vehicle being worked on:

Year _____ Make _____ VIN _____

Model _____

Model and type of transmission _____

Procedure

Task Completed

1. Park the vehicle on a level surface.   ☐

2. Describe the condition of the fluid (color, condition, smell).

   _____

   _____

3. What is indicated by the fluid's condition?

   _____

   _____

4. Connect a tachometer to the engine. If the vehicle has one in the instrument panel, there is no need to connect another one. Explain how you connected the tachometer:

   _____

   _____

5. Place the tachometer on the inside of the vehicle so that it can be easily read.   ☐

6. Mark the face of the tachometer with a grease pencil at the recommended maximum rpm for this test. What is the maximum rpm? _____

7. Check the engine's coolant level.   ☐

8. Block the front wheels and set the parking brake.   ☐

9. Place the transmission in PARK and allow the engine and transmission to warm.   ☐

Conduct a Stall Test (continued)

10. Place the gear selector into the gear that is recommended for this test. What is the recommended gear? _____

☐ 11. Press the throttle pedal to the floor with your right foot and firmly press the brake pedal with your left. Hold the brake pedal down.

12. When the tachometer needle stops rising, note the engine speed and let off the throttle. What was the highest rpm during the test? _____

13. Place the gearshift into neutral and allow the engine to run at 1000 rpm for at least one minute. What is the purpose of doing this?

_____

_____

☐ 14. If noise is heard from the transmission during the stall test, raise the vehicle on a hoist.

☐ 15. Use the stethoscope to determine if the noise is from the transmission or the torque converter.

16. What are your conclusions from this test?

_____

_____

_____

_____

Instructor's Response _____

_____

_____

_____

Job Sheet 21

21

Name _____ Date _____

Servicing an Automatic Transmission Oil Pump

Upon completion of this job sheet, you should be able to inspect, measure, and replace the components of an oil pump assembly.

ASE Correlation

This job sheet is related to the ASE Automatic Transmission and Transaxle Test's Content area: *Off-Vehicle Transmission and Transaxle Repair; Oil Pump and Converter.*
Task: Inspect, measure, and replace oil pump assembly and components.

Tools and Materials

Machinist Dye   Feeler Gauge

Describe the transmission being worked on and the vehicle the transmission is from:

Year _____ Make _____ Model _____

Model _____

Model and type of transmission _____

Procedure

Task Completed

1. With the oil pump removed from the transmission, remove the front pump bearing race, front clutch thrust washer, gasket and O-ring. Inspect the pump bodies, pump shaft, and ring groove areas. Record your findings.

   _____

   _____

   _____

2. Mount the pump on a stand and unbolt and separate the pump bodies.   ☐

3. Mark the gears with machinist bluing ink or paint before removing them so that the gears will remain in the same relationship during reassembly. What did you use to mark them and where did you mark them?

   _____

   _____

4. Inspect the gears and all internal surfaces for defects and visible wear. Record your findings.

   _____

   _____

5. With the pump mounted on the stand, with a feeler gauge measure between the outer gear and the crest in the pump housing crest. What tolerance do the specifications call for? What did you measure?

_____

_____

6. Measure between the outer gear teeth and the crescent. What tolerance do the specifications call for? What did you measure?

_____

_____

7. Place the pump flat on the bench and with a feeler gauge and straightedge, measure between the gears and the pump cover. What tolerance do the specifications call for? What did you measure?

_____

_____

8. Measure the clearance between the C-ring and the ring groove. What tolerance do the specifications call for? What did you measure?

_____

_____

9. Using the stand to center the pump, torque the securing bolts to specifications. What is the specified torque?

_____

_____

10. Replace all O-rings and gaskets.

Instructor's Response _____

_____

_____

General Hydraulic System Service

Upon completion and review of this chapter, you should be able to:

❏ Remove and install the valve body from common transmissions.

❏ Check valve body mating surfaces.

❏ Inspect and measure valve body bores, springs, sleeves, retainers, brackets, check balls, and screens.

❏ Inspect and replace valve body spacers and gaskets.

❏ Check and/or adjust valve body bolt torque.

❏ Inspect and replace the governor cover and seals.

❏ Inspect, adjust, repair, and replace the governor sleeve, valve, weights, springs, retainers, and gear.

Valve Body Removal

If the pressure test indicated that there is a problem associated with the valves in the valve body, a thorough disassembly, cleaning in fresh solvent, careful inspection, and the freeing up and polishing of the valves may correct the problem. Sticking valves and sluggish valve movements are caused by poor maintenance, the use of the wrong type of fluid, and/or overheating the transmission. The valve body of most transmissions can be serviced when the transmission is in the vehicle; however, it is typically serviced when the transmission has been removed for other repairs.

Typically, to remove a valve body from a transmission while it is still in the vehicle, begin by draining the fluid and removing the oil pan. Then disconnect the manual and throttle lever assemblies. Carefully remove the detent spring and screw assemblies. Loosen and remove the valve body screws (Figure 7-1). Before lowering the valve body and separating the assembly, hold the assembly with the valve body on the bottom and the transfer and separator plates on top. Holding the assembly in this way will reduce the chances of dropping the steel balls that are located in the valve body (Figure 7-2). Lower the valve body and note where these steel balls are located in the valve body; remove them and set them aside, along with the various screws.

> ✔ **SERVICE TIP:** To avoid having to spend hours crawling on the floor of the shop looking for lost parts, place your hand or fingers over spring-loaded valves or plugs when removing them from the valve body.

Removal of a valve body from a transmission on a bench is similar. Make sure you have the transmission positioned so that the valve body can be lowered in such a way that you can keep the check balls in place.

There are many different designs and configurations used in today's transmissions. Each of them requires unique steps for removal and installation. It is extremely important that you refer to the service manual for these specifics. To illustrate these variables, valve body removal and installation procedures for some common transmissions follows, as well as a description of the components that are included in the valve body assembly.

Chrysler Transaxles

Chrysler's 41TE transaxle is found in many of their cars and mini-vans. The valve body assembly of these units is comprised of the valve body, a transfer plate, and a separator plate. The valve

Basic Tools

Basic hand tools

Cleaning pan

OSHA-approved air nozzle

Cleaning solution

Classroom Manual
Chapter 7, page 204

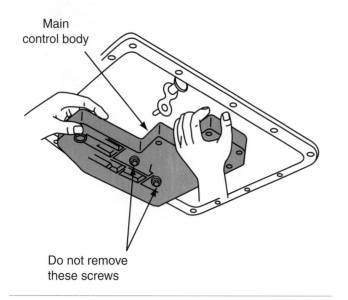

Main control body

Do not remove these screws

Figure 7-1 When removing a valve body from the transmission, remove only the bolts that are necessary to lower the valve body. Also, be careful not to lose any springs and check balls.

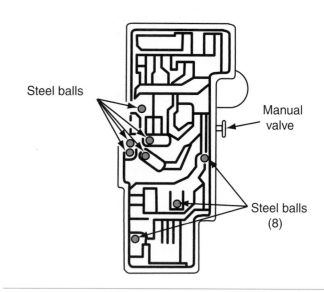

Steel balls

Manual valve

Steel balls (8)

Figure 7-2 Location of manual shift valve and check balls in a typical valve body.

The **solenoid/ pressure switch** is a single unit that consists of various solenoids and pressure switches.

The **manual shaft** is connected to the gearshift linkage.

body contains many different valves and check balls (Figure 7-3). All of these control fluid flow to the TCC, the **solenoid/pressure switch** assembly, and the various hydraulic apply devices.

To remove the valve body, move the manual valve lever clockwise into low gear. Loosen and remove the valve body retaining bolts. Then, using a screwdriver, push the park rod rollers out of the guide bracket (Figure 7-4). Then remove the valve body from the housing. Carefully pull the valve body assembly from the case. The **manual shaft** is attached to the valve body and must be carefully pulled from its bore.

Once removed, the valve body can be disassembled. This begins with the removal of the TR sensor assembly (Figure 7-5). The accumulator retaining plate and accumulator are then removed.

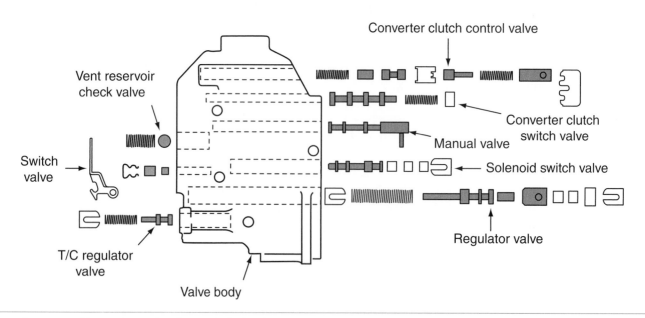

Converter clutch control valve

Vent reservoir check valve

Switch valve

Manual valve

Converter clutch switch valve

Solenoid switch valve

T/C regulator valve

Valve body

Regulator valve

Figure 7-3 The various valves in a typical Chrysler 41TE.

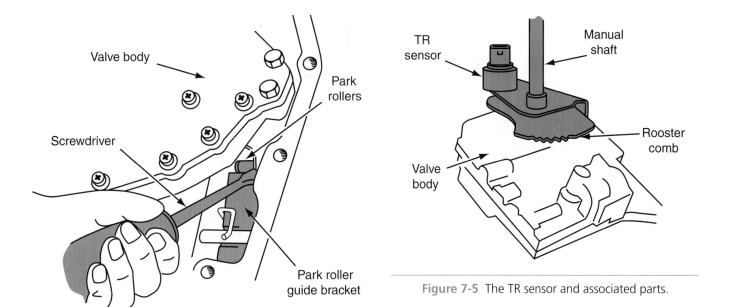

Figure 7-4 Location of the park rod rollers and guide.

Figure 7-5 The TR sensor and associated parts.

The transfer plate can now be unbolted and separated from the valve body. While separating these units, pay attention to the check balls. It is very possible that a loose check ball will drop out.

Remove the oil screen and overdrive clutch check valve from the separator plate. Then remove the separator plate (Figure 7-6). Then remove the thermal valve and check balls from the valve body. Keep track of their location and count the balls as they are removed. Disassemble, inspect, and clean the valve body components as needed.

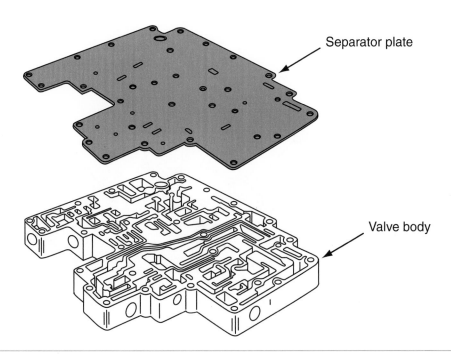

Figure 7-6 A valve body with its separator plate.

To install the valve body, guide the park rod rollers into the guide bracket while positioning the valve body. Install the valve body-to-case mounting bolts and tighten them to specifications.

Ford Transmissions

The 4R70W transmission is commonly found in Ford pickups and SUVs. Removing the valve body from this transmission is a rather simple process. The valve body is bolted directly to the transmission case and few items need to be removed to remove it. The electrical connectors to the various solenoids mounted to or around the valve need to be disconnected. Once they are, the wiring harness can be removed by disconnecting it from its main connector. After the harness is removed, the shift and TCC solenoids can be unbolted and removed. Then remove the retaining bolt for the manual valve detent lever spring and then the spring (Figure 7-7). Now unbolt the valve body from the transmission case. Remove and discard the pump outlet screen.

Remove the separator plate and discard the gaskets. Note the location of the check balls (Figure 7-8) and remove them before cleaning the valve body. Once the valve body is serviced, install the check balls and the separator plate. Install a new pump outlet screen and gaskets during reassembly.

To ensure proper alignment of the valve body, this transmission has two alignment bolts. Position the valve body gasket and valve body onto the case by using these alignment bolts as a guide. Loosely install all of the valve body retaining bolts. Then install the manual valve detent lever spring and tighten it in place. Before doing this, make sure the drive pin of the detent lever is correctly positioned in the manual valve.

Now tighten the valve body retaining bolts according to the specified sequence and to the correct torque. Install the shift and TCC solenoids and reconnect the wiring harness to the main connector and to the solenoids.

General Motors' Transmissions

GM's 4L60-E transmission is used in a wide variety of RWD vehicles. Like the valve body in a Ford transmission, the valve body of a 4L60 is rather straightforward. Although the procedures are similar, the 4L60 requires a different sequence and, due to its design, additional steps. Photo Sequence 11 covers the procedure for removing and installing a valve body when the 4L60 is still in the vehicle.

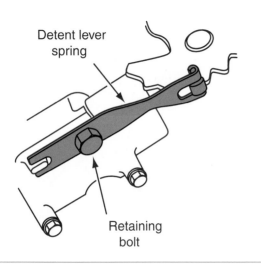

Figure 7-7 The detent lever spring for the manual valve.

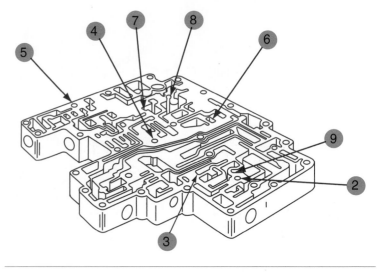

Figure 7-8 Location of the check balls in a 4R70W transmission.

Photo Sequence 11
Removing and Installing a Valve Body

P11-1 Position the vehicle on a lift and loosen the transmission pan to drain the fluid.

P11-2 After the fluid has drained, remove the pan and the transmission filter.

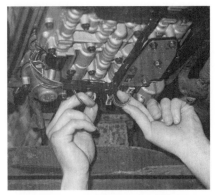

P11-3 Allow the fluid to drip off the transmission, then disconnect the electrical connections to the various switches and solenoids mounted to or around the valve body.

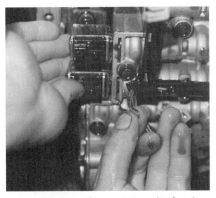

P11-4 Remove the retaining clip for the PWM solenoid, and then do the same for the shift solenoids.

P11-5 Remove the TCC solenoid.

P11-6 Remove the retaining bolts or clips for the electrical harness. Move the harness off to the side of the transmission.

P11-7 Now remove the pressure switch retaining bolts and switch.

P11-8 Now unbolt and remove the transmission fluid pressure switch assembly. Carefully inspect this assembly for any damage or debris.

P11-9 Unbolt and remove the manual detent spring.

(continued)

P11-10 Loosen and remove the remaining valve body mounting bolts.

P11-11 Lower the valve body just enough to disconnect the manual valve's linkage, then lower the valve body away from the transmission and remove it.

P11-12 Unbolt and remove the accumulator assembly.

P11-13 Before reinstalling the valve body, check the fit of the new gasket to the transmission.

P11-14 Reinstall the accumulator.

P11-15 Position the valve body on the case and loosely install some of the mounting bolts. Then, install the manual detent spring.

P11-16 Install the transmission fluid pressure switch assembly. Do not tighten the mounting bolts.

P11-17 Position the electrical harness and loosely install the retaining bolts.

P11-18 Install the TCC solenoid but do not tighten the mounting bolts.

Photo Sequence 11 (cont'd.)
Removing and Installing a Valve Body

P11-19 Now tighten all of the valve body mounting bolts according to the specified order and to the specified torque.

P11-20 Install the various solenoids and switches to the valve body. Then tighten all retaining bolts to the specified torque.

P11-21 Reconnect the electrical connectors to the various solenoids and switches.

P11-22 Install a new fluid filter.

P11-23 Install the transmission pan with a new gasket and tighten the retaining bolts to specifications.

P11-24 Fill the transmission with fluid, following the correct procedure and using the correct fluid.

The removal process begins with disconnecting the electrical connections to the various switches and solenoids mounted to or around the valve body (Figure 7-9). The wiring harness cannot be removed until the TCC solenoid has been removed. The order in which the solenoids are removed is important. Some retaining bolts and brackets are covered by the solenoids. Remove the PWM solenoid to gain access to the retaining bolts for the TCC solenoid. Once the TCC solenoid is removed, the electrical harness retaining bolts can be removed.

Now remove the pressure switch retaining bolts and switch. Now unbolt and remove the **transmission fluid pressure switch** assembly (Figure 7-10). This unit needs to be carefully inspected and checked for any damage or debris.

The manual detent spring can now be unbolted and removed. Once this is off, the remaining valve body mounting bolts can be removed. Once loosened, the valve body should be slightly lowered and the manual valve's linkage disconnected. This will allow for removal of the valve body.

The **transmission fluid pressure switch** assembly contains five different pressure switches and is connected to five different hydraulic circuits.

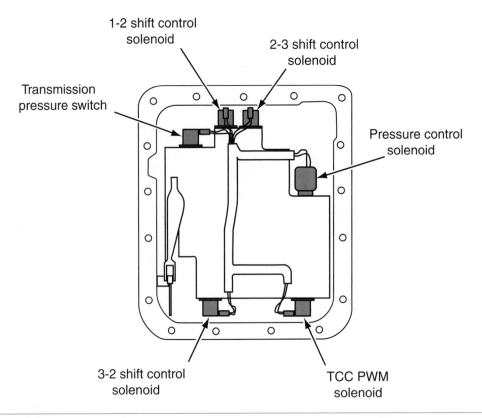

Figure 7-9 The various solenoids and switches located on or near the valve body of a 4L60-E transmission.

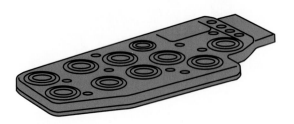

Figure 7-10 A transmission fluid pressure (TFP) switch assembly for a 4L60-E.

Installation is done in the reverse order of removal. Again it is important that the solenoids be installed according to the sequence. All of the valve body mounting bolts should be loosely installed as components are installed, until all of the bolts are in their proper location. Then the bolts should be tightened in the specified order and to the specified torque.

Honda Transaxles

The valve body assembly in this unique transaxle is comprised of a main valve body, the regulator valve body, the servo valve body, and the accumulator body (Figure 7-11). The main valve body contains the pump gears and the manual, modulator, shift, servo control, TCC control, and cooler check valves. The shift valves are controlled by fluid flow directed by one of the three shift solenoids.

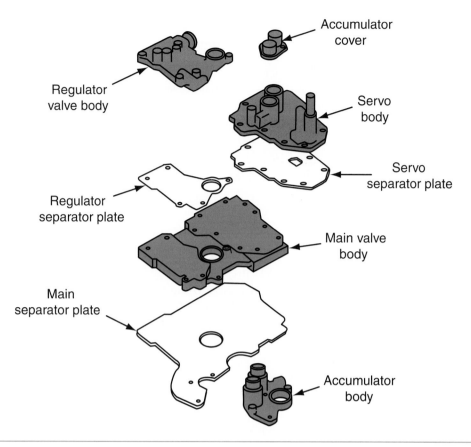

Figure 7-11 The complete valve body assembly for a late-model Honda transaxle.

The regulator valve body contains the regulator, TCC timing, and relief valves. The servo valve body contains additional shift valves, two forward speed accumulators, and the servo valve. The accumulator valve body contains two forward speed accumulators and a lubrication check valve.

Transmission fluid from the regulator passes through the manual valve to the different shift and control valves. Oil feed pipes and internal passages are used to move the fluid to the various clutches in the transmission.

To remove this valve body, the ATF feed pipes must be removed first. Then the ATF strainer and servo detent plate should be unbolted and removed. Now press down on the accumulator cover while loosening its mounting bolts. Loosen the bolts in a staggered pattern. The cover is spring loaded and the threads in the servo valve body will strip if the cover is not evenly removed.

Now the servo valve body, servo separator plate, accumulator valve body, and regulator valve body can be removed. After these are removed, the stator shaft and shaft stop should be removed. Then unhook the detent spring from the detent arm and remove the detent arm shaft, detent arm, and control shaft.

Note the location of the cooler check ball and spring and remove them. Once these are removed, the main valve body can be unbolted and removed. While removing the valve body, be prepared to catch the TCC control valve and spring. These are held in position by the valve body.

Remove the pump gears, noting which side of the gears faces up. Now the main separator plate and its dowels can be removed. Clean and inspect all parts. Installation of the valve body assembly follows the reverse order of the removal procedure.

Valve Body Service

CUSTOMER CARE: If while you are disassembling and inspecting the valve body you discover signs of poor transmission maintenance or abuse, make note of it. When you next talk to the customer, explain what you found and what the resulting problem was. Make sure you explain this in an understanding and non-offensive way.

If previous tests suggest a problem with only one or two valves, start your inspection at those valves. Doing this will not only save you time, but will also reduce the chance of something being misplaced or ruined during a total disassembly. If the transmission had heavily contaminated fluid, the entire valve body should be inspected and cleaned, or replaced.

Disassembly

A valve body contains many valves. These valves are typically held in their bores by a plug or cover plate. These must be removed to gain access to the valves. The cover plates are bolted or screwed to the valve body (Figure 7-12). Plugs can be held in place in a number of different ways. All of them can be released by pressing the plug slightly into the bore (Figure 7-13).

Begin disassembly by removing the manual shift valve from the valve body. Then, remove the pressure regulator retaining screws while keeping one hand around the spring retainer and adjusting screw bracket. Remove the pressure regulator valve. Then remove all of the valves and springs from the valve body. Make sure that you keep all springs and other parts with their associated valves.

It is important that you somehow keep track of the position of the springs in relationship to the valves. You can draw diagrams on a piece of paper, which can serve as a quick reference during reassembly. You can also use a Polaroid or some other type of instant camera to photograph the valve body with its channel plate exposed and the check balls in place. This will make an excellent reference for reassembly.

SERVICE TIP: Another trick is to use the cardboard sheet included in every gasket set. Fold this sheet in one-inch pleats like an accordion and lay it on your bench with the slick side of the cardboard facing up. Then follow these steps:

1. Take the valves and springs out of the valve body and lay them in the different grooves in the sequence they were removed.
2. Clean the valve body castings.
3. Clean the valves and springs. Do not put them back on the cardboard; rather, put them directly into their bores in the valve body.

Classroom Manual
Chapter 7, page 196

It has been a long while since carburetors were installed on engines, but **carburetor cleaner** is readily available and is quite effective for cleaning aluminum parts.

Cleaning and Inspection

Remove the check balls and springs. Note the exact location of each and count them as they are removed. Compare your count to the number specified in the service manual. This will ensure that all of them have been removed and some hidden or stuck ones will not be lost during cleaning. Never use a magnet to remove the check balls. The balls may become magnetized and tend to collect particles in the fluid. Continue disassembly by removing the rest of the valves and springs.

After all of the valves and springs have been removed from the valve body, soak the valve body and separator and transfer plates in mineral spirits for a few minutes. Some rebuild shops soak the valve body and its associated parts in **carburetor cleaner** or lacquer thinner, then the parts are washed off with water. Thoroughly clean all parts and make sure all passages within the valve body are clear and free of debris. Carefully blow-dry each part individually with dry compressed air. Never wipe the parts of a valve body with a rag or paper towel. Lint from either will collect in the valve body passages and cause shifting problems.

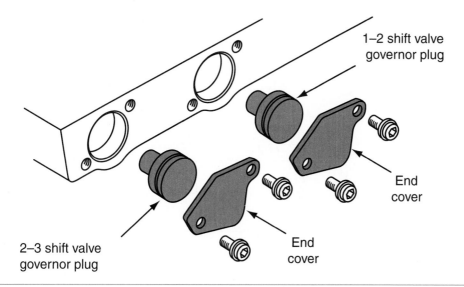

Figure 7-12 Examples of the covers used to retain valves and their springs in the bores of a valve body.

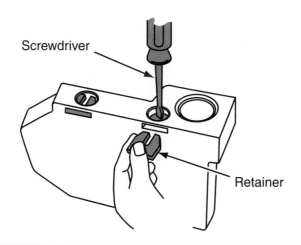

Figure 7-13 To remove or install the retainer for a valve retaining plug, the plug is depressed and the retainer pulled out or inserted into its groove.

 WARNING: Wear safety glasses when drying parts with compressed air. The overspray of solvent and other cleaners can damage your eyes and lead to blindness.

 **CAUTION:** Do not leave valve bodies or other parts submerged in carburetor cleaner longer than five minutes.

Check the separator plate for scratches or other damage (Figure 7-14). Scratches or score marks can cause oil to bypass correct oil passages and result in system malfunction. If the plate is defective in any way, it must be replaced. Check the oil passages in the upper and lower valve bodies for varnish deposits, scratches, or other damage that could restrict the movement of the valves. Make sure all fluid drain back openings are clear and have no varnish or dirt buildup. Check all of the threaded holes and related bolts and screws for damaged threads, and replace as needed.

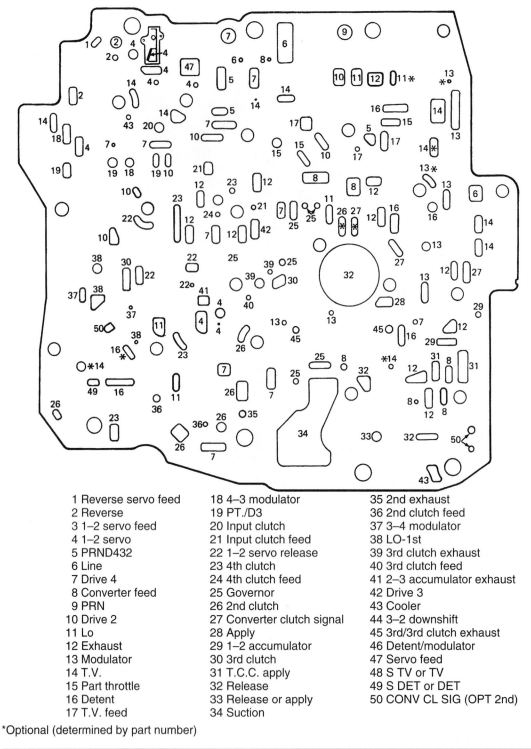

1 Reverse servo feed	18 4–3 modulator	35 2nd exhaust
2 Reverse	19 PT./D3	36 2nd clutch feed
3 1–2 servo feed	20 Input clutch	37 3–4 modulator
4 1–2 servo	21 Input clutch feed	38 LO-1st
5 PRND432	22 1–2 servo release	39 3rd clutch exhaust
6 Line	23 4th clutch	40 3rd clutch feed
7 Drive 4	24 4th clutch feed	41 2–3 accumulator exhaust
8 Converter feed	25 Governor	42 Drive 3
9 PRN	26 2nd clutch	43 Cooler
10 Drive 2	27 Converter clutch signal	44 3–2 downshift
11 Lo	28 Apply	45 3rd/3rd clutch exhaust
12 Exhaust	29 1–2 accumulator	46 Detent/modulator
13 Modulator	30 3rd clutch	47 Servo feed
14 T.V.	31 T.C.C. apply	48 S TV or TV
15 Part throttle	32 Release	49 S DET or DET
16 Detent	33 Release or apply	50 CONV CL SIG (OPT 2nd)
17 T.V. feed	34 Suction	

*Optional (determined by part number)

Figure 7-14 Typical bores of a separator plate. (Courtesy of General Motors Corporation, Service Operations)

If the valve body has an oil strainer, make sure it is clean. This can be checked by pouring clean ATF through the strainer and observing the fluid flow out of the strainer. If there is poor flow, the strainer should be replaced.

Examine each valve for nicks, burrs, and scratches. If the valve lands have worn areas, this means the valve has been rubbing in its bore. Make sure each valve properly fits into its respective bore. To do this, hold the valve body vertically and install an unlubricated valve into its bore. Let the valve fall of its own weight into the valve body until the valve stops. Then, place your finger over the valve bore and turn the valve body over. The valve should again drop by its own weight. If the valve moves freely under these conditions, it will operate freely with fluid pressure. Repeat this test on all valves.

The problem may be corrected by polishing the valve lands if the valves are made of steel. If the valves are made of aluminum and are damaged or worn, the valve body should be replaced. To polish a valve, use a polishing stone, crocus cloth, or very fine sandpaper (600 or higher grit). Many good technicians allow the sandpaper to soak in ATF before using it to polish a valve. Evenly rotate the valve on the polishing material, making sure this does not round the edges of the valve. Do not use anything that is coarse on the valve, and polish the lands only enough to ensure free movement in the bore. After the valve is polished, clean it.

An alternative to using a polishing stone or crocus cloth is the use of toothpaste or a mixture of ATF and a very fine abrasive cleaning paste. Place the paste in your hand, then wrap your hand around the valve, and move it back and forth, letting the abrasives remove the burr or nick.

After using any of these polishing methods to remove a nick or burr, the valve must be thoroughly cleaned to remove all of the cleaning and abrasive materials. After the valve has been recleaned, it should be tested in its bore again.

If the valve cannot be cleaned well enough to move freely in its bore, the valve body should be replaced. Individual valve body parts are usually not available. Individual valves are lapped to a particular valve body and, therefore, if any parts need to be replaced, the entire valve body must be replaced.

The valve bores need attention as well. Roll up a half sheet of ATF-soaked 600+ grit sandpaper. Insert the paper into the valve's bore. Turn the roll of paper so that it unrolls and expands to the size of the bore. Then twist the paper while moving it in and out of the bore. Then clean the entire valve body in solvent and dry it with compressed air.

Although it is desirable to have the valves move freely in their bores, excessive wear is also a problem. There should never be more than 0.001-inch clearance between the valve and its bore. If the bore or valve is worn, the entire valve body needs to be replaced.

> ✔ **SERVICE TIP:** Make sure you include the manual valve while you are inspecting and cleaning the valves. It is easy to overlook this valve because it is often removed so early during the disassembly of the valve body and it won't be inserted into the valve body until the valve body is installed into the transmission case.

Check each spring for signs of distortion. If any spring is damaged, the valve body should be replaced. A good way to check the springs is to lay them on their side and roll them. If they roll true, they are not distorted. If they wobble, they are distorted.

With a straightedge laid across the sealing surface of the valve body, use a feeler gauge to check its flatness. The size of gauge that fits between the straightedge and the surface of the valve body measures the amount the surface is warped or distorted. If it is slightly warped, it can be flat filed. Be very careful when doing this. Keep the file flat and always file in one direction. If the surface is warped beyond repair, the valve body must be replaced.

Reassembly

After the valve body has been cleaned, it should be allowed to air dry. Once dry it should be dipped into a pan of clean ATF. While the valve body is soaking in ATF, locate the installation specifications in the service manual.

> ✔ **SERVICE TIP:** When assembling a valve body, always have your notes, photograph, and the correct service manual close for quick reference. Many valves and springs look similar; however, each has its own bore and purpose (Figure 7-15).

Special Tools

Polishing stone
Crocus cloth

An Arkansas stone is the name for a common type of polishing stone.

Special Tools

Sheet of 600+ grit sandpaper

Sandpaper abrasiveness is rated by grit—the higher the number, the finer the grit.

Special Tools

Straightedge
Feeler gauge set

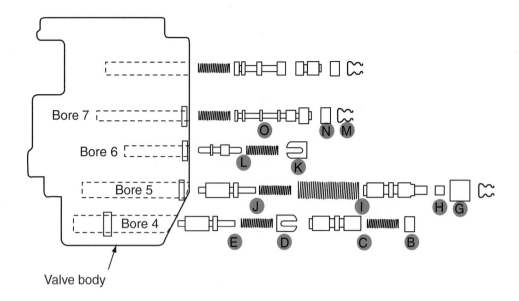

Bore 4	Bore 5	Bore 6	Bore 7
Ⓐ Spring retainer plate	Ⓕ Clip	Ⓚ Retainer plate	Ⓜ Clip
Ⓑ Bore plug	Ⓖ Sleeve	Ⓛ TV limit valve and spring	Ⓝ Bore plug
Ⓒ Orifice control valve and spring	Ⓗ Plug		Ⓞ 1–2 shift valve and spring
Ⓓ Spring retainer plate 1	Ⓘ 3–4 shift valve and spring		
Ⓔ 2–3 capacity modulator valve and spring	Ⓙ 3–4 TV modulator valve and spring		

Figure 7-15 Typical location of some of the valves and springs in a valve body.

Lubricate all parts with clean ATF. Then, install the valves and associated springs into their bores. It is important that you place the valve retaining plugs or caps in the correct bore and in the correct direction (Figure 7-16). Once the cap is in position, carefully depress it with a small screwdriver and install the retaining clip.

Install all check balls and springs in their correct location. If you have any doubts as to where they should be placed, refer to the service manual. Count the check balls as you install them to make sure you have inserted all of them.

Before beginning to install the valve body, check the new valve body gasket to make sure it is the correct one by comparing it to the old gasket. If the gasket appears to be the correct one, lay it over the separator plate and hold it up to a light, making sure no oil holes are blocked (Figure 7-17). Also check to make sure the gasket seals off the worm tracks and won't allow the fluid to go where it shouldn't go. Then install the bolts to hold valve body sections together and the valve body to the case. Tighten the bolts to the torque specifications to prevent valve body warpage and possible leakover. Over-torquing can also cause the bores to distort, which would not allow the valves to move freely once the valve body is tightened to the transmission case.

Many transmissions use bolts of various lengths to secure the valve body to the case. It is

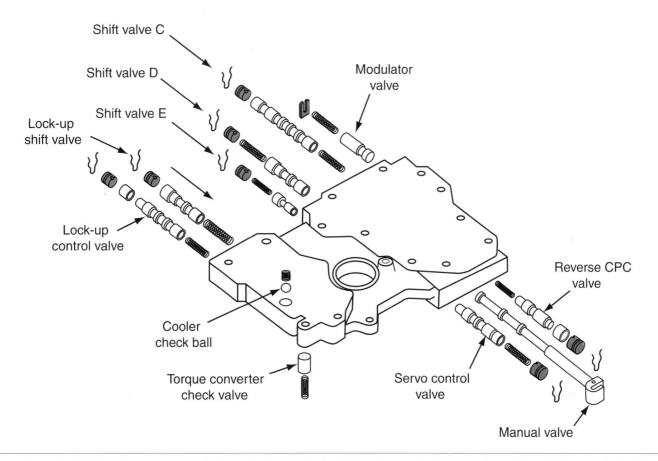

Figure 7-16 The plugs used to retain the valves in a valve body have designated bores and must face in the correct direction.

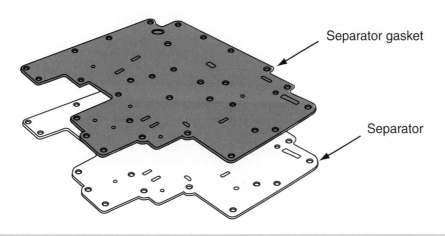

Figure 7-17 Always check a new valve body gasket by laying it over the separator plate and making sure it does not block any holes. If the gasket is correct, use clean ATF to hold it in place during assembly.

extremely important that the correct length bolt be used in each bore. This is so important that service manuals list the size and location of the mounting bolts (Figure 7-18).

> **CAUTION:** The primary cause of valve sticking is the over-tightening of the valve body bolts. This is especially true of aluminum valve bodies. Always be careful when handling a valve body; they are very precise components.

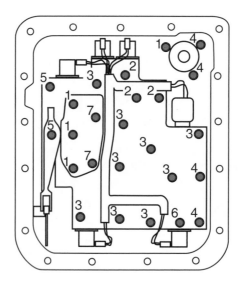

Each numbered bolt location corresponds to a specific bolt size and length, as indicated by the following:

1. M6 X 1.0 X 65.0
2. M6 X 1.0 X 54.4
3. M6 X 1.0 X 47.5
4. M6 X 1.0 X 35.0
5. M8 X 1.25 X 20.0
6. M6 X 1.0 X 12.0
7. M6 X 1.0 X 18.0

Figure 7-18 When installing the valve body mounting bolts, make sure you use the correct bolt size and length in the correct location. Refer to the service manual for guidance.

On some Chrysler transmissions, as well as a few others, the spring for the pressure regulator valve is adjustable. Adjustments are made by using an Allen wrench and machinist's rule (Figure 7-19).

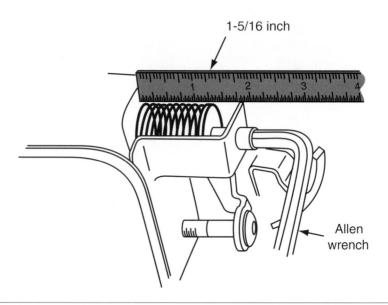

Figure 7-19 On some transmissions, the pressure regulator is adjustable. For the transmission shown in this figure, an Allen wrench is used to adjust a screw until it is at the specified distance. At this point the pressure regulator should be correctly adjusted.

Governor Service

If the pressure tests suggested that there was a governor problem, it should be removed, disassembled, cleaned, and inspected. Some governors are mounted internally and the transmission must be removed to service it. Other governors can be serviced by removing the extension housing or oil pan, or by detaching an external retaining clamp and then removing the unit (Figure 7-20).

Improper shift points are typically caused by a faulty governor or governor drive gear system. However, some electronically controlled transmissions do not rely on the hydraulic signals from a governor; rather, they rely on the electrical signals from sensors. Sensors, such as speed and load sensors, signal to the transmission's computer when gears should be shifted. Faulty electrical components and/or loose connections can also cause improper shift points.

Service to the governor, on many transmissions, can be done with the transmission still in the vehicle. If the transmission has a shaft-mounted governor, the governor is driven directly by the output shaft and can be accessed by removing the extension housing from the transmission. Some transaxles require complete disassembly of the transaxle to access the governor. Other transmissions may have a protrusion off the side of the extension housing that contains the governor. These governors are typically driven by a gear and are accessible by removing the governor cover from the protrusion.

Classroom Manual
Chapter 7, page 211

Basic Operation

When the forward drive range is selected by the driver, line pressure flows from the manual valve to the governor valve (Figure 7-21). When the output shaft begins to rotate, the weights of the governor start to move out. As the weights move out, the governor's port opens, allowing fluid flow.

The fluid flow acts against the shift valve spring. Once the pressure can overtake the spring tension, the shift valve moves, causing an upshift. If governor pressure never builds enough to overcome the spring, the shift valve won't move. If the governor pressure is slow to build, delayed shifts will result. And, if governor pressure is higher than normal, early shifts will take place.

Disassembly

To disassemble a typical gear-driven governor, remove the governor cover by carefully prying it out of its bore. Once the cover has been removed, remove the primary governor valve from its bore in the governor housing. Then remove the secondary valve retaining pin, secondary valve spring, and valve (Figure 7-22).

Special Tools

Soft-jawed vise

Feeler gauge set

Assortment of pin drifts

Electric drill

Assortment of drill bits

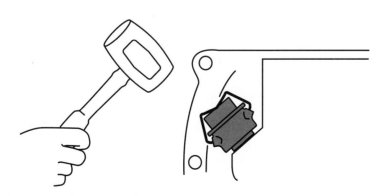

Figure 7-20 Some governor assemblies are contained in a separate housing and retained by a cover and external retaining clamp.

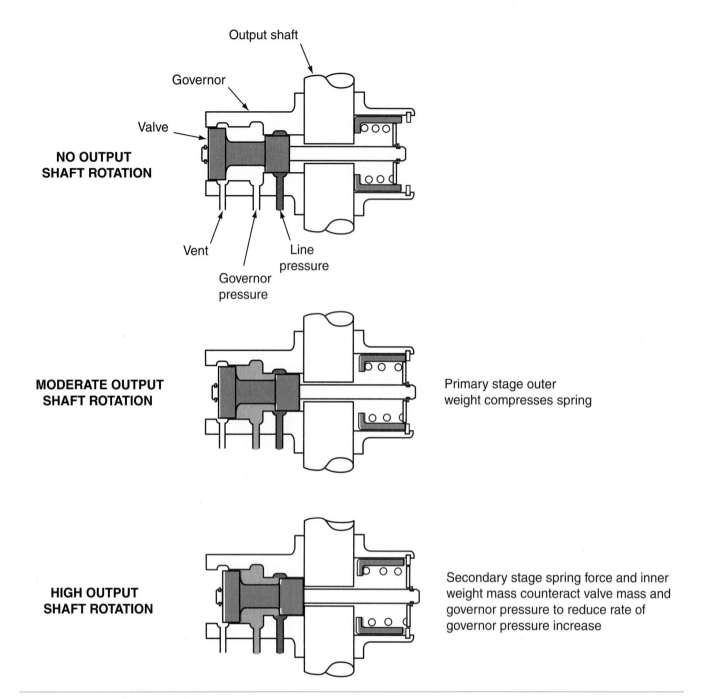

NO OUTPUT SHAFT ROTATION

Output shaft

Governor

Valve

Vent

Governor pressure

Line pressure

MODERATE OUTPUT SHAFT ROTATION

Primary stage outer weight compresses spring

HIGH OUTPUT SHAFT ROTATION

Secondary stage spring force and inner weight mass counteract valve mass and governor pressure to reduce rate of governor pressure increase

Figure 7-21 The action of a governor.

To disassemble a shaft-mounted governor after the extension housing and it has been removed, cut and remove the retaining pins in the primary weights and thrust cap. Then remove the thrust cap, weights, springs, and valve. Place the remaining assembly in a soft-jawed vise and drive out the retaining pin for the drive gear. Then press off the gear.

Prior to disassembling the governor, it is wise to check the action of the governor valve by moving the weights. With the weights held to the shaft, the exhaust port of the valve should be open. The amount the port is opened can be measured with a feeler gauge. Typically the port should be open at least 0.020 inches. With the weights held in their fully extended position, the exhaust port should be closed and the inlet port opened. Again, the amount the port is opened can be measured with a feeler gauge. Typically the port should be open at least 0.020 inches.

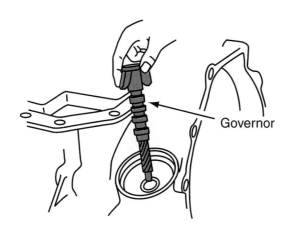

Figure 7-22 After the retaining cover has been removed, the governor assembly can be pulled out of its bore.

While moving the weights, pay attention to their movement. They should move freely and return to their rest position without much effort.

Thoroughly clean and dry governor parts using the same method as is used for valve body parts. Test the valve in its bore in the governor housing; it should move freely in the bore without sticking or binding. Also check the valve for any signs of burning or scoring and replace it, if necessary. Inspect the springs for a loss of tension and burning marks and replace if necessary. Make sure you check the ports of the governor for any buildups that may restrict fluid flow.

Reassembly

To reassemble a shaft-mounted governor, place the spring around the secondary valve and insert it into the secondary valve bore. Then insert the retaining pin into the governor housing pinholes. Now, install the primary valve into the governor housing. The governor cover should then be driven into place with a new seal. Make sure you lubricate the seal with ATF before driving the cover into the bore.

CAUTION: Never interchange components of the primary and secondary governors. Also note that the flat faces of the primary valve must face outward when it is installed.

To reassemble a shaft-driven governor, use a press and install the drive gear. Many rebuilders recommend that a new hole be bored for the gear retaining pin. This provides for secure mounting of the gear to the governor. If you need to or choose to do this, set the assembly in a soft-jawed vise. Position the governor so you can drill a new hole about 90 degrees away from the original bore. The new hole should be slightly smaller than the outside diameter of the retaining pin. Normally the hole should have an inside diameter of 1/8 inch. After the hole has been bored, insert a new retaining pin to secure the gear. Thoroughly clean the assembly before you continue to assemble and install it. Then install the weights, springs, valve, and thrust cap. Insert new retaining pins through the thrust cap and weights and crimp both ends of the pins to prevent them from working out.

If the governor assembly was removed from the governor support and parking gear, be sure to tighten the bolts to specifications with a torque wrench. After assembly, install the governor and torque the bolts to specifications. Over-tightening can cause the valve to stick. Some transmissions use a drive ball on the output shaft that locks the governor to the shaft (Figure 7-23). Make sure it is in place when installing the governor.

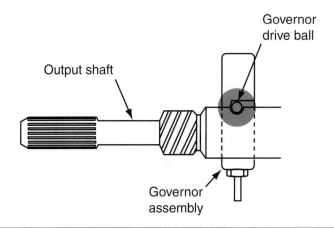

Figure 7-23 Governor drive ball in output shaft.

A customer with a late-model Honda complains that his automatic transmission seems to start out in the wrong gear when drive is selected. The technician takes the car for a road test to verify the complaint. She observed and verified the complaint. It did seem like the vehicle started out in second or third gear and had very poor and unsafe acceleration. But once the car got going, the transmission seemed to work fine.

Following the normal diagnostic routine, she found nothing all that unusual. So she took the car out again on a road test, this time with pressure gauges installed. She jotted down the test results and returned to the shop to search published information about this particular transmission.

She found that on Honda AS, AK, CA, and F4 model transmissions that are governor controlled, a complaint of no reverse and/or wrong gear starts can be caused by either stuck valves in the valve body or high governor pressure at a stop. The information further stated that if the pressure gauge reads less than 2 psi when the vehicle is at a stop, the problem is sticky valves in the valve body. If the gauge reads 2 psi or more, then high governor pressure is the problem. Either the governor valve is sticking or something else is allowing a higher pressure to leak into the governor circuit.

Since she experienced more than 2 psi, she knew the problem was high governor pressure. She then removed the governor and used 800-grit sandpaper to polish the valve. Then she reassembled and installed the governor. She kept the pressure gauges connected and took a road test. She found that the car accelerated as it should and found the pressure at a stop to be normal.

There is a small tube that must be installed into the governor shaft. This tube separates line pressure from governor pressure, and if it is left out, line pressure will flow directly into the governor circuit, causing high governor pressure and the same problem.

Terms To Know

Carburetor cleaner	Solenoid/pressure switch	Transmission fluid pressure switch
Manual shaft		

ASE-Style Review Questions

1. *Technician A* says all parts of the valve body should be soaked in mineral spirits before reassembling. *Technician B* says a lint-free rag is a must when wiping down valves.
Who is correct?
   A. A only
   B. B only
   C. Both A and B
   D. Neither A nor B

2. While removing scratches in a valve:
Technician A uses a flat file to remove the scratch. *Technician B* uses a sand blaster or glass bead machine to polish the surface of the valve.
Who is correct?
   A. A only
   B. B only
   C. Both A and B
   D. Neither A nor B

3. *Technician A* says over-torqueing the hold-down bolts of the valve body can cause the valves to stick in their bore.
Technician B says flat filing the surface of the valve body will allow the valve body to seal properly and will therefore allow the valves to move more freely in their bores.
Who is correct?
   A. A only
   B. B only
   C. Both A and B
   D. Neither A nor B

4. While assembling a valve body after cleaning it:
Technician A lubricates all parts with clean ATF. *Technician B* says it is important that the valve retaining plugs or caps be placed in the correct bore and in the correct direction.
Who is correct?
   A. A only
   B. B only
   C. Both A and B
   D. Neither A nor B

5. After soaking valve body parts in mineral spirits:
Technician A wipes the parts off with a paper towel. *Technician B* blow-dries each part individually with compressed air.
Who is correct?
   A. A only
   B. B only
   C. Both A and B
   D. Neither A nor B

6. When reassembling a shaft-driven governor:
Technician A drills new holes on the weights and inserts a new retaining pin through the holes.

Technician B drills a new hole for the gear retaining pin.
Who is correct?
   A. A only
   B. B only
   C. Both A and B
   D. Neither A nor B

7. *Technician A* says all transmissions rely on the hydraulic signals from a governor to determine shift timing.
Technician B says sensors, such as speed and load sensors, send a signal to the computer to verify the governor's command to upshift.
Who is correct?
   A. A only
   B. B only
   C. Both A and B
   D. Neither A nor B

8. *Technician A* says that if diagnosis suggests a problem with only one or two valves, visual inspection of the valve body should start at those valves.
Technician B says that if the transmission had heavily contaminated fluid, the entire valve body should be inspected and cleaned, or replaced.
Who is correct?
   A. A only
   B. B only
   C. Both A and B
   D. Neither A nor B

9. While checking the action of a governor:
Technician A says that with the weights held away from the shaft, the exhaust port of the valve should be open.
Technician B says the weights should move freely and return to their rest position without much effort.
Who is correct?
   A. A only
   B. B only
   C. Both A and B
   D. Neither A nor B

10. *Technician A* says that if a valve cannot be cleaned well enough to move freely in its bore, the valve body should be replaced.
Technician B says that if there is even the slightest bit of damage or varnish buildup in a valve's bore, the entire valve body must be replaced.
Who is correct?
   A. A only
   B. B only
   C. Both A and B
   D. Neither A nor B

ASE Challenge Questions

1. *Technician A* says that if the governor pressure is slow to build, early upshifts will result.
   Technician B says that if line pressure is higher than normal, early shifts will take place.
   Who is correct?
   - **A.** A only
   - **B.** B only
   - **C.** Both A and B
   - **D.** Neither A nor B

2. Which of the following is *not* a common cause for sticking valves and sluggish valve movements?
   - **A.** Overtorqued valve body bolts.
   - **B.** A faulty pressure control solenoid.
   - **C.** The use of the wrong type of fluid.
   - **D.** Overheating the transmission.

3. *Technician A* says the valve body of all transmissions and transaxles can be serviced with the transmission still in the vehicle, providing that surrounding components are removed first.
   Technician B says the governor in all transmissions and transaxles can be serviced with the transmission still in the vehicle, providing that surrounding components are removed first.
   Who is correct?
   - **A.** A only
   - **B.** B only
   - **C.** Both A and B
   - **D.** Neither A nor B

4. *Technician A* says oil feed pipes are used to move the fluid to the various clutches and apply devices in a transmission.
   Technician B says shift and TCC solenoids are often mounted directly to the valve body.
   Who is correct?
   - **A.** A only
   - **B.** B only
   - **C.** Both A and B
   - **D.** Neither A nor B

5. *Technician A* says that if aluminum valves are scored or otherwise damaged, the individual valve or entire valve body should be replaced.
   Technician B says problems rarely result from excessive clearance between the valve and its bore.
   Who is correct?
   - **A.** A only
   - **B.** B only
   - **C.** Both A and B
   - **D.** Neither A nor B

Job Sheet 22

22

Name _____ Date _____

Service a Valve Body

Upon completion of this job sheet, you should be able to disassemble, clean, inspect, and re-assemble a valve body.

ASE Correlation

This job sheet is related to the ASE Automatic Transmission and Transaxle Test's Content Area: *In-Vehicle Transmission and Transaxle Repair.*
Tasks: Inspect valve body mating surfaces, bores, valves, springs, sleeves, retainers, brackets, check balls, screens, spacers, and gaskets; replace as necessary.
Check and adjust valve body bolt torque.

Tools and Materials

An automatic transmission or transaxle on a bench   Supply of clean solvent
Compressed air and air nozzle   Lint-free shop towels
Inch-pound torque wrench   Service manual
Measuring calipers

Describe the transmission being worked on:

Model and type of transmission _____

Year _____ Make _____ VIN _____

Model _____

Procedure

Task Completed

1. Remove the valve body attaching screws from the transmission. Start at the outside bolts and work toward the center if the service manual does not give specific instructions. ☐

2. Remove the valve body assembly and place in it a container of clean solvent. ☐

3. Remove the end plates and covers from the assembly. ☐

4. In the space below, draw out a simplified view of the valve body. Note the location of all the valves, check balls, and springs. ☐

5. Begin to disassemble the unit. Lay all the parts on a clean surface in the order in which they were disassembled. This will help during the reassembly of the valve body. Be careful with the check balls; they may not be all the same size. Some transmissions use slightly larger or smaller check balls in circuits. How many check balls were in the valve body? Were they all the same size?

_____

_____

_____

_____

6. Remove all the gaskets from the assembly. Place them aside. Do not throw them away. They will be needed for comparison when installing new gaskets.

7. Thoroughly clean the main body and plates of the assembly. Do not dry the valve body with anything that would leave lint. Allow the unit to air dry or dry it off with compressed air.

8. Inspect the valve body for damage and cracks. Also check the flatness of the body and the plates. Describe your findings.

_____

_____

_____

_____

9. What are the valves made of? _____

10. Check all the valves for free movement in their bores. Also check the valves for wear, scoring, and signs of sticking. Describe your findings and recommendations.

_____

_____

_____

_____

11. Correctly reinstall the valves and springs in their bores.

12. Replace the end plates or covers. Install the retaining screws by hand, then tighten them to the specific torque. The specified torque is _____.

13. Install the check balls into their proper location.

14. Compare the new gaskets with the old and install the correct new ones on the valve body.

15. Align the spring-loaded check balls and all other parts as needed.

16. Place the gasket on top of the transfer plate. Align the gasket holes with the transfer plate and the bores in the valve body.

17. Position the valve body on the transmission. Align the parking and other internal linkages. Then install the retaining screws by hand.

18. Tighten the screws to the specified torque. Tighten the screws in the center first, then work to the outside of the valve body. The specified torque is _____.

19. Summarize this job sheet.

_____

_____

_____

_____

_____

Instructor's Response _____

_____

_____

_____

Job Sheet 23

23

Name _____ Date _____

Servicing Governors

Upon completion of this job sheet, you should be able to inspect, repair, and replace a governor assembly.

ASE Correlation

This job sheet is related to the ASE Automatic Transmission and Transaxle Test's Content Area: *In Vehicle Transmission and Transaxle Repair.*
Task: Inspect, repair, and replace governor assembly.

Tools and Materials

Torque wrench

Describe the vehicle that was assigned to you:

Year _____ Make _____ Model _____

VIN _____ Engine type and size _____

Transmission type and model _____

Procedure

Task Completed

1. If the pressure tests suggested that there was a governor problem, it should be removed, disassembled, cleaned, and inspected. Some governors are mounted internally and the transmission must be removed to service the governor. Others can be serviced by removing the extension housing or oil pan, or by detaching an external retaining clamp and then removing the unit. How did you remove yours?

   _____

   _____

   _____

2. To disassemble a typical governor, remove the primary governor valve from its bore in the governor housing. ☐

3. Remove the secondary valve retaining pin, secondary valve spring, and valve. ☐

4. Thoroughly clean and dry these parts. ☐

5. Test each valve in its bore in the governor housing; they should move freely in their bores without sticking or binding. Record your findings.

   _____

   _____

6. Check the valves for any signs of burning or scoring and replace the valve body, if necessary. Record your findings.

   _____

   _____

Servicing Governors (continued)

7. Inspect the springs for a loss of tension and burning marks and replace, if necessary. Record your findings.

_____

_____

☐   8. To reassemble the governor, place the spring around the secondary valve, and insert it into the secondary valve bore.

☐   9. Insert the retaining pin into the governor housing pinholes.

☐   10. Install the primary valve into the governor housing.

11. If the governor assembly was removed from the governor support and parking gear, be sure to tighten the bolts to specifications with a torque wrench. After assembly, install the governor and torque the bolts to specifications. What are the specifications?

_____

_____

Instructor's Response _____

_____

_____

Gear and Shaft Service

Upon completion and review of this chapter, you should be able to:

❏ Inspect, measure, and replace thrust washers and bearings.

❏ Inspect and replace bushings.

❏ Inspect and measure a planetary gear assembly and replace parts as necessary.

❏ Inspect and replace shafts.

❏ Remove and install the various types of shaft seals.

❏ Inspect, repair, or replace transaxle drive chains, sprockets, gears, bearings, and bushings.

❏ Inspect and replace parking pawl, shaft, spring, and retainer.

❏ Inspect, measure, repair, adjust, or replace transaxle final drive components.

CHAPTER 8

Basic Tools

Basic mechanic's tool set

Clean lint-free rags

Appropriate service manual

Clean ATF

Thrust Washers, Bushings, and Bearings

 SERVICE TIP: All bushings and thrust washers (Figure 8-1) should be pre-lubed during transmission assembly.

Thrust Washers and Bearings

The purpose of a thrust washer is to support a **thrust load** and keep parts from rubbing together, preventing premature wear on parts such as planetary gearsets (Figure 8-2) and transfer or countershaft shaft assemblies (Figure 8-3). Selective thrust washers come in various thicknesses to take up clearances and adjust endplay. **Selective thrust washers** are used in many different locations. Some transmissions have a selective thrust washer between the front and rear multiple friction disc assemblies (Figure 8-4). Thrust washers and thrust bearings are used wherever rotating parts must have their endplay maintained. To control the endplay of non-rotating parts, selective shims or spacers are used.

Flat thrust washers and bearings have a fixed thickness and are used throughout a transmission. Typically these are numbered by the manufacturer for easy identification (Figure 8-5). Thrust washers should be inspected for scoring, flaking, and wear through to the base material. Flat thrust washers should also be checked for broken or weak tabs (Figure 8-6). These tabs are critical for holding the washer in place. On metal type flat thrust washers, the tabs may appear cracked at the bend of the tab; however, this is a normal appearance due to the characteristics of the materials used to manufacture them. Plastic thrust washers will not show wear unless they are damaged. The only way to check their wear is to measure the thickness with a micrometer and compare them to a new part. All damaged and worn thrust washers and bearings should be replaced.

Proper thrust washer thicknesses are important to the operation of an automatic transmission. After following the recommended procedures for checking the endplay of various components, refer to the manufacturer's chart for the proper thrust plate thickness for each application.

Use a petroleum jelly type lubricant to hold thrust washers in place during assembly. This will keep them from falling out of place, which will affect endplay. Besides petroleum jelly, there are special greases designed just for automatic transmission assembly that may work fine.

CAUTION: Under no circumstances should you use white lube or chassis lube. These greases will not mix in with the fluid and can plug up orifices, passages, and even hold check balls off their seats.

Thrust load is a load placed in parallel with the center of an axis.

Thrust washers are often referred to as thrust plates.

The use of the name **selective thrust washer** means the thrust washer for this application is available in different thicknesses. The correct thrust washer must be selected to provide the correct endplay or clearance.

Thrust washers are not included in an overhaul kit because they are select fit. Even if the existing thrust washers are the correct size, they should be replaced with new ones if they are pitted or scored.

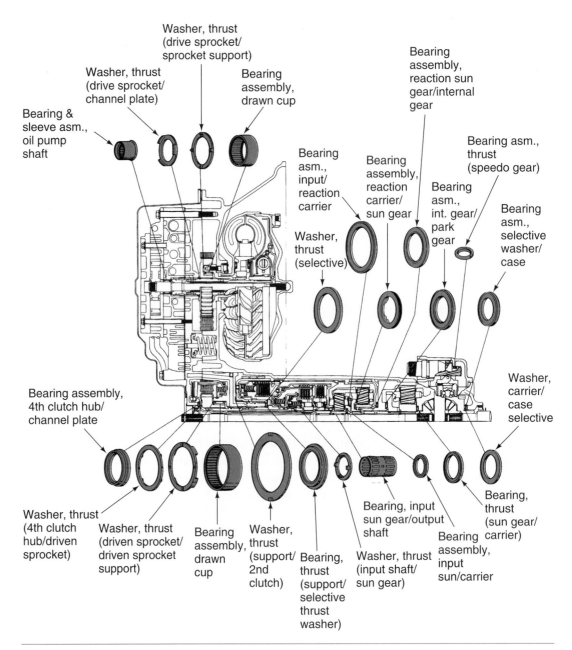

Figure 8-1 Location of thrust washers and bearings in a typical transaxle. (Courtesy of General Motors Corporation, Service Operations)

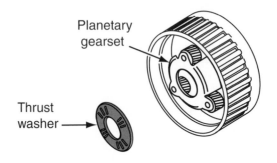

Figure 8-2 The purpose of a thrust washer is to support a thrust load and keep parts such as planetary gearsets from rubbing together.

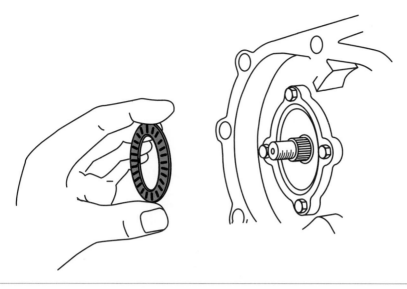

Figure 8-3 Thrust washers are also used in transfer or countershaft shaft assemblies.

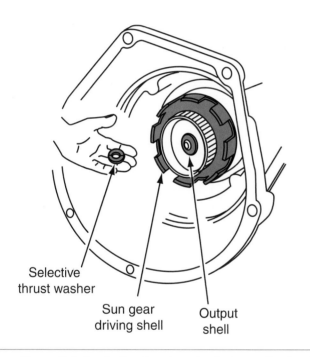

Selective thrust washer

Sun gear driving shell

Output shell

Figure 8-4 A selective thrust washer between the front and rear multiple friction disc assembly.

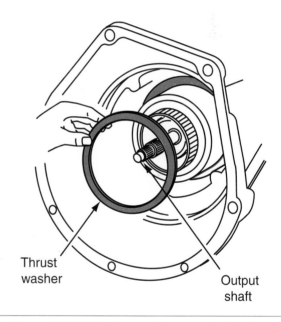

Thrust washer

Output shaft

Figure 8-5 This is the #11 thrust washer for this transaxle. Flat thrust washers and bearings are numbered by the manufacturer for easy identification.

Bearings

All bearings should be checked for roughness before and after cleaning. Carefully examine the inner and outer races, and the rollers, needles (Figure 8-7), or balls for cracks, pitting, **etching**, or signs of overheating.

Sprag and roller clutches should be inspected in the same way as bearings. Check their operation by attempting to rotate them in both directions (Figure 8-8). If working properly, they will allow rotation in one direction only. Also, visually inspect each spring of the clutch unit.

Etching is a discoloration or removal of some material caused by corrosion or some other chemical reaction.

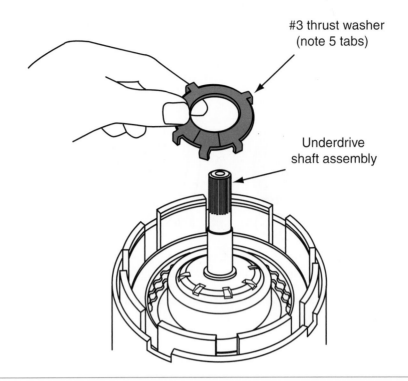

Figure 8-6 A thrust washer with tabs. These tabs must be carefully inspected for cracks or other damage.

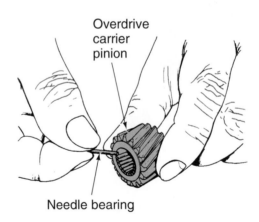

Figure 8-7 Needle bearings are often located inside the small pinion gears. Coating the inside of the gear with petroleum jelly before the needle bearings are positioned will help to keep them in place. (Courtesy of General Motors Corporation, Service Operations)

Bushings

Bushings should be inspected for pitting and scoring. Always check the depth that bushings are installed to and the direction of the oil groove, if so equipped, before you remove them. Many bushings that are used in the planetary gearing and output shaft areas have oiling holes in them. Be sure to line these up correctly during installation or you may block off oil delivery and destroy the geartrain. If any damage is evident on the bushing, it should be replaced.

Bushing wear can be visually checked as well as checked by observing the lateral movement of the shaft that fits into the bushing. Any noticeable lateral movement indicates wear and the

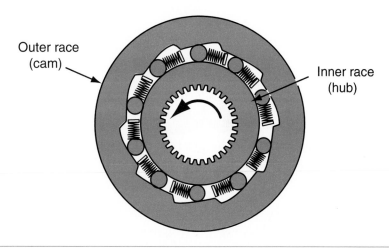

Figure 8-8 To check the action of a one-way clutch, hold the inner race and rotate the clutch in both directions. The clutch should rotate smoothly in one direction and lock in the other direction.

bushing should be replaced. The amount of clearance between the shaft and the bushing can be checked with a wire-type feeler gauge. Insert the wire between the shaft and the bushing; if the gap is greater than the maximum allowable, the bushing should be replaced. Normally, bushings must fit the shafts they ride on with about a 0.0015- to 0.003-inch clearance. You can check this fit by measuring the inside diameter of the bushing and the outside diameter of the shaft with a dial caliper or micrometer (Figure 8-9). This is a critical fit throughout the transmission and especially at the converter drive hub, where 0.005 inches is the desired fit.

Special Tools

Wire-type feeler gauge set

Micrometer

Bushing pullers and drivers

Hydraulic press

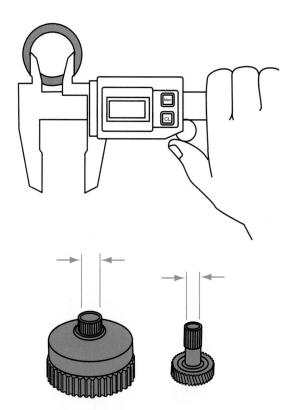

Figure 8-9 The inside diameter of bushings should be measured for wear with a telescoping gauge and a micrometer.

Figure 8-10 A typical bushing removal and installation tool.

Most bushings are press-fit into a bore. To remove them they are driven out of the bore with a properly sized bushing tool. Some bushings can be removed with a slide hammer fitted with an expanding or threaded fixture that grips the inside of the bushing. Another way to remove bushings is to carefully cut one side of the bushing and collapse it. Once collapsed, the bushing can be easily removed with a pair of pliers. Small-bore bushings that are located in areas where it is difficult to use a bushing tool can be removed by tapping the inside bore of the bushing with threads that match a selected bolt that fits into the bushing. After the bushing has been tapped, insert the bolt and use a slide hammer to pull the bolt and bushing out of its bore.

Whenever possible, all new bushings should be installed with the proper bushing driver and a press. The use of these tools prevents damage to the bushing and allows for proper seating of the bushing into its bore (Figure 8-10).

Planetary Gear Assemblies

Classroom Manual
Chapter 8, page 249

Special Tools

Feeler gauge set

Micrometer

The purpose of inspection at this point is to try to eliminate the possibility of putting noise into a newly rebuilt unit. This would result in a costly comeback, so close inspection of the planetary gearset is a must. All planetary gear teeth should be inspected for chips or stripped teeth. Any gear that is mounted to a splined shaft needs the splines checked for mutilated or shifted splines. Helical gears have many advantages over straight-cut gears, such as providing low operating noise, but you must check the endplay of the individual gears during your inspection. The helical cut of the gears makes them thrust to one side during operation. This can put a lot of load on the thrust washers and may wear them beyond specification. Checking these was discussed earlier, but particular attention should be given to the planetary carriers.

Look first for obvious problems like blackened gears or pinion shafts. These conditions indicate severe overloading and require that the carrier be replaced. Occasionally the pinion gear and shaft assembly can be replaced individually. When looking at the gears themselves, a bluish condition can be a normal condition, as this is part of a heat-treating process used during manufacture. Check the planetary pinion gears for loose bearings. Check each gear individually by rolling it on its shaft to feel for roughness or binding of the needle bearings. Wiggle the gear to be sure it is not loose on the shaft. Looseness will cause the gear to whine when it is loaded. Also, inspect the gears' teeth for chips or imperfections, as these will also cause whine.

Check the gear teeth around the inside of the front planetary ring gear. Check the fit between the front planetary carrier to the output shaft splines. Remove the snap ring and thrust washer from the front planetary ring gear (Figure 8-11). Examine the thrust washer and the outer splines of the front drum for burrs and distortion. The rear clutch friction discs must be able to slide on these splines during engagement and disengagement. With the snap ring removed, the front planetary carrier can be removed from the ring gear. Check the planetary carrier gears for endplay by placing a feeler gauge between the planetary carrier and the planetary pinion gear (Figure 8-12). Compare the endplay to specifications. On some Ravigneaux units, the clearance at both ends of the long pinion gears must also be checked and compared to specifications.

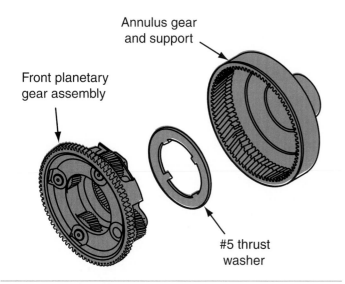

Figure 8-11 The planetary assembly, its thrust washer, and annulus gear should be inspected for signs of abnormal wear.

Figure 8-12 The clearance between the pinion gears and the planetary carrier should be checked and compared to specifications.

Check the splines of the sun gear. Sun gears should have their inner bushings inspected for looseness on their respective shafts. Also check the fit of the sun shell to the sun gear (Figure 8-13). The shell can crack where the gear mates with the shell. The sun shell should also be checked for a bell-mouthed condition where it is tabbed to the clutch drum. Any variation from a true round should be considered junk and should not be used. Look at the tabs and check for the best fit into the clutch drum slots. This involves trial fitting the shell and drum at all the possible combinations and marking the point where they fit the tightest. A snug fit here will eliminate bell mouthing due to excess play at the tabs. It can also reduce engagement noise in reverse, second, and fourth gears. This excess play allows the sun shell tabs to strike the clutch drum tabs as the transmission shifts from first to second or when the transmission is shifted into reverse.

The gear carrier should have no cracks or other defects. Replace any abnormal or worn parts. Check the thrust bearings for excessive wear, and if required, correct the input shaft thrust clearance by using a washer with the correct thickness. Determine the correct thickness by measuring the thickness of the existing thrust washer and comparing it to the measured endplay. Now move the gear back and forth to check its endplay. Some shop manuals will give a range for this check, but if none can be found you can figure about 0.007 to 0.025 inch as an average amount. All the pinions should have about the same endplay.

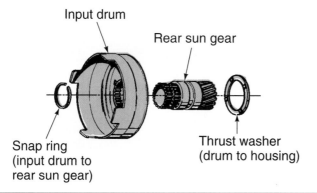

Figure 8-13 The fit of all drums onto the splines of their mating shafts should be checked. (Courtesy of General Motors Corporation, Service Operations)

Shafts

Classroom Manual
Chapter 8, page 273

Carefully examine the area on all shafts that rides in a bushing, bearing, or seal. Check the entire length of the shaft for signs of overheating and other damage. If the shaft has gears, the teeth of the gears should be inspected for damage and breaks. Also inspect the splines for wear, cracks, or other damage (Figure 8-14). A quick way to determine spline wear is to fit the mating splines and check for lateral movement.

Shafts are checked for scoring in the areas where they ride in bushings. As the shaft is a much harder material than the bushings, any scoring on the shaft would indicate a lack of lubrication at that point. The affected bushing should appear worn into the backing metal. Because shaft-to-bearing fit is critical to correct oil travel throughout the transmission, a scored shaft should be replaced. Lubricating oil is carried through most shafts; therefore, an internal inspection for debris is necessary. A blocked oil delivery hole can starve a bushing, resulting in a scored shaft. The internal oil passage of a shaft may not be able to be visually inspected and only observation during cleaning will give an indication of the openness of the passage. Washing the shaft passage out with a solvent and possibly running a piece of small-diameter wire through the passage will dislodge most particles. Be sure to check that the ball that closes off the end of the shaft, if the shaft is so equipped, is securely in place. A missing ball could be the cause of burned planetary gears and scored shafts due to a loss of oil pressure. Any shaft that has an internal bushing should be inspected, as described earlier. Replace all defective parts as necessary.

Shafts should be checked for wear in the ring groove area. This is especially critical if the groove accommodates a metal ring. Make sure there is no step wear in the groove and that the sides and bottom are square. Also make sure the groove is not too wide for the ring. If a 0.005-inch feeler gauge will fit in the groove with the ring in place, the groove is worn and the shaft should be replaced.

Input and output shafts can be solid, drilled, or tubular. The solid and drilled shafts are supported by bushings, so the bushing journals of the shafts should be free of noticeable wear at these points. Small scratches can be removed with 320-grit emery paper. Grooved or scored shafts require replacement. The splines should not show any sign of waviness along their length. Check drilled shafts to be sure the drilled portion is open and free of any foreign material. Wash out the shaft with solvent and run a small-diameter wire through the shaft to dislodge any particles. After running the wire through the opening, wash out the shaft once more and blow it out with compressed air.

If the shaft has a check ball, such as the 4L60 turbine shaft (Figure 8-15), be certain the ball seats in the correct direction. This particular check ball controls oil flow direction to the converter.

Special Tools

Feeler gauge set
320-grit sandpaper
Cleaning solvent
OSHA-approved air nozzle

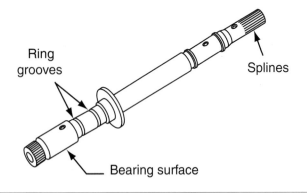

Figure 8-14 All shafts, including their splines and ring grooves, should be carefully inspected for wear or other damage.

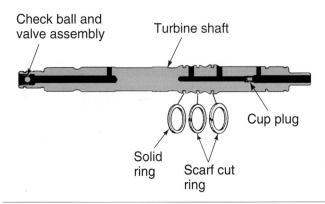

Figure 8-15 If the turbine shaft is fitted with a check ball, make sure it is able to seat and unseat. (Courtesy of General Motors Corporation, Service Operations)

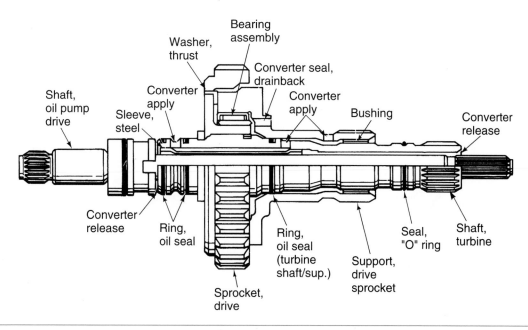

Figure 8-16 Some transmission shafts support another shaft through the bushings fitted to the inside diameter of one shaft. These bushings should be carefully inspected. (Courtesy of General Motors Corporation, Service Operations)

Some shafts have a ball pressed into one end to block off one end of the shaft. This is used to hold oil in the shaft so the oil is diverted through holes in the side of the shaft. These holes supply oil to bushings, one-way clutches, and planetary gears. If the ball does not fully block the end of the shaft, oil pressure can be lost, causing failure of these components. Some shafts may be used to support another shaft (Figure 8-16), as in the GM 4L30. The output shaft uses the rear of the input shaft to center and support itself. The small bushing found in the front end of the output shaft should always be replaced on this transmission during rebuild. If the input shaft pilot is worn or scored, a replacement shaft will be necessary.

All hubs, drums, and shells should be carefully examined for wear and damage. Especially look for nicked or scored band application surfaces on drums, worn or damaged lug grooves in clutch drums, worn splines, and burned or scored thrust surfaces. Minor scoring or burrs on band application surfaces can be removed by lightly polishing the surface with a 600-grit crocus cloth. Any part that is heavily scored or scratched should be replaced.

Parking Pawl

● **CUSTOMER CARE:** Always remind your customers that they should not rely totally on the Park gear selector position when parking the vehicle. The parking brake should be set in addition to fully placing the selector into Park. This is especially of concern when the engine is running. Many accidents have happened because the transmission slipped out of Park while the engine was running.

The parking pawl assembly (Figure 8-17) can be inspected after the transmission is disassembled or, on some transmissions, while the transmission is still in the vehicle. Examine the engagement lug on the pawl; make sure it is not rounded off. If the lug is worn, it will allow the pawl

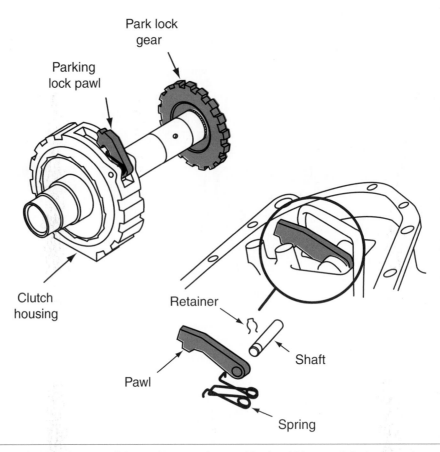

Figure 8-17 Each part of the parking pawl assembly should be carefully inspected.

to possibly slip out or not fully engage in the parking gear. Most parking pawls pivot on a pin; this also needs to be checked to make sure there is no excessive looseness at this point. The spring that pulls the pawl away from the parking gear must also be checked to make sure it can hold the pawl firmly in place. Also check the position and seating of the spring to make sure it will remain in that position during operation.

The pushrod or operating shaft (Figure 8-18) must provide the correct amount of travel to engage the pawl to the gear. Make sure the shaft is not bent or that the pivot hole in the internal shift linkage is not worn oblong. Also, make sure the bushing or sleeve that supports the manual shaft is in good condition (Figure 8-19).

Any components found unsuitable should be replaced. It should be noted that the components that make up the parking lock system are the only parts holding the vehicle in place when parked. If they do not function correctly, the car may roll or even drop into reverse when the engine is running, causing an accident or injury. Replace any questionable or damaged parts.

 WARNING: A careful inspection of the parking pawl assembly is essential to avoid possible injury, death, and/or lawsuits.

Before installing the rear extension housing, assemble the parking pin, washer, spring, and pawl. Be sure they are assembled properly. Then install the housing and tighten the bolts to specifications.

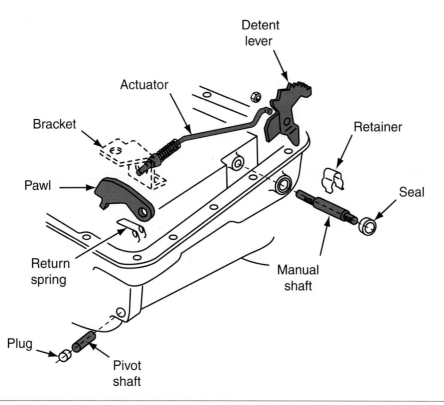

Figure 8-18 The pushrod or operating shaft for the parking pawl must provide the correct amount of travel to engage the pawl to the gear. Make sure the shaft is not bent or that the pivot hole in the internal shift linkage is not worn oblong.

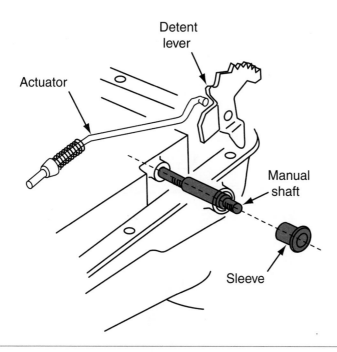

Figure 8-19 A typical sleeve for the operating rod of a parking pawl assembly.

Drive Chains

Classroom Manual
Chapter 8, page 270

Special Tools

Paint stick

Machinist's rule

Bushing pullers and
  drivers

Hydraulic press

The drive chains used in some transaxles should be inspected for side play and stretch. These checks are made during disassembly and should be repeated as a double check during reassembly. Chain deflection is measured between the centers of the two sprockets. Typically very little deflection is allowed.

Deflect the chain inward on one side until it is tight (Figure 8-20). Mark the housing at the point of maximum deflection. Then deflect the chain outward on the same side until it is tight (Figure 8-21). Again mark the housing in line with the outside edge of the chain at the point of maximum deflection. Measure the distance between the two marks. If this distance exceeds specifications, replace the drive chain.

Check each link of the chain by pushing and pulling the links away from the pin that holds them together. All of the links should move very little and each move the same amount. It is important to realize that a chain is only as strong as its weakest link. Check each link carefully.

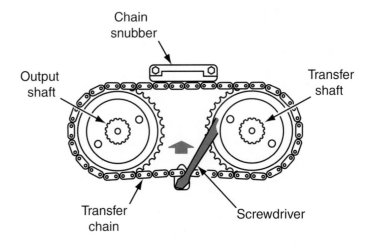

Figure 8-20 While measuring the slack of the drive chain, outwardly deflect the chain, and make a mark to that point of deflection.

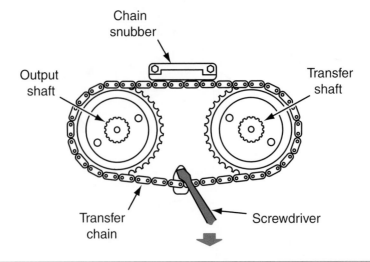

Figure 8-21 Continue measuring the slack of the drive chain by deflecting the chain inwardly. Mark the point of deflection. The distance between the outward and inward marks is the amount of chain slack.

Be sure to check for an identification mark on the chain during disassembly. These can be painted or dark-colored links, and may indicate either the top or the bottom of the chain, so be sure you remember which side was up.

The sprockets should be inspected for tooth wear and for wear at the point where they ride. If the chain was found to be too slack, it may have worn the sprockets in the same manner that engine timing gears wear when the timing chain stretches. A slightly polished appearance on the face of the gears is normal.

Chain deflection is commonly referred to as chain slack.

Bearings and Bushings

The bearings and bushings used on the sprockets need to be checked for damage. The radial needle thrust bearings must be checked for any deterioration of the needles and cage. The running surface in the sprocket must also be checked, as the needles may pound into the gear's surface during abusive operation. The bushings should be checked for any signs of scoring, flaking, or wear. Replace any defective parts.

Typically, the bearings and bushings are removed with a puller (Figure 8-22) and installed with a driver and a press (Figure 8-23).

The removal and installation of the chain drive assembly of some transaxles requires that the sprockets be spread slightly apart (Figure 8-24). The key to doing this is to spread the sprockets just the right amount. If they are spread too far, they will not be easy to install or remove.

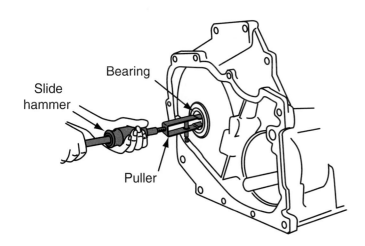

Figure 8-22 The bearings and bushings in a chain drive assembly should be removed with a puller.

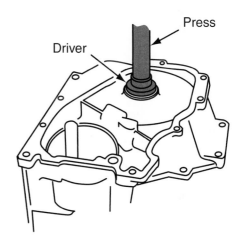

Figure 8-23 The bearings and bushings in a chain drive assembly should be installed with a driver and a hydraulic press.

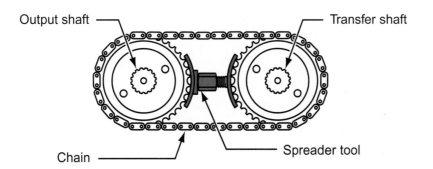

Figure 8-24 To remove and install some drive chains, they must be slightly spread apart with a special tool.

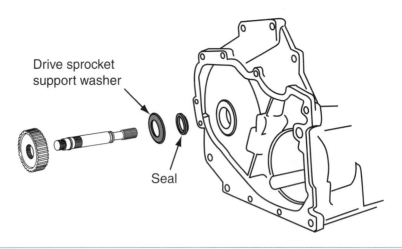

Figure 8-25 The shafts and gears of a chain drive assembly have numerous seals that must be replaced whenever the unit is disassembled.

The shafts and gears of the assembly have numerous seals and thrust washers (Figure 8-25). The seals must be replaced whenever the unit is disassembled.

Transfer Gears

Classroom Manual
Chapter 8, page 268

Some transaxles use gears, instead of a drive chain, to move or transfer the output of the transmission to the final drive unit (Figure 8-26). The shafts and gears in these transaxles must be carefully inspected and replaced if they are damaged.

To remove and install the transfer shaft gear, a holding tool must be used to stop the transfer gear from turning while loosening or tightening the retaining nut (Figure 8-27). The nut is typ-

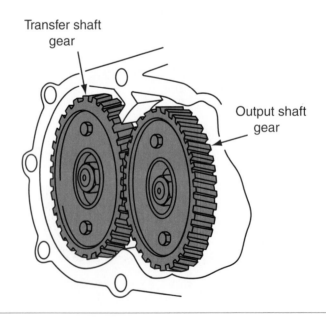

Figure 8-26 The transfer shafts and gears must be carefully inspected and replaced if they are damaged.

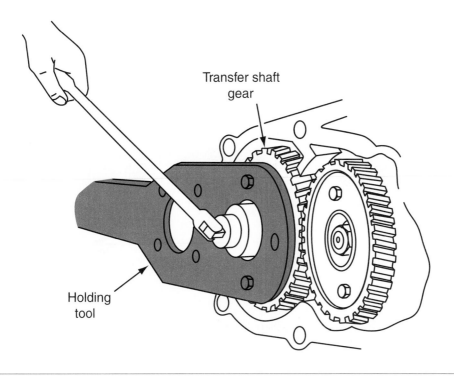

Figure 8-27 To remove and install the transfer shaft gear, a holding tool must be used to stop the transfer gear from turning while loosening or tightening the retaining nut.

ically tightened to 200 ft.-lbs. To remove the transfer shaft from the transaxle case, the retaining snap ring must be removed (Figure 8-28). Then the shaft can be pulled out with its bearing (Figure 8-29). The shaft's bearing must be pressed on and off the shaft.

A puller must be used to remove the transfer gear from the output shaft (Figure 8-30), after its retaining bolt has been removed.

Special Tools

Bushing pullers and drivers

Hydraulic press

Dial indicator

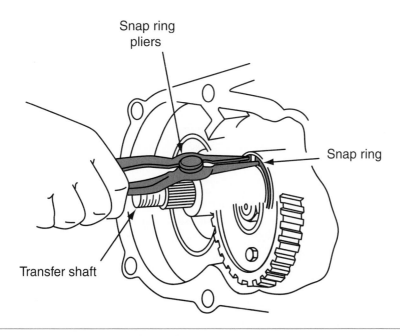

Figure 8-28 To remove the transfer shaft from the transaxle case, the retaining snap ring must be first removed.

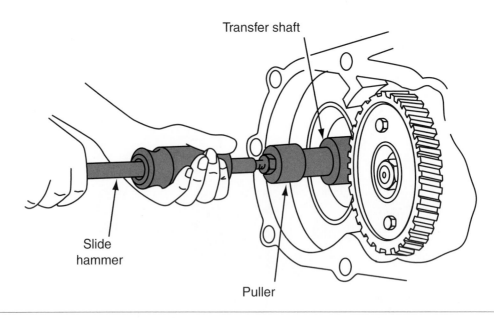

Figure 8-29 The transfer shaft can be pulled out with its bearing with a suitable puller tool.

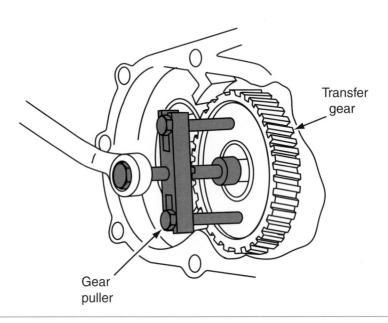

Figure 8-30 A puller must be used to remove the transfer gear from the output shaft, after its retaining bolt has been removed.

Behind the transfer shaft gear is a selective shim used to provide correct meshing of the teeth of the transfer shaft gear and the output shaft transfer gear (Figure 8-31) and to control transfer shaft endplay. Endplay is measured with a dial indicator. To ensure good contact with the indicator's plunger and the end of the transfer shaft, Chrysler recommends that a steel check ball be placed between the plunger tip and the shaft (Figure 8-32). To help hold the ball in place during the check, coat the ball in heavy grease. If the endplay is not within specifications, select a thrust washer with the correct thickness and install it behind the transfer gear.

The output shaft should also be checked for endplay. If the endplay is incorrect, a different size shim should be installed behind the output shaft transfer gear (Figure 8-33). Since the gear is splined to the shaft, a thrust bearing is not needed here. After endplay has been corrected and

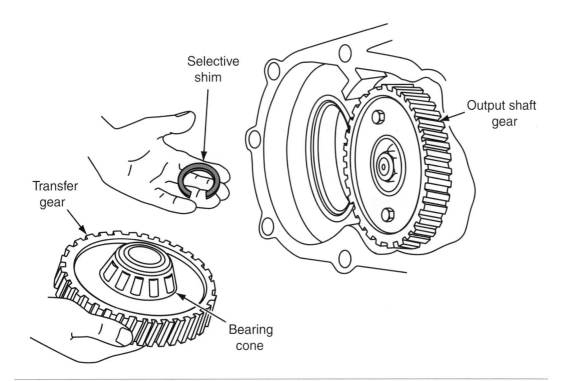

Figure 8-31 Behind the transfer shaft gear is a selective shim used to provide correct meshing of the teeth of the transfer shaft gear and the output shaft transfer gear and to control transfer shaft endplay.

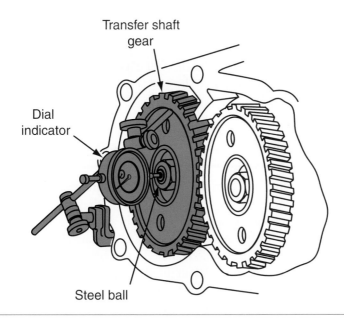

Figure 8-32 Chrysler recommends that a steel check ball coated in heavy grease be placed between the plunger tip and the shaft to ensure good contact with the indicator's plunger and the end of the transfer shaft.

after the output shaft gear has been re-installed, the turning torque of the shaft should be measured. If the turning torque is too high, a slightly thicker shim should be installed. If the turning torque is too low, a slightly thinner shim should be installed. The bearings for the transfer and output shafts are pressed on and off the shafts.

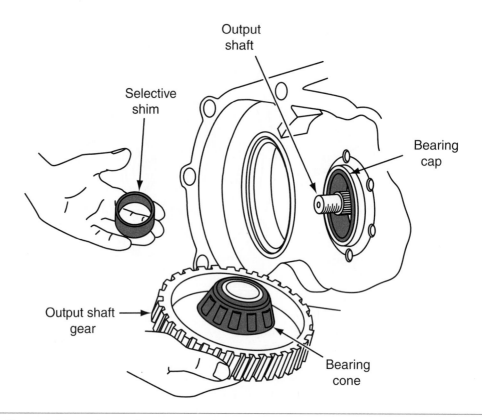

Figure 8-33 A selective shim is installed behind the output shaft transfer gear to control output shaft endplay.

It is very important that the transfer gears be tightened to specifications. The torque setting not only ensures that the gears will remain on the shafts, but it also maintains the correct bearing adjustments.

Final Drive Units

Classroom Manual
Chapter 8, page 267

Special Tools

Inch-pound torque wrench

Transaxle final drive units should be carefully inspected. Examine each gear, thrust washer, and shaft for signs of damage. If the gears are chipped or broken, they should be replaced. Also inspect the gears for signs of overheating or scoring on the bearing surface of the gears.

Final drive units may be helical gear or planetary gear units. The helical type should be checked for worn or chipped teeth, overloaded tapered roller bearings, and excessive differential side gear and spider gear wear. Excessive play in the differential is a cause of engagement clunk (Figure 8-34). Be sure to measure the clearance between the side gears and the differential case and to check the fit of the spider gears on the spider gear shaft. Proper clearances can be found in the appropriate shop manual. It is possible that the side bearings of some final drive units are preloaded with shims (Figure 8-35). Select the correct size shim to bring the unit into specifications. With a torque wrench, measure the amount of rotating torque. Compare your readings against specifications (Figure 8-36).

If the bearing preload and endplay is fine, as is the condition of the bearings, the parts can be reused. However, always install new seals during assembly. It should be noted that these bearings function the same as RWD rear axle side bearings and you should set the preload to the specifications for a new bearing. Used bearings should be set to the amount found during teardown or about one half the preload of a new bearing.

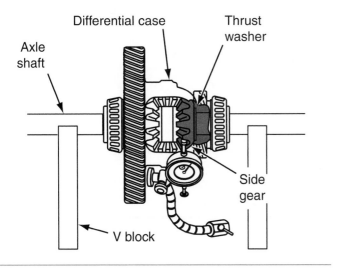

Figure 8-34 The backlash of the side gears in a final drive unit should be checked prior to assembling the transaxle.

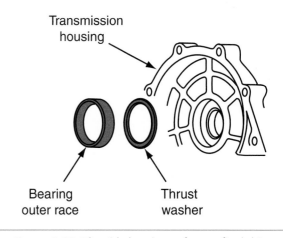

Figure 8-35 The side bearings of some final drive units are preloaded with selective shims.

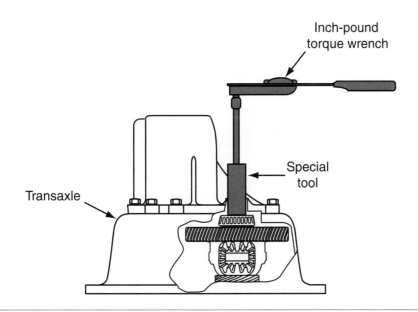

Figure 8-36 Using an inch-pound torque wrench, the turning torque of the transaxle assembly should be checked after assembly.

Planetary gear-type final drives (Figure 8-37) are also checked for the same differential case problems that the helical type would encounter. The planetary pinion gears need to be checked for looseness or roughness on their shafts and for endplay (Figure 8-38). Any problems found normally result in the replacement of the carrier as a unit, because most pinion bearings and shafts are not sold as separate parts. Again, specifications for these parts are found in the shop manual. Photo Sequence 12 covers the procedure for servicing a planetary gear-type final drive unit.

Planetary gear-type final drives, like helical final drives, are available in more than one possible ratio for a given type of transaxle, so care should be taken to ensure that the same gear ratios are used during assembly. This is not normally a problem when overhauling a single unit;

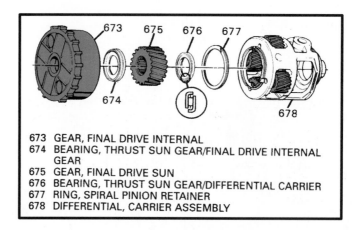

673 GEAR, FINAL DRIVE INTERNAL
674 BEARING, THRUST SUN GEAR/FINAL DRIVE INTERNAL GEAR
675 GEAR, FINAL DRIVE SUN
676 BEARING, THRUST SUN GEAR/DIFFERENTIAL CARRIER
677 RING, SPIRAL PINION RETAINER
678 DIFFERENTIAL, CARRIER ASSEMBLY

Figure 8-37 Planetary type final drives should be checked in the same basic way as helical type final drives. (Courtesy of General Motors Corporation, Service Operations)

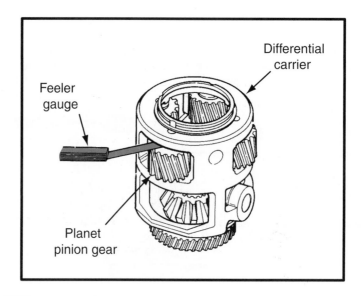

Figure 8-38 The endplay of the planetary pinion gears should be checked with a feeler gauge. (Courtesy of General Motors Corporation, Service Operations)

however, in a shop where many transmissions are being repaired, it is possible to mix up parts, causing problems during the rebuild.

Adjusting Side Gear Endplay

Special Tools

Adapter kit for transaxle

Dial indicator

0 to 1-inch micrometer

Most of the procedures for servicing FWD final drive units are the same as for RWD differential units. Because FWD units typically use the end of the output shaft as the drive pinion gear, all normal pinion shaft adjustments are not needed. Ring gear and side bearing adjustments are still necessary. These adjustments are typically made with the differential case assembled and out of the transaxle case.

The procedure shown in Photo Sequence 12 is typical for measuring and adjusting the side gear endplay in a transaxle's final drive. Always refer to your service manual before proceeding to make these adjustments on a transaxle.

Photo Sequence 12
Servicing Planetary Gear-Type Final Drive Units

P12-1 Place the final drive unit into the transaxle's oil pan. Doing this will lessen the chances of losing bearings while disassembling the unit. Make sure the pan is clean.

P12-2 With a pin punch and hammer, remove the differential pinion shaft retaining ring.

P12-3 Remove the retaining ring from the end of the output shaft. This ring must not be reused!

P12-4 Pull the output shaft from the differential carrier.

P12-5 Remove the final drive sun gear.

P12-6 Remove the differential pinion shaft retaining pin, the differential pinion shaft, pinion gears, and thrust washers and place them aside. Keep these parts together and in the order and orientation in which they were assembled.

P12-7 Remove the differential side gears and thrust washers.

P12-8 Using a screwdriver, carefully remove the final drive carrier retaining ring.

P12-9 Remove the planet pinion pins, then the planet pinion gears, thrust washers, needle bearings, and the bearing spacers.

(continued)

P12-10 Inspect the needle bearings, thrust washers, pinion gears, and planet pinion pins.

P12-11 Remove the final drive sun gear to carrier thrust bearing.

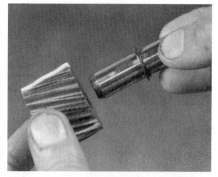

P12-12 To begin reassembly, install the pinion gear needle bearing spacer, thrust washer, and pinion gear onto the planet pinion gear pin.

P12-13 Install the needle bearings, one at a time, into the top and bottom of the planet pinion gear. Use transjel or petroleum jelly to help hold the bearings in position.

P12-14 Coat the sun gear to final drive carrier thrust washer with transjel or petroleum jelly. Then install the thrust bearing onto the final drive carrier.

P12-15 Separate the pinion pin from the gear. Assemble the planet pinion gear thrust washers and gears into the final drive carrier. Make sure the gears face in the same direction as they faced when they were removed. Also be careful not to let the bearings fall out.

P12-16 Install the planet pinion gear pins into the carrier. Then, install the final drive carrier retaining ring.

P12-17 With a feeler gauge, check the endplay of the pinion gears. If the endplay is too low, the planet pinion assembly must be removed and the correct thickness thrust washer installed. If the endplay is too much, the differential assembly must be replaced.

P12-18 Install the differential side gears and thrust washers into the carrier.

P12-19 Apply transjel or petroleum jelly onto the thrust washers for the pinion gears, and then install them into the carrier.

P12-20 Align the bores of the pinion gears with the bore in the carrier for the pinion pin, and then install the pin. Make sure the hole in the pivot pin, for its retaining pin, is aligned with the retaining pin bore.

P12-21 Install the pinion pin's retaining pin.

P12-22 Insert the sun gear into the carrier, making sure the gear is placed in the correct direction.

P12-23 Install the output shaft into the final drive assembly.

P12-24 Install a new retaining snap ring onto the end of the output shaft.

1. Install the correct adapter into the differential bearings (Figure 8-39).
2. Mount the dial indicator to the ring gear with the plunger resting against the adapter.
3. With your fingers or a screwdriver, move the ring gear up and down (Figure 8-40).
4. Record the measured endplay.
5. Measure the old thrust washer with the micrometer.
6. Install the correct sized thrust washer (Figure 8-41).
7. Repeat the procedure for the other side.

Adjusting Bearing Preload

The following procedure is typical for the measurement and adjustment of the differential bearing preload in a transaxle. Always refer to your service manual before proceeding to make these adjustments on a transaxle.

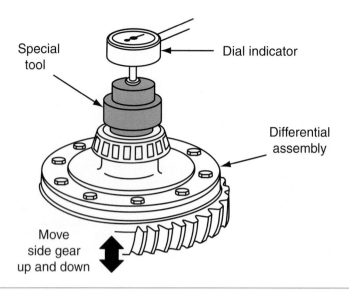

Figure 8-39 Tool setup for measuring side gear endplay.

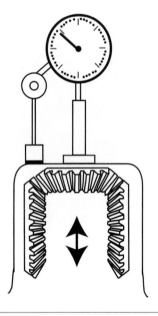

Figure 8-40 Move the ring gear up and down and observe the dial indicator.

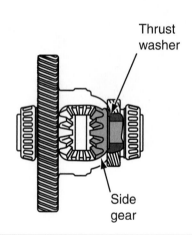

Figure 8-41 Location of thrust washers.

Special Tools

Set of gauging shims

Dial indicator

Fresh lubricant

Inch-pound torque
wrench

1. Remove the bearing cup and existing shim from the differential bearing retainer.
2. Select a gauging shim that will allow for 0.001 to 0.010 inch endplay.
3. Install the gauging shim into the differential bearing retainer (Figure 8-42).
4. Press in the bearing cup.
5. Lubricate the bearings and install them into the case.
6. Install the bearing retainer.

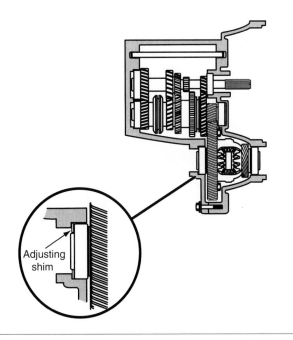

Figure 8-42 Typical location of preload shim.

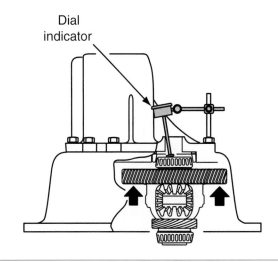

Figure 8-43 Setup for measuring differential bearing preload.

7. Tighten the retaining bolts.

8. Mount the dial indicator with its plunger touching the differential case (Figure 8-43).

9. Apply medium pressure in a downward direction while rolling the differential assembly back and forth several times.

10. Zero the dial indicator.

11. Apply medium pressure in an upward direction while rotating the differential assembly back and forth several times.

12. The required shim to set preload is the thickness of the gauging shim plus the recorded endplay.

13. Remove the bearing retainer, cup, and gauging shim.

14. Install the required shim.

15. Press the bearing cup into the bearing retainer.

16. Install the bearing retainer and tighten the bolts.

17. Check the rotating torque of the transaxle (Figure 8-44). If this is less than specifications, install a thicker shim. If the torque is too great, install a slightly thinner shim.

18. Repeat the procedure until the desired torque is reached.

The side bearings of most final drive units must be pulled off and pressed onto the differential carrier (Figure 8-45). Always be sure to use the correct tools for removing and installing the bearings.

The ring gear of many transaxles is riveted to the differential carrier. The rivets must be drilled, then driven out with a hammer and drift to separate the ring gear from the carrier (Figure 8-46). To install a new ring gear to the carrier, nuts and bolts are used (Figure 8-47). These nuts and bolts must be of the specified hardness and should be tightened in steps and to the specified torque.

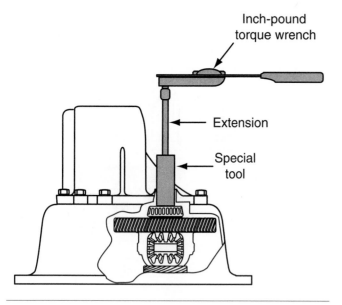

Figure 8-44 Using an inch-pound torque wrench to check the turning torque of a transaxle.

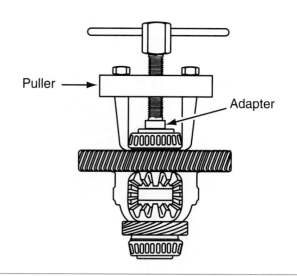

Figure 8-45 Typical setup for removing side bearings from a differential case.

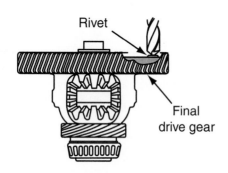

Figure 8-46 Ring gear rivets must be drilled and driven out to separate the ring gear from the differential case.

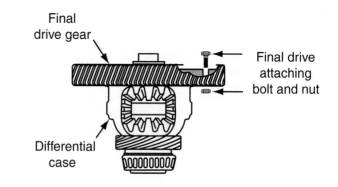

Figure 8-47 The ring gear should be fastened to the differential case with special nuts and bolts.

The final drives of most transaxle differential carriers are fitted with a speedometer gear pressed onto the carrier and under one side bearing (Figure 8-48). These gears are pulled off and pressed onto the carrier.

Final Drive Bearing Replacement

The final drive unit is either positioned in the transaxle case or in a separate housing mounted to the transaxle case (Figure 8-49). Regardless of its location, it is supported by bearings. Normally, tapered roller bearings are used. These bearings are removed with pullers (Figure 8-50) and installed with a press and driver (Figure 8-51). When replacing the bearings, make sure the bearing seats are free from nicks and burrs. Those defects will not allow the bearing to seat properly and will give false endplay measurements. If the bearings were replaced, endplay and gear clearances must be checked and corrected before installing the final drive unit into the housing.

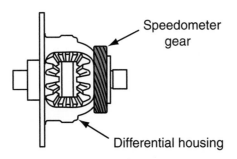

Figure 8-48 The location of the speedometer gear on the differential case.

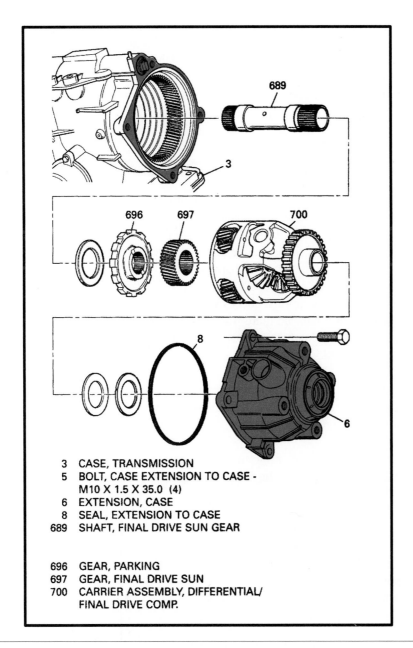

3   CASE, TRANSMISSION
5   BOLT, CASE EXTENSION TO CASE -
        M10 X 1.5 X 35.0 (4)
6   EXTENSION, CASE
8   SEAL, EXTENSION TO CASE
689   SHAFT, FINAL DRIVE SUN GEAR

696   GEAR, PARKING
697   GEAR, FINAL DRIVE SUN
700   CARRIER ASSEMBLY, DIFFERENTIAL/
        FINAL DRIVE COMP.

Figure 8-49 Some final drive units are housed in a separate housing bolted to the transaxle case. It is important that the seals be replaced during assembly. (Courtesy of General Motors Corporation, Service Operations)

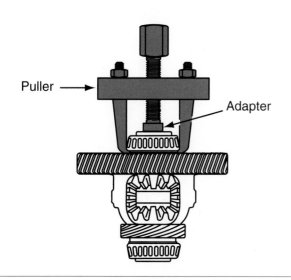

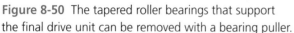

Figure 8-50 The tapered roller bearings that support the final drive unit can be removed with a bearing puller.

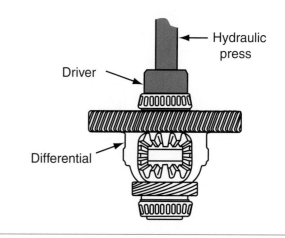

Figure 8-51 The bearings that support the final drive unit should be installed with a hydraulic press and driver.

CASE STUDY

A customer had his 1998 Ford Taurus towed into the shop. He stated that nothing happened when the transmission was in any gear. The technician looked things over and attempted to verify the complaint. Sure enough, the car wouldn't move in forward or in reverse. He did notice a slight change in noise when he moved the lever into drive and reverse, so he knew something was happening.

He proceeded to do the normal checks with the scan tool and found nothing unusual except for the codes that said there was no signal from the output sensor. That didn't alarm him because he knew there wouldn't be a signal since the car wouldn't move. He then checked the fluid. The level was correct and smelled fine.

He then raised the car and checked the front axle shafts to see if they were stuck. He could turn the wheels easily; in fact, he noticed how easy it was to rotate them. But he wasn't sure what to make of that.

He then ran a pressure test and found the pressure to be normal at idle speed and observed pressure changes as he selected drive and reverse. From this he knew it was not a hydraulic problem.

Since the problem didn't appear to be electronic or hydraulic, he knew it had to be mechanical. Thinking about what he had observed, he came to the conclusion that the problem must be in the input or the output of the transaxle. He strongly suspected the output because of the change of noise he noticed. Giving it more thought, he was convinced the problem was the output because of the little effort it took to rotate the wheels.

He proceeded to pull the transaxle and began his disassembly and inspection. He carefully checked the drive chain and sprocket assembly and found no evidence of a problem. He then pulled the output shaft from the transaxle. As soon as he had it out, he saw the problem; the splines were stripped and well rounded off.

He proceeded to overhaul the transmission and to clean every part thoroughly. He replaced the output shaft and transfer gear, then put the transaxle back together.

After the transaxle was in, he put it in drive and it moved. During the road test he found no problems. He told the customer it was now working fine and the customer was pleased.

Terms To Know

Etching   Selective thrust washers   Thrust load

ASE-Style Review Questions

1. While inspecting a planetary gearset:
   Technician A says blackened pinion shafts indicate severe overloading.
   Technician B says bluish gears indicate overheating.
   Who is correct?
   A. A only
   B. B only
   C. Both A and B
   D. Neither A nor B

2. *Technician A* says loose planetary pinion gear bearings will cause the gear to whine when it is loaded.
   Technician B says damaged teeth on a planetary pinion gear will cause the gear to whine.
   Who is correct?
   A. A only
   B. B only
   C. Both A and B
   D. Neither A nor B

3. *Technician A* removes bushings by carefully cutting one side of the bushing and collapsing it. Once collapsed, the bushing can be easily removed with a pair of pliers.
   Technician B removes small-bore bushings by tapping the inside bore of the bushing with threads that match a selected bolt, which fits into the bushing. After tapping the bushing, the bolt is inserted and a slide hammer used to pull the bolt and bushing out of its bore.
   Who is correct?
   A. A only
   B. B only
   C. Both A and B
   D. Neither A nor B

4. *Technician A* says bushings can be heated with a torch to remove them easily.
   Technician B says bushings can be removed with a slide hammer and the correct attachment.
   Who is correct?
   A. A only
   B. B only
   C. Both A and B
   D. Neither A nor B

5. While checking a planetary gearset:
   Technician A says the end clearance of the pinion gears should be checked with a feeler gauge.
   Technician B says the end clearance of the long pinions in a Ravigneaux gearset should be checked at both ends.
   Who is correct?
   A. A only
   B. B only
   C. Both A and B
   D. Neither A nor B

6. *Technician A* says most FWD final drive units require that ring and pinion backlash and pinion depth be set to specifications.
   Technician B says ring gear and side gear bearing adjustments are required on all FWD final drive units.
   Who is correct?
   A. A only
   B. B only
   C. Both A and B
   D. Neither A nor B

7. *Technician A* says a blocked oil delivery passage will cause a shaft to score.
   Technician B says that if a shaft is fitted with a check ball and the check ball does not seat properly, low oil pressure will result.
   Who is correct?
   A. A only
   B. B only
   C. Both A and B
   D. Neither A nor B

8. *Technician A* says thrust washers should be inspected for scoring, flaking, and wear through to the base material.
   Technician B says plastic thrust washers will not wear unless they are damaged.
   Who is correct?
   A. A only
   B. B only
   C. Both A and B
   D. Neither A nor B

9. While inspecting a parking pawl assembly:
   Technician A says the pivot pin must fit loosely in its bracket and tightly in the pawl.
   Technician B says the engagement lug on the pawl must be square in order to fully engage into the parking gear.
   Who is correct?
   A. A only
   B. B only
   C. Both A and B
   D. Neither A nor B

10. While checking the endplay of a transfer gear assembly:
    Technician A says that if the endplay of the output transfer shaft is incorrect, a different size thrust washer should be installed behind the output shaft transfer gear.
    Technician B says that after the endplay of the output shaft has been corrected, the turning torque of the shaft should be measured. If the turning torque is too high, a slightly thinner shim should be installed. If the turning torque is too low, a slightly thicker shim should be installed.
    Who is correct?
    A. A only
    B. B only
    C. Both A and B
    D. Neither A nor B

ASE Challenge Questions

1. While servicing a final drive unit:
   Technician A checks gear and bearing endplay whenever new bearings are installed in the unit.
   Technician B reuses the bearings, seals, and thrust washers if the bearing preload and endplay are fine, as is the condition of the bearings.
   Who is correct?
   A. A only
   B. B only
   C. Both A and B
   D. Neither A nor B

2. *Technician A* says a drive chain that is too loose should be shortened by removing a pair of links in the chain.
   Technician B says the drive sprockets should be replaced if the gear teeth are polished or show any other signs of wear.
   Who is correct?
   A. A only
   B. B only
   C. Both A and B
   D. Neither A nor B

3. *Technician A* checks bushing wear by observing the lateral movement of the shaft that fits into the bushing. Any noticeable lateral movement indicates wear and therefore the bushing should be replaced.
   Technician B checks for bushing wear by measuring the inside diameter of the bushing and the outside diameter of the shaft with a dial caliper or micrometer.
   Who is correct?
   A. A only
   B. B only
   C. Both A and B
   D. Neither A nor B

4. *Technician A* drills out the ring gear retaining rivets to remove the ring gear from the carrier.
   Technician B installs the ring gear nuts and bolts of the specified hardness.
   Who is correct?
   A. A only
   B. B only
   C. Both A and B
   D. Neither A nor B

5. *Technician A* says all hubs, drums, and shells should be carefully examined for wear and damage.
   Technician B says minor scoring or burrs on band application surfaces of a drum can be removed by lightly polishing the surface with a 200-grit crocus cloth.
   Who is correct?
   A. A only
   B. B only
   C. Both A and B
   D. Neither A nor B

Job Sheet 24

24

Name _____ Date _____

Checking Thrust Washers, Bushings, and Bearings

Upon completion of this job sheet, you should be able to inspect, measure, and replace thrust washers and bearings and inspect the bushings in a transmission/transaxle.

ASE Correlation

This job sheet is related to the Automatic Transmission and Transaxles Test's Content Area: *Off-Vehicle Transmission and Transaxle Repair, Gear Train, Shafts, Bushings, and Case.*
Tasks: Inspect, measure, and replace thrust washers and bearings and inspect bushings; replace as needed.

Tools and Materials

Wire-type feeler gauge set   Bushing driver set
Bushing puller tool

Describe the vehicle being worked on:

Year _____ Make _____ Model _____

VIN _____ Engine type and size _____

Model and type of transmission _____

Procedure

1. The best time to inspect thrust washers, bearings, and bushings is during disassembly. The bushings should be inspected for pitting and scoring. Describe their condition.

   _____

   _____

2. Check the depth that bushings are installed to and the direction of the oil groove, if so equipped, before you remove them. Many bushings that are used in the planetary gearing and output shaft areas have oiling holes in them. Be sure to line these up correctly during installation or you may block off oil delivery and destroy the geartrain. Describe their condition.

   _____

   _____

3. Observe the lateral movement of the shaft that fits into the bushing. Any noticeable lateral movement indicates wear and the bushing should be replaced. Describe your findings.

   _____

   _____

Checking Thrust Washers, Bushings, and Bearings (continued)

4. The amount of clearance between the shaft and the bushing can be checked with a wire-type feeler gauge. Insert the wire between the shaft and the bushing; if the gap is greater than the maximum allowable, the bushing should be replaced. What are the specifications for this gap and how do they compare to your measurement?

_____

_____

5. Measure the inside diameter of the bushing and the outside diameter of the shaft with a Vernier type caliper or micrometer. Compare the two and state your conclusions.

_____

_____

6. Most bushings are press-fit into a bore. Remove them by driving them out of the bore with a properly sized bushing tool. Some bushings can be removed with a slide hammer fitted with an expanding or threaded fixture that grips the inside of the bushing. Another way to remove bushings is to carefully cut one side of the bushing and collapse it. Once collapsed, the bushing can be easily removed with a pair of pliers. What did you use to remove the bushing?

_____

_____

☐ 7. Small-bore bushings located in areas where it is difficult to use a bushing tool can be removed by tapping the inside bore of the bushing with threads that match a bolt that fits into the bushing. After the bushing has been tapped, insert the bolt and use a slide hammer to pull the bolt and bushing out of its bore.

☐ 8. All new bushings should be pre-lubed during transmission assembly and installed with the proper bushing driver. Make sure they are not damaged and are fully seated in their bores.

9. The purpose of a thrust washer is to support a thrust load and keep parts from rubbing together. Selective thrust washers come in various thicknesses to take up clearances and adjust shaft endplay. Flat thrust washers and bearings should be inspected for scoring, flaking, and wear through to the base material. Describe their condition.

_____

_____

10. Flat thrust washers should also be checked for broken or weak tabs. These tabs are critical for holding the washer in place. On metal-type flat thrust washers, the tabs may appear cracked at the bend of the tab; however, this is a normal appearance. Describe their condition.

_____

_____

11. Only damaged plastic thrust washers will show wear. The only way to check their wear is to measure the thickness and compare it to a new part. Describe their condition.

_____

_____

12. Proper thrust washer thicknesses are important to the operation of an automatic transmission. Always follow the recommended procedure for selecting the proper thrust plate. Use a petroleum jelly type lubricant to hold thrust washers in place during assembly.

☐

> **CAUTION:** Never use white lube or chassis lube. These greases will not mix with the fluid and can plug up orifices, passages, and hold check balls off their seats.

13. All bearings should be checked for roughness before and after cleaning.

☐

14. Carefully examine the inner and outer races, and the rollers, needles, or balls for cracks, pitting, etching, or signs of overheating. Describe their condition.

_____

_____

15. Give a summary of your inspection.

_____

_____

_____

Instructor's Response _____

_____

_____

_____

Job Sheet 25

25

Name _____   Date _____

Servicing Oil Delivery Seals

Upon completion of this job sheet, you should be able to inspect oil delivery seal rings, ring grooves, and sealing surface areas.

ASE Correlation

This job sheet is related to the Automatic Transmission and Transaxles Test's Content Area: *Off-Vehicle Transmission and Transaxle Repair, Gear Train, Shafts, Bushings, and Case.*
Tasks: Inspect oil delivery seal rings, ring grooves, and sealing surface areas.

Tools and Materials

Petroleum jelly   Crocus cloth
Seal driver tools   Feeler gauge set

Procedure Guidelines

❏ Three types of seals are used in automatic transmissions: O-ring and square-cut (lathe-cut), lip, and sealing rings. These seals are designed to stop fluid from leaking out of the transmission and to stop fluid from moving into another circuit of the hydraulic circuit.

❏ O-ring and square-cut seals are used to seal non-rotating parts. When installing a new O-ring or square-cut seal, coat the entire surface of the seal with assembly lube or petroleum jelly. Make sure you don't stretch or distort the seal while you are working it into its holding groove. After a square-cut seal is installed, double check it to make sure it is not twisted. The flat surface of the seal should be parallel with the bore. If it is not, fluid will easily leak passed the seal.

❏ Lip seals that are used to seal a shaft typically have a metal flange around their outside diameter. The shaft rides on the lip seal at the inside diameter of the seal assembly. The rigid outer diameter provides a mounting point for the lip seal and is pressed into a bore. Once pressed into the bore, the outer diameter of the seal prevents fluid from leaking into the bore, while the inner lip seal prevents leakage past the shaft.

❏ Piston lip seals are set into a machined groove on the piston. This type of lip seal is not housed in a rigid metal flange. They are designed to be flexible and provide a seal while the piston moves up and down. While the piston moves, the lip flexes up and down. The most important thing to keep in mind while installing a lip seal is to make sure the lip is facing the correct direction. The lip should always be aimed toward the source of pressurized fluid. If installed backward, fluid under pressure will easily leak past the seal. Also remember to make sure the surfaces to be sealed are clean and not damaged.

❏ Teflon or metal sealing rings are commonly used to seal servo pistons, oil pump covers, and shafts. These rings may be designed to provide for a seal, but they may also be designed to allow a controlled amount of fluid leakage. Sealing rings are either solid rings or are cut. Cut sealing rings are of one of three designs: open-end, butt-end, or locking end.

❏ Solid sealing rings are made of a Teflon-based material and are never reused. To remove them, simply (but carefully) cut the seal after it has been pried out of its groove. Installing a new solid sealing ring requires special tools. These tools allow you to stretch the seal while pushing it into position. Never attempt to install a solid seal without the proper tools. Because these seals are soft, they are easily distorted and damaged.

❏ Open-end sealing rings fit loosely into a machined groove. The ends of the rings do not touch when they are installed. This type of ring is typically removed and installed with a pair of snap ring pliers. The ring should be expanded just enough to move it off or onto the shaft.

❏ Butt-end sealing rings are designed so that their ends butt up or touch each other once the seal is in place. This type of seal can be removed with a small screwdriver. The blade of the screwdriver is used to work the ring out of its groove. To install this type of ring, use a pair of snap ring pliers and expand the ring to move it into position.

❏ Locking-end rings may have hooked ends that connect or have ends that are cut at an angle to hold the ends together. These seals are removed and installed in the same way as butt-end rings. After these rings are installed, make sure the ends are properly positioned and touching.

❏ All seals should be checked in their own bores prior to installation. They should be slightly smaller or larger (±3%) than their groove or bore. If a seal is not the proper size, find one that is. Do not assume that because a particular seal came with the overhaul kit it is the correct one.

❏ Never install a seal when it is dry. The seal should slide into position and allow the part it seals to slide into it. A dry seal is easily damaged during installation.

❏ Install only genuine seals recommended by the manufacturer of the transmission.

Describe the vehicle being worked on:

Year _____ Make _____ Model _____

VIN _____ Engine type and size _____

Transmission type and model _____

Task Completed

☐

Procedure

1. Before installing seals, clean the shaft and/or bore area.

2. Carefully inspect these areas for damage. File or stone away any burrs or bad nicks and polish the surfaces with a fine crocus cloth, then clean the area to remove the metal particles. Describe your findings.

_____

_____

☐

3. Lubricate the seal, especially any lip seals, to ease installation.

4. All metal sealing rings should also be checked for proper fit. Since these rings seal on their outer diameter, the seal should be inserted in its bore and should feel tight there. If the seal has some form of locking ends, these should be interlocked prior to trying the seal in its bore. Describe your findings.

_____

_____

5. Check the fit of the sealing rings in their shaft groove. Describe your findings.

_____

_____

6. Check the side clearance of the ring by placing the ring into its groove and measuring the clearance between the ring and the groove with a feeler gauge. Describe your findings.

_____

_____

7. While checking the clearance, look for nicks in the grooves and for evidence of groove taper or stepping. Describe your findings.

_____

_____

8. Use the correct driver when installing a seal and be careful not to damage the seal during installation.

☐

Instructor's Response _____

_____

_____

_____

Job Sheet 26

26

Name _____ Date _____

Servicing Planetary Gear Assemblies

Upon completion of this job sheet, you should be able to inspect and measure planetary gear assemblies.

ASE Correlation

This job sheet is related to the Automatic Transmission and Transaxles Test's Content Area: *Off-Vehicle Transmission and Transaxle Repair, Gear Train, Shafts, Bushings, and Case.*
Tasks: Inspect and measure planetary gear assembly (includes sun, ring gear, thrust washers, planetary gears, and carrier assembly); replace as needed.

Tools and Materials

Snap ring pliers
Feeler gauge set

Describe the vehicle being worked on:

Year _____ Make _____ Model _____

VIN _____ Engine type and size _____

Model and type of transmission _____

Procedure

1. The planetary gears used in automatic transmissions are the helical type gear and all gear teeth should be inspected for chips or stripped teeth. Describe their general condition before disassembling the gearset.

   _____

   _____

2. Any gear that is mounted to a splined shaft needs the splines checked for mutilation or shifted splines. Record your findings.

   _____

   _____

3. Note any discoloration of the parts and explain the cause for it.

   _____

   _____

4. Check the planetary pinion gears for loose bearings. Record your findings.

   _____

   _____

5. Check each gear individually by rolling it on its shaft to feel for roughness or binding of the needle bearings. Wiggle the gear to be sure it is not loose on the shaft. Looseness will cause the gear to whine when it is loaded. Record your findings.

_____

_____

6. Inspect the gears' teeth for chips or imperfections, as these will also cause whine. Record your findings.

_____

_____

7. Check the gear teeth around the inside of the front planetary ring gear. Record your findings.

_____

_____

8. Check the fit between the front planetary carrier to the output shaft splines. Record your findings.

_____

_____

9. Remove the snap ring and thrust washer from the front planetary ring gear. Record your findings.

_____

_____

10. Examine the thrust washer and the outer splines of the front drum for burrs and distortion. Record your findings.

_____

_____

11. With the snap ring removed, the front planetary carrier can be removed from the ring gear. Check the planetary carrier gears for endplay by placing a feeler gauge between the planetary carrier and the planetary pinion gear. Compare the endplay to specifications. Record your findings.

_____

_____

12. Check the splines of the sun gear. Record your findings.

_____

_____

13. Sun gears should have their inner bushings inspected for looseness on their respective shafts. Record your findings.

_____

_____

Servicing Planetary Gear Assemblies (continued)

14. Check the fit of the sun shell to the sun gear and inspect the shell for cracks, especially at the point where the gears mate with the shell. Record your findings.

_____

_____

15. Check the sun shell for a bell-mouthed condition where it is tabbed to the clutch drum. Any variation from a true round should be considered junk and should not be used. Record your findings.

_____

_____

16. Look at the tabs and check for the best fit into the clutch drum slots. This involves trial fitting the shell and drum at all the possible combinations and marking the point where they fit the tightest. Record your findings.

_____

_____

17. Check the gear carrier for cracks and other defects. Record your findings.

_____

_____

18. Check the thrust bearings for excessive wear and, if required, correct the input shaft thrust clearance by using a washer with the correct thickness. Record your findings.

_____

_____

19. To determine the correct thickness, measure the thickness of the existing thrust washer and compare it to the measured endplay. All the pinions should have about the same endplay. Record your findings.

_____

_____

20. Replace all defective parts and reassemble the gearset.   ☐

Instructor's Response _____

_____

_____

Job Sheet 27

27

Name _____ Date _____

Servicing Internal Transaxle Drives

Upon completion of this job sheet, you should be able to inspect the transaxle drive, link chains, sprockets, gears, bearings, and bushings.

ASE Correlation

This job sheet is related to the Automatic Transmission and Transaxles Test's Content Area: *Off-Vehicle Transmission and Transaxle Repair, Gear Train, Shafts, Bushings, and Case.*
Tasks: Inspect transaxle drive, link chains, sprockets, gears, bearings, and bushings; perform necessary action and inspect, measure, repair, adjust or replace transaxle final drive components.

Tools and Materials

Marking tool
Machinist's rule

Describe the vehicle being worked on:

Year _____ Make _____ Model _____

VIN _____ Engine type and size _____

Model and type of transmission _____

Procedure

Chain Drives

Task Completed

1. This inspection is done with the transaxle on a bench and partially disassembled and should be repeated as a double check during reassembly. Begin by checking chain deflection between the centers of the two sprockets. Deflect the chain inward on one side until it is tight. ☐

2. Mark the housing at the point of maximum deflection. ☐

3. Then deflect the chain outward on the same side until it is tight. ☐

4. Again mark the housing in line with the outside edge of the chain at the point of maximum deflection. ☐

5. Measure the distance between the two marks. If this distance exceeds specifications, replace the drive chain. Describe your findings.

_____

_____

6. Be sure to check for an identification mark on the chain during disassembly. The mark can be a painted or dark-colored link, and may indicate either the top or the bottom of the chain, so be sure you remember which side was up. How was your chain marked?

_____

_____

7. The sprockets should be inspected for tooth wear and for wear at the point where they ride. If the chain was found to be too slack, it may have worn the sprockets in the same manner that engine timing gears wear when the timing chain stretches. A slightly polished appearance on the face of the gears is normal. Describe your findings.

_____

_____

8. Check the bearings and bushings used on the sprockets for damage. Describe your findings.

_____

_____

9. The radial needle thrust bearings must be checked for any deterioration of the needles and cage. Describe your findings.

_____

_____

10. The running surface in the sprocket must also be checked, as the needles may pound into the gear's surface during abusive operation. Describe your findings.

_____

_____

11. The bushings should be checked for any signs of scoring, flaking, or wear. Describe your findings.

_____

_____

12. Based on the above, what parts need to be replaced?

_____

_____

_____

_____

Final Drive Units

1. Final drive units may be helical gear or planetary gear units. A careful inspection of the assembly is done with the transaxle disassembled. The helical type should be checked for worn or chipped teeth, overloaded tapered roller bearings, and excessive differential side gear and spider gear wear. Describe your findings.

_____

_____

2. Measure the clearance between the side gears and the differential case. Compare your measurement to specifications. Describe your findings.

_____

_____

3. Check the fit of the spider gears on the spider gear shaft. Describe your findings.

_____

_____

4. Check the assembly's endplay. How do your measurements compare to specifications?

_____

_____

5. What is used to preload the side bearings on this transaxle?

_____

_____

6. With a torque wrench, measure the amount of rotating torque. Compare your readings against specifications. Describe your findings.

_____

_____

7. If the bearing preload and endplay are fine, as is the condition of the bearings, the parts can be reused. However, always install new seals during assembly.

☐

8. Planetary-type final drives are checked for the same problems as helical types. Check for worn or chipped teeth, overloaded tapered roller bearings, and excessive differential side gear and spider gear wear. Describe your findings.

_____

_____

9. The planetary pinion gears need to be checked for looseness or roughness on their shafts and for endplay. Describe your findings.

_____

_____

10. Check the endplay of the assembly. How did you do this and how do your measurements compare to specifications?

_____

_____

11. What is used to preload the side bearings on this transaxle?

_____

_____

12. With a torque wrench, measure the amount of rotating torque. Compare your readings against specifications. Describe your findings.

_____

_____

13. What are your conclusions about the final drive unit?

_____

_____

_____

_____

Instructor's Response _____

_____

_____

_____

Friction and Reaction Unit Service

Upon completion and review of this chapter, you should be able to:

- ❏ Inspect and replace bands and drums.

- ❏ Adjust bands, internally and externally.

- ❏ Inspect and replace servo including bore, piston, seals, pin, spring, and retainers and repair or replace as necessary.

- ❏ Inspect an accumulator bore, piston, seals, spring, and retainers while the transmission is in or out of the vehicle.

- ❏ Inspect and service clutch drum, piston, check balls, springs, retainers, seals, and friction and pressure plates.

- ❏ Measure and adjust clutch pack clearance.

- ❏ Air test the operation of the clutch pack and servo assemblies.

- ❏ Inspect and service roller and sprag clutch, races, rollers, sprags, springs, cages, and retainers.

Friction and Reaction Units

An automatic transmission achieves its various speed gears and reverse gear through holding or driving different members of the planetary gearset. Transmissions use compound gears, which are controlled by compound friction and reaction devices. Although diagnosis, based on symptoms, can lead you to the device that isn't holding or driving properly, recognition of those suspected parts becomes difficult when the transmission is apart. Clutch and brake application charts are a must during diagnosis. They are also a must during transmission overhaul.

As an illustration of the complexities involved with part recognition, let us take a brief look at a Chrysler 41TE transaxle (Figure 9-1). This is a four-speed unit with electronic controls that control the hydraulic system. The hydraulic system is typical in that it directly controls the transaxle's mechanical components. Although many of the friction and reaction devices are controlled by hydraulics, they are very mechanical and can be considered part of the mechanical system of the transmission.

The input clutch assembly of the 41TE is a good example of a compound friction unit. This assembly houses the underdrive, overdrive, and reverse clutches. These are the three different input clutches for the transmission. All of them are housed together in the input clutch assembly (Figure 9-2). These clutches direct the input to a compound planetary gearset.

The underdrive clutch is applied in first, second, and third gears. When applied, the underdrive hub drives the rear sun gear. The overdrive clutch is applied in third and overdrive gears. When the overdrive clutch is applied, it drives the front planet carrier. The underdrive clutch and the overdrive clutch are applied in third gear, giving two inputs to the gearset, which provides for direct drive. The reverse clutch is only applied in reverse and drives the front sun gear assembly.

The 41TE also relies on two additional clutches: the 2nd/4th gear and Low/Reverse units (Figure 9-3). These clutches hold a member of the planetary gearset, while the input clutches drive others. The holding clutches are placed at the rear of the transmission case.

The 2nd/4th clutch is engaged in second and fourth gears and grounds the front sun gear to the case. The Low/Reverse clutch is applied in park, reverse, neutral, and first gear. When this clutch is engaged, the front planet carrier and rear ring gear assembly are grounded to the case.

The planetary gearset in the 41TE is located between the input clutch assembly and the rear of the transaxle. This compound gearset does not have a common member; rather, the two planetary units are connected in tandem, with the front carrier connected to the rear ring gear and the rear carrier connected to the front ring gear.

Basic Tools

Mechanic's basic tool set

Appropriate service manual

Torque wrench

Clean ATF

Clean drain pan

Lint-free rags

Classroom Manual
Chapter 9, page 281

Underdrive occurs when there is speed reduction through the gears.

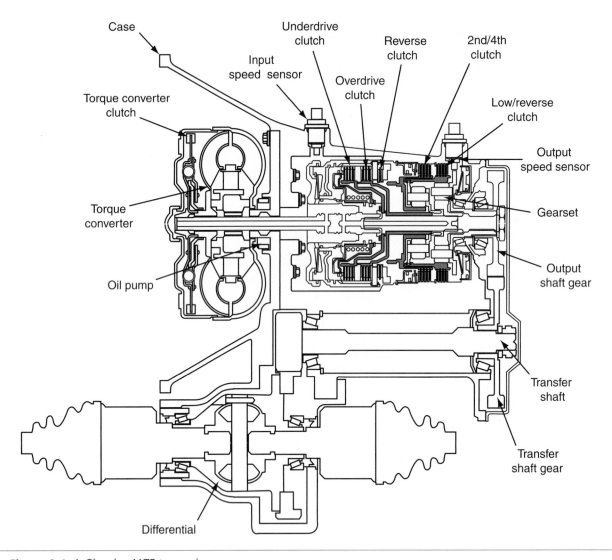

Figure 9-1 A Chrysler 41TE transaxle.

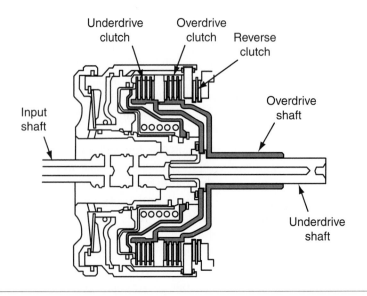

Figure 9-2 The input clutch assembly of a 41TE.

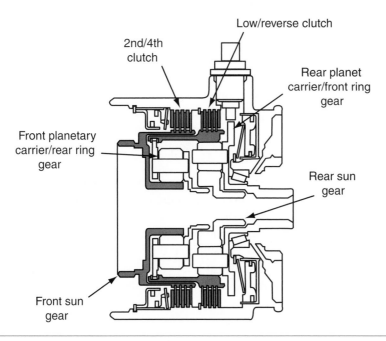

Figure 9-3 The holding clutches for a 41TE.

During inspection and disassembly of the friction units and their hub or drum, pay close attention to which planetary gearset member each is attached. Keep in mind that some planetary members are connected to each other and, as one member rotates, it rotates the other at the same speed. Also beware that sometimes the drum or housing for a clutch is also the drum for a brake band (Figure 9-4). Often this means that the same planetary member can be driven or held to achieve different gears.

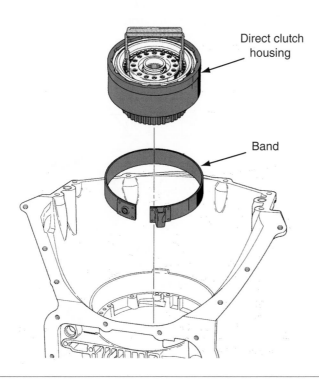

Figure 9-4 The drum for this clutch also serves as the clamping surface for the band. The tool shown in the clutch assembly is used to pull the assembly out of the transmission housing.

Band Service

Classroom Manual
Chapter 9, page 283

Servicing bands and their components includes inspection of the bands as well as the drums that the bands are wrapped around. Before the introduction of overdrive automatic transmissions, most bands operated in a free condition during most driving conditions. This means the band was not applied in the cruising gear range. However, many overdrive automatic transmissions use a band in the overdrive cruise range, which puts an additional load on the band and subsequently causes additional wear on the band. For this reason, a thorough inspection of the bands is very important (Figure 9-5).

The bands in a transmission are either single or double wrap, depending on the application. Both types can be the heavy-duty cast iron type or the normal strap type. The frictional material used on clutches and bands is quite absorbent. This characteristic can be used to tell if there is much life left in the lining. Simply squeeze the lining with your fingers to see if any fluid appears. If fluid appears, this tells you the lining can still hold fluid and has some life left in it. It is hard to tell exactly how long the band will last, but at least you have an indication that it is still useable. Strap or flex-type bands should never be twisted or flattened out. This may crack the lining and lead to flaking of the lining.

Inspection

Band failures found during overhaul are easy to spot. Look for chipping, cracks, burn marks, glazing, and nonuniform wear patterns and flaking. If any of these defects are apparent, the band should be replaced.

Also inspect brake band frictional material for wear. If the linings show wear, carefully check the band struts (Figure 9-6), levers, and anchors for wear. Replace any worn or damaged parts. Look at the linings of heavy-duty bands to see if the lining is worn evenly. A twisted band will show tapered wear on the lining. If the frictional material is blackened, this is caused by an excessive buildup of heat. High heat may weaken the bonding of the lining and allow the lining to come loose from the metal portion of the band. On double-wrap bands, check both segments of the lining for cracks and other damage. Make sure you carefully check the band lugs for damage as well.

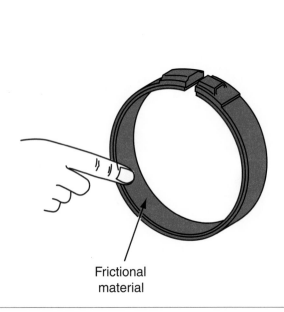

Frictional material

Figure 9-5 The frictional material of the bands must be carefully inspected.

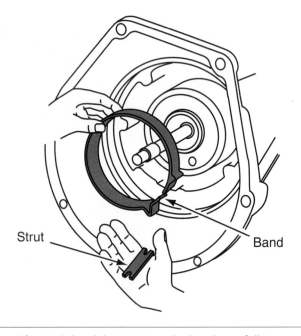

Strut

Band

Figure 9-6 While removing the band, carefully inspect the band strut.

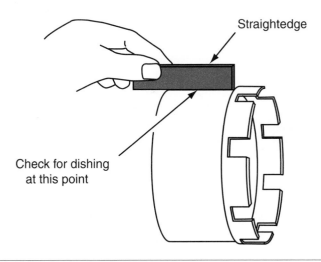

Straightedge

Check for dishing at this point

Figure 9-7 Check the flatness of the drum with a straightedge.

The drum surface should be checked for discoloration, scoring, glazing, and distortion. The drums will be either iron castings or steel stampings. Cast iron drums that are not scored can generally be restored to service by sanding the running surface with 180-grit emery paper in the drum's normal direction of rotation. A polished surface is not desirable on cast iron drums.

The surface of the drum must be flat (Figure 9-7). This is not usually a problem with a cast iron drum, but it can affect the stamped steel-type drum. It is possible for the outer surface of the drum to dish outward during its normal service life. This is a common problem on the GM 4L60 and should be inspected on any transmission that has a stamped steel band surface. Check the drum for flatness across the outer surface where the band runs. Any dishing here will cause the band to distort as it attempts to get a full grip on the drum. Distortion of the band weakens the bond of the frictional material to the band and will cause early failure due to flaking of the frictional lining. A dished stamped steel drum should be replaced. Check the service manual for maximum allowable tolerances.

Band Adjustments

After the band assembly has been installed in the transmission housing and around its drum, the band needs to be adjusted. Band adjustment is also part of a "**transmission tune-up**" on some models. Many transmissions have provisions for externally adjusting the band running clearance (Figure 9-8). Some transmissions have no provisions for band adjustment other than selectively sized servo apply pins and struts.

To set the running clearance on transmissions that have an adjustment screw, loosen the locknut on the adjuster screw and back it off about five turns. Backing off the locknut allows for tightening the screw to a specified torque, which simulates a fully applied band (Figure 9-9). The amount of required torque varies with different manufacturers, but it gives the same result.

After torquing, the adjuster screw is backed off a number of turns as specified by the manufacturer. To hold the adjustment, the locknut is generally tightened to 30–35 ft.-lb., while the adjuster screw is held stationary. The timing of band application has a lot to do with how the shift feels to the driver. This is one of the reasons there are so many different tightening torques on the various transmissions. The torque setting and number of turns the adjuster is backed off provides the proper clearance and grip for the many different types of bands used. Additionally, the pitch of the threads on the adjuster screws varies even on the same type of transmission. This is why you cannot use a single adjustment sequence for all transmissions.

Not all screw-type band adjusters are on the exterior of the transmission case. Torqueflite transmissions have the low-reverse band adjustment inside the oil pan. This requires the removal

A **transmission tune-up** usually includes a fluid and filter change, an inspection, and the adjustment of bands, if possible.

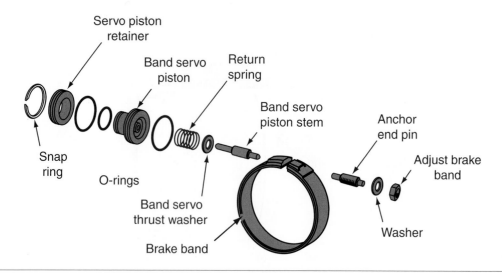

Figure 9-8 A complete band/servo assembly with an adjusting screw.

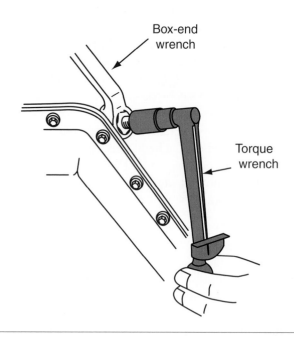

Figure 9-9 Adjusting a band with a torque wrench and box-end wrench.

Adjusting the bands may or may not be part of a scheduled maintenance program. Some manufacturers recommend periodic adjustments; others recommend adjustment only after the transmission has been overhauled. Some transmissions do not have a band adjustment.

of the pan to make band adjustments. This may seem inconvenient, but since the low-reverse band is normally applied at idle conditions, not much lining wear occurs. Therefore, the low-reverse band does not require adjustment as often as the intermediate band, and it can be done during a fluid and filter change when the pan is removed.

As you can see, band adjustment is not a difficult task. However, gaining access to the adjuster screw may require the removal of some other linkages, wiring, or even cooler lines that could be in your way. Be sure any components you moved out of the way are restored to their original position after the band adjustment is finished.

CAUTION: It is important that band adjustment procedures be followed exactly as outlined by the manufacturer. Serious damage to the transmission can result if the procedure is not followed.

Servo and Accumulator Service

On some transmissions, the servo and accumulator assemblies are serviceable with the transmission in the vehicle (Figure 9-10). Others require the complete disassembly of the transmission. Internal leaks at the servo or clutch seal cause excessive pressure drops during gear changes.

Before disassembling a servo or any other component, carefully inspect the area to determine the exact cause of the leakage. Do this before cleaning the area around the seal. Look at the path of the fluid leakage and identify other possible sources. These sources could be worn gaskets, loose bolts, cracked housings, or loose line connections.

Inspect the outside area of the seal. If it is wet, determine if the oil is leaking out or if it is merely a lubricating film of oil. When removing the servo, continue to look for the causes of the leak. Check both the inner and outer parts of the seal for wet oil, which means leakage. When removing the seal, inspect the sealing surface, or lips, before washing. Look for unusual wear, warping, cuts and gouges, or particles embedded in the seal.

Band servos and accumulators are basically pistons with seals in a bore held in position by springs and retaining snap rings (Figure 9-11). Remove the retaining rings and pull the assembly from the bore for cleaning. Cast iron seal rings may not need replacement, but rubber and elastomer seals should always be replaced (Figure 9-12). Most experienced automatic transmission technicians will replace all rings during a rebuild, just to be safe.

Also check the condition of the pins, piston, and springs (Figure 9-13). Look for signs of wear and damage. Also check the fit of the pins and pistons in the case. The bores in the case should not allow the pins and pistons to wobble. If they do wobble in the bore, either they are worn or the bores in the case are worn.

Accumulators

Begin the disassembling of an accumulator by removing the accumulator plate snap ring. After removing the accumulator plate, remove the spring and accumulator pistons. If rubber seal rings are installed on the piston, replace them whenever you are servicing the accumulator. Lubricate the new accumulator piston ring and carefully install it on the piston. Lubricate the accumulator cylinder walls and install the accumulator piston and spring. Then reinstall the accumulator plate and retaining snap ring.

Classroom Manual
Chapter 9, page 305

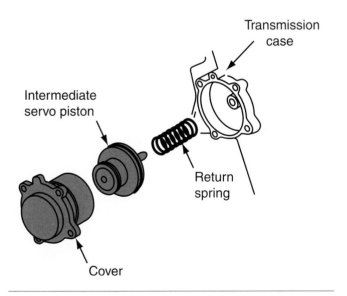

Figure 9-10 The servos in some transmissions are serviceable while the transmission is in the vehicle and are contained in its own bore.

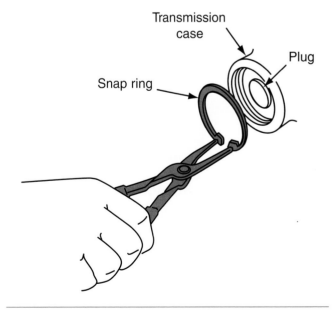

Figure 9-11 Servos and accumulators are normally retained by a snap ring.

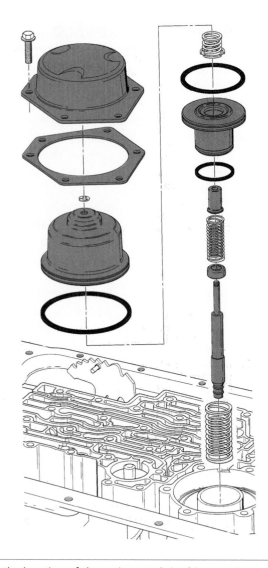

Figure 9-12 Note the location of the various seals in this servo/accumulator assembly.

Many accumulator pistons can be installed upside down. This results in free travel of the piston or too much compression of the accumulator spring. Note the direction of installation of the piston during the teardown process, as you will not always find a good picture when you need to reinstall the accumulator. Because the movement of the accumulator has an effect on shift feel, correct installation is critical. It is common for manufacturers to mate servo piston assemblies with accumulators. This takes up less space in the transmission case and, because they have the same basic shape, can reduce some of the machining during manufacture.

Servos

A servo is disassembled in a similar fashion. The servo's piston, spring, piston rod, and guide should be cleaned and dried. Check the servo piston for cracks, burrs, scores, and wear. Servo pistons may be made of either aluminum or steel. Aluminum pistons should be carefully checked for cracks and their fit on the guide pins. Cracked pistons will allow for a pressure loss and being loose on the guide pin may allow the piston to bind in its bore. Whether it is a steel or aluminum piston, the seal groove should be free of nicks or any imperfection that might pinch or bind the seal. Clean up any problems with a small file or scraper.

Classroom Manual
Chapter 9, page 286

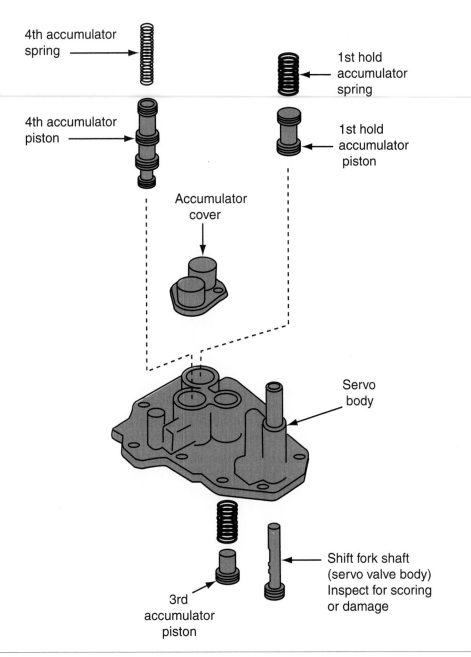

4th accumulator spring

1st hold accumulator spring

4th accumulator piston

1st hold accumulator piston

Accumulator cover

Servo body

Shift fork shaft (servo valve body) Inspect for scoring or damage

3rd accumulator piston

Figure 9-13 Another design of a servo/accumulator assembly.

Seals

Most original equipment servo seals are the Teflon type. These seals will exhibit no feeling of drag in the bore due to the slipperiness of the Teflon. A majority of replacement transmission gasket sets will supply cast iron hook-end seal rings in place of the Teflon seals. When they are installed, the cast iron seals will have a noticeable drag as the servo piston is moved through the bore. This is not a problem, and in some cases may even improve operation.

Check the cast iron seal rings to make sure they are able to turn freely in the piston groove. These seal rings are not typically replaced unless they are damaged, so carefully inspect them.

Some servo pistons have a molded-on rubber seal. This type is usually replaced during overhaul since the rubber is subject to the same deterioration as other seals of similar construction.

Inspection

Inspect the servo or accumulator spring for possible cracks. Also check where the spring rests against the case or piston. The spring may wear a groove into the aluminum, so be sure the piston or case material has not worn too thin. Some servos utilize a steel chafing plate in order to eliminate this problem.

Inspect the servo cylinder for scores or other damage. Move the piston rod through the piston rod guide and check for freedom of movement. Check the band servo components for wear and scoring. Replace all other components as necessary, then reassemble the servo assembly.

Assembly

Multipart servo/accumulators must be assembled in the correct sequence. Incorrect assembly can result in dragging bands and harsh shifting. If the servo pin has Teflon seals, these will need to be cut off with a knife and a new Teflon seal installed with the correct sleeve and sizing tools.

When reassembling the servo, lubricate the seal ring with ATF and carefully install it on the piston rod. Lubricate and install the piston rod guide with its snap ring into the servo piston. Then install the servo piston assembly, return spring, and piston guide into the servo cylinder. Some servos are fitted rubber lip seals that should be replaced (Figure 9-14). Lubricate and install the new lip seal. On spring-loaded lip seals, make sure that the spring is seated around the lip and that the lip is not damaged during installation.

Selective Servo Apply Pins

Classroom Manual
Chapter 9, page 288

Servo apply pins are called selective when they are available in a variety of lengths. The correct length must be used to prevent transmission damage and ensure proper band operation.

Transmissions without a band adjusting screw use **selective servo apply pins** that maintain the correct clearance between the band and the drum. This type of servo apply pin must be checked for correct length during overhaul to ensure proper stroking action of the servo piston as well as shift timing. The travel needed to apply a band relates to the timing of band application. Although the valving and orificing determines true shift timing, the adjustment of the servo pin completes the job by providing the correct clearance of the band to the drum. To select the correct apply pin, most transmissions require the use of special tools. The tools are used to simulate a fully applied band and allow the technician to determine the correct length pin.

To select the servo apply pin in a 4L80-E transmission, two special tools and a torque wrench are needed. A gauge pin (Figure 9-15) is placed into the bore for the servo. Then the checking tool (Figure 9-16) is positioned over the bore with its hex nut facing the linkage for the parking pawl. The checking tool is fastened with two servo cover bolts that are torqued to a specified amount. The hex nut on the checking tool is then torqued to 25 ft.-lb. and the exposed part of the

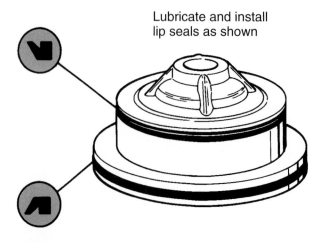

Lubricate and install
lip seals as shown

Figure 9-14 Proper installation of servo piston seals.

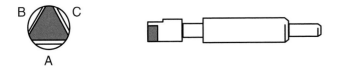

Figure 9-15 A special gauge pin that is placed into the bore for the servo to determine which selective servo pin should be used.

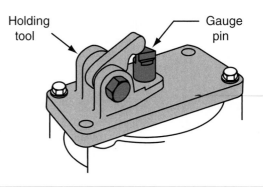

Figure 9-16 The checking tool is positioned over the bore and gauge pin with its hex nut facing the linkage for the parking pawl.

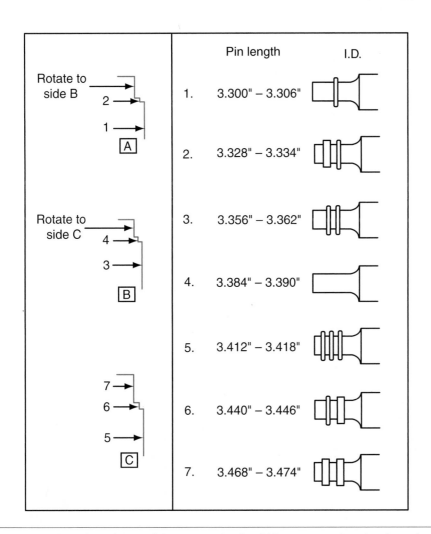

	Pin length	I.D.
Rotate to side B	1. 3.300" – 3.306"	
	2. 3.328" – 3.334"	
	3. 3.356" – 3.362"	
Rotate to side C	4. 3.384" – 3.390"	
	5. 3.412" – 3.418"	
	6. 3.440" – 3.446"	
	7. 3.468" – 3.474"	

Figure 9-17 The exposed part of the gauge pin should be compared to the chart given in the service manual to determine the correct pin length.

pin gauge is studied to determine the proper apply pin length. A chart in the service manual (Figure 9-17) relates the appearance of the exposed part of the pin gauge to the correct pin length.

If a replacement band, drum, or even case has been used, an apply pin check must be made. This is the only way to be certain you have the correct length pin. Since these pins are selective, you will have to start by checking the length of the pin already in the transmission. If it is too long or too short, a new length pin will be necessary. These pins are available at the parts departments of most dealerships and suppliers for automatic transmission rebuilders.

Classroom Manual
Chapter 9, page 292

Many transmission rebuilders replace all friction discs and steel plates as a part of the overhaul process. This can actually save time because it eliminates the inspection time and there is no doubt about whether any discs or plates were of questionable condition.

Multiple-Friction Disc Assemblies

Two types of multiple-friction disc assemblies are found in transmissions: rotating drum (Figure 9-18) and case grounded (Figure 9-19). Both are serviced in the same way but require slightly different inspection procedures.

 SERVICE TIP: New friction discs should not be used with used steel plates unless the steel plates are deglazed.

All friction disc and steel plate packs are held in place by snap rings. These snap rings may be selective in thickness and must be kept with the clutch pack during disassembly. It is common to use the same diameter snap ring on more than one clutch in a single transmission. However, the rings can be of differing thicknesses. This thickness variation can be used to set the clearance in the clutch pack. The snap rings may also have a distinct shape that will be effective only when used in the correct groove.

An example of complete disassembly, inspection, assembly, and clearance checks is shown in Photo Sequence 13. Always refer to the manufacturer's recommendations for a particular transmission.

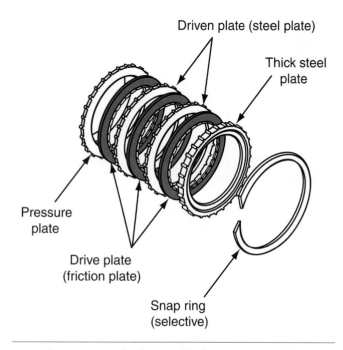

Figure 9-18 Clutch assembly from a rotating drum type clutch.

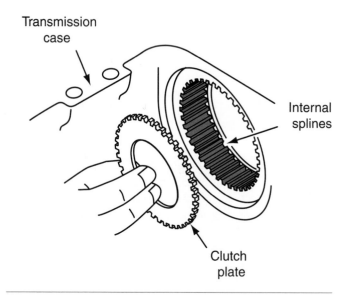

Figure 9-19 A case ground clutch assembly.

Photo Sequence 13
Proper Procedure for Disassembling, Inspecting, Assembling, and Clearance Checking Direct Clutch

P13-1 Set the direct clutch on the bench.

P13-2 Pry out the snap ring.

P13-3 Remove the pressure plate and the clutch pack assembly.

P13-4 Install the piston compressor and compress the piston.

P13-5 Remove the snap ring.

P13-6 Remove the piston compressor, the spring retainer assembly, and the piston.

P13-7 Install new seals on the piston.

P13-8 Check the movement of the check ball.

P13-9 Lubricate the outer piston seal with petroleum jelly.

(continued)

Proper Procedure for Disassembling, Inspecting, Assembling, and Clearance Checking Direct Clutch

P13-10 Reinstall the piston.

P13-11 Place the piston spring retainer into the drum.

P13-12 Install the compressor and snap ring.

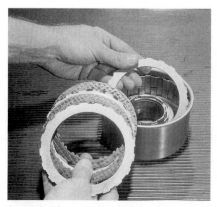

P13-13 Alternately install friction and steel discs.

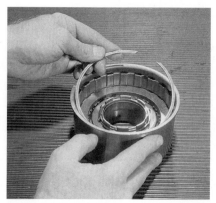

P13-14 Install the pressure plate and snap ring.

P13-15 Check the clearance with a feeler gauge.

P13-16 Air test the assembly.

Disassembly

Using a screwdriver, remove the large clutch retaining plate snap ring and remove the thick steel clutch pressure plate (Figure 9-20). Now remove the clutch pack. Some clutch packs have a wave snap ring that must be looped through the clutch pack until the steel plates can be removed.

Once the snap ring is removed, the backing plate will be the first steel plate you will take out. This plate, like the snap ring, may be a selective part. Since the backing plate is thicker than the other steel plates in the clutch pack, it is easy to spot. Now remove the remainder of the friction discs and steel plates, keeping them in the order in which they were in the drum (Figure 9-21).

Using a clutch spring compressor tool, compress the clutch return springs. Then using snap ring pliers, remove the clutch hub retainer snap ring, retaining plate, and springs. Most retainers have small tabs that prevent you from removing the snap ring without using a compressor (Figure 9-22).

Special Tools

Snap ring pliers

Feeler gauge

OSHA-approved air nozzle

If the return spring assembly is installed upside down, the unit will pass an air check but will not apply the clutches.

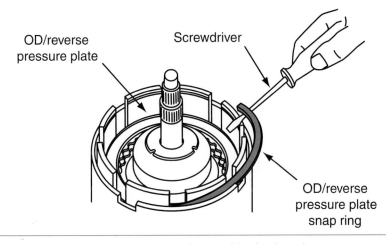

Figure 9-20 Removing large snap ring to disassemble clutch pack.

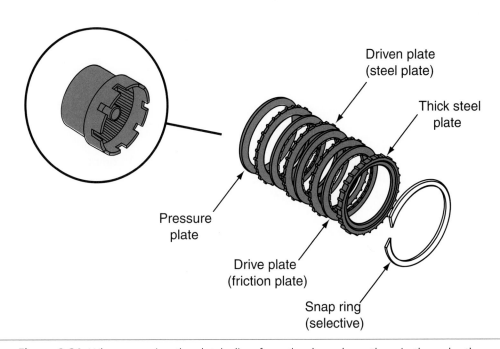

Figure 9-21 When removing the clutch discs from the drum, keep them in the order they were originally in.

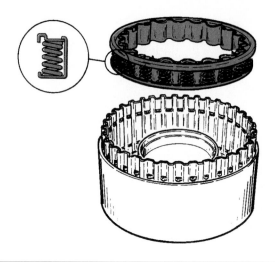

Figure 9-22 Retainer ring for captive return springs.

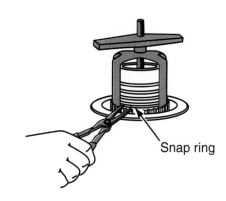

Figure 9-23 A spring compressor tool for a multiple-friction disc assembly in a GM 4T80E transaxle.

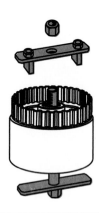

Figure 9-24 A spring compressor tool for a multiple-friction disc assembly in Honda transaxles.

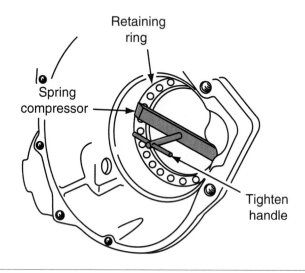

Figure 9-25 A spring compressor tool for a multiple-friction disc assembly that is set into the transmission housing.

There are many types of clutch spring compressor tools available. The many different locations and depths of the clutch springs in their drums dictate having more than one simple compressor (Figures 9-23, 9-24, and 9-25).

With the compressor installed on the spring retainer (Figure 9-26), compress the spring and retainer just enough to allow the snap ring to be removed. Pushing the retainer down too much may bend or distort it. Make sure the snap ring is not partially caught in its groove before releasing the spring compressor. Also be very careful when releasing the spring compressor tool; some springs have very high tensions and can injure you if they get a chance to fly out.

 WARNING: Be careful when compressing the springs. Careless procedures can allow the springs and retainer to fly into your face.

With the retainer removed, a single large coil spring or multiple small coil springs will be exposed. Note the number and placement of the multiple springs for assembly.

If no coil springs were encountered when the clutch pack was removed, then a Belleville or disc-type return spring is used (Figure 9-27). This type of clutch will have a heavy steel plate with

Classroom Manual
Chapter 9, page 296

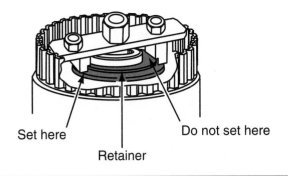

Set here

Retainer

Do not set here

Figure 9-26 Proper positioning of the spring compressor will prevent damage to the piston and seals.

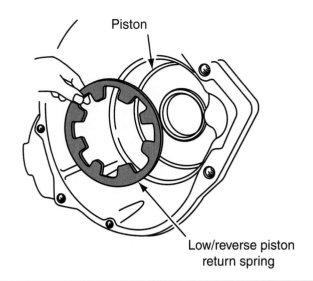

Piston

Low/reverse piston return spring

Figure 9-27 A clutch assembly fitted with a Belleville spring.

one rounded side in the bottom of the clutch pack. This heavy plate is a pressure plate. The released position of these springs has very little or no outward force on the retaining snap ring; therefore, the snap ring can be removed without the use of a compressor. The snap ring for this application could be selective, so remember which snap ring goes where.

Some clutches can have two or more snap rings, so pay attention to their placement in the clutch. Additionally, a wavy snap ring is sometimes used to retain the Belleville spring. This is used to give some cushion to the application of the clutch.

For easy removal of the piston from its drum, mount the clutch on the oil pump. Then lift the piston out of the bore. If it will not come out, the drum can be inverted and slammed squarely against a hardwood bench top. This will dislodge a tight piston, but it will not damage the drum.

Another way to remove the piston is to charge the apply circuit of the clutch with compressed air. Use an air nozzle with a rubber tip and apply air pressure to pop out the piston from the drum. Air is blown in either at the feed hole in the drum or by placing the drum on the clutch support and using its normal feed passage (Figure 9-28). Wrap a rag around the clutch drum to catch the piston and the fluid overspray. Make sure your fingers are out of the way of the moving piston.

Belleville-type return springs are usually found in the forward clutch.

 WARNING: Air pressure should be reduced to 25–30 lb. to avoid expelling the piston at a high rate of speed, which could cause injury.

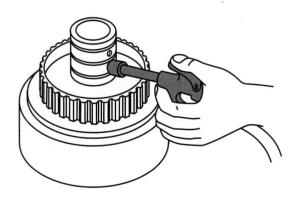

Figure 9-28 Use an air nozzle with a rubber tip and apply air pressure at the feed hole in the drum to pop out the piston from the drum.

Pistons located in the transmission case can be removed by blowing compressed air into the apply passage in the case.

With the piston removed, take note of the types of seals used and their position. This is important with lip seals, as the lip will face the direction the fluid comes from. Lip seals installed in the wrong direction will not hold pressure. Seals are generally replaced during overhaul. The reuse of old seals is not recommended.

After the clutch pack has been disassembled, check the clutch drum, clutch housing, or transmission housing for defects. The splined area in the transmission housing should be carefully inspected for broken or chipped splines and other damage. If these splines are damaged, the case should be replaced. Clutch drums and housings should be checked for worn or damaged bushings. These should be replaced if they show signs of wear or damage. The splines inside the drum should be inspected for broken or chipped splines and other damage. If these splines are damaged, the drum or housing should be replaced. The outside surface of the housing should be inspected for signs of overheating and other defects. The flatness of the surface should also be checked by placing a straightedge across the surface (Figure 9-29). Look straight at the housing and the straightedge. If light shines between these two surfaces, the surface is distorted and the housing should be replaced.

Inspection and Cleaning

Once a clutch assembly has been taken apart, you may wish to inspect the clutch components or continue to disassemble the remainder of the clutch units in the transmission. If you choose the latter, make sure you keep the parts of each clutch separate from the others.

Clean the components of the clutch assembly. Make sure all clutch parts are free of any residue of varnish, burned disc facing material, or steel filings. Take special care to wash out any foreign material from the inside of drums and the hub disc splines. If left in, the material can be washed out by the fresh transmission fluid and sent through the transmission. This can ruin the rebuild.

The clutch splines must be in good shape with no excessively rounded corners or shifted splines. Test their fit by trial fitting each new clutch disc on the splines. Move the discs up and down the splines to check for binding. If they bind, this can cause dragging of the discs during a time when they should be free floating. Replace the hubs if the discs drag during this check. Check the spring retainer. It should be flat and not distorted at its inner circumference. Check all springs for height, cracks, and straightness (Figure 9-30). Any springs that are not the correct height or that are distorted should be replaced. Many retainers have the springs attached to them by crimping. This speeds up production at the assembly line. Turning this type of retainer upside down is a

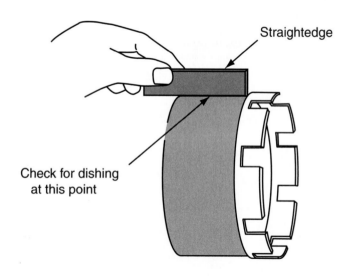

Figure 9-29 Checking the flatness of a housing with a straightedge.

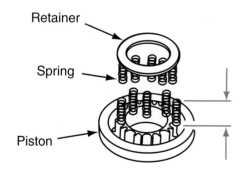

Figure 9-30 Check the height of each one of the return springs.

quick check of spring length. Closely examine the Belleville spring for signs of overheating or cracking, and replace it if it is damaged.

The steel plates should be checked to be sure they are flat and not worn too thin. Check all **steels** against the thickest one in the pack or a new one. Most steels will have an identifying notch or mark on the outer tabs. If the plates pass inspection, remove the polished surface finish and the steels are ready for reuse. The steel plates should also be checked for flatness by placing one plate on top of the other and checking the space on the inside and outside diameters. Clutch plates must not be warped or cone-shaped. Also, check the steel plates for burning and scoring and for damaged driving lugs. Check the grooves inside the clutch drum and check the fit of the steel plates, which should travel freely in the grooves.

Close inspection of the friction discs is simple. The discs will show the same types of wear as bands will. Disc facings should be free of chunking, flaking, and burned or blackened surfaces. Discs that are stripped of their facing have been overheated and subjected to abuse. In some cases, the friction discs and steel plates can be welded together. This occurs when the facing comes off the disc due to a loosening of the facing's bonding because of extreme heat. As the facing comes off, metal-to-metal contact is made between the discs and steel plates. The friction involved as the clutch tries to engage causes them to fuse together. This may lock the clutch in an engaged condition. Depending on which clutch is affected, driveability problems can include: drives in NEUTRAL, binds up in REVERSE (forward clutch seized), starts in direct drive, binds up in second (high-reverse clutch seized), and other problems that are not common.

If the discs do not show any signs of deterioration, squeeze each disc to see if fluid is still trapped in the facing material. If fluid comes to the surface, the disc is not glazed. Glazing seals the surface of the disc and prevents it from holding fluid. Holding fluid is basic to proper disc operation. It allows the disc to survive engagement heat, which would burn the facing and cause glazing. Fluid stored in the frictional material cools and lubricates the facings as it transfers heat to the steel plates and also carries heat away as some oil is spun out of the clutch pack by centrifugal force. This helps avoid the scorching and burning of the disc.

Clutch discs must not be charred, glazed, or heavily pitted. If a disc shows signs of flaking frictional material or if the frictional material can be scraped off easily, replace the disc.

A black line around the center of the friction surface also indicates that the disc should be replaced. Examine the teeth on the inside diameter of each friction disc for wear and other damage (Figure 9-31).

The steel plates are commonly called **steels** by the trade.

Classroom Manual
Chapter 9, page 292

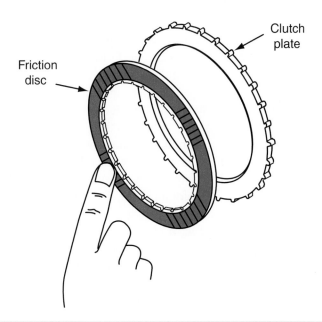

Figure 9-31 Carefully examine the clutch discs.

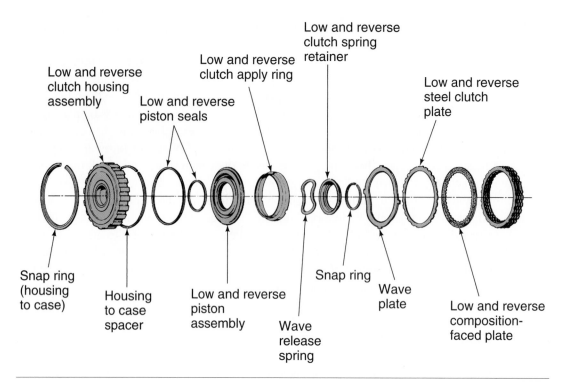

Low and reverse clutch spring retainer

Low and reverse clutch housing assembly

Low and reverse clutch apply ring

Low and reverse steel clutch plate

Low and reverse piston seals

Snap ring (housing to case)

Housing to case spacer

Low and reverse piston assembly

Wave release spring

Snap ring

Wave plate

Low and reverse composition-faced plate

Figure 9-32 A clutch assembly with a waved plate. (Courtesy of General Motors Corporation, Service Operations)

Classroom Manual
Chapter 9, page 295

Wave plates are used in some clutch assemblies to cushion the application of the clutch (Figure 9-32). These should be inspected for cracks and other damage. Never mix wave plates from one clutch assembly to another. As an aid in assembly, most wave plates have identifying marks.

The clutch pistons (Figure 9-33) are checked for cracks, warpage, and fit in their bores. Also check the check ball and the check ball bore for damage. Carefully examine the seal ring grooves and inside diameter of the piston for cracks, nicks, and burrs. Groove wear can be accelerated by excessive pump pressures. The excess pressure forces the seal rings against the sides of the grooves so hard that fluid cannot get between the ring and groove to lubricate the ring. If the bores of the clutch are severely grooved, a stuck pressure regulator valve could be the problem.

The reverse side of the pump cover is the clutch support that incorporates seal rings for the fluid circuits leading to the clutch drums. The seal rings fit loosely into the grooves in the clutch support and rely on pump oil pressure to push them against the side of the groove to make the seal. The seal rings should be checked for side play in the grooves and for proper fit into the drum. Check the grooves for burrs, step wear, or pinched groove conditions. It should be noted that these seals rotate with the drum. Any condition that hinders rotation will cause the ring seals to bind, resulting in drum wear. This can destroy the drum if it is not corrected.

Carefully inspect aluminum pistons, which may have hairline cracks that will cause pressure leakage during use. This could cause clutch slipping as a result of lost hydraulic force pushing the piston against the discs. If burned discs were found in the clutch pack, be sure to check for cracks in the piston.

Stamped steel pistons have replaced most aluminum pistons because they are cheaper to produce. Aluminum pistons require a casting process followed by machining, whereas stamped pistons can be made by the thousands and may only require a simple spot-weld operation to prepare them for use. Stamped pistons show cracking more easily than aluminum. Also, look for any separation where spot welding is used on these pistons.

It is common to find annular check balls in clutch pistons (Figure 9-34). These balls allow for an air release from the bore while it is being filled with ATF and for a quick release of pres-

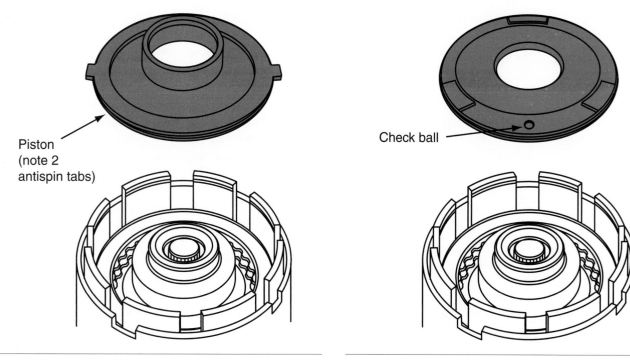

Figure 9-33 A clutch piston with anti-spin tabs.

Piston
(note 2
antispin tabs)

Check ball

Figure 9-34 A clutch piston with an annular check ball.

sure when the clutch is released. Inspect each check ball to be sure it is free in its bore. Even after a thorough cleaning in solvent, the check balls may not be free. A fine wire can be pushed into the bore, followed by a spray of carburetor cleaner, to remove any stubborn deposits. The bore should then be blown out with compressed air.

A check ball may also be located in the drum and should be checked. However, it is often very difficult to use the same cleaning techniques as used on pistons. A quick way to determine if the check ball is free is to shake the clutch drum to hear the relief check ball rattle. If the check ball does not rattle, replace the drum.

The ability of the check ball to seal is also important. To check how well the ball seats in its bore, pour clean solvent into the bore. Observe the other end of the bore. If fluid leaks out, the ball is not seating and the piston or drum should be replaced.

Examine the outside surface of the drum for glazing. Glazing can be removed with emery cloth. Also check the drum's cylinder walls for deep scratches and nicks.

Inspect the front clutch bushing for wear and scores. If the bushing is worn, replace it. Also inspect any bushings found in the clutch drums for excessive wear, scoring, or looseness in the drum bore. Replace as needed.

Clutch Pack Reassembly

Begin assembly of a clutch unit by gathering the new seals and other new parts that may be necessary. Prior to installation, all clutch discs and bands are to be soaked in the type of transmission fluid that will be used in the transmission. The minimum soak time is 15 minutes. Be sure that all discs are submerged in the fluid and that both sides are coated.

CAUTION: Use caution when installing the piston to prevent damage to the seals. Be careful not to stretch the seals during installation.

Before fitting the new rubber seals to the piston, check them against the old seals. This will ensure correct sizing and shape of the new seal. Most overhaul kits include more seals than are required to complete the job. This is because changes in transmission design may dictate the use of a seal or gasket of different design or size. Therefore, both the old design seal and the new design seal are included in the kit. The rebuilder must check to be sure the correct seal is being used. This also holds true for cast iron and Teflon ring seals.

 SERVICE TIP: Using the hook-end, cast iron, high-reverse, stator support seals from a C-4 on a C-5 will result in no reverse and no high gear because the inner bore of the C-5 clutch drum is larger than that of the C-4. Doing this would be building in a failure that, if not detected during an air check, would require the removal of the transmission to correct.

Once the correct seals are chosen from the kit, they can be installed on the piston (Figure 9-35). Remember to position lip seals so they face the same direction that the fluid pressure comes into the drum.

Seals should be lubricated with automatic transmission fluid, trans-gel, or petroleum jelly. Never use chassis lube, "white lube," or motor oil. These will not melt into the transmission fluid as it heats up, but will clog filters and orifices or cause valves and check balls to stick.

Manufacturers switched to Teflon seals for these positions because they helped reduce the wear at the bore in the drum. During an overhaul, they may be replaced with hook-end-type steel rings.

Care must be taken to be sure the ends of scarf-cut Teflon seals are installed correctly. These rings must also be checked for fit into the drum. They should have a snug, but not too tight, fit into the bore.

Some manufacturers use an endless type of Teflon seal. These seals must be installed with special sleeves and pushing tools to avoid overstretching the seal. The seals are first pushed over the installing sleeve to their location in the groove, then a sizing tool is slipped over the sleeve and seal ring to fit the seal to the groove (Figure 9-36).

CAUTION: Never assemble a clutch assembly when it is dry. Always lubricate its components thoroughly with clean ATF.

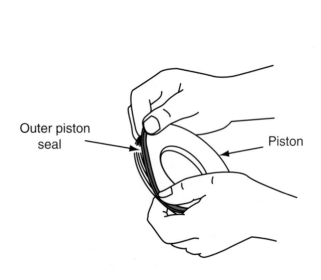

Figure 9-35 Install new seals around the outside of the piston before reassembling the clutch.

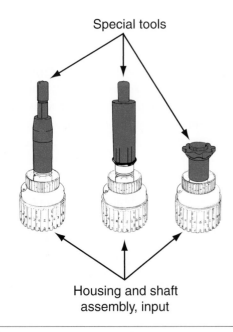

Figure 9-36 Tools and procedure for installing solid seals onto an input shaft.

Assemble the piston, being careful not to allow the seal to kink or become damaged during installation. The piston can now be installed. Several methods may be used to aid piston installation. Lathe-cut seals can be helped into their bores by using a thin feeler gauge mounted on a handle. These are available from most automotive tool suppliers. Pistons with lathe-cut seals are installed by positioning the piston in the bore of the clutch drum. Then slowly work the piston down in the bore until resistance is felt. Using the feeler-type seal installer, work your way around the outer circumference of the seal, using a downward action followed by a clockwise pulling motion as you push the seal back into the groove in the piston.

Occasionally, a piston will not allow access to the outer seal area. A large chamfered edge is at the top of the seal bore to allow the seal to be worked into the bore without the help of any special tools. The piston can be installed by rotating the piston as you push down. Use even pressure to avoid binding the piston or cocking the piston in its bore. Uneven pressure can also cause the ring seal to be pushed out or to tear.

> ☑ **SERVICE TIP:** Some rebuilders use a wax stick to coat lathe-cut seals for installation. This is available under the trade name Door Ease. Its original use was to stop squeaks on rubber door bumpers and latches. If Door Ease is used, do not coat the drum bore with ATF. The Door Ease works fine by itself.

Pistons with lip seals require a more delicate installation. The lips can be bent back or torn unless proper caution is taken during installation. The basic shape of a lip seal makes it necessary to use an installation tool (Figure 9-37). The lip must be pushed back toward the piston body in order to allow the seal to enter the bore. Lip seals will often stick in snap ring grooves as you try to slip the piston and seals into the drum. Piano wire installers can be used to roll lip seals back away from the snap ring grooves or the bore of the drum (Figure 9-38). The round cross section of the wire prevents cutting or tearing the seal lip during installation.

While holding the piston as squarely in the drum as you can, work the tool around the lip to allow the seal to enter the bore or around the center of the drum. Do not apply too much

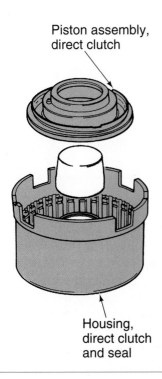

Piston assembly,
direct clutch

Housing,
direct clutch
and seal

Figure 9-37 Using a seal protector to install a piston into a drum without damaging the seals. (Courtesy of General Motors Corporation, Service Operations)

Some rebuilders use an old credit card in place of the feeler gauge. Its wide surface and plastic edge will not cut into the rubber seals.

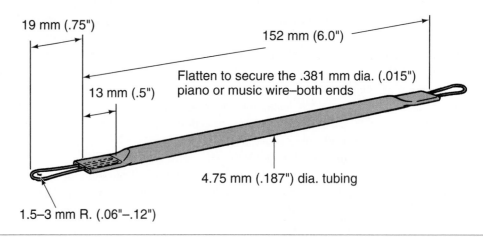

Figure 9-38 Piano wire-type piston installation tool. (Courtesy of General Motors Corporation, Service Operations)

downward force as the seal lip is worked into the bore. The piston will fall into place after the lip is fully inserted.

Pistons with multiple seals require special care to avoid damaging the other seals as you work on one seal. Multiple-seal piston installation is made simple by using plastic ring seal installers. These rings compress the seals back into the piston grooves so they will not hang up. Two installers are often used at the same time.

Regardless of the type of tool used to install the piston seals, always take your time to avoid tearing or rolling the new seals during installation.

When working with stamped steel pistons, install the ring spacer on the top side of the piston, making sure it has the correct thickness for the application.

Once the piston seals enter the bore, push the piston all the way down until it stops. Then lift the piston up slightly. Push it back down to the bottom of the bore to be sure it is all the way down. After the piston is installed, rotate the piston by hand to ensure that there is no binding.

Reassemble the springs and retainer after the piston has been installed. Place the single or multiple spring set onto the piston using the pockets provided to locate them. Be sure the loose spring sets are spaced as they were during teardown. Set the retainer plate over the springs. Make sure the retainer is facing in the correct direction. On some transmissions, if the retaining ring and springs are faced in the wrong direction, the clutch assembly will work well during an air check but it will not apply the clutches during use. Captive spring retainers are simply set on the piston and require no other setup.

Position the spring compressor on the retainer plate and compress the springs. Be careful not to allow the retainer to catch in the snap ring groove while compressing the spring. This will bend or distort the plate, making it unsuitable for use. Remember to compress the springs only enough to get the snap ring into its groove. Use snap ring pliers to expand or contract the snap ring. Once the snap ring is installed and fully seated, release the compressor.

Clutches with a Belleville return spring may not require a compressor for assembly. The Belleville spring is merely laid on the piston and centered in the bore. A wire ring is sometimes inserted on the piston where the Belleville spring touches the piston. This is used to prevent the steel spring from chafing the aluminum piston. If left out, there will be too much endplay and piston damage will occur.

The large snap ring that retains the Belleville spring is now inserted in its groove. This snap ring may be either flat or of wavy construction. There may also be a plastic spacer ring used under the snap ring to center the spring. Be certain the snap ring seats firmly against the drum. Then install the pressure plate on top of the Belleville spring.

> Loose springs are used mostly with aluminum pistons.

Continue clutch assembly by stacking the clutch pack. Begin by installing the dish plate, with the dish facing outward. If a wavy plate or cushion spring is used, it will normally be installed next to the piston. Alternately stack the steels and friction discs until the correct number of plates has been installed. Place the backing plate (the thickest steel plate) on top and install the retainer ring in its groove. Assemble the remaining clutch packs.

☑ **SERVICE TIP:** Used steels can be deglazed by sanding their faces with 320-grit emery paper, which can produce a dull surface on the plates.

As mentioned before, all models of a transmission do not always have the same number of discs and steels in their clutches. Most overhaul kits supply enough discs and steels to rebuild all models. They are likely to have more discs and steels than are required. This is why a technician should always note how many discs and steels are in each pack while disassembling the transmission.

Trying to fit all of the supplied discs and steels may result in clutches with no freeplay or no room for the snap ring. If these problems come up while stacking the pack, refer to a service manual. It may list the number of discs and steels for the model you are building. After the plates are installed, position the retainer plate and install the retaining snap ring. Proceed by measuring the clearance between the plate and snap ring.

■ **CAUTION:** Steel plates and friction discs from the same transmission may look the same, but may have different thicknesses. Compare the new plates and discs to the ones removed from each clutch during teardown. Incorrect disc and steel thicknesses can cause buildup headaches, if you are not aware of this fact. The new disc facings should match the type that was removed in both thickness and grooving, if correct clutch engagement is to be maintained.

☑ **SERVICE TIP:** There are three different direct clutch counts for the A500 and A518. The only reliable way to determine the correct number of clutches for the transmission you are working on is to measure the snap ring groove height in the drum. If the distance from the groove to the top of the drum is 0.485 inch, there should be 5 clutches and 4 steels. If the measurement is 0.35 inch, there should be 6 clutches and 5 steels. If the measurement is 0.1 inch, 8 clutches and 7 steels should be installed.

Clearance Checks

The clearance check of a multiple-disc pack is critical for correct transmission operation. Excessive clearance causes delayed gear engagements, while too little clearance causes the clutch to drag. Adjusting the clearance of multiple-disc packs can be done with the large outer snap ring in place.

With the clutch pack and pressure plate installed, use a feeler gauge to check the distance between the pressure plate and the outer snap ring (Figure 9-39). Clearance can also be measured between the backing plate and the uppermost friction disc. If the clutch pack has a waved snap ring, place the feeler gauge between the flat pressure plate and the wave of the snap ring farthest away from the pressure plate. Compare the distance to specifications. Attempt to set pack clearance to the smallest dimension shown in the chart.

Clearances can also be checked with a dial indicator and hook tool (Figure 9-40). The hook tool is used to raise one disc from its downward position and the amount that it is able to move is recorded on the dial indicator. This represents the clearance.

Another way to measure clearance is to mount the clutch drum on the clutch support and use 25–35 psi of compressed air through the oil pump body channels to charge the clutch. Clearance can be measured by mounting a dial indicator so that it reads backing plate movement as the air forces the piston to apply the clutch (Figure 9-41).

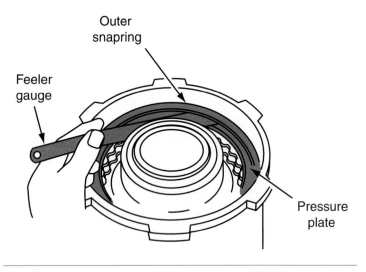

Figure 9-39 Measuring clutch clearance with a feeler gauge.

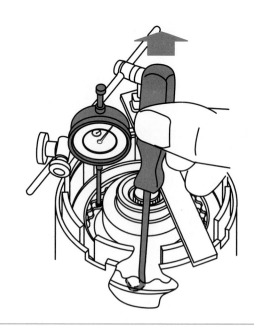

Figure 9-40 Using a hook tool to measure clutch clearance.

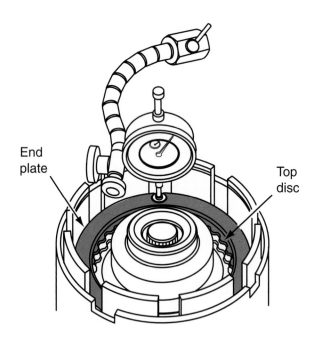

	Service limit	
1st	0.65–0.85 mm	(0.026–0.033 in.)
2nd	0.65–0.85 mm	(0.026–0.033 in.)
3rd	0.40–0.60 mm	(0.016–0.024 in.)
4th	0.40–0.60 mm	(0.016–0.024 in.)
Low-hold	0.80–1.00 mm	(0.031–0.039 in.)

Figure 9-41 Dial indicator set-up for measuring clutch clearances.

P/N	Plate no.	Thickness mm (in)
22551–PX4–003	1	2.1 (0.082)
22552–PX4–003	2	2.2 (0.086)
22553–PX4–003	3	2.3 (0.090)
22554–PX4–003	4	2.4 (0.094)
22555–PX4–003	5	2.5 (0.098)
22556–PX4–003	6	2.6 (0.102)
22557–PX4–003	7	2.7 (0.106)
22558–PX4–003	8	2.8 (0.110)
22559–PX4–003	9	2.9 (0.114)

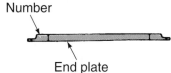

Number

End plate

Figure 9-42 A typical chart of the various pressure plate thicknesses available to correct clutch clearances.

If the clearance is greater than specified, install a thicker snap ring to take up the clearance. If the clutch clearance is insufficient, install a thinner snap ring.

Another way to adjust clutch clearance is to vary the thickness of the clutch pressure plate. By using a pressure plate of the desired thickness, you can obtain adequate clutch clearance (Figure 9-42).

Proper clearance can also be achieved through the use of special steel clutch plates. These are available from some manufacturers of transmission overhaul kits.

If specifications are not available, a general clutch pack clearance can be used. Allow 0.01 inch clearance for each friction disc in the clutch pack. For example, if the pack has four friction discs, set the minimum clearance to 0.04 in. Since this is the minimum clearance, some transmission specialists then add 0.03 in. to the clearance. Therefore, in the example given, the maximum clearance would be 0.07 in.

Air Testing

After the clearance of the clutch pack is set, perform an air test on each clutch. This test will verify that all the seals and check balls in the hydraulic component are able to hold and release pressure.

Air checks can also be made with the transmissions assembled. This is the absolute best way to check the condition of the circuit, because there are very few components missing from the circuit. The manufacturers of different transmissions have designed test plates that are available to test different hydraulic circuits (Figure 9-43). Testing with the transmission assembled also allows for testing of the servos.

To test a clutch assembly, install the oil pump assembly with its reaction shaft support over the input shaft and slide it into place on the front clutch drum. When the clutch drums are mounted on the oil pump, all components in the circuit can be checked. If the clutch cannot be checked in this manner, blocking off apply ports with your finger and applying air pressure through the other clutch apply port will work.

✔ **SERVICE TIP:** FWD transaxles that utilize a chain to connect the torque converter output to the transaxle input shaft do not have a common shaft for the clutches and the oil pump. Therefore, it is impossible to test the drums from these transaxles while they are mounted on the oil pump. Instead, the drums should be tested while they are mounted on the front sprocket support. These are tested the same way as any other drum.

The 4T60-E and AX4S transmissions are common transaxles that have a chain link and an off-center input shaft.

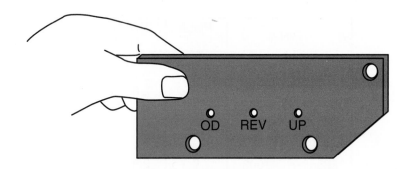

Figure 9-43 A manufacturer's air pressure test plate.

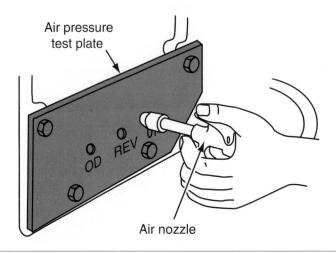

Figure 9-44 To test a clutch, apply air to the hole in the test plate designated for that clutch.

To conduct an air test with the transmission assembled, invert the entire assembly and place it in an open vise or transmission support tool. Pour clean ATF into the circuits that will be tested. Try to get as much fluid into the circuit as you can. Then, install the air test plate. On many transmissions it is necessary to remove the valve body in order to mount the test plate. Then, air test the circuit using the test hole designated for that clutch (Figure 9-44). Be sure to use low pressure compressed air (25–35 psi) to avoid damage to the seals. Higher pressures may blow the rubber seals out of the bore or roll them on the piston.

While applying air pressure, you may notice some escaping air at the metal or Teflon seal areas. This is normal, but not necessarily desirable, as these seals have a controlled amount of leakage designed into them. Basically, the less a seal leaks, the better seal it is. There should be no air escaping from the piston seals. The clutch should apply with a dull but positive thud. It should also release quickly without any delay or binding. Examine the check ball seat for evidence of air leakage.

Only use compressed air that is free of dirt and moisture.

Clutch Volume Index

Many electronically controlled transmissions regulate shift feel by regulating the volume of fluid used to apply or release a clutch or brake. The computer monitors gear ratio changes by monitoring the input and output speed of the transmission. By comparing the signal from the turbine or input speed sensor to the signal from the output speed sensor, the computer can determine the operating gear ratio. The computer also monitors the **Clutch Volume Index (CVI)**.

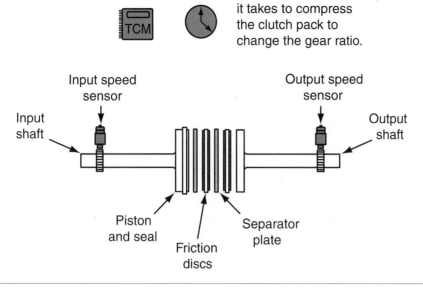

The TCM times how long it takes to compress the clutch pack to change the gear ratio.

Figure 9-45 Things considered by the system's computer (TCM) when it calculated CVI.

CVIs represent the volume of fluid needed to compress a clutch pack or apply a brake band. Based on its constant calculation of gear ratio, the computer can monitor how long it takes to change a gear (Figure 9-45). If the time is too long, more fluid is sent to the apply device. The volume of the fluid required to apply the friction elements is continuously updated as part of the transmission's adaptive learning. As friction material wears, the volume increases.

One-Way Clutch Assemblies

One-way clutches are holding devices. They allow a member to rotate in one direction only. They may be used to ground a member or effectively hold a member by preventing it from rotating in the direction it will tend to rotate in. One-way clutches are also used to transmit torque from one member to another. In these cases, the inner race is splined to a member of the gearset. The outer race is splined to another member. An example of this setup is shown in Figure 9-46. The inner race of this overdrive one-way clutch is splined to the coast clutch cylinder, which is splined to the overdrive sun gear. The one-way clutch's outer race is splined to the overdrive ring gear. When

CVIs can be monitored with a scan tool and the readings compared to charts given in the service manual. Keep in mind that the **Clutch Volume Index (CVI)** lets you know how much fluid is needed to apply or release a friction component. When it is out of spec, a problem is indicated.

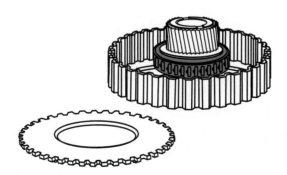

Figure 9-46 The inner race of this overdrive one-way clutch is splined to the coast clutch cylinder, which is splined to the overdrive sun gear; the outer race is splined to the overdrive ring gear.

the overdrive ring gear has a counterclockwise torque from the vehicle itself, the gearset attempts to rotate the sun gear clockwise. The attempted counterclockwise rotation of the ring locks the one-way clutch, which in turn locks the ring gear with the sun gear. This action provides for a direct drive through the gearset.

When the overdrive gear is selected, the coast clutch cylinder is held by the inner race of the one-way clutch, which in turn holds the overdrive sun gear. This causes the planet gear assembly to walk around the sun and drive the ring gear clockwise with an overdrive. During coast, the one-way clutch releases and allows the ring gear to rotate faster than the sun gear.

Because they are purely mechanical in nature, one-way clutches are relatively simple to inspect and test. The durability of these clutches relies on constant fluid flow during operation. If a one-way clutch has failed, a thorough inspection of the hydraulic fluid feed circuit to the clutches must be made to determine if the failure was due to fluid starvation. The rollers and sprags ride on a wave of fluid when they overrun. Since most of these clutches spend the majority of their time in the overrunning state, any loss of fluid can cause rapid failure of the components.

Sprags, by design, produce the fluid wave effect as they slide across the inner and outer races, making them somewhat less prone to damage. Rollers, due to their spinning action, tend to throw off fluid, which allows more chance for damage during fluid starvation. During the check of the hydraulic circuit, take a look at the feed holes in the races of the clutch. Use a small-diameter wire and spray carburetor cleaner or brake cleaner to be certain the feed holes are clear. Push the wire through the feed holes and spray the cleaner into them. Blowing through them with compressed air after cleaning is recommended.

One-way clutches should be inspected for wear or damage, spline damage, surface finish damage, and damaged retainers and grooves. Some clutch units have a plastic race that needs to be carefully inspected for chafing or other damage.

Roller Clutches

Classroom Manual
Chapter 9, page 289

Roller clutches should be disassembled to inspect the individual pieces (Figure 9-47). The surface of the rollers should have a smooth finish with no evidence of any flatness. Likewise, the race should be smooth and show no sign of Brinelling, as this indicates severe impact loading. This condition may also cause the roller clutch to "buzz" as it overruns.

All rollers and races that show any type of damage or surface irregularities should be replaced. Check the folded springs for cracks, broken ends, or flattening out. All of the springs from

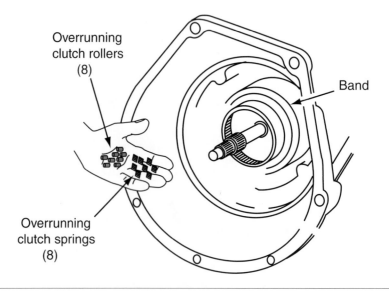

Figure 9-47 Inspect all parts of a roller-type one-way clutch.

a clutch assembly should have approximately the same shape. Replace all distorted or otherwise damaged springs. The cam race of a roller clutch may show the same Brinelled wear due to impact overloading. The cam surface, like the smooth race, must be free of all irregularities.

Sprag Clutches

Sprag clutches (Figure 9-48) cannot easily be disassembled; therefore, a complete and thorough inspection of the assembly is necessary. Pay particular attention to the faces of the sprags. If the faces are damaged, the clutch unit should be replaced. Sprags and races with scored or torn faces are an indication of dry running and require the replacement of the complete unit and an inspection of the lubrication system.

Classroom Manual
Chapter 9, page 289

Installation

Once the one-way clutches are ready for installation, verify that they overrun in the proper direction. In some cases, it is possible to install these clutches backwards, which would cause them to overrun and lock in the wrong direction. This would result in some definite driveability problems. One way to make sure you have installed the clutch in the correct direction is to determine the direction of lock-up before installing the clutch (Figure 9-49), then study the transmission's powerflow and match the direction of the clutch with it. Most one-way clutches have some kind of marking that indicates in which direction the clutch should be set (Figure 9-50).

● **CUSTOMER CARE:** Some of the components discussed in this chapter are not typically replaced when rebuilding a transmission and may not have been included in the cost estimate given to the customer. Make sure the customer is aware of all extra costs before they come in to pick up their vehicle after service.

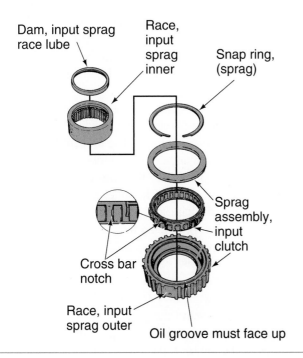

Figure 9-48 A typical sprag one-way clutch assembly. (Courtesy of General Motors Corporation, Service Operations)

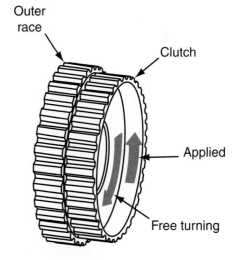

The overrun clutch hub must turn clockwise, but not counterclockwise.

Figure 9-49 One way to make sure you have installed the clutch in the correct direction is to determine the direction of lock-up before installing the clutch. The operation of a one-way clutch is checked in the same way.

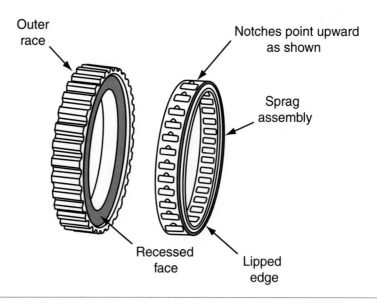

Outer race

Notches point upward as shown

Sprag assembly

Recessed face

Lipped edge

Figure 9-50 Most one-way clutches have some marking that indicates in which direction the clutch should be set.

CASE STUDY

A car dealer called to see if the Automotive Department at the college could help with a diagnostic problem. His customer had a new Ford pickup. The owner complained of a buzzing noise while driving. The technician at the dealership disassembled the transmission and installed an overhaul kit. The next day, the customer picked up his truck. He returned the next day, slightly annoyed, with the same complaint of buzzing while driving. The same technician once again disassembled the transmission and installed another overhaul kit. This time the technician road tested the truck after installing the kit. The noise was still there. By this time, both the owner of the truck and the technician were not very happy. After two "rebuilds," the transmission still had the same noise.

The transmission was brought to the college and was checked out on the transmission "dyno." In the forward ranges, the buzzing was only apparent in intermediate and high gear ranges. There was no buzzing in low or reverse gear ranges. The students, using a power flow chart, determined that the only geartrain member not in motion in low and reverse was the rear carrier. Because the rear carrier is held by the one-way clutch, we decided to disassemble the transmission to check the clutch. The inner race of the roller clutch had a flat spot from an apparent machining mishap during manufacture. Every time a roller went past the flat spot, it clicked into the flat, producing the buzz at road speeds. We installed a new roller inner race, which was supplied by the dealer, and, like magic, the buzz was gone.

Why were students able to determine the problem while an experienced technician could not find it? Very simply, the technician did not understand the basic theories of transmission operation. The road test very clearly showed that the buzzing occurred only in second and third gears. A basic knowledge of the Simpson geartrain and knowing how to read and use a flow chart would have helped the technician diagnose the problem. Unfortunately, this technician, like many others, only knew how to "put a kit in it" and hope that the noise would go away. Just as a point of interest, according to the service manager, the technician loaded up her tools and quit when she learned that students, armed with a power flow chart and a shop manual, had diagnosed and repaired the problem that had her stumped.

Terms to Know

Clutch Volume Index (CVI)   Steels   Transmission tune-up
Selective servo apply pins

ASE-Style Review Questions

1. While visually inspecting a roller-type one-way clutch:
   Technician A replaces the clutch because the rollers have a smooth finish.
   Technician B replaces the clutch because the race shows signs of brinnelling.
   Who is correct?
   A. A only   **C.** Both A and B
   B. B only   **D.** Neither A nor B

2. *Technician A* says most original equipment servo seals are of the Teflon type.
   Technician B says a majority of replacement transmission gasket sets will supply cast iron hook end seal rings, in place of the Teflon seals, and these are acceptable replacements.
   Who is correct?
   A. A only   **C.** Both A and B
   B. B only   **D.** Neither A nor B

3. Which of the following is *not* a good reason to replace a band?
   A. The band is chipped.
   B. The friction material is glazed.
   C. The friction material is soaked with fluid.
   D. The friction material has random wear patterns.

4. While conducting an air test on a transaxle:
   Technician A notices escaping air at the metal or Teflon seals. He proceeds to replace them.
   Technician B says the clutch should apply with a slight delay, then a dull thud.
   Who is correct?
   A. A only   **C.** Both A and B
   B. B only   **D.** Neither A nor B

5. While inspecting a clutch assembly that has an aluminum piston:

Technician A carefully inspects the piston for hairline cracks that will cause pressure leakage, resulting in clutch slippage.
Technician B says that if burned discs were found in the clutch pack, the piston should be checked for defects and cracks.
Who is correct?
   A. A only   **C.** Both A and B
   B. B only   **D.** Neither A nor B

6. *Technician A* says all adjustable bands have their locknut and adjusting screw on the outside of the transmission case.
   Technician B says some bands do not have adjusting screws.
   Who is correct?
   A. A only   **C.** Both A and B
   B. B only   **D.** Neither A nor B

7. *Technician A* says transmissions originally equipped with Teflon seals must be refitted with Teflon seals during an overhaul.
   Technician B says a press is needed for the installation of Teflon seals.
   Who is correct?
   A. A only   **C.** Both A and B
   B. B only   **D.** Neither A nor B

8. While assembling a multiple-friction disc pack:
   Technician A alternately stacks the steels and friction discs until the correct number of plates has been installed.
   Technician B installs all of the discs contained in the overhaul kit and corrects the clearance with a selective pressure plate or snap ring.
   Who is correct?
   A. A only   **C.** Both A and B
   B. B only   **D.** Neither A nor B

9. *Technician A* says an air test can be used to check servo action.
Technician B says an air test can be used to check for internal fluid leaks.
Who is correct?
   A. A only
   B. B only
   C. Both A and B
   D. Neither A nor B

10. While checking clutch discs:
Technician A says the steel plates should be replaced if they are worn flat.
Technician B says the friction discs should be squeezed to see if they can hold fluid. If they hold fluid and look okay, they are serviceable.
Who is correct?
   A. A only
   B. B only
   C. Both A and B
   D. Neither A nor B

ASE Challenge Questions

1. *Technician A* says friction discs that have blackened surfaces have been subject to overheating.
Technician B says that if the friction facing has been burned or worn off, the discs can become welded together and can cause the vehicle to creep when the transmission is in neutral.
Who is correct?
   A. A only
   B. B only
   C. Both A and B
   D. Neither A nor B

2. *Technician A* says excessive groove wear in a clutch piston can be caused by excessive pump pressures.
Technician B says excessive wear, scoring, and grooving in the bore of a clutch can be caused by a stuck pressure regulator valve.
Who is correct?
   A. A only
   B. B only
   C. Both A and B
   D. Neither A nor B

3. *Technician A* says incorrect assembly of servos and accumulators can result in dragging bands and harsh shifting.
Technician B says internal leaks at the servo or clutch seal will cause excessive pressure drops during gear changes.
Who is correct?
   A. A only
   B. B only
   C. Both A and B
   D. Neither A nor B

4. *Technician A* says that if a one-way clutch has failed, a thorough inspection of the hydraulic fluid feed circuit to the clutch must be made to determine if the failure was due to fluid starvation.
Technician B says that if the race of a roller-type one-way clutch has random hard spots, it is likely that the roller clutch will "buzz" as it overruns.
Who is correct?
   A. A only
   B. B only
   C. Both A and B
   D. Neither A nor B

5. While inspecting a servo:
Technician A says the servo needs new seals because both the inside and outside of the cover seal are wet.
Technician B says that whenever the servo is disassembled, all rubber, cast-iron, and Teflon seals should be replaced.
Who is correct?
   A. A only
   B. B only
   C. Both A and B
   D. Neither A nor B

Job Sheet 28

28

Name _____ Date _____

Inspecting Apply Devices

Upon completion of this job sheet, you should be able to inspect various apply devices of a transmission.

ASE Correlation

This job sheet is related to the ASE Automatic Transmission and Transaxle Test's Content Area: *Off-Vehicle Transmission and Transaxle Repair; Friction and Reaction Units.*
Tasks: Inspect clutch drum, piston, check-balls, springs, retainers, seals, and friction and pressure plates; replace as needed. Inspect roller and sprag clutch, races, rollers, sprags, springs, cages, and retainers; replace as needed.

Tools and Materials

Compressed air and air nozzle   Lint-free shop towels
Supply of clean solvent   Service manual

Describe the vehicle being worked on:

Year _____ Make _____ Model _____

VIN _____ Engine type and size _____

Model and type of transmission _____

Procedure

Task Completed

1. Disassemble the transmission into major units. Set each unit aside until this job sheet refers to it. Describe any problems you encountered while disassembling the transmission. Be sure to follow the procedures given in the appropriate service manual while taking the transmission apart.

   _____

   _____

   _____

   _____

2. Clean each overrunning clutch assembly in fresh solvent. Allow the assemblies to air dry.   ☐

3. Check the rollers and sprags for signs of wear or damage. Describe your findings.

   _____

   _____

4. Check the springs for distortion, distress, and damage. Describe your findings.

   _____

   _____

Inspecting Apply Devices (continued)

5. Check the inner race and the cam surfaces for scoring and other damage. Describe your findings.

_____

_____

6. Check the condition of the snap rings. Describe your findings.

_____

_____

7. Summarize the condition of the overrunning clutch units.

_____

_____

_____

_____

☐ **8.** Wipe the transmission's bands clean with a dry, lint-free cloth.

9. Check the bands for damage, wear, distortion, and lining faults. Describe your findings.

_____

_____

10. Inspect the band apply struts for distortion, and other damage. Describe your findings.

_____

_____

11. Summarize the conditions of the bands.

_____

_____

_____

_____

☐ **12.** Clean servo and accumulator parts in fresh solvent and allow them to air dry.

13. Inspect their bores for scoring and other damage. Describe their condition.

_____

_____

14. Check the piston and piston rod for wear, nicks, burrs, and scoring. Describe your findings.

_____

_____

15. Inspect all springs for damage and distortion. Describe their condition.

_____

_____

Inspecting Apply Devices (continued)

16. Check the mating surfaces between the piston and the walls of their bores for scoring, wear, nicks, and other damage. Describe your findings.

_____

_____

17. Check the movement of each piston in its bore. Describe that movement.

_____

_____

18. Check the fluid passages for restrictions and clean out any dirt present in the passages. Describe your findings.

_____

_____

19. Summarize the condition of the servos and accumulators in this transmission.

_____

_____

_____

_____

Instructor's Response _____

_____

_____

Job Sheet 29

29

Name _____ Date _____

Checking and Overhauling a Multiple-Friction Disc Assembly

Upon completion of this job sheet, you should be able to air test clutch packs and servos; and disassemble, inspect, and reassemble a multiple-friction disc assembly.

ASE Correlation

This job sheet is related to the ASE Automatic Transmission and Transaxle Test's Content Area: *Off-Vehicle Transmission and Transaxle Repair; Friction and Reaction Units.*
Tasks: Measure clutch pack clearance; adjust as needed. Air test operation of clutch and servo assemblies.

Tools and Materials

Feeler gauge set
Special tools for the assigned transmission
Emery cloth
Crocus cloth
Clean solvent

Lint-free shop towels
OSHA-approved air nozzle
Air test plate
Service manual

Describe the transmission being worked on:

Model and type of transmission _____

Vehicle the transmission is from:

Year _____ Make _____ Model _____

VIN _____ Engine type and size _____

Procedure

Task Completed

1. With the transmission disassembled, set aside each clutch pack for inspection. ☐

2. Disassemble each clutch pack. Keep each assembly separate from the others. Record the number of steel and friction plates in each assembly.

   _____

   _____

   _____

3. Check the steel plates for discoloration, scoring, distortion, and other damage. Describe their condition:

   _____

   _____

   _____

4. If the steel plates are in good condition, rough up the shiny surface with emery (80-grit) cloth. ☐

Checking and Overhauling a Multiple-Friction Disc Assembly (continued)

5. Inspect the friction plates for wear, damage, and distortion. Describe their condition.

_____

_____

_____

☐ 6. Soak all new and reusable friction plates in ATF for at least 15 minutes before reassembling the clutch pack.

7. Check the pressure plate for discoloration, scoring, and distortion. Describe its condition:

_____

_____

_____

8. Inspect the coil springs for distortion and damage. Describe their condition:

_____

_____

_____

☐ 9. Inspect the Belleville spring for wear and distortion. Make sure to check the inner fingers. Describe your findings:

10. Check the clutch drums for damaged lugs, grooves, and splines. Also check them for score marks. Describe their condition:

_____

_____

_____

11. Check the movement of the check balls in the drums. Describe your findings:

_____

_____

_____

12. Summarize the condition of the clutch packs:

_____

_____

_____

13. Reassemble the clutch packs with the required new parts and with new seals. Follow the recommended procedure for doing this.

_____

_____

_____

14. Insert a feeler gauge between the pressure plate and the snap ring. What is the measured clearance? _____

15. What is the specified clearance? _____

16. If the measured clearance is not the same as the specified clearance, what should you do?

_____

_____

_____

17. Do what is necessary to correct the clearance. ☐

18. Tip the clutch assembly on its side and insert a feeler gauge between the pressure plate and the adjacent friction plate. What is the measured clearance? _____

19. What is the specified clearance? _____

20. If the measured clearance is not the same as what was specified, what should you do?

_____

_____

21. Do what is necessary to correct the clearance. ☐

22. After the clearance is set, check the service manual for the proper procedure for air testing the clutch pack. Summarize the procedure.

_____

_____

_____

23. Place the pack or the partially assembled transmission in a vise or on a stand. ☐

24. Apply low air pressure to the designated test port. Pay attention to the sound of the air leaking and the activation of the disc pack. Describe what you heard.

_____

_____

25. Release the air and listen for the release of the pack. Describe what happened.

_____

_____

_____

26. Based on the above, what are your conclusions?

_____

_____

_____

27. When the transmission is assembled, install the air test plate to the designated area on the transmission. What does the plate attach to?

_____

_____

28. What components can be checked with the test plate?

_____

_____

_____

29. Apply low pressure to each of the test ports and record the results.

_____

_____

_____

30. What is indicated by the results of this test?

_____

_____

_____

Instructor's Response _____

_____

_____

Rebuilding Common Transmissions

Upon completion and review of this chapter, you should be able to:

❏ Disassemble, service, and reassemble a Chrysler Torqueflite 36RH transmission.

❏ Disassemble, service, and reassemble a Chrysler 41TE transaxle.

❏ Disassemble, service, and reassemble a Chrysler 42LE transaxle.

❏ Disassemble, service, and reassemble a Ford 4R44E transmission.

❏ Disassemble, service, and reassemble a Ford 4EAT/Mazda GF4A-EL transaxle.

❏ Disassemble, service, and reassemble a Ford AX4S transaxle.

❏ Disassemble, service, and reassemble a GM 4L60 transmission.

❏ Disassemble, service, and reassemble a GM 3T40 transaxle.

❏ Disassemble, service, and reassemble a GM 4T60 transaxle.

❏ Disassemble, service, and reassemble a Honda F4 transaxle.

❏ Disassemble, service, and reassemble a Nissan L4N71B/E4N71B transmission.

❏ Disassemble, service, and reassemble a Toyota A541E transaxle.

This chapter contains many photo sequences that show the step-by-step procedures for overhauling common transmissions and transaxles. Although a particular model of transmission was chosen for the photographs, variations of the same transmission can be used. The procedures in this chapter are for educational purposes and do not include the many changes that a different model would require.

These sequences do not include detailed procedures for rebuilding subassemblies, such as the valve body and clutch packs. Detailed information on these, as well as other subassemblies, is given in other chapters of this manual.

Chrysler Transmissions

Chrysler Corporation introduced the Torqueflite transmission in 1956. This transmission was the first modern three-speed automatic transmission with a torque converter and the first to use the Simpson two-planetary compound geartrain. Nearly all Torqueflite-based transmissions and transaxles use a rotor-type oil pump and all use a Simpson geartrain. Late-model Torqueflites have a one-piece aluminum housing and a bolt-on extension housing.

There are two basic versions of the three-speed Torqueflite transmission: the 30RH (old A-904) and the 36RH (old A-727). The difference between the two is the intended use. The 36RH is designed for heavier-duty use.

The 42RE is a commonly found four-speed transmission and is used in pick-ups and SUVs. The 42RE is based on the 36H, which is a three-speed transmission. The primary difference between the two transmission models is that the 42RE has an additional planetary gearset at the rear of the compound planetary gearset. Hydraulic control of the first three forward speeds is the same as the 36RH. Overdrive is controlled by a direct clutch, an overdrive clutch, and an overdrive one-way clutch. The output from the compound gearset passes through the overdrive planetary

Basic Tools

Mechanic's basic tool set

Appropriate service manual

Transmission holding fixture

Torque wrench

Rubber-tipped air nozzle

Pan of clean ATF

Feeler gauge

Classroom Manual
Chapter 10, page 312

Special Tools

Dial indicator and holding fixture

Clutch compressor tool

Oil pump puller

Seal remover/installer

gearset. In all gears, except overdrive, torque flows through the direct clutch to the output shaft. When overdrive is selected, the torque passes through the ring gear to the carrier of the overdrive planetary gearset.

The 42RE and 36RH are nearly identical, as are the service procedures for both. Like all transmissions, modifications have been made each year to include more electronic controls. This is true of the 36RH and especially true of the 42RE. Photo Sequence 14 goes through the procedure for overhauling a 36RH transmission. These overhaul procedures are similar to those for other Chrysler transmissions.

Photo Sequence 14
Typical Procedure for Overhauling a 36RH Transmission

PS14-1 Check and record the endplay of the input shaft, then unbolt and remove the oil pan and gasket.

PS14-2 Unbolt and remove the oil filter. Then unbolt and remove the valve body by first disconnecting the parking lock rod from the manual lever and unbolting the TCC solenoid.

PS14-3 Remove the accumulator piston spring and lift the piston from the case.

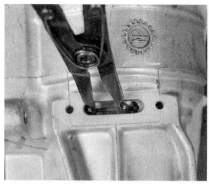

PS14-4 Remove the parking lock rod. Then unbolt and remove the extension housing. There is a cover plate in the bottom of the extension housing that once removed will allow access to the output shaft's snap ring. This ring must be expanded while pulling the extension housing off the case.

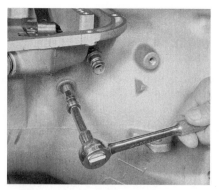

PS14-5 Tighten the front band adjustment screw to prevent damage to the discs during disassembly.

PS14-6 Remove the oil pump bolts and pull the oil pump out of the case with special pullers.

PS14-7 Loosen the front band adjustment screw and remove the band strut and anchor. Then remove the band.

PS14-8 Remove the retaining or snap ring, then the front clutch assembly, with the input shaft, from the case.

PS14-9 Remove front planetary gearset.

PS14-10 Loosen the rear band adjustment screw, then remove the band strut and band. Then remove the output shaft with the governor attached to it.

PS14-11 Compress the return spring for the front servo and remove the retaining snap ring.

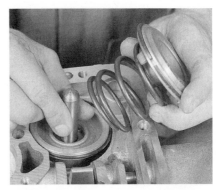

PS14-12 Remove the servo assembly from the case.

PS14-13 Compress the return spring for the rear servo and remove the retaining snap ring.

PS14-14 Remove servo assembly from the case.

PS14-15 Inspect all servo piston seals.

PS14-16 Reinstall the servo assemblies after careful inspection.

PS14-17 Inspect the planetary gearset and drive shells. Replace components as needed.

PS14-18 Inspect the bands and their struts and anchors. Replace as needed.

PS14-19 Disassemble and inspect the clutch units. Replace parts as needed. Remember to allow the discs to soak in ATF before assembling the clutch pack.

PS14-20 Replace the pistons' oil seal rings in the clutch assemblies.

PS14-21 Reassemble the clutch packs, and then check the clearances.

PS14-22 Unbolt the stator support from the oil pump housing. Separate the support from the oil pump.

PS14-23 Check the wear of the gears and the pump itself. Replace parts as needed.

PS14-24 Replace the pump seals and reassemble the pump. Torque the bolts to specifications.

PS14-25 Disassemble, clean, and inspect the valve body.

PS14-26 Bolt valve body together and tighten bolts to specifications.

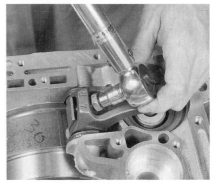

PS14-27 Install the output shaft with the governor into the case. Then install the rear band and anchor assembly and tighten the adjustment screw.

PS14-28 Install the rear planetary gearset and drum. Make sure all thrust washers are installed in their correct locations.

PS14-29 Install the front clutch assembly and the front band assembly, and then tighten the adjustment screw.

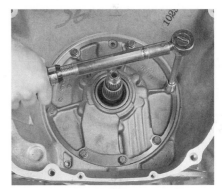

PS14-30 Assemble servos with new seals. Then install the oil pump and torque the bolts in the proper sequence and to the proper torque.

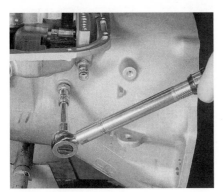

PS14-31 Adjust both bands according to specifications.

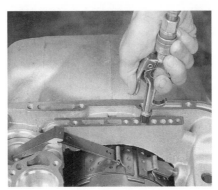

PS14-32 Air test the entire transmission before installing the oil filter and oil pan.

Special Tools

Dial indicator and
holding fixture

Clutch compressor
tool

Oil pump puller

Seal remover/installer

Output shaft holding
tool

Chrysler Torqueflite Transaxles

In 1978, the basic Torqueflite transmission was modified for use as a FWD transaxle. Torqueflite transaxles contain the same basic parts as the 36RH and 42RE transmissions, with the addition of a transfer shaft, final drive gears, and differential unit.

Many different transaxle models have been used since then. The 41TE is commonly found in mini-vans and some late-model cars. This is a four-speed unit with electronic controls. The compound planetary gearset in the 41TE does not have a common member; rather, the two planetary units are connected in tandem, with the front carrier connected to the rear ring gear and the rear carrier connected to the front ring gear.

To control the forward gear ranges, the 41TE relies on an input clutch assembly, a 2nd/4th clutch, and a Low/Reverse clutch. The input clutch assembly houses the three different input clutches: the underdrive, overdrive, and reverse clutches.

The underdrive clutch is applied in first, second, and third gears. When applied, the underdrive hub drives the rear sun gear. The overdrive clutch is applied in third and overdrive gears. When the overdrive clutch is applied, it drives the front planet carrier. The underdrive clutch and the overdrive clutch are applied in third gear, giving two inputs to the gearset, which provides for direct drive. The reverse clutch is only applied in reverse and drives the front sun gear assembly.

The 2nd/4th gear and Low/Reverse units are holding devices and are placed at the rear of the transmission case. The 2nd/4th clutch is engaged in second and fourth gears and grounds the front sun gear to the case. The Low/Reverse clutch is applied in park, reverse, neutral, and first gear. When this clutch is engaged, the front planet carrier and rear ring gear assembly is grounded to the case. The overhaul procedure for this transaxle is shown in Photo Sequence 15.

Photo Sequence 15
Typical Procedure for Overhauling a 41TE Transaxle

PS15-1 Remove all electrical switches from the outside of the transaxle case.

PS15-2 Remove the solenoid assembly and gasket.

PS15-3 Remove the torque converter.

Photo Sequence 15 (cont'd.)
Typical Procedure for Overhauling a 41TE Transaxle

PS15-4 Loosen and remove the oil pan's bolts. Then remove the oil pan.

PS15-5 Remove the oil filter.

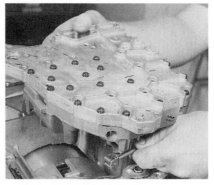

PS15-6 Loosen and remove the valve body attaching bolts. Then push the park rod rollers away from the guide bracket and remove the valve body assembly. Pull straight up on the valve body as the manual shaft is attached to it.

PS15-7 Remove the retaining snap ring for the accumulator, then remove the accumulator assembly.

PS15-8 Remove the oil pump seal, using the correct seal puller.

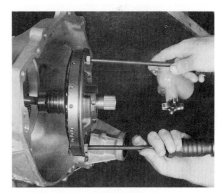

PS15-9 Loosen and remove the oil pump's attaching bolts. Then, using the correct puller, remove the oil pump. While pulling the pump, push in on the input shaft.

PS15-10 Pull out the input shaft and clutch assembly.

PS15-11 Remove and discard the oil pump gasket. Then remove the caged needle bearing assembly from the input shaft.

PS15-12 Remove the front sun gear assembly.

PS15-13 Twist and pull on the front carrier and rear ring gear assembly to remove it.

PS15-14 Remove the rear sun gear with its thrust bearing.

PS15-15 Using the proper clutch compressor, remove the 2–4 clutch snap ring and clutch retainer. Mark the alignment of the retainer with the return spring located below it. Then remove the return spring.

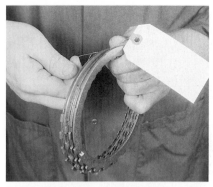

PS15-16 Remove the clutch pack and tag it as the 2–4 pack.

PS15-17 Remove the low/reverse tapered snap ring from the case. Follow the recommended sequence while prying the snap ring out of the case.

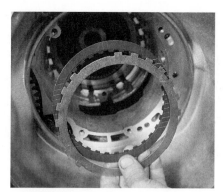

PS15-18 Remove the low/reverse reaction plate and one friction disc.

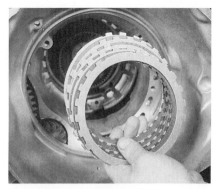

PS15-19 Remove the low/reverse reaction plate flat snap ring, and then remove the clutch pack. Tag the pack as the low/ reverse clutch.

PS15-20 Loosen and remove the rear cover bolts. Then remove the rear cover.

PS15-21 Remove the transfer shaft gear nut and washer. Then, using a gear puller, remove the transfer gear.

PS15-22 Remove the transfer gear selective shim and the bearing retainer. Then remove the transfer shaft bearing snap ring.

PS15-23 Remove the transfer shaft and the output shaft gear bolt and washer. Then, using a gear puller, remove the output gear and selective shim.

PS15-24 Remove the rear carrier assembly.

PS15-25 After careful inspection of all parts, begin reassembly by installing the rear carrier assembly.

PS15-26 Install the output gear, selective shim, washer, and bolt. While holding the shaft with a special holding tool, tighten the bolt to specifications.

PS15-27 Install the transfer shaft.

PS15-28 Install the transfer shaft bearing snap ring, selective shim, and bearing retainer. Then install the transfer gear, washer, and nut. While holding the shaft with a special tool, tighten the nut to specifications.

PS15-29 Apply 1/8-inch bead of RTV on the rear cover. Then install the rear cover and tighten the rear cover bolts to specifications.

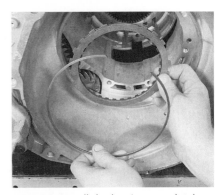

PS15-30 Install the low/reverse clutch pack, reaction plate flat snap ring, friction disc, and low/reverse reaction plate. Then install the low/reverse tapered snap ring into the case. Follow the recommended sequence while installing the snap ring into the case.

Photo Sequence 15 (cont'd.)
Typical Procedure for Overhauling a 41TE Transaxle

PS15-31 Install the 2–4 clutch pack snap ring and return spring.

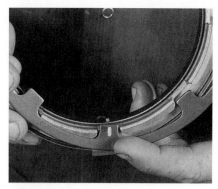

PS15-32 Using the proper clutch compressor, install the 2–4 clutch snap ring and clutch retainer. Make sure the alignment marks on the spring and retainer are matched.

PS15-33 Install the rear sun gear with its thrust bearing. Then, install the front carrier and rear ring gear assembly by twisting and pushing on it.

PS15-34 Install the front sun gear assembly.

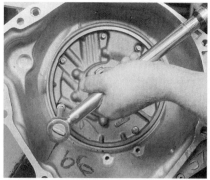

PS15-35 Install the new oil pump gasket. Then, install the oil pump. Tighten the oil pump's attaching bolts to specifications.

PS15-36 Install the oil pump seal, using the correct seal installing tool.

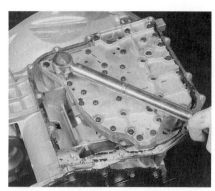

PS15-37 Install the accumulator assembly and retaining snap ring. Then install the valve body assembly. Push the park rod rollers away from the guide bracket while positioning the valve body. Install and tighten the valve body retaining bolts to specifications.

PS15-38 Install a new oil filter. Then, install the oil pan with a new gasket. Tighten the oil pan bolts to specifications.

PS15-39 Install the torque converter, the solenoid assembly with a new gasket, and all electrical switches on the outside of the transaxle case.

With the introduction of new mid-size FWD cars in 1993, Chrysler introduced the 42LE, which is a longitudinally mounted transaxle. Although this transaxle is based on the 41TE and in most ways is very similar to the 41TE, its location in the vehicle requires different overhaul procedures. The 42LE uses a chain drive to transfer the output of the planetary gearset to a hypoid-type final drive unit. As a result, its appearance and construction is different. Therefore, Photo Sequence 16 is included to show the step-by-step procedures for overhauling this transaxle.

Classroom Manual
Chapter 10, page 320

Special Tools

Dial indicator and
  holding fixture

Clutch compressor
  tool

Oil pump puller

Seal remover/installer

Output shaft holding
  tool

Photo Sequence 16
Typical Procedure for Overhauling a 42LE Transaxle

PS16-1 Remove all electrical switches from the outside of the transaxle case.

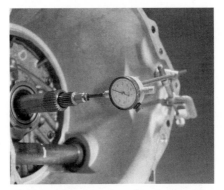

PS16-2 With the transaxle standing on end and the torque converter removed, measure the endplay of the transaxle.

PS16-3 Remove the electrical wiring harness.

PS16-4 Loosen and remove the oil pan's bolts. Then remove the oil pan and the pan gasket.

PS16-5 Remove the oil filter.

PS16-6 Loosen and remove the valve body attaching bolts. Then remove the valve body assembly. Care should be taken not to lose the overdrive and underdrive accumulator springs, which may fall out when removing the valve body.

PS16-7 Remove the retaining snap ring for the accumulators, and then remove the accumulator assemblies. Be careful to keep the parts of each accumulator separated from one another.

PS16-8 Unbolt and remove the solenoid assembly from the valve body.

PS16-9 Remove the snap ring and the long stub shaft from the transaxle.

PS16-10 Index the inner and outer differential adjusters at the case.

PS16-11 Remove the outer lock bracket and back out the adjuster one complete revolution.

PS16-12 Place the seal protector on the shaft and carefully remove the differential cover.

PS16-13 Remove the differential carrier and ring gear assembly.

PS16-14 Remove the inner adjuster lock bracket, and then remove the inner adjuster.

PS16-15 Remove the rear chain cover.

PS16-16 Remove the chain snubber guide.

PS16-17 Remove the snap rings and wave washers from the output and transfer shafts.

PS16-18 Install the special chain spreader tool between the chain sprockets and remove the sprockets as a unit.

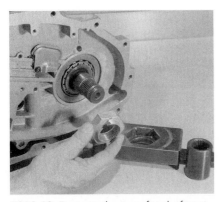

PS16-19 Remove the transfer shaft nut, rear cone, rear cup, oil baffle, rear shim, transfer shaft, and transfer shaft seals. Follow the procedures in the service manual for removing the staked sections of the flange nut. Use a special tool to hold the nut while you rotate the transfer shaft to loosen the nut.

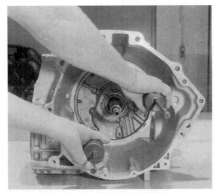

PS16-20 Remove the oil pump attaching bolts, then using the correct puller, remove the oil pump. While pulling the pump, push in on the input shaft.

PS16-21 Remove and discard the oil pump gasket. Then remove the bypass valve from the case.

PS16-22 Pull out the input shaft and clutch assembly.

PS16-23 Remove the caged needle bearing from the input shaft. Note its orientation as it comes off so it can be properly installed during reassembly.

PS16-24 Remove the front sun gear assembly and thrust washer.

PS16-25 Twist and pull on the front carrier and rear annulus gear assembly to remove it. Then remove the rear sun gear with its thrust bearing.

PS16-26 Using the proper clutch compressor, remove the 2–4 clutch snap ring and clutch retainer. Mark the alignment of the retainer with the return spring located below it.

PS16-27 Remove the 2–4 clutch retainer and return spring.

PS16-28 Remove the lower reaction plate.

PS16-29 Remove the clutch pack and tag it as the 2–4 pack.

PS16-30 Remove the low/reverse tapered snap ring from the case. Follow the recommended sequence while prying the snap ring out of the case.

PS16-31 Remove the low/reverse reaction plate.

PS16-32 Remove the low/reverse reaction plate flat snap ring and the clutch pack. Tag the pack as the low/reverse clutch.

PS16-33 Use an arbor press to press the output shaft from the case.

Typical Procedure for Overhauling a 42LE Transaxle

PS16-34 After careful inspection of all parts, begin reassembly by pressing the output shaft into the case with an arbor press.

PS16-35 Install the low/reverse clutch pack and the reaction plate flat snap ring.

PS16-36 Install the low/reverse reaction plate.

PS16-37 Install the low/reverse tapered snap ring according to the procedure outlined in the service manual.

PS16-38 Install the 2–4 clutch pack.

PS16-39 Install the lower reaction plate.

PS16-40 Install the return spring and 2–4 clutch retainer.

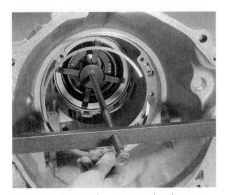

PS16-41 Using the proper clutch compressor, install the snap ring for the 2–4 clutch retainer.

PS16-42 Install the rear sun gear with its thrust washer and then seat the rear annulus gear and front carrier into its proper position.

PS16-43 Install the front sun gear.

PS16-44 Install the front sun gear thrust washer on the front sun gear.

PS16-45 Install the caged needle bearing on the input shaft so that the inside diameter elbow flange is seated against the rear sun gear. Use petroleum jelly to hold it in place.

PS16-46 Install the input shaft clutch assembly.

PS16-47 Install a new oil pump gasket on the oil pump.

PS16-48 Then, install the bypass valve and oil pump. Tighten the oil pump's attaching bolts to specifications.

PS16-49 Install the transfer shaft nut, rear cone, rear shim, transfer shaft seal, oil baffle, rear cup, transfer shaft, and new shaft seal. Use the special tool to hold the nut while you turn the transfer shaft to tighten the nut. Follow the procedures in the service manual for staking the nut.

PS16-50 Install the chain and sprockets as a unit, using the special chain spreader tool.

PS16-51 Install the wave washers and snap rings on the output and transfer shafts.

PS16-52 Install the chain snubber guide.

PS16-53 Install the rear chain cover and torque the mount bolts to specifications.

PS16-54 Install a new O-ring onto the inner adjuster. Lube the inner adjuster threads and O-ring with gear oil and reinstall according to the index marks made during disassembly. Then install the inner adjuster and lock bracket. Torque the lock bracket mount bolt to specifications.

PS16-55 Install the differential carrier and ring gear assembly.

PS16-56 Install the differential cover being careful to protect the seals using the special tool. Torque the mounting bolts to the specified torque.

PS16-57 Install the outer adjuster locknut and torque it to specifications.

PS16-58 Install the long stub shaft and snap ring.

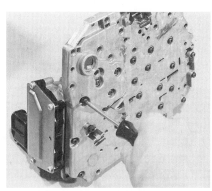

PS16-59 Install the solenoid assembly onto the valve body.

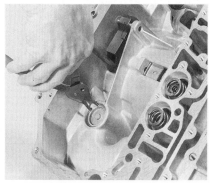

PS16-60 Install the accumulator assemblies and retaining snap rings.

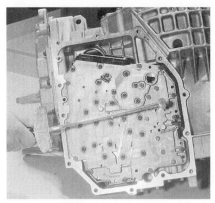

PS16-61 Then install the valve body assembly. Install and tighten the valve body retaining bolts to specifications.

PS16-62 Install a new oil filter.

PS16-63 Then, install the oil pan with a new gasket. Tighten the oil pan bolts to specifications.

PS16-64 Install and connect the wiring harness.

PS16-65 Install the torque converter and all electrical switches on the outside of the transaxle case.

Classroom Manual
Chapter 10, page 322

Special Tools

Dial indicator and holding fixture

Clutch compressor tool

Oil pump puller

Oil pump alignment tool

Seal remover/installer

Tapered punch

Ford Motor Company Transmissions

Ford Motor Company began to use the Simpson geartrain with the introduction of the C-4 transmission in 1964. Previous transmissions were based on the Ravigneaux design, as are some of their current models. The most commonly used transmissions in cars and trucks are the 4R70W, 4R100, and 4R44E. Another commonly used transmission is a heavy-duty version of the 4R44E, referred to as the 4R55E. The 4R44E (old A4LD) is a Simpson gearset-based four-speed transmission. It has full electronic controls and relies on three multiple-friction disc packs, three brake bands, and two one-way clutches to provide the various gear ratios. The overhaul procedures for this transmission are shown in Photo Sequence 17.

Photo Sequence 17
Typical Procedure for Overhauling a 4R44E Transmission

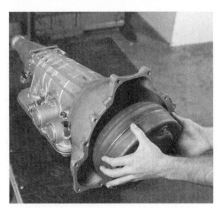

PS17-1 Remove the torque converter.

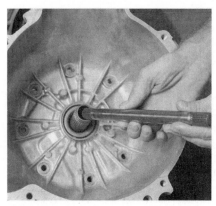

PS17-2 Remove the input shaft. Note that the two splined ends of the shaft are different.

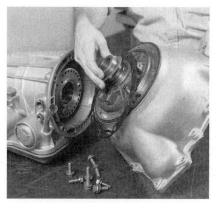

PS17-3 Remove the bell-housing retaining bolts, and then the bell housing and oil pump as an assembly. Also remove the thrust washer and gasket.

PS17-4 Separate the oil pump from the bell housing.

PS17-5 Remove the steel plate from the housing.

PS17-6 Remove the oil pan bolts, then the oil pan.

PS17-7 Remove the retaining bolt for the filter screen. Remove the filter screen, then remove the detent spring under the screen.

PS17-8 Disconnect the two wires at the TCC solenoid.

PS17-9 Remove the valve body retaining bolts. While moving the valve body away, unlock and remove the selector lever connecting link. Then remove the valve body and gasket. Note and mark the valve body bolts; they are of different lengths.

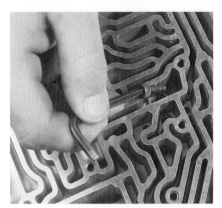

PS17-10 Remove the Allen head retaining bolt holding the center support from within the case's worm tracks.

PS17-11 Remove the bolts retaining the extension housing to the case, then remove the extension housing, parking pawl, and pawl spring.

PS17-12 Remove the governor retaining bolts, then the governor.

PS17-13 Loosen the overdrive band locknut and back off adjustment screw. Discard the locknut.

PS17-14 Remove the anchor and apply struts.

PS17-15 Lift out the clutch assembly and band. Mark the band as "overdrive" and label the anchor end of the band.

PS17-16 Lift out the overdrive one-way clutch and planetary assembly.

PS17-17 Remove the center support retaining snap ring.

PS17-18 Remove the overdrive apply lever and shaft. Then remove the overdrive control bracket from the valve body side of the case. The overdrive apply lever shaft is longer than the intermediate apply lever shaft.

PS17-19 Remove and mark the thrust washer from the top of the center support.

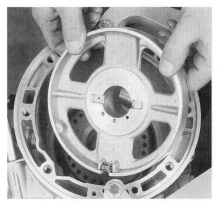

PS17-20 Carefully remove the center support bearing by prying on it evenly and upwardly.

PS17-21 Remove and mark the thrust washer from below the center support.

PS17-22 Loosen the intermediate band locknut and back off the adjusting screw. Discard the locknut. Remove the anchor and apply struts. Then remove the reverse/high and forward clutch assembly.

PS17-23 Remove the intermediate band. Mark the band as "intermediate" and label the anchor end of the band.

PS17-24 Remove the forward planetary gear assembly with its thrust washer. Mark the thrust washer.

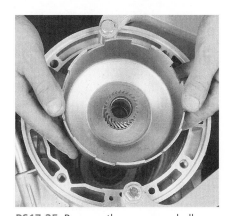

PS17-25 Remove the sun gear shell.

PS17-26 Remove the planet and ring gear assemblies.

PS17-27 Remove the large snap ring from the case to release the reverse planetary gear assembly.

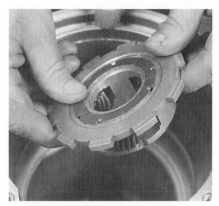

PS17-28 Remove the reverse planetary gear assembly with its thrust washer. Mark the thrust washer.

PS17-29 Remove the small snap ring on the output shaft and remove the output shaft ring gear.

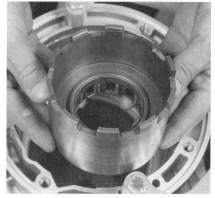

PS17-30 Remove the low/reverse drum and one-way clutch assembly.

PS17-31 Remove the low/reverse servo from the valve body side of the case.

PS17-32 Remove the low/reverse band. Then remove and mark the thrust washer.

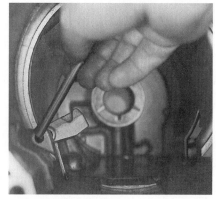

PS17-33 Remove the intermediate band apply lever and shaft.

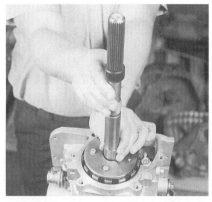

PS17-34 Turn the transmission so that the output shaft points upward. Then lift out the output shaft.

PS17-35 Remove the park gear/collector body assembly from the rear of the case. Then remove and mark the thrust washer.

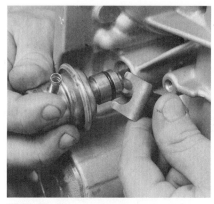

PS17-36 Remove the vacuum modulator and throttle valve retaining bolt, and then remove the modulator, throttle valve actuator rod, and throttle valve.

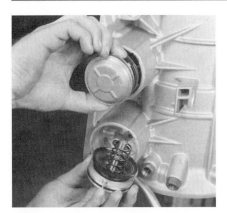

PS17-37 Remove the intermediate and overdrive servos' cover snap rings, covers, pistons, and springs. Mark the covers as to which is the overdrive servo cover.

PS17-38 Remove the neutral/safety switch and kickdown lever nut, lever, and O-ring.

PS17-39 Remove the linkage centering pin, manual lever nut, manual lever, internal kickdown lever and park pawl rod, and detent plate assembly. Then remove the lever shaft oil seal.

PS17-40 Remove the TCC and 3–4 shift solenoid connector. To remove the connector, a tab on the outside of the case must be depressed while the connector is pulled.

PS17-41 Before installing the center support into the case, install new high clutch seals onto the support hub. You must size these seals. Failure to do so will cause the seals to be cut or rolled over

during installation. Use the overdrive brake drum for sizing. Carefully rotate the center support while inserting it into the drum housing. Apply a liberal amount of petroleum jelly to the center support hub and seals.

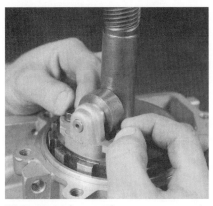

PS17-42 Make sure the center support is seated fully into the overdrive drum. Set the assembly aside to allow the assembly to stand for several minutes; this will allow the seals to seat in their grooves.

PS17-43 Position the thrust washer in the rear of the case. Install the collector body and output shaft. Then carefully install the governor onto the collector body. Tighten the bolts to specifications.

PS17-44 Position the thrust washer into the case from the front. Then install the low/reverse brake drum, output shaft ring gear, and snap ring. Position the reverse planet assembly and thrust washers.

PS17-45 Install the snap ring into the drum to hold the planet assembly. Then install the low/reverse band as marked at disassembly. Replace the piston's O-ring and install the low/reverse servo piston assembly to hold the band in position.

PS17-46 Replace the piston's O-ring and install the intermediate servo piston assembly, piston cover, and snap ring. Replace the piston's O-ring and install the overdrive servo piston assembly, piston cover, and snap ring.

PS17-47 Install the intermediate servo apply lever and shaft into the case. Then install the complete forward clutch and reverse/high clutch assemblies. Rotate the transmission so that the output shaft points downward. Install the intermediate band, apply strut, and anchor strut. Temporarily install the input shaft.

PS17-48 Check the transmission's rear endplay with a depth gauge to determine the amount of space between the thrust washer surface of the overdrive center support and the intermediate brake drum. Select the proper sized thrust washer required to obtain endplay specifications. Install the thrust washer and retain it with petroleum jelly. Note: If the average reading during this check is outside of specifications, this indicates improper reassembly, missing parts, or that some of the parts are out of specifications. Correct the problem before continuing reassembly. If within specifications, remove the depth gauge and input shaft.

PS17-49 Insert the input shaft with its short splines facing down, through the center support, and into the splines of the forward clutch cylinder. Carefully place the center support into the case. Do not seat it into the intermediate brake drum, but make sure it is in line with the retaining bolt hole in the worm tracks.

PS17-50 Without applying force on the center support, gently wiggle the input shaft, allowing the center support to slide into the intermediate brake drum with its own weight. Once fully seated, position the thrust washer on top of the center support.

Photo Sequence 17 (cont'd.)
Typical Procedure for Overhauling a 4R44E Transmission

PS17-51 Install the snap ring to retain the center support. The ends of the snap ring should be positioned in the wide shallow cavity located in the five o'clock position. Then install the Allen head bolt that retains the center support to the case. Note: Two types of snap rings are used. One has no notches and the other has a notch on its outer and inner diameters. The outer diameter notch should be positioned on the left.

PS17-52 Install the sun gear and support into the overdrive planet assembly and one-way clutch. Make sure the needle bearing race is centered inside the planetary assembly.

PS17-53 Install the overdrive planet assembly and one-way clutch into the case. Install the overdrive drum assembly, overdrive bracket, apply lever, and shaft. Then install the overdrive band, apply strut, and anchor strut.

PS17-54 Make sure the needle bearing race in the overdrive planetary assembly is centered and the overdrive clutch is fully seated. Then place the selective washer on top of the overdrive clutch drum and temporarily install the pump assembly (without its gasket) into the case.

PS17-55 Using a dial indicator, check the endplay. If the endplay exceeds limits, replace the selective washer with one that will allow for proper endplay. After the endplay is correct, remove the oil pump and the selective thrust washer.

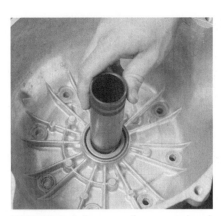

PS17-56 Install a new oil pump seal and position the separator plate onto the converter housing. Position the pump assembly onto the separator plate and converter housing. Install the retaining bolts finger tight.

PS17-57 With the recommended tool, align the pump in the converter housing to prevent seal leakage, pump gear breakage, or bushing failure.

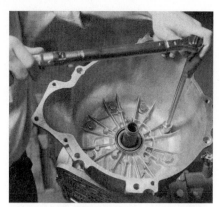

PS17-58 Install and tighten the retaining bolts. Then remove the alignment tool. Install the input shaft into the pump and install the converter into the pump gears. Rotate the converter to check for free movement. Then remove the converter and input shaft.

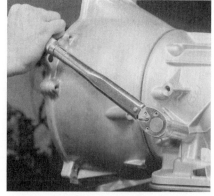

PS17-59 Coat the converter housing gasket with petroleum jelly and position it onto the transmission case. Install the seal ring onto the converter housing and position the selected thrust washer onto the rear of the pump. With the converter housing and pump aligned to the transmission case, install the retaining bolts with new washers and tighten them to specifications.

PS17-60 Install a new locknut on the overdrive band adjusting screw. Tighten the adjusting screw to 10 ft.-lb. and back it off exactly two turns. Then tighten the locknut to 35–45 ft.-lb. Repeat this procedure for the intermediate band. Perform air pressure tests to ensure proper transmission operation.

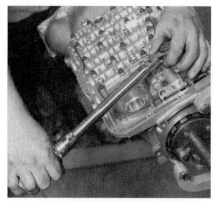

PS17-61 Install the shift lever oil seal. Then install the internal shift linkage, external manual control lever, and centering pin. Then tighten the nut to specifications. Install the O-ring, kickdown lever, and retaining nut. Install the neutral/safety switch. Install the converter clutch solenoid connector throttle valve, rod, vacuum diaphragm, retaining clamp, and bolt.

PS17-62 Using tapered punches, align the valve body to the separator plate and gasket. Use petroleum jelly to hold the gasket in place, then tighten the retaining bolts to specifications. Attach and lock the link to the manual valve. Carefully ease the valve body into the case and install and tighten the retaining bolts. Make sure the correct length bolts are installed in their proper location.

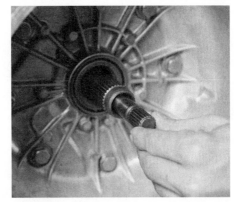

PS17-63 Connect the converter clutch solenoid wires and 3–4 shift solenoid wires. Then install the reverse servo piston assembly and spring. Make sure the piston rod is correctly seated into the reverse band apply end. Then install the correct servo rod and cover, using a new servo cover gasket. Then tighten the attaching bolts. Install new O-rings on the filter screen. Then install the filter screen and the oil pan with a new gasket. Tighten the pan bolts to specifications in two steps.

PS17-64 Install the parking pawl and return spring into the extension housing. Then install the extension housing with a new gasket. Make sure the parking pawl rod is fully seated. Then install and tighten the extension housing bolts. Finish assembly by installing the input shaft and torque converter. Make sure the torque converter is properly seated.

Ford Transaxles

Ford's 4EAT is also referred to as the GF4A-EL when used in a Mazda vehicle. This transaxle is an electronically controlled Ravigneaux-based four-speed unit. There are basically two different models of this transaxle. The major difference in models is the gear ratio of the final drive unit. This transaxle is used in many FWD vehicles and uses five multiple-disc clutches, two one-way clutches, and one band to provide the different gear ratios. This type of transaxle is equipped with a rotor-type oil pump driven by the torque converter through a drive shaft. Photo Sequence 18 covers the overhaul procedures for this transaxle.

Classroom Manual
Chapter 10, page 329

Special Tools

Dial indicator and various holding fixtures

Clutch spring compressor tool

Oil pump puller

Various seal removers and drivers

Photo Sequence 18
Typical Procedure for Overhauling a Ford 4EAT/Mazda GF4A-EL Transaxle

PS18-1 Remove torque converter, dipstick, oil filler tube, and breather hose from transmission case, and pull out the oil pump drive shaft.

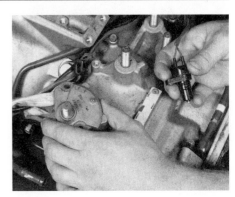

PS18-2 Remove the vehicle speed pulse generator and inhibitor switch.

Photo Sequence 18 (cont'd.)
Typical Procedure for Overhauling a Ford 4EAT/Mazda GF4A-EL Transaxle

PS18-3 Remove the oil pan and gasket. Examine the residue in the pan. Then, remove the oil strainer and O-ring.

PS18-4 Remove the valve body cover and gasket.

PS18-5 Disconnect the solenoid connectors and ATF temperature sensor, and then remove the coupler assembly.

PS18-6 Remove the valve body assembly.

PS18-7 Remove the oil pump and gasket.

PS18-8 Remove the clutch assembly by removing the turbine shaft snap ring, and pulling out the drum and clutch assembly.

PS18-9 Remove the small sun gear and one-way clutch.

PS18-10 Remove the band, then its anchor and strut.

PS18-11 Remove the piston stem from the band servo.

Photo Sequence 18 (cont'd.)
Typical Procedure for Overhauling a Ford 4EAT/Mazda GF4A-EL Transaxle

PS18-12 Remove the servo.

PS18-13 Remove the remaining one-way clutch assembly and the carrier hub assembly.

PS18-14 Remove the retaining snap ring, then remove the internal gear from the output shell.

PS18-15 Remove the remaining clutch assembly.

PS18-16 Separate the transaxle housing.

PS18-17 Remove the output shell from the output shaft.

PS18-18 Remove the manual shaft and shift plate.

PS18-19 Remove the snap ring and the parking assist lever, and then remove the parking pawl assembly.

PS18-20 Remove the differential assembly.

PS18-21 Remove the bearing housing.

PS18-22 Then remove the idler gear from the case.

PS18-23 Inspect the oil pump and all other parts, and repair or replace as necessary. Make sure to install all new rubber oil seals.

PS18-24 Install idler gear by tapping it into place.

PS18-25 Install the selected shim and bearing race into the bearing housing, then install the bearing housing.

PS18-26 Install the differential unit.

PS18-27 Install the parking pawl assembly. Then install the parking assist lever and snap ring.

PS18-28 Install the manual shaft and shift plate.

PS18-29 Install the output shell to the output gear and install the thrust bearing onto the output shell.

PS18-30 Fasten the converter housing to the transaxle case. Tighten the bolts gradually and to specifications.

PS18-31 Install the turbine shaft and 3–4 clutch assembly.

PS18-32 Install the internal gear to the output shell. Then install the retaining snap ring.

PS18-33 Install the carrier hub assembly.

PS18-34 Install the one-way clutch.

PS18-35 Install the band servo assembly into the case.

PS18-36 Install the band assembly into the case and adjust it so that it is fully expanded.

PS18-37 Install remaining one-way clutch and small sun gear.

PS18-38 Install the clutch assembly.

PS18-39 Check and correct for total endplay.

PS18-40 Adjust the band.

PS18-41 Install a new O-ring, oil filter, and pan gasket. Then install oil pan and tighten bolts to specifications.

PS18-42 Install valve body. Make sure to tighten bolts to specifications and according to the order prescribed by the manufacturer.

PS18-43 Install the solenoid and inhibitor switch with their wiring harness.

PS18-44 Install a new gasket, then install and tighten down the valve body cover.

PS18-45 Install the vehicle speed pulse generator.

PS18-46 Install dipstick tube and oil filler tube assemblies. Then install the oil pump drive shaft and torque converter.

The AX4S (old AXOD) four-speed transaxle is used in many FWD vehicles and is commonly found in Taurus, Windstar, and Sable models. It relies on two simple planetary units that operate in tandem. The planet carriers of each planetary unit are locked to the other planetary unit's ring gear. Each planetary set has its own sun gear and set of planet pinion gears. The AXOD uses four multiple-friction disc assemblies, two band assemblies, and two one-way clutches to control the operation of the planetary gearset. All the multiple-disc packs are applied hydraulically and released by several small coil springs when hydraulic pressure is diverted from the clutch's piston. The overhaul procedure for this transaxle is shown in Photo Sequence 19.

Classroom Manual
Chapter 10, page 365

Special Tools

Dial indicator and
  holding fixture

Clutch compressor
  tool

Oil pump puller

Seal remover/installer

Servo cover
  compressor tool

Depth micrometer
  and fixture

Seal protector

Photo Sequence 19
Typical Procedure for Overhauling an AX4S Transaxle

PS19-1 Remove the torque converter and mount the transaxle in a vertical position on a holding fixture. Then drain the fluid.

PS19-2 Move the transaxle to a horizontal position. Remove the two governor cover bolts, and then remove the cover and seal. Discard seal. Lift the governor, speedometer drive gear assembly, and bearing from the case.

PS19-3 Remove the bolts from the overdrive servo cover. Then mark the alignment of the cover and remove it, the piston assembly, and spring. Discard the O-ring from the cover. Then remove the low-intermediate servo cover bolts. Mark the alignment of the cover and remove it, the piston assembly, and spring. Remove and discard the gasket.

PS19-4 Remove the neutral safety switch bolts and remove the switch.

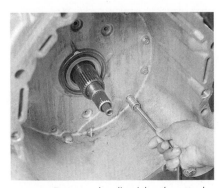

PS19-5 Remove the dipstick tube attaching bolt and pull the tube from the case. Then, remove the chain cover bolts from inside the torque converter housing.

PS19-6 Remove the valve body cover bolts, valve body cover, and gasket. Discard the gasket.

Typical Procedure for Overhauling an AX4S Transaxle

PS19-7 Unplug the electrical connectors from the pressure switches and solenoid. Use both hands to do this and do not pull on the wires; pull on the connector. Compress the tabs on both sides of the five-pin bulkhead connector from inside the chain cover and remove the connector and wiring.

PS19-8 Using a 9-mm wrench on the flats of the manual shaft, rotate the shaft clockwise so that the manual valve is all the way in. Note the location, according to length, of the oil pump and valve body bolts. Then remove the bolts. Do not remove the two bolts that hold the oil pump and valve body together. The oil pump cover bolts should also not be removed at this time.

PS19-9 Push in the TV plunger. Pull the pump and valve body assembly outward. Rotate the valve body clockwise and remove the manual valve link from the manual valve and disconnect it from the detent lever. Now remove the oil pump and valve body assembly.

PS19-10 Remove the throttle valve bracket bolts and remove the bracket. Pull the oil pump drive shaft out of the case and remove and discard the Teflon seals from the shaft.

PS19-11 Place the transaxle in a vertical position. Then remove and discard the circlip for the left-hand output shaft. Then remove the shaft's seal with a seal remover.

PS19-12 Mark the location and length of the chain cover bolts. Then remove them and the chain cover. Discard the gasket.

PS19-13 Mark and remove the metal accumulator springs from the chain cover.

PS19-14 Simultaneously lift out both sprockets with the chain assembly. Mark and remove the thrust washers from the drive and driven sprocket supports. Inspect the drive sprocket support bearing to determine if it needs replacing.

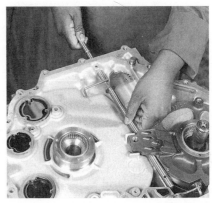

PS19-15 Remove the lockpin and roll pins from the manual shaft. Be careful not to damage the machined surfaces. Slide the shaft out of the case, and then pry the seal from the case.

PS19-16 Use a straightedge and note whether the machined bolt hole surface of the driven sprocket support is above or below the case's machined surface. Remember this for reassembly. Then remove the driven sprocket support assembly. It may be necessary to back out the reverse clutch anchor bolt.

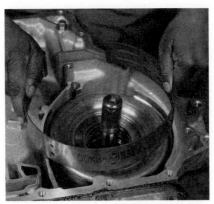

PS19-17 Remove the Teflon seals from the support assembly. Mark and remove the thrust washers and needle bearing. Then remove the plastic overdrive band retainer and the overdrive band.

PS19-18 Lift the front sun gear and shell assembly out of the case.

PS19-19 Remove the oil pan cover bolts and the cover. Discard the gasket. Remove the reverse apply tube/oil filter bolt, bracket, and oil filter screen.

PS19-20 Remove the tube retaining bracket bolts and brackets. *Note:* For complete transaxle disassembly, the reverse apply tube *must* be removed prior to removing the reverse clutch. The rear lube tube must also be removed and the seal replaced whenever the differential is removed.

PS19-21 Remove the park rod abutment bolts. Remove the roll pin for the parking pawl shaft. Then, using a magnet, remove the parking pawl shaft, park pawl, and return spring.

PS19-22 Place the transaxle in a horizontal position. Grasp the outer diameter of the reverse clutch cylinder with your fingertips and slide the assembly out of the case.

PS19-23 Place the transaxle in a vertical position. Grasp the front planetary shaft and lift out both the front and rear planetary assemblies. Lift out the low-intermediate drum and sun gear assembly. Then, remove the low-intermediate band.

PS19-24 Remove the snap ring for the final drive assembly from the case using a screwdriver inserted through the side of the case. Lift out the final drive assembly using the output shaft.

PS19-25 Remove the final drive ring gear, thrust washer, and needle bearing. Remove and discard the rear lube tube seal by tapping it toward the inside of the case. Then, install the converter oil seal and right-hand output shaft seal using a seal installer.

PS19-26 Place the transaxle in a vertical position. Install the needle bearing over the case boss with its flat side facing up and outer lip facing down.

PS19-27 Install the final drive ring gear with its external splines up. Lightly tap the ring gear to fully seat it in the case.

PS19-28 Reassemble the governor drive gear, differential assembly, final drive sun gear, parking gear, needle bearings, rear planetary support, and thrust washer.

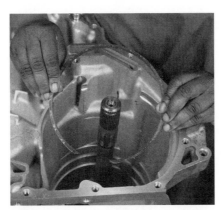

PS19-29 Lower the final drive assembly into the case. Install the snap ring and align it with the low/intermediate band anchor pin.

PS19-30 Mount a dial indicator with its plunger on the end of the output shaft. Check end clearance. If the clearance is not within specifications, replace the thrust washer with one that will bring the clearance to specifications. Available thicknesses are: 0.045–0.049" (1.15–1.25 mm) Orange; 0.055–0.059" (1.40–1.50 mm) Purple; and 0.064–0.069" (1.65–1.75 mm) Yellow.

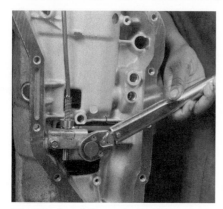

PS19-31 Install the park pawl, return spring, park pawl shaft, and locator pin. Make sure the park pawl engages the park gear and returns freely. Install the park rod actuating lever and park rod into the case. Install the park rod abutment and start the abutment bolts. Push in the park pawl and locate the rod between the pawl and the abutment.

PS19-32 Using a 3/8" drift, gently install the lube tube seal flush against the rear case support. Install the low/intermediate band and align the anchor pin pocket with the anchor pin. Install the low/intermediate drum and sun gear assembly.

PS19-33 Reassemble the ring gear and shell assembly, rear planetary, needle bearing, front planetary, and snap ring. Carefully slide the planetary assembly over the output shaft.

PS19-34 Lower the reverse clutch into the case and start clutch plate engagement. Align the clutch cylinder anchor pin pocket with the anchor pin case hole. Use the intermediate clutch hub to complete clutch plate engagement and fully seat the reverse clutch. Rotate the planet with the hub to engage splines.

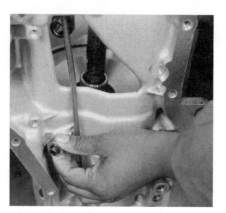

PS19-35 Start the reverse anchor pin bolt, but do not tighten. Reassemble the forward, direct, and intermediate clutch assembly. Lower the assembly into the case. Align the shell and sun gear splines with those in the forward planetary.

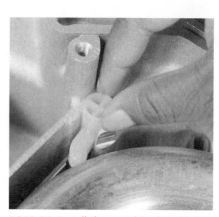

PS19-36 Install the overdrive band into the case. Install the plastic retainer with the crosshairs facing up.

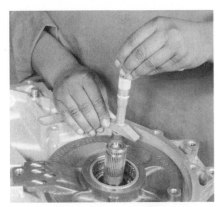

PS19-37 Check the drive sprocket end clearance to determine required thrust washer thicknesses. To do this, you must first determine if the machined bolt hole surfaces on the driven sprocket support are above or below the case machined surface. If they are above the surface, place a depth micrometer on the machined bolt hole surface and measure the distance to the case's machined surface. If they are below the case's surface, place the depth micrometer on the case's surface and measure the distance to the machined bolt hole surface. Install the correct thrust washer. Repeat this procedure for the drive sprocket.

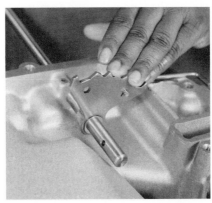

PS19-38 Tap the seal for the manual shaft into the case. Start the manual shaft through the seal and slide the manual detent lever onto the shaft. Then slide the shaft through the park rod actuating lever and tap it into the bore in the case. Install a new lock pin through the case hole. Make sure it is aligned with the groove in the shaft. Install new roll pins.

PS19-39 Align the tabs of the thrust washers and install them onto the drive and driven sprocket supports. Lubricate and install the cast iron sealing ring onto the input shaft. Install the chain over the sprockets.

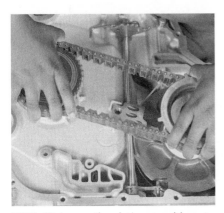

PS19-40 Lower the chain assembly onto sprocket supports. Rotate the supports while lowering the chain assembly to ensure that the sprockets are fully seated on the supports.

PS19-41 Install and align the thrust washers into the chain cover. Install a new chain cover gasket onto the cover. Then install the correct accumulator springs in their proper location. Carefully apply downward pressure, to overcome accumulator spring pressure and start two of the chain cover bolts. Position these bolts 180 degrees apart.

PS19-42 Start the remaining chain cover bolts and tighten them in sequence and to specifications. The input shaft should now have little endplay and should rotate freely. If it does not rotate freely, remove the chain cover and check for a damaged cast iron seal.

Typical Procedure for Overhauling an AX4S Transaxle

PS19-43 Tighten the park rod abutment bolts, reverse anchor pin bolt, and locknut to specifications. Lightly tap the lube tubes until they are fully seated in their bores. Install the tube retaining brackets.

PS19-44 Install the O-rings onto the oil filter and press the filter into the case. Install the reverse apply tube/oil filter bracket and bolt. Install the oil pan with a new gasket. Tighten the bolts to specifications.

PS19-45 Install new Teflon seals onto the pump drive shaft and install the shaft.

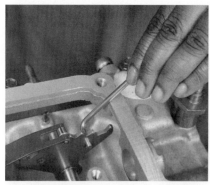

PS19-46 Install the TV bracket and tighten bracket bolts to specifications. Connect the manual valve link to the detent lever.

PS19-47 Start the oil pump and valve body assembly over the pump shaft and connect the manual valve link to the manual valve.

PS19-48 Check the alignment of the valve body to the pump and make sure the gasket is the correct one for the application. Then, install the oil pump and valve body assembly.

PS19-49 Install the bulkhead connector and other electrical connectors. Install the neutral start switch. With the manual shaft in neutral, align the switch using a 0.089" drill bit. Then tighten the switch to specification.

PS19-50 Install the valve body cover with a new gasket. Then tighten the bolts to specification. Install the dipstick tube grommet and dipstick tube in the case.

General Motors Transmissions

Classroom Manual
Chapter 10, page 328

Most General Motors transmissions and transaxles are based on the Simpson geartrain. Many different versions of the same basic design can be found. For many designs, the differences lie in the electronic controls and their intended use. The 4L60-E is used in most full-size GM cars and trucks and in some heavy-duty and high-performance vehicles. It is the four-speed unit that features full electronic control. The available gear ratios are provided by two planetary gearsets. The ring gear of one gearset is splined to the carrier of the other. One gearset is referred to as the input gearset, the other is the reaction gearset. To control gearset activity, five multiple-friction disc assemblies, one sprag one-way clutch, one roller one-way clutch, and a band are used. The 4L60-E has a one-piece case casting that incorporates the bell housing. The extension housing is a separate casting bolted to the rear of the case. The overhaul procedures for this transmission are outlined in Photo Sequence 20.

Special Tools

Dial indicator and holding fixture

Clutch compressor tool

Oil pump puller

Seal remover/installer

Servo cover compressor tool

Output shaft support tool

Photo Sequence 20
Typical Procedure for Overhauling a 4L60 Transmission

PS20-1 Remove the torque converter.

PS20-2 Mount the transmission in a holding fixture and position it so the oil pan is facing up. Position a servo cover compressor on two oil pan bolts. Then, compress the servo cover and remove the retaining ring, servo cover, and O-ring.

PS20-3 Remove the 2–4 servo assembly. Servo pin length should be checked prior to disassembly. This helps identify any wear on the 2–4 band and/or reverse input drum.

PS20-4 Remove the governor cover and O-ring, then remove the governor assembly.

PS20-5 Remove the speed sensor assembly or the speedometer driven gear. This type of transmission will be equipped with one or the other.

PS20-6 Remove the extension housing bolts, then pull the housing away from the case.

PS20-7 Remove the oil pan bolts, pan, and gasket.

PS20-8 Remove the oil filter.

PS20-9 Remove the solenoid retaining bolts. Then remove the solenoid and O-ring.

PS20-10 Remove the wiring harness. Note the location and position for reference during reassembly.

PS20-11 Loosen and remove the valve body and auxiliary valve body attaching bolts. Mark the location of each bolt, as they are different lengths. Also mark the location of any check balls you encounter.

PS20-12 Carefully remove the 1–2 accumulator cover retaining bolts, cover and pin assembly, piston, seal, and spring.

PS20-13 Remove spacer plate, and note the location of the check balls and filter.

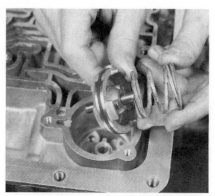

PS20-14 Remove the 3–4 accumulator spring, piston, and pin. Note the location of the spacer plate, gasket, check balls, and filters.

PS20-15 Remove oil pump retaining bolts. Then using the correct puller, remove the oil pump.

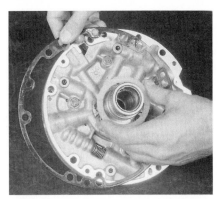

PS20-16 Remove the oil pump seal and gasket. Then remove the reverse input clutch-to-pump thrust washer from the pump.

PS20-17 Remove the reverse and input clutch assembly by lifting it out with the input shaft.

PS20-18 Remove the 2–4 band anchor pin. Then remove the band from the case.

PS20-19 Remove the input sun gear. Install an output shaft support tool onto the output shaft, then remove the input carrier to output shaft retaining ring.

PS20-20 Remove the input carrier and output shaft. Then remove the input carrier thrust washer from the reaction carrier shaft.

PS20-21 Remove the input ring gear and the reaction carrier shaft.

PS20-22 Remove the reaction sun shell and thrust washer. Then, remove the sun shell to clutch race thrust washer and support-to-case retaining ring.

PS20-23 Remove the spring retainer from the low/reverse support, then remove the reaction sun gear, low/reverse clutch race, clutch rollers, support assembly, and reaction carrier assembly.

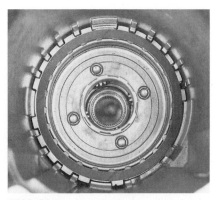

PS20-24 Remove the low/reverse clutch assembly.

PS20-25 Remove the reaction ring gear and bearing assembly. Then remove the reaction ring gear support-to-case bearing assembly.

PS20-26 Remove the parking lock bracket retaining bolts and the lock bracket.

PS20-27 Remove the parking pawl shaft, parking pawl, and return spring by using a screw extractor to remove the shaft plug from the case.

PS20-28 Using the correct spring compressor, compress the low/reverse clutch spring retainer and remove the retaining ring and spring assembly.

PS20-29 Remove the low/reverse piston assembly by applying air to the apply passage.

PS20-30 Remove the manual shaft nut, shaft, and retainer. Then remove the parking lock actuator assembly and inner detent lever.

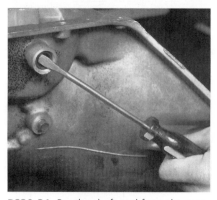

PS20-31 Pry the shaft seal from the case.

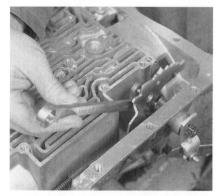

PS20-32 Install a new shaft seal into the case. Then install the parking lock actuator assembly and detent lever. Install the manual shaft, nut, and retainer.

PS20-33 Install new seals onto the low/reverse piston and coat them with petroleum jelly.

PS20-34 With the transmission in a vertical position, install the piston into the transmission case. Make sure the piston is properly aligned and fully seated.

PS20-35 Install the clutch springs. Using the spring compressor, compress the springs past the ring groove in the case. Then install the retaining ring.

PS20-36 Coat the bearing assembly with petroleum jelly and install it into the case. Then install the reaction ring gear and support.

PS20-37 Install the bearing onto the support, then install the oil deflector and reaction carrier assembly into the case.

PS20-38 Install the clutch pack with the correct number of plates.

PS20-39 Install the low/reverse support into the case, then install the inner race by pushing down while rotating it until it is fully engaged.

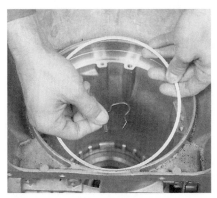

PS20-40 Install the spring retainer, then install the low/reverse retainer ring.

PS20-41 Install the snap ring onto the reaction sun gear, then install the sun gear into the reaction carrier.

PS20-42 Install the thrust washer onto the low/reverse clutch race and install the reaction sun gear shell onto the sun gear.

PS20-43 Install the thrust washer onto the reaction sun gear shell, making sure the thrust washer tangs are positioned in the slots in the shell.

PS20-44 Install the input ring gear and reaction carrier shaft into the sun gear shell. The carrier shaft splines must engage with the reaction carrier.

PS20-45 Install the thrust washer onto the reaction carrier shaft. Then install the output shaft into the transmission case.

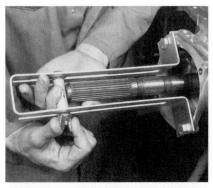

PS20-46 Install the output shaft support tool so that the shaft is positioned as upward as possible. Then install the input carrier assembly onto the output shaft.

PS20-47 Install a new retaining ring onto the output shaft, then remove the output shaft support tool.

PS20-48 Install the input sun gear.

PS20-49 Install the selective thrust washer and bearing assembly onto the input housing.

PS20-50 Install new input shaft seals, and then size them according to the procedures given in the service manual.

PS20-51 Position the reverse input assembly onto the input clutch assembly. Then install the reverse and input clutch assemblies into the case as a single unit. Align the 3–4 clutch plates of the input assembly with the input ring gear.

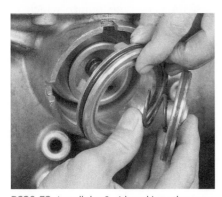

PS20-52 Install the 2–4 band into the case. Align the anchor pin end with the case pin hole and install the anchor pin into the case. Make sure the anchor pin lines up with the end of the band. Then install the 2–4 servo assembly into the case and index the apply pin onto the end of the band. Install the servo cover with a new O-ring.

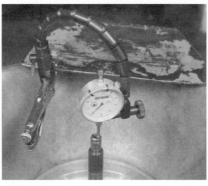

PS20-53 Check endplay. Install the correct thickness of thrust washer at the rear of the oil pump, and then install the oil pump after installing a new pump O-ring and gasket. Make sure all holes, especially the filter and pressure regulator holes, are lined up.

PS20-54 Install the retaining bolts and torque them to specifications.

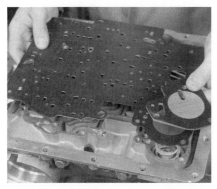

PS20-55 Install the 3–4 accumulator piston pin into the case. Then install the piston seal and piston onto the pin. Install the accumulator spring, check balls, filters, retainer, and the spacer plate and gasket.

PS20-56 Install the 1–2 accumulator spring, oil seal ring, piston, cover, and bolts. Tighten the bolts to specifications.

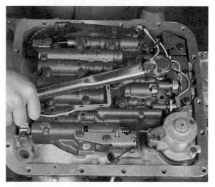

PS20-57 Install the valve body and auxiliary valve body. Then install the retaining bolts and tighten them to specifications.

PS20-58 Install the speedometer gear or speed sensor rotor onto the output shaft. Then install a new O-ring into the output shaft sleeve. Install the extension housing and torque the bolts to specifications. Using a seal installer, install a new oil seal in the extension housing.

PS20-59 Install the speedometer driven gear or the speed sensor into the extension housing. Then install all external electrical connectors. Install the manual shift lever and the torque converter.

Classroom Manual
Chapter 10, page 332

Special Tools

Dial indicator and
  holding fixture

Clutch compressor
  tool

Oil pump puller

Seal remover/installer

Servo cover
  compressor tool

Output shaft support
  tool

General Motors Transaxles

The 3T40 is a three-speed automatic transaxle designed for light-duty use. The torque converter does not directly drive the transaxle; rather, the turbine shaft drives two sprockets and a chain, which transfer engine torque to the geartrain. The 3T40 uses a vane-type pump, which is mounted to the valve body and is driven indirectly through a drive shaft by the torque converter. It also uses three multiple-friction disc assemblies, one band, and a single one-way roller clutch to provide the various gear ratios. Each clutch is applied hydraulically and released by several small coil springs. The recommended overhaul procedure for this transaxle is shown in Photo Sequence 21.

Photo Sequence 21
Typical Procedure for Overhauling a 3T40 Transaxle

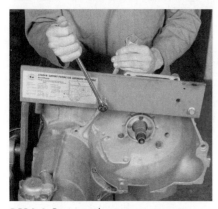

PS21-1 Remove the torque converter. Place the transaxle on a holding fixture and position the unit so that the right axle end is down to drain the fluid. Be careful when using this type of fixture; it is easy to cut your finger off if it is not assembled correctly.

PS21-2 Remove the speed sensor housing bolts, housing, and oil seal. Then, remove the speed sensor, speed sensor rotor, and governor assembly. Or remove the speedometer driven gear mount retainer bolt and retainer, then pull the driven gear from the governor housing. Remove the governor cover bolts, cover, and O-ring.

PS21-3 Remove the oil pan bolts, and then remove the oil pan and oil strainer.

PS21-4 Remove the reverse oil pipe retaining bracket-to-servo cover bolt. Remove the retaining servo cover bolts and lift off the intermediate servo cover and gasket. Then remove the intermediate servo assembly.

PS21-5 Remove the snap ring and the intermediate band apply pin from the intermediate servo piston. Then remove the third accumulator exhaust valve and spring. *Note:* The apply pin is selective size.

PS21-6 Disconnect the low/reverse oil pipe, oil pipe seal backup washer, and O-ring. Then remove the low/reverse cup plug.

PS21-7 Remove the dipstick stop and parking lock bracket from above the parking pawl and parking pawl actuator rod.

PS21-8 Rotate the final drive unit until both ends of the output shaft retaining ring are showing.

PS21-9 With a C-ring removal tool, loosen the ring from the output shaft. Rotate the shaft 180 degrees and remove the ring from the shaft. Then remove the output shaft.

PS21-10 Remove the valve body cover and mounting bolts. Then remove the throttle lever and bracket assembly retaining bolts from the valve body. Lift off the throttle lever and bracket assembly with the throttle valve cable link.

PS21-11 Remove all auxiliary valve body bolts except the lower left bolt. This bolt is only removed when it is necessary to separate the auxiliary valve body from the main valve body. Remove the remaining control valve bolts and lift the control valve and oil pump assembly away from the valve body.

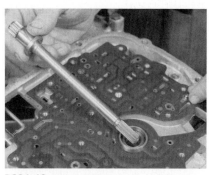

PS21-12 Remove the check ball from the direct clutch passage on the spacer plate. Then lift out the oil pump drive shaft.

PS21-13 Remove the spacer plate and gaskets. Then remove the other check balls from the case cover. Mark the location of all check balls.

PS21-14 Using the correct tools and setup, measure the input shaft-to-case cover endplay. The endplay is adjusted by a selective snap ring on the input shell.

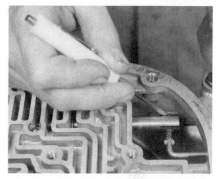

PS21-15 Disconnect the manual valve rod from the manual valve.

PS21-16 Remove the transmission case cover bolts and separate the case cover from the case. Once separated, lay the case cover with the 1–2 accumulator facing up. This prevents the loss of the 1–2 accumulator pin.

PS21-17 Remove the 1–2 accumulator spring, piston, and gasket. Then remove the case cover-to-drive sprocket thrust washer and driven sprocket thrust bearing assembly.

PS21-18 Remove and discard the O-ring from the input shaft, then remove the drive sprocket, driven sprocket, and chain as an assembly with the selective fit thrust washers.

PS21-19 Remove the detent lever-to-manual shaft pin and the manual shaft-to-case pin. Then pull the manual lever from the case. Remove the manual valve rod, detent lever assembly, and park lock actuator rod.

PS21-20 Remove the driven sprocket support and thrust washer from the direct clutch assembly.

PS21-21 Remove the intermediate band anchor hole plug, then remove the intermediate band assembly.

PS21-22 Remove the input shaft with the direct and forward clutch assembly. As the shaft is being pulled out, separate the clutch assemblies. Then remove the thrust washers and the input ring gear.

PS21-23 Remove the input carrier assembly, the thrust washers for the input ring and sun gears, the sun gear, and the input drum.

PS21-24 Measure sun gear-to-input drum endplay to check the fit of selective snap ring. The snap ring is positioned on the reaction sun gear. If the endplay exceeds specifications, replace the snap ring with one of the correct thickness during reassembly.

PS21-25 Measure the low/reverse clutch housing-to-low roller clutch race thrust washer endplay. This washer is located between the clutch housing and the one-way clutch assembly. If the endplay exceeds specifications, replace it with one of the correct thickness during reassembly.

PS21-26 Remove the reaction sun gear, then remove the low/reverse clutch housing-to-case snap ring. Now with the correct tool, remove the low/reverse housing.

PS21-27 Remove the low reverse clutch housing-to-case spacer ring from its groove in the case. Then lift the final drive sun gear shaft and the reaction gearset assembly out of the case.

PS21-28 Remove the roller clutch and reaction carrier assembly from the final drive sun gear shaft.

PS21-29 Measure the final-drive-to-case endplay to select the correct thickness of thrust washer for reassembly. The thrust washer is located between the differential carrier and the carrier case.

PS21-30 Remove the final drive ring gear spacer-to-case snap ring. Then, using the appropriate tool, pull the final drive unit from the case.

PS21-31 Remove the final drive differential-to-case thrust washer and the differential carrier-to-case thrust roller bearing from the case.

PS21-32 Prior to assembly, lubricate all bearings, bushings, and seals with clean ATF. Install the correct sized differential and carrier thrust washers. Use petroleum jelly to keep them in place.

PS21-33 Install the final drive ring gear spacer. Make sure the opening in the spacer aligns with the parking pawl opening in the case.

PS21-34 Install the final drive spacer-to-case snap ring. Then install the reaction sun gear into the case.

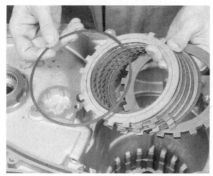

PS21-35 Install the low/reverse backing plate and clutch plates into the case. Then install the housing-to-case spacer ring.

PS21-36 Install the low/reverse clutch housing into the case. Make sure the oil feed bores in the housing line up with those in the case.

PS21-37 Install the selective snap ring onto the reaction sun gear, then install the sun gear onto the final drive sun gear shaft. Rotate the reaction sun gear while pushing down on the low/reverse clutch housing until the housing drops below the snap ring groove in the case.

PS21-38 Install the thick low/reverse clutch housing-to-case snap ring. Then install the input drum onto the reaction sun gear. Install the tanged thrust washers for the input sun and ring gears. The thrust washer for the ring gear is larger than the one for the sun gear.

PS21-39 Install the input pinion carrier onto the input sun gear. Then put the ring gear over the carrier.

PS21-40 Install the input shaft-to-input ring gear thrust washer onto the forward and direct clutch assembly. Then install the clutch assembly into the case, making sure they are fully seated.

PS21-41 Install the intermediate band. Install the anchor hole plug.

PS21-42 Install the driven sprocket support-to-direct clutch housing thrust washer, then install the support.

PS21-43 Install the manual shaft and parking lock actuator rod into the case. Then install the detent lever on the manual shaft and push the manual shaft into place. Install the manual shaft-to-detent lever and the manual shaft-to-case retaining pins.

PS21-44 Assemble the drive and driven sprockets with their chain link. Then install the assembly with the appropriate thrust washers.

PS21-45 Install the case cover-to-driven sprocket roller bearing thrust washer. Make sure it is facing the correct direction.

PS21-46 Install the 1–2 accumulator piston and spring into the case. Then install the inner and outer case-to-cover gaskets and case cover. Install and tighten the case cover bolts according to specifications.

PS21-47 Connect the manual valve rod to the manual valve. Install the check balls into the case cover.

PS21-48 Install the check ball into the direct clutch passage on the spacer plate. Then install the oil pump shaft into the bore in the case cover.

PS21-49 Install the control valve body and tighten the bolts to specifications. Then install the valve body wiring harness. Connect the lever link for the TV bracket and install the bracket onto the valve body.

PS21-50 Install the valve body cover with a new cover. Tighten bolts to specifications.

PS21-51 Rotate the transaxle so that the oil pan is facing up. Then install the output shaft into the transaxle. Rotate the final drive to allow the retaining ring to be installed in its groove in the output shaft.

PS21-52 Install the parking lock bracket and dipstick stop. Then install a new low/reverse oil pipe seal assembly, O-ring back-up washer, and the O-ring for the end of the oil pipe. Install the oil pipe and retainer bracket.

PS21-53 Install the intermediate servo piston assembly with new seals. Install the third accumulator exhaust valve and spring into the check valve bore.

PS21-54 Install the reverse oil pipe bracket-to-oil pipe and servo cover, the servo cover, and the servo cover bolts. Install a new oil strainer and O-ring.

PS21-55 Install the oil pan with a new oil pan gasket and tighten the bolts to specifications.

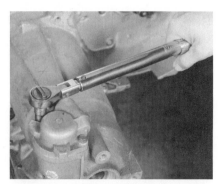

PS21-56 Install the governor assembly with a new O-ring on its cover. Install the speedometer gear or speed sensor rotor into the governor cover. Then tighten the governor cover bolts to specifications. Install the torque converter.

The geartrain of the 4T60 is based on two simple planetary gearsets operating in tandem. The combination of the two planetary units functions much like a compound unit. The two tandem units do not share a common member; rather, certain members are locked together or are integral with each other. The front planet carrier is locked to the rear ring gear and the front ring gear is locked to the rear planet carrier. The transaxle houses a third planetary unit which is used only as the final drive unit and not for overdrive.

The 4T60 uses a variable displacement vane-type pump, and four multiple-friction disc assemblies, two bands, and two one-way clutches to provide the various gear ranges. One of the one-way clutches is a roller clutch, the other is a sprag. The multiple-disc packs are released by several small coil springs when hydraulic pressure is diverted from the clutch's piston. Photo Sequence 22 covers the overhaul procedures for this transaxle.

Classroom Manual
Chapter 10, page 355

Special Tools

Dial indicator and
holding fixture

Clutch compressor
tool

Oil pump puller

Seal remover/installer

Servo cover
compressor tool

Output shaft support
tool

Photo Sequence 22
Typical Procedure for Overhauling
a 4T60 Transaxle

PS22-1 Remove the torque converter, then place the transaxle in a holding fixture. Remove the speedometer sensor and governor assembly.

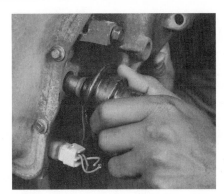

PS22-2 Remove the bottom oil pan, oil filter, modulator, and modulator valve.

PS22-3 Remove the accumulator cover with governor feed and return pipes and accumulator pistons, gaskets, retainers, and pipes from their bores.

PS22-4 Mark the cover for the reverse servo so you know where it came from. Then, remove the cover by applying pressure to it and removing the retaining ring. Then, remove the servo assembly from the case.

PS22-5 Mark the cover for the 1–2 servo so you know where it came from. Then, apply pressure to the cover and remove the servo cover retaining ring. Then remove the cover and servo assembly.

PS22-6 Remove the side cover bolts, nuts, and washers. Then remove the side covers and gaskets. Disconnect and remove the wiring harness to the pressure switches, solenoid, and case connector.

PS22-7 Remove the TV lever, linkage, and bracket from the valve body assembly. Remove the pump assembly cover bolts, then the pump cover. Remove the servo pipe retainer bolt, retainer plate, mounting bolts, and valve body.

PS22-8 Remove the oil reservoir weir. Then mark the location of and remove the check balls between the spacer plate and the valve body and between the channel plate and the spacer plate.

PS22-9 Disconnect the manual valve link from the manual valve. Place the detent lever in the park position and remove the retaining clip. Then remove the channel plate with its gaskets.

PS22-10 Remove the oil pump drive shaft. Remove the input clutch accumulator and converter clutch piston assemblies. Remove the fourth clutch plates and the apply plate, then remove the fourth clutch's thrust bearing, hub, and shaft.

PS22-11 Rotate the final drive unit until both ends of the output shaft retaining ring are showing.

PS22-12 With a C-ring removal tool, loosen the ring from the output shaft. Rotate the shaft 180 degrees and remove the ring from the shaft. Then remove the output shaft.

PS22-13 Remove and discard the O-ring from the input shaft, then remove the drive sprocket, driven sprocket, and chain as an assembly with the selective fit thrust washers. Place the chain assembly on a bench so that it is in the same direction and position it was in the transmission.

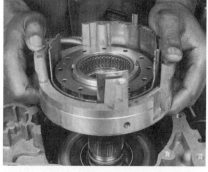

PS22-14 Remove the driven sprocket support and thrust washer from between the sprockets and the channel plate. Remove the scavenging scoop and driven sprocket support with the second clutch thrust washer.

PS22-15 Using the correct tool, remove the second clutch and input shaft clutch housings as an assembly. Remove the reverse band.

PS22-16 Remove the thrust washers. Then measure the endplay of the input shaft clutch housing. Select the correct thickness of thrust washer and set aside for reassembly.

PS22-17 Remove the input clutch sprag assembly, third clutch assembly, and the input sun gear. Then remove the reverse reaction drum, input carrier assembly, and thrust washer.

PS22-18 Remove the reaction carrier, reaction sun gear/drum assembly, and forward band.

PS22-19 Remove the reaction sun gear thrust bearing and final drive sun gear shaft.

PS22-20 Check final drive endplay and select the correct size thrust washer for the unit.

PS22-21 Using the proper tool, remove the final drive assembly and selective thrust washers and bearings.

PS22-22 Clean and inspect the transaxle case and all of the transaxle components. Then position the correct size thrust washers and bearings onto the final drive assembly and install the unit. Petroleum jelly can be used to hold the washers and bearings in place while positioning the unit.

PS22-23 Install the final drive sun gear shaft through the final drive ring gear; the splines must engage with the parking gear and the final drive sun gear.

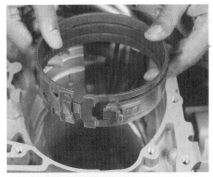

PS22-24 Install the forward band into the case, making sure the band is aligned with the anchor pin.

PS22-25 Install the reaction sun gear-to-final drive ring gear thrust bearing. Then assemble the reaction sun gear and drum assembly onto the final drive ring gear.

PS22-26 Check the endplay of the carriers in the reaction planetary gearset. Then install the thrust washer and carrier assembly into the case. Rotate the carrier until the pinions engage with the reaction sun gear.

PS22-27 Install the input carrier with its thrust bearing into the case. Then install the reverse reaction drum, making sure its splines engage with the input carrier.

PS22-28 Inspect the roller and sprag clutches, then put the spacer onto the sun gear, followed by the input sprag retainer, sprag assembly, and roller clutch. Make sure the input sprag and third roller clutch hold and freewheel in opposite directions while holding the input sun gear.

PS22-29 Lubricate the inner seal on the input clutch's piston, then install the seal and assemble the input piston into the input housing. Install a new O-ring onto the input shaft.

PS22-30 Install the spring retainer and guide into the piston, then install the third clutch piston housing into the input housing. Using the proper compressor, install the retaining snap ring.

PS22-31 Install the inner seal for the third clutch, then install the third clutch piston into the housing. Compress the spring retainer and install the retaining snap ring.

PS22-32 Install the wave plate, and then install the correct number of clutch plates in the correct sequence. Install the input clutch backing plate and the retaining snap ring. Air check the operation of the clutch.

PS22-33 Assemble the second clutch piston in the housing. Install the apply ring, return spring, snap ring, and wave plate. Then assemble the correct number of clutch plates in the correct sequence. Install the backing plate and snap ring, then air check the clutch's operation.

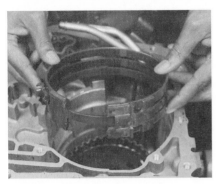

PS22-34 Install the thrust washers. Using the correct tool, install the second clutch and input shaft clutch housings as an assembly. Install the reverse band.

PS22-35 Install the scavenging scoop and driven sprocket support with the second clutch thrust washer. Install the driven sprocket support and thrust washer between the sprockets and the channel plate.

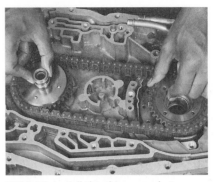

PS22-36 Install the drive sprocket, driven sprocket, and chain as an assembly.

PS22-37 Install the output shaft. Start the C-ring onto the output shaft, then rotate the shaft 180 degrees and fully seat the ring onto the shaft.

PS22-38 Install the fourth clutch's thrust bearing, hub, and shaft. Then install the fourth clutch plates and the apply plate. Install the input clutch accumulator and converter clutch piston assemblies. Install the oil pump drive shaft.

PS22-39 Install the channel plate with new gaskets; connect the manual valve link to the manual valve. Place the detent lever in the park position and install the retaining clip.

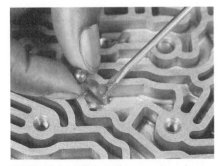

PS22-40 Install the oil reservoir weir. Install the check balls in their proper location between the spacer plate and the valve body and between the channel plate and the spacer plate.

PS22-41 Install the servo pipe retainer bolt, retainer plate, mounting bolts, and valve body. Install the pump assembly cover bolts, then the pump cover. Tighten the bolts to specifications. Install the TV lever, linkage, and bracket onto the valve body assembly.

PS22-42 Connect and install the wiring harness to the pressure switches, solenoid, and case connector. Then install the side covers and gaskets. Install the side cover bolts, nuts, and washers, and then tighten them to specifications.

PS22-43 Install the 1–2 servo cover, servo assembly, and the retaining ring.

PS22-44 Install the reverse servo cover, servo assembly, and the retaining ring.

PS22-45 Install the accumulator cover with governor feed and return pipes and accumulator pistons, gaskets, retainers, and pipes into their bores.

PS22-46 Install the bottom oil pan, oil filter, modulator, and modulator valve.

PS22-47 Install the torque converter, speedometer sensor, and governor assembly.

Classroom Manual

Chapter 10, page 368

Special Tools

Dial indicator and holding fixture

Clutch compressor tool

Seal remover/installer

Gear puller

Small chisel

Punch and drift set

Honda Transaxles

The Honda CA, F4, and G4 transaxles are used in many Honda and Acura cars. Other models of this unique transaxle are similar in operation but may have a third shaft. This third shaft was added to provide for engine braking. These transaxles do not use planetary gearsets to provide for the different gear ranges. Constant-mesh helical and square-cut gears are used in a manner similar to that of a manual transmission.

These transaxles have a mainshaft and countershaft on which the gears ride. To provide the four forward and one reverse gear, different pairs of gears are locked to the shafts by hydraulically controlled clutches. Reverse gear is obtained through the use of a shift fork that slides the reverse gear into position. Four multiple-disc clutches, the sliding reverse gear, and a one-way clutch are used to control the gears. Photo Sequence 23 outlines the overhaul procedure for this unique transaxle.

Photo Sequence 23
Typical Procedure for Overhauling a Honda F4 Transaxle

PS23-1 With the transaxle on a holding fixture, loosen and remove the bolts securing the right side cover. Then remove the side cover.

PS23-2 Lock the mainshaft using the mainshaft holding tool. Engage the parking brake pawl with the parking gear and then loosen and remove the mainshaft and countershaft locknuts. *Note:* The mainshaft locknut has left-hand threads.

PS23-3 Remove the thrust washer, thrust needle bearing, and first gear needle bearing.

PS23-4 Remove the mainshaft first gear.

PS23-5 Remove the parking pawl, shaft, stop pin, and spring.

PS23-6 Install a gear puller and remove the parking gear and countershaft first gear as a unit.

PS23-7 Remove the reverse idler bearing holder assembly.

PS23-8 Remove the parking shaft arm and spring.

PS23-9 Remove the throttle control lever.

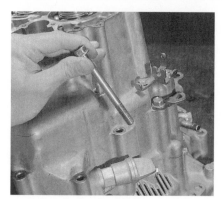

PS23-10 Loosen and remove the transmission housing mounting bolts.

PS23-11 Install the transmission housing puller on the countershaft and screw the puller bolt against the countershaft to separate the housing.

PS23-12 Remove the housing.

PS23-13 Remove the countershaft reverse gear collar and reverse gear.

PS23-14 Remove the countershaft fourth gear, needle bearing, distance collar, reverse gear selector hub, reverse gear selector, and reverse gear shift fork.

PS23-15 Remove the mainshaft and countershaft as an assembly.

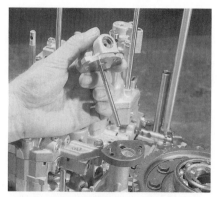

PS23-16 Remove the governor housing, pipe, and separator plate.

PS23-17 Remove the governor shaft, holder, and gear assembly.

PS23-18 Remove the second/third accumulator cover. The cover is spring loaded, so press down on the cover while unscrewing the bolts in a star pattern to prevent stripping the threads in the accumulator housing.

PS23-19 Remove the second/third accumulator springs.

PS23-20 Remove the bolts that secure the fourth accumulator cover and spring, and then remove the accumulator cover. *Note:* This cover is spring loaded, therefore you should be careful not to strip the threads in the servo body while removing or installing the cover bolts.

PS23-21 Remove the modulator valve body and separator plate.

PS23-22 Remove the bolts that secure the servo body, and then remove the servo body and separator plate.

PS23-23 Remove the servo secondary valve body. Be careful not to lose the two steel balls, ball springs, timing accumulator piston and spring, and secondary filter.

PS23-24 Remove the LC-shift valve body and separator plate.

PS23-25 Remove the regulator valve body.

PS23-26 Remove the stator shaft, shaft arm, and stop pin from the main valve body.

PS23-27 Remove the cotter pin, washer, rollers, and pin from the manual valve.

PS23-28 Remove the main valve body. Be careful not to lose the three steel balls, check ball springs, torque converter check valve, and check valve spring.

PS23-29 Remove the oil pump gears, gear shaft, check valve, check valve spring, and separator plate.

PS23-30 Remove the ATF filter screen.

PS23-31 Remove the differential assembly from the torque converter housing.

PS23-32 Carefully inspect and clean all gears and other parts of each sub-assembly. Make sure you keep the sub-assembly parts separated form one another. Replace any defective or worn parts. Then install the differential assembly into the torque converter housing.

PS23-33 Install a new ATF filter screen.

PS23-34 Install the main separator plate, pump gears, pump gear shafts, check valve, and check valve spring.

PS23-35 Install the main valve body. Make sure the pump drive gear and the pump shaft rotate smoothly. Be careful not to lose the three steel balls, check ball springs, torque converter check valve, and check valve spring. Tighten the valve body bolts to specifications.

PS23-36 Install the manual valve pin, rollers, washer, and cotter pin.

PS23-37 Install the stator shaft, shaft arm, and stop pin in the main valve body.

PS23-38 Install the pressure regulator valve body.

PS23-39 Install the LC-shift valve separator plate and body. Tighten the bolts to specifications.

PS23-40 Install the servo secondary valve body. Make certain the two steel balls, ball springs, timing accumulator piston and spring, and secondary filter are in the proper position.

PS23-41 Install the servo valve separator plate, valve body, and mounting bolts. Tighten to specifications.

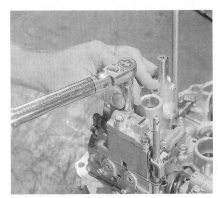

PS23-42 Install the modulator valve body and separator plate. Torque the retaining bolts to specifications.

PS23-43 After replacing the O-ring, install the fourth accumulator spring cover.

PS23-44 Install the second/third accumulator springs.

PS23-45 Using the handle of a hammer, compress the second/third accumulator springs by pressing down on the cover while you tighten the accumulator retaining bolts.

PS23-46 Install the servo valve body and mounting bolts. Tighten to specifications.

PS23-47 Install the governor shaft, holder, and gear assembly.

PS23-48 Install the governor housing, pipe, and separator plate.

PS23-49 Install the governor housing and tighten the bolts to specifications. Bend up the tabs on the lock plates with a punch.

PS23-50 Install the mainshaft and countershaft as a unit.

PS23-51 Install the needle bearings, countershaft fourth gear, distance collar, reverse selector hub, and reverse selector with the shift fork on the countershaft.

PS23-52 Install the countershaft reverse gear collar and reverse gear.

PS23-53 Install the reverse shift fork and tighten the bolt to specifications.

PS23-54 Install the reverse idler gear and needle bearing into the housing. Align the spring pin of the control shaft with the transmission housing groove. Then align the torque converter housing and a new gasket with the transmission housing.

PS23-55 Install the transmission housing bolts and torque them to specifications in two steps.

PS23-56 Install the throttle control lever assembly and tighten it to specifications.

PS23-57 Install the parking brake shift arm and spring, and then tighten it to specifications.

PS23-58 Install the reverse idler bearing and holder and tighten to specifications.

PS23-59 Install the countershaft first gear needle bearing.

PS23-60 Install the parking gear and countershaft first gear.

PS23-61 Engage the parking pawl with the parking gear and check for proper alignment.

PS23-62 Install new countershaft and mainshaft locknuts and tighten to specifications. Stake them to their shafts using a steel punch. Tighten to specifications.

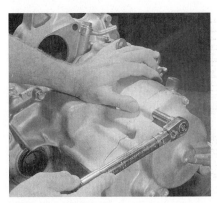

PS23-63 Install and torque the end cover mounting bolts to specifications. Install the torque converter, making sure it is fully seated.

CUSTOMER CARE: Because this model uses a sliding reverse gear, damage to the transaxle can occur if the car is towed or moved while the transaxle's gear selector is in NEUTRAL but the transaxle is still mechanically in reverse gear. A servo normally is used to delay forward gear engagement after the transaxle has been in reverse. Simply shifting into NEUTRAL may not disengage reverse gear. Anytime the vehicle is to be moved, start the engine and place the transaxle into DRIVE, then shift it into NEUTRAL. Turn off the engine and the vehicle is ready to be moved.

Special Tools

Dial indicator and holding fixture

Clutch compressor tool

Oil pump puller

Seal remover/installer

Classroom Manual
Chapter 10, page 340

Nissan Motor Company Transmissions

Widely used Nissan RWD transmissions are the 3N71B, which is a three-speed unit, and the L4N71B/E4N71B series, which provide four forward gears through the use of a Simpson gearset and an additional planetary unit mounted in front of the Simpson gearset. The primary difference between the "L" and the "E" is that the E-model provides electronic control for shifting and torque converter clutch operation. Nissan transaxles are similar to other transaxles covered in this chapter.

These transmissions use four multiple friction disc assemblies, two servos and bands, and a one-way clutch to provide for the different ranges of gears. All the clutches except the low/reverse unit are released by several small coil springs. The low/reverse unit utilizes a Belleville-type spring for greater clamping pressures. Photo Sequence 24 covers the typical overhaul procedures for this model transmission. Always follow the recommended procedure for the transmission or transaxle you are working on.

Photo Sequence 24
Typical Procedure for Overhauling a Nissan L4N71B Transmission

PS24-1 Place the transmission on a suitable workbench. Check endplay and record your readings.

PS24-2 Remove the torque converter.

PS24-3 Unscrew and remove the electrical kickdown solenoid and O-ring.

PS24-4 Unscrew and remove the throttle modulator valve, its diaphragm rod and O-ring.

PS24-5 Remove the speedometer drive assembly, with gear and O-ring.

PS24-6 Remove the oil pan and inspect its contents. An analysis of foreign material will provide clues regarding the types of problems to look for during the overhaul procedure. Check the service manual for specific details.

PS24-7 Remove valve body from the case.

PS24-8 Remove manual valve from the valve body to prevent the valve from dropping out.

PS24-9 Back off the servo piston stem locknut and tighten piston stem snug to prevent the front clutch drum from dropping out when removing the front pump.

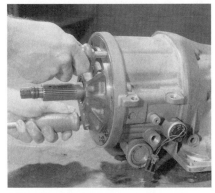

PS24-10 Use the correct puller and remove the front pump from the case.

PS24-11 Remove front clutch thrust washer and bearing race.

PS24-12 Remove the overdrive servo cover.

Photo Sequence 24 (cont'd.)
Typical Procedure for Overhauling a Nissan L4N71B Transmission

PS24-13 Loosen the overdrive servo locknut, and back off the overdrive band's servo piston stem to release the band.

PS24-14 Remove the front and rear clutch assemblies. Note the positions of the front pump thrust washers and rear thrust washer.

PS24-15 Remove the brake band strut and overdrive brake band.

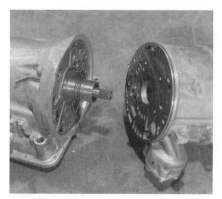

PS24-16 Remove the overdrive housing.

PS24-17 Remove the extension housing. Take care not to lose the parking pawl, spring, pin, and parking actuator.

PS24-18 Remove the governor valve assembly.

PS24-19 Remove the front drum support and input shaft.

PS24-20 Remove the second brake band strut and brake band.

PS24-21 Remove the front clutch and planetary gear pack.

PS24-22 Remove the output shaft snap ring.

PS24-23 Remove the output shaft.

PS24-24 Remove the snap ring, connecting drum, and one-way clutch assembly.

PS24-25 Use a screwdriver to remove the large retaining snap ring. Then remove the low and reverse brake assembly.

PS24-26 Install the low and reverse brake assembly starting with the steel dished plate and then alternating steel and friction plates.

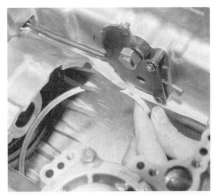

PS24-27 Install the retaining plate and snap ring. Check for proper clearance between the snap ring and retaining plate. Select the proper thickness of retaining plate that will give the correct ring-to-plate clearance if the measurement does not meet the specified limits. Check the operation of the low and reverse brake by using compressed air.

PS24-28 Install the one-way clutch assembly, connecting drum, and snap ring.

PS24-29 Install the governor needle bearing, thrust washer, output shaft, and oil distributor in the case.

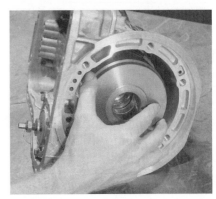

PS24-30 Install the planetary gear pack and front clutch assembly.

PS24-31 Install the snap ring on the output shaft.

PS24-32 Install the second brake band, band strut. and band servo. Lubricate the O-ring seals with ATF.

PS24-33 Lubricate the drum support gasket with ATF and install the drum support into the housing.

PS24-34 Install the governor valve assembly and torque the retaining bolts to specifications.

PS24-35 Install the parking actuator and parking pawl assemblies in the extension housing. Then install the extension housing. Torque its mounting bolts to specifications.

PS24-36 Install the front drum support gasket and new O-ring onto the overdrive housing. Gently tap it into place using a rubber mallet. Make sure the bolt holes are aligned.

PS24-37 Install the needle bearing race and direct clutch thrust washer. Then install the overdrive brake band, strut, and servo assembly. Lubricate the servo O-ring before installing it.

PS24-38 Install the overdrive pack on the drum support.

PS24-39 Adjust the overdrive band. Tighten the piston stem to specifications. Then back off two full turns. Tighten the servo locknut to specifications. Test the overdrive servo with compressed air.

PS24-40 Install the overdrive servo cover and tighten to specifications.

PS24-41 Adjust the second brake band. Tighten the piston stem to specifications, and then back off three full turns.

PS24-42 Secure the piston stem while you tighten the servo locknut to specifications.

PS24-43 Install the oil pump bearing and thrust washer.

PS24-44 Install the oil pump assembly, using the special tool. Be sure to lubricate the gasket and O-ring with ATF. Align the mounting bolt holes in the pump housing with the bolt holes in the overdrive housing.

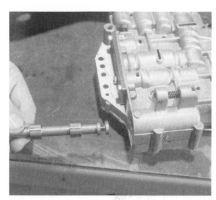

PS24-45 Lubricate and install the manual valve into the control valve body.

PS24-46 Install the control valve body and tighten the valve body attaching bolts to the specified torque.

PS24-47 Install the oil pan gasket, oil pan retaining bolts. Tighten the bolts to specifications, while using a star pattern to tighten them.

PS24-48 Install the speedometer drive assembly with its drive gear and lubricated O-ring.

Photo Sequence 24 (cont'd.)
Typical Procedure for Overhauling a Nissan L4N71B Transmission

PS24-49 Install the vacuum diaphragm, diaphragm rod, and O-ring.

PS24-50 Install the kickdown (or downshift) solenoid and O-ring.

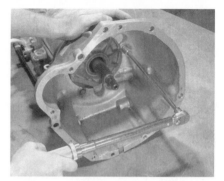

PS24-51 Install the torque converter housing. Torque the mounting bolts to specifications in a star pattern.

Classroom Manual
Chapter 10, page 342

Special Tools

Dial indicator and
 holding fixture
Clutch compressor
 tool
Oil pump puller
Air nozzle
Seal remover/installer
Gear pullers and
 drivers
Small chisel
Punch and drift set

Toyota Transaxles

Toyota led the way with electronically controlled transmissions and since the first model was released, they have continued to refine that basic model. That base model, the A-140E transaxle, was a three-speed transmission with an overdrive assembly. The basis for operation is a Simpson gearset in line with a single overdrive planetary gearset. The transaxles uses four multiple-disc clutches, two band and servo assemblies, and three one-way clutches to provide the various gear ranges. The A541E transaxle is a revised copy of the A140E. The biggest change was in the electronic controls, which now had an adaptive learning capability. This transaxle has been used in many different Toyota and Lexus models, including the Toyota Camry, Toyota Avalon, and the Lexus ES-300. This transaxle uses six multiple-disc clutches, three one-way clutches, and one brake band. The operation of the transaxle and the lockup converter is totally controlled by the PCM. However, a throttle pressure cable is used to mechanically modulate line pressure. Photo Sequence 25 covers the typical overhaul procedures for this model transmission.

Photo Sequence 25
Typical Procedure for Overhauling a Toyota A541E Transaxle

PS25-1 With the transaxle mounted in a stand and its external electrical components, oil pan, and linkages removed, remove the upper cover of the transaxle case.

PS25-2 Remove the oil pipe bracket.

Photo Sequence 25
Typical Procedure for Overhauling a Toyota A541E Transaxle

PS25-3 Remove the oil strainer.

PS25-4 Remove the retaining bolts for the detent spring, and then remove the spring.

PS25-5 Unbolt and remove the retaining bolts for the manual valve body and remove it. Then remove the retaining bolt for the oil pipes, and then remove the pipes.

PS25-6 Disconnect the electrical connectors to the solenoids, then remove the connector clamp.

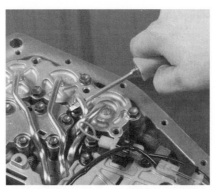

PS25-7 Remove the oil pipes' retaining bolts, then pry out the apply pipes.

PS25-8 Unbolt the valve body retaining bolts, and then disconnect the throttle cable from the valve body while removing the valve body.

PS25-9 Remove the throttle cable solenoid wiring.

PS25-10 Remove the accumulator pistons and springs.

PS25-11 Measure the piston stroke for the second coast brake. Use a divider to transfer the distance to a vernier caliper or micrometer for measurement.

PS25-12 Remove the second coast brake piston's retaining ring and use air pressure to remove the second coast brake piston.

PS25-13 Position the transaxle so that it is upright, then remove the oil pump with a puller.

PS25-14 Remove the direct and forward clutch assembly.

PS25-15 Separate the direct clutch from the forward clutch.

PS25-16 Remove the guide for the second/coast brake band.

PS25-17 Remove the second/coast brake band.

PS25-18 Remove the ring gear for the front planetary gearset.

PS25-19 Remove the front planetary gearset with its bearings.

PS25-20 Remove the sun gear and sun gear input drum.

Photo Sequence 25 (cont'd.)
Typical Procedure for Overhauling a Toyota A541E Transaxle

PS25-21 Remove the second brake drum, and then remove the second brake piston return spring, and the #1 one-way clutch.

PS25-22 Remove the friction discs and plates from the housing.

PS25-23 Remove the snap ring and the #2 one-way clutch.

PS25-24 Then remove the thrust washer and the rear planetary ring gear.

PS25-25 Remove the flange, discs, and plates of the first/reverse brake.

PS25-26 Rotate the transaxle and remove the retaining bolts for the overdrive unit.

PS25-27 Remove the overdrive cover.

PS25-28 Then pull the overdrive planetary gearset from the transaxle.

PS25-29 Remove the output gear and nut.

PS25-30 Remove the retaining ring, output shaft, and bearing race.

PS25-31 Install the output shaft with a new race and retaining ring.

PS25-32 Install the output gear with a new nut and tighten the nut to specifications.

PS25-33 Stake the output gear's retaining nut to the output shaft.

PS25-34 Remove the overdrive clutch pack by compressing the first/reverse brake piston return spring and remove the snap ring. Then remove the piston.

PS25-35 Remove the parking lock pawl bracket and guide.

PS25-36 Remove the manual shaft and seal.

PS25-37 Loosen the final drive and axle seal covers.

PS25-38 Remove the covers, then remove and service the differential unit.

PS25-39 Begin reassembly by installing the differential and drive axle unit. Make sure the unit is properly set up.

PS25-40 Install new seals and gaskets, then position the final drive housing and tighten the bolts to specifications.

PS25-41 Install the parking pawl lock, pawl bracket and guide, and the manual shaft. Use a new seal and collar when doing this. Then check the operation of the parking lock pawl.

PS25-42 Install the first/reverse brake piston into the case and install the piston return spring and retaining snap ring.

PS25-43 Install the overdrive brake band and overdrive planetary gearset into the case.

PS25-44 Install the overdrive cover with a new gasket. Tighten the retaining bolts to specifications.

PS25-45 Check the endplay of the intermediate shaft. If it is within specs, coat the thrust washer with petroleum jelly and install the rear planetary gearset.

PS25-46 Install the discs, plates, and flange for the first/reverse brake. Then install the retaining snap ring.

PS25-47 Check the operation of the brake with compressed air.

Photo Sequence 25 (cont'd.)
Typical Procedure for Overhauling a Toyota A541E Transaxle

PS25-48 Check the clearance of the disc pack with a feeler gauge.

PS25-49 Install the #2 one-way clutch into the case with the shiny side of the flange facing out.

PS25-50 Install the second/coast brake band guide.

PS25-51 Then coat the thrust washer for the #1 one-way clutch and install it with the clutch.

PS25-52 Install the discs and flange for the second brake, then install the snap ring.

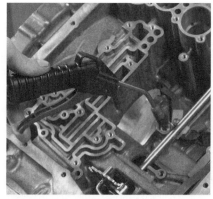

PS25-53 Check the operation of the second brake with compressed air.

PS25-54 Coat the thrust washer with petroleum jelly, then install the sun gear with its drum.

PS25-55 Turn the drum clockwise until it seats into the #1 one-way clutch.

PS25-56 Install a new O-ring onto the intermediate shaft, then install the front planetary gearset.

PS25-57 Coat the thrust washers with petroleum jelly and assemble the forward and direct clutch assemblies.

PS25-58 Install the second/coast brake band and the forward and direct clutch assembly.

PS25-59 Position the transaxle with the oil pump opening up. Coat the new pump O-ring with ATF and install it on the pump.

PS25-60 Install the pump into the transmission housing.

PS25-61 Check the endplay of the input shaft.

PS25-62 Install the piston for the second/coast brake. Use a new snap ring to retain it.

PS25-63 Identify the correct accumulator and piston spring for each of the accumulator bores.

PS25-64 Install the accumulator pistons and springs. Use a new gasket and torque the bolts to specifications.

PS25-65 Connect the wires to the solenoids and connect the throttle cable.

PS25-66 Put the oil pipes into their proper position and then mount the valve body to the case. Install and tighten the retaining bolts to specifications.

PS25-67 Connect the solenoids to the electrical harness. Then install the manual valve body and detent spring. Be sure to tighten the retaining bolts to specifications.

PS25-68 Install and properly tighten the oil pipe bracket and retaining bolts.

PS25-69 Install the oil strainer.

PS25-70 Complete assembly by installing the upper cover of the transaxle case and its external electrical components, oil pan, and linkages.

CASE STUDY

A very irate customer had his late-model Honda towed back to the shop. Just three weeks had passed since the transmission had been rebuilt and now it had failed again.

The technician who had done the overhaul was assigned the car. He conducted a visual inspection and found nothing obviously wrong with the electrical system or the transaxle itself. However, the fluid had a burned smell. He conducted an oil pressure test and found extremely low pressures. He suspected that the pump or pressure regulator had failed.

He pulled the transaxle and checked the pressure regulator and oil pump. He found the pump's gears seized together. This was the first Honda transaxle he had ever rebuilt and he was disappointed that it had failed. Wondering how this could happen, he reviewed the service manual and technical service bulletins. In bold print in the service manual he found the answer. He had failed to correctly align the oil pump shaft during assembly. Based on his experiences with other transmissions he had rebuilt, he hadn't known this was critical and had assumed that the shaft would only fit one way.

He aligned the shaft, replaced the oil pump gears, and reassembled the transmission. He road tested the car and conducted a pressure test. Everything was fine. He then gave the car back to the customer, after apologizing, knowing that the transaxle was now right. The customer appreciated his honesty and left with confidence that the transmission was okay. The technician learned a lesson: always refer to and follow the directions given in the service manual.

ASE-Style Review Questions

1. *Technician A* says some seals must be cut to size before they are installed.
   Technician B says square-cut seals are designed to roll over when a part is fit over them.
   Who is correct?
   A. A only   C. Both A and B
   B. B only   D. Neither A nor B

2. While assembling a transmission:
   Technician A coats the steel clutch discs with petroleum jelly.
   Technician B soaks the friction discs in clean ATF before installing them.
   Who is correct?
   A. A only   C. Both A and B
   B. B only   D. Neither A nor B

3. *Technician A* says endplay is often corrected by selective snap rings.
   Technician B says endplay is often corrected by selective thrust washers.
   Who is correct?
   A. A only   C. Both A and B
   B. B only   D. Neither A nor B

4. *Technician A* says all Chrysler transmissions and transaxles are Simpson gear based.
   Technician B says the 36RH and 42LE are longitudinally mounted and are very similar in construction and operation. The major difference between the two is that one is a transaxle.
   Who is correct?
   A. A only   C. Both A and B
   B. B only   D. Neither A nor B

5. *Technician A* says all transaxles that have a drive chain use the chain to transfer transmission output to the final drive unit.
   Technician B says tandem planetary gearsets do not have a common member; rather, a member of one gearset is connected to a member of the other gearset. Sometimes there are two of these connections.
   Who is correct?
   A. A only   C. Both A and B
   B. B only   D. Neither A nor B

6. A transaxle slips in all gear ranges. Which of the following is the *least* likely cause?
   A. A defective one-way   C. Faulty electronic
       clutch   controls
   B. Low fluid level   D. A clogged fluid filter

7. While assembling a transaxle:
   Technician A reuses all seals unless they are damaged.
   Technician B lubricates all seals and bearings with clean bearing grease before installing them.
   Who is correct?
   A. A only   C. Both A and B
   B. B only   D. Neither A nor B

8. A transmission abruptly makes unwanted downshifts at high speeds. Which of the following is the *most* likely cause?
   A. Throttle cable out of   C. Linkage out of
       adjustment   adjustment
   B. Defective oil pump   D. Sticking valves in
                               the valve body

9. While diagnosing incorrect shift points;
Technician A says a disconnected shift solenoid could be the cause.
Technician B says a dirty valve body could be the cause.
Who is correct?
A. A only
B. B only
C. Both A and B
D. Neither A nor B

10. While checking transmission endplay:
Technician A measures the movement of the shaft with a dial indicator.
Technician B uses a clutch compressor tool to get the maximum movement reading.
Who is correct?
A. A only
B. B only
C. Both A and B
D. Neither A nor B

ASE Challenge Questions

1. The customer complains of harsh automatic downshifts.
Technician A says the anticlunk spring may be broken or positioned incorrectly.
Technician B says line pressure may be entering the governor assembly.
Who is correct?
A. A only
B. B only
C. Both A and B
D. Neither A nor B

2. All of the following may cause a rough initial engagement in forward and reverse EXCEPT:
A. Excessive backlash in final drive/differential assembly
B. Retarded ignition timing
C. Missing check ball
D. Leaking transmission oil filter

3. During disassembly, the low/reverse band is found to be very worn with some frictional material missing.
Technician A says the damage may be the result of improper band adjustment.
Technician B says high oil pump pressure may be the cause.

Who is correct?
A. A only
B. B only
C. Both A and B
D. Neither A nor B

4. The 3–4 gear switch winding tested open.
Technician A says this would prevent the torque converter from lockup in fourth gear.
Technician B says the PCM may place the system in default under this condition.
Who is correct?
A. A only
B. B only
C. Both A and B
D. Neither A nor B

5. The vehicle experiences an intermittent second gear start.
Technician A says a bad one-way clutch may be the cause.
Technician B says low governor pressure may be the cause.
Who is correct?
A. A only
B. B only
C. Both A and B
D. Neither A nor B

Job Sheet 30

30

Name _____ Date _____

Identification of Special Tools and Procedures

Upon completion of this job sheet, you should be able to use the service manual to determine what special tools and procedures are required to correctly service a particular transmission or transaxle.

Tools and Materials

Appropriate service manual

Procedure

Your instructor will assign you to a particular transmission or transaxle. You will use the service manual to identify and describe the special tools recommended for the complete overhaul of the transmission. You will also identify any tools your shop has that are acceptable substitutes for the factory-recommended special tools. After you have assembled the list of special tools, go through the steps for complete overhaul and identify those procedures that are more than remove and install types (such as checking input shaft endplay).

1. Describe the transmission that was assigned to you:

   Model and type of transmission _____

   Describe the vehicle it is from: Year _____ Make _____

   Model _____ VIN _____

2. What service manual or electronic database are you using for this assignment? Be specific.

   _____

   _____

3. List the special tools referenced in the overhaul section of the manual for this transmission. Include in your list the part or subsystem the tool is used on.

   _____

   _____

   _____

   _____

   _____

   _____

   _____

   _____

   _____

   _____

   _____

Identification of Special Tools and Procedures (continued)

4. List the required special tools that are available in your shop.

_____

_____

_____

_____

_____

_____

_____

_____

_____

5. List the available tools that are acceptable substitutes for the special tools you don't have in your shop.

_____

_____

_____

_____

_____

_____

_____

_____

6. Review the overhaul procedures and describe the special procedures that must be known in order to properly service the transmission assigned to you. Include in your description the parts or subsystems that require this procedure. Do not simply say, "check endplay." Describe what needs to be done to check the endplay.

_____

_____

_____

_____

_____

_____

_____

_____

Instructor's Response _____

_____

Appendix

ASE Practice Examination

Final Exam Automatic Transmission/Transaxle A2

1. Which of the following is the *least* likely cause for a buzzing noise from a transmission?
   A. Improper fluid level or condition
   B. Defective oil pump
   C. Defective flexplate
   D. Damaged planetary gearset

2. A vehicle experiences engine flare in low gear only.
   Technician A says the torque converter lockup clutch is slipping.
   Technician B says the transmission oil pump is not providing the required pressure.
   Who is correct?
   A. A only
   B. B only
   C. Both A and B
   D. Neither A nor B

3. The results of a pressure test are being discussed:
   Technician A says low idle pressure may be caused by a defective exhaust gas recirculation system.
   Technician B says low Neutral and Park pressures may indicate a fluid leakage past the clutch and servo seals.
   Who is correct?
   A. A only
   B. B only
   C. Both A and B
   D. Neither A nor B

4. The vehicle creeps in neutral.
   Technician A says a too high engine idle speed could be the cause.
   Technician B says a too tight clutch pack may be the problem.
   Who is correct?
   A. A only
   B. B only
   C. Both A and B
   D. Neither A nor B

5. *Technician A* says over-torque valve body fasteners may cause a lack of engine braking in manual low.
   Technician B says a lack of engine braking in manual third may be caused by a bad overrunning clutch.
   Who is correct?
   A. A only
   B. B only
   C. Both A and B
   D. Neither A nor B

6. The transmission's output shaft and its sealing components are being discussed:
   Technician A says the shaft and all of its sealing components must be replaced if nicks and scratches are found in the shaft's sealing area.
   Technician B says all of the shaft's seals and rings must be replaced during a rebuild.
   Who is correct?
   A. A only
   B. B only
   C. Both A and B
   D. Neither A nor B

7. The vehicle will only upshift to second at full throttle. This could be caused by any of the following EXCEPT:
   A. Clogged oil passages
   B. Low fluid level
   C. Bad clutch pack
   D. Open upshift switch

8. The vehicle will not move in any gear.
   Technician A says a misadjusted T.V. cable could be the cause.
   Technician B says leakage at the oil pump and/or valve body could cause this condition.
   Who is correct?
   A. A only
   B. B only
   C. Both A and B
   D. Neither A nor B

9. Sensors are being discussed:
   Technician A says most speed sensors are ac generators.
   Technician B says most speed sensors use a stationary magnet, rotor, and a voltage sensor.
   Who is correct?
   A. A only
   B. B only
   C. Both A and B
   D. Neither A nor B

10. *Technician A* says the PCM monitors the amount of voltage generated by the speed sensor to calculate the vehicle's speed.
    Technician B says the output of a speed sensor is pulsed as an on/off voltage signal when displayed on a DSO.
    Who is correct?
    A. A only
    B. B only
    C. Both A and B
    D. Neither A nor B

11. Shift solenoids are being discussed:
    Technician A says engine flare during upshifts may be caused by high resistance in the solenoid's windings.

Technician B says delay shifts may be caused by open solenoid windings.
Who is correct?

A. A only   **C.** Both A and B
B. B only   **D.** Neither A nor B

12. When diagnosing the cause of no engine braking during manual low operation:
Technician A checks for a defective oil pump.
Technician B suspects a damaged drive link.
Who is correct?

A. A only   **C.** Both A and B
B. B only   **D.** Neither A nor B

13. The customer complains of transmission noises in all gears except Park and Neutral.
Technician A says the oil pump may be the cause.
Technician B says the drive chain or sprocket may be the source of the noise.
Who is correct?

A. A only   **C.** Both A and B
B. B only   **D.** Neither A nor B

14. Converter installation is being discussed:
Technician A says converter depth must be measured before it is removed from the transmission.
Technician B says the converter must engage the turbine (input) shaft, stator shaft, and pump drive to be properly installed.
Who is correct?

A. A only   **C.** Both A and B
B. B only   **D.** Neither A nor B

15. A vehicle with an EAT exhibits erratic shifting. The speed sensor is suspect.
Technician A says the PM sensor can be checked on the vehicle with a voltmeter.
Technician B says there should be a specified resistance between the leads when checked with an ohmmeter.
Who is correct?

A. A only   **C.** Both A and B
B. B only   **D.** Neither A nor B

16. The vehicle will crank in Reverse or Drive, but not in any other gear.
Technician A says the Park/Neutral switch is open.
Technician B says the T.V. linkage is misadjusted.
Who is correct?

A. A only   **C.** Both A and B
B. B only   **D.** Neither A nor B

17. While diagnosing the cause of sluggish acceleration:
Technician A suspects a faulty governor.

Technician B checks the condition of the fluid.
Who is correct?

A. A only   **C.** Both A and B
B. B only   **D.** Neither A nor B

18. Which of the following is the *most* likely cause of slipping in all forward gear ranges?

A. Faulty governor   **C.** Faulty band or clutch
B. Clogged oil filter   **D.** Sticking valve in the valve body

19. A vehicle with a newly rebuilt EAT shifts directly to third from low. The most probable cause in this instance is:

A. 1–2 servo is defective   **C.** Manual linkages are out of adjustment
B. 2–3 shift valve is sticking   **D.** Throttle linkage is out of adjustment

20. While diagnosing the cause of no forced downshifts during full throttle operation:
Technician A suspects a misadjusted manual linkage.
Technician B suspects a dirty valve body and valves.
Who is correct?

A. A only   **C.** Both A and B
B. B only   **D.** Neither A nor B

21. Which of the following is the *least* likely cause for the transmission not upshifting and operating only in first gear?

A. Improper fluid level or condition   **C.** Defective oil pump
D. Faulty governor
B. Faulty valve body

22. The vehicle exhibits a scraping or grating noise in every gear position anytime engine torque is changed up or down.
Technician A says this could be caused by a damaged planetary gearset.
Technician B says cracks around the flexplate mounting holes would cause this noise.
Who is correct?

A. A only   **C.** Both A and B
B. B only   **D.** Neither A nor B

23. A MAP sensor test reveals below specified voltage signals at all engine speeds.
Technician A says an EAT may have late upshifts based on the signals.
Technician B says a vacuum modulator controlled T.V. pressure will cause late upshifts based on the signals.
Who is correct?

A. A only   **C.** Both A and B
B. B only   **D.** Neither A nor B

24. The transmission will not shift and has high line pressure during testing. This could be caused by:

    A. A stuck regulator valve

    B. High governor pressure

    C. Low T.V. pressure

    D. A faulty manual lever position sensor

25. The valve body bores are being discussed:

    Technician A says the bores should be measured with small bore (hole) gauges.

    Technician B says the bores need only be cleaned if the valves can be polished.

    Who is correct?

    A. A only

    B. B only

    C. Both A and B

    D. Neither A nor B

26. During disassembly, some extension-to-case bolts were found to have aluminum strips in the threads.

    Technician A says the extension housing should be replaced.

    Technician B says to clean and repair the internal threads with a thread insert.

    Who is correct?

    A. A only

    B. B only

    C. Both A and B

    D. Neither A nor B

27. *Technician A* says use two flare nut wrenches to disconnect transmission cooler lines.

    Technician B says use vice grip pliers to hold a cooler line that is twisting as the flare nut is turned.

    Who is correct?

    A. A only

    B. B only

    C. Both A and B

    D. Neither A nor B

28. Air checking the transmission is being discussed:

    Technician A says using a special adapter plate is necessary on some transmissions.

    Technician B says a special style blow gun should be used.

    Who is correct?

    A. A only

    B. B only

    C. Both A and B

    D. Neither A nor B

29. During inspection, the valve body mounting surface of the case is found to be slightly warped.

    Technician A says the problem may be corrected by using a fine file on the valve body.

    Technician B says the case must be replaced if the mounting surface for the valve body is warped.

    Who is correct?

    A. A only

    B. B only

    C. Both A and B

    D. Neither A nor B

30. A bushing was nicked and scratched as the shaft was being removed. There is no other apparent wear or damage.

    Technician A says use fine emery cloth to repair the damage.

    Technician B says sanding or polishing the bushing may cause the shaft to wobble in the bushing bore.

    Who is correct?

    A. A only

    B. B only

    C. Both A and B

    D. Neither A nor B

31. While diagnosing the cause of gear slipping in second gear only:

    Technician A suspects a worn or damaged clutch assembly.

    Technician B suspects a damaged drive link.

    Who is correct?

    A. A only

    B. B only

    C. Both A and B

    D. Neither A nor B

32. While conducting a pressure test:

    Technician A says the cause of low pressure in all operating ranges could be a clogged filter.

    Technician B says the cause of high pressure in all operating ranges could be a defective throttle valve.

    Who is correct?

    A. A only

    B. B only

    C. Both A and B

    D. Neither A nor B

33. *Technician A* says pump side clearance is determined using a straightedge and feeler gauges.

    Technician B says the pump's gear side clearance may be measured with a dial indicator.

    Who is correct?

    A. A only

    B. B only

    C. Both A and B

    D. Neither A nor B

34. *Technician A* says the transmission fluid level may be checked hot or cold in Park with the engine running on some vehicles.

    Technician B says dark color with a burned smell may indicate a transmission rebuild may be needed.

    Who is correct?

    A. A only

    B. B only

    C. Both A and B

    D. Neither A nor B

35. There are traces of transmission fluid in the engine coolant.

    Technician A says the leak may be coming from a poor line connection to the radiator.

    Technician B says a cracked cooler line may be the source of the leak.

    Who is correct?

    A. A only

    B. B only

    C. Both A and B

    D. Neither A nor B

36. *Technician A* says when removing the chain, input gear, and output gear from certain transaxles, a special tool must be used to spread the gears. *Technician B* says a special tool must be used to draw the gears together in order to remove the chain from this type of transaxle.
Who is correct?
A. A only
B. B only
C. Both A and B
D. Neither A nor B

37. The customer complains of delayed upshifts during hard acceleration.
Technician A says the throttle valve linkage adjustment is the most probable cause.
Technician B says a clogged catalytic converter may be the cause.
Who is correct?
A. A only
B. B only
C. Both A and B
D. Neither A nor B

38. A shudder is noticeable at speeds around 40–55 mph in third or fourth gear.
Technician A says the TCC solenoid may be partially blocked.
Technician B says TCC oil pressure may be too high.
Who is correct?
A. A only
B. B only
C. Both A and B
D. Neither A nor B

39. While diagnosing the cause of no torque converter clutch engagement on a transmission that seems to shift fine:
Technician A suspects a damaged clutch pressure plate.
Technician B suspects a severely worn input clutch.
Who is correct?
A. A only
B. B only
C. Both A and B
D. Neither A nor B

40. Inspection of the torque converter is being discussed:
Technician A says to replace the converter if the transmission's front pump is damaged.
Technician B says to reweld or repair loose converter drive lugs.
Who is correct?
A. A only
B. B only
C. Both A and B
D. Neither A nor B

41. *Technician A* says transaxle turning torque may be checked by turning the axle with a torque wrench. *Technician B* says the turning torque is measured by turning the final drive's axle (side) gear with a torque wrench.

Who is correct?
A. A only
B. B only
C. Both A and B
D. Neither A nor B

42. An ohmmeter is being used to check several shift solenoids:
Technician A says high resistance in shift solenoids will cause engine flare during upshifts.
Technician B says an infinity reading means the solenoid's windings have more resistance than the meter can measure.
Who is correct?
A. A only
B. B only
C. Both A and B
D. Neither A nor B

43. *Technician A* says some OBD-II systems have a TCC break-in mode or routine.
Technician B says this OBD-II mode must be manually activated using a scan tool.
Who is correct?
A. A only
B. B only
C. Both A and B
D. Neither A nor B

44. Transmission/transaxle removal is being discussed:
Technician A says some electrical components of the ABS system must be disconnected at the wheels on RWD vehicles.
Technician B says the converter access plate must be removed before removing a transmission or transaxle.
Who is correct?
A. A only
B. B only
C. Both A and B
D. Neither A nor B

45. Which of the following is the *least* likely cause for ATF coming out of the transmission vent or the dipstick tube?
A. Defective pressure regulator valve
B. Plugged drain holes in the valve body
C. Fluid at too high of a level
D. Defective oil cooler

46. Transmission/transaxle installation is being discussed:
Technician A says to always replace the pilot bearing/bushing for smooth torque converter movement on the flexplate.
Technician B says the converter drive hub should be lubricated with ATF to help prevent damage to the pump's front seal.
Who is correct?
A. A only
B. B only
C. Both A and B
D. Neither A nor B

47. While diagnosing the cause of transmission overheating:

    Technician A checks for contaminated fluid.
    Technician B suspects a damaged flexplate.
    Who is correct?

    A. A only
    B. B only
    C. Both A and B
    D. Neither A nor B

48. *Technician A* says slippage in first gear can only be caused by a faulty vehicle speed sensor.
    Technician B says a defective one-way clutch can cause slippage in all gears.
    Who is correct?

    A. A only
    B. B only
    C. Both A and B
    D. Neither A nor B

49. *Technician A* says to clean the valve body in carburetor cleaner, rinse in water, and blow dry with compressed air.

    Technician B says valves that are dragging in their bores should be polished with fine emery cloth until the drag is removed.
    Who is correct?

    A. A only
    B. B only
    C. Both A and B
    D. Neither A nor B

50. Transmission oil pumps are being discussed:
    Technician A says high line pressure would result if the relief valve spring is broken or missing.
    Technician B says the relief valve is designed to restrict the fluid flow until operating pressures are achieved.
    Who is correct?

    A. A only
    B. B only
    C. Both A and B
    D. Neither A nor B

APPENDIX

Automatic Transmission Special Tool Suppliers

APD Transmission Parts
  Atlanta, GA

Automatic Transmission Rebuilders Association
  Ventura, CA

Baum Tools Unlimited, Inc.
  Longboat Key, FL

Big A Auto Parts, APC Inc.
  Houston, TX

Carquest Corp.
  Tarrytown, NY

Hastings Manufacturing
  Hastings, MI

KD Tools, Danaher Tool Group
  Lancaster, PA

Kent-Moore, Div. SPX Corp.
  Warren, MI

Lisle Corp.
  Clarinda, IA

MacTools
  Washington Courthouse, OH

MATCO Tool Co.
  Stow, OH

NAPA Hand/Service Tools
  Lancaster, PA

OTC, Div. SPX Corp.
  Owatonna, MN

Parts Plus
  Memphis, TN

Snap-on Tools Corp.
  Kenosha, WI

Transtar Industries, Inc.
  Cleveland, OH

APPENDIX

Metric Conversions

	to convert these	to these,	multiply by:
TEMPERATURE	Centigrade Degrees	Fahrenheit Degrees	1.8 then + 32
	Fahrenheit Degrees	Centigrade Degrees	0.556 after − 32
LENGTH	Millimeters	Inches	0.03937
	Inches	Millimeters	25.4
	Meters	Feet	3.28084
	Feet	Meters	0.3048
	Kilometers	Miles	0.62137
	Miles	Kilometers	1.60935
AREA	Square Centimeters	Square Inches	0.155
	Square Inches	Square Centimeters	6.45159
VOLUME	Cubic Centimeters	Cubic Inches	0.06103
	Cubic Inches	Cubic Centimeters	16.38703
	Cubic Centimeters	Liters	0.001
	Liters	Cubic Centimeters	1000
	Liters	Cubic Inches	61.025
	Cubic Inches	Liters	0.01639
	Liters	Quarts	1.05672
	Quarts	Liters	0.94633
	Liters	Pints	2.11344
	Pints	Liters	0.47317
	Liters	Ounces	33.81497
	Ounces	Liters	0.02957
WEIGHT	Grams	Ounces	0.03527
	Ounces	Grams	28.34953
	Kilograms	Pounds	2.20462
	Pounds	Kilograms	0.45359
WORK	Centimeter Kilograms	Inch-Pounds	0.8676
	Inch-Pounds	Centimeter-Kilograms	1.15262
	Meter Kilograms	Foot-Pounds	7.23301
	Foot-Pounds	Newton-Meters	1.3558
PRESSURE	Kilograms/Square Centimeter	Pounds/Square Inch	14.22334
	Pounds/Square Inch	Kilograms/Square Centimeter	0.07031
	Bar	Pounds/Square Inch	14.504
	Pounds/Square Inch	Bar	0.06895

Glossary
Glosario

Abrasion Wearing or rubbing away of a part.
Abrasión El desgaste o consumo por rozamiento de una parte.

Acceleration An increase in velocity or speed.
Aceleración Un incremento en la velocidad.

Accumulator A device used in automatic transmissions to cushion the shock of shifting between gears, providing a smoother feel inside the vehicle.
Acumulador Un dispositivo que se usa en las transmisiones automáticas para suavizar el choque de cambios entre las velocidades, así proporcionando una sensación más uniforme en el interior del vehículo.

Adhesives Chemicals used to hold gaskets in place during the assembly of an engine. They also aid the gasket in maintaining a tight seal by filling in the small irregularities on the surfaces and by preventing the gasket from shifting due to engine vibration.
Adhesivo Los productos químicos que se usan para sujetar a los empaques en una posición correcta mientras que se efectua la asamblea de un motor. También ayuden para que los empaques mantengan un sello impermeable, rellenando a las irregularidades pequeñas en las superficies y previniendo que se mueva el empaque debido a las vibraciones del motor.

Aeration The process of mixing air into a liquid.
Aireación El proceso de mezclar el aire en un líquido.

ALDL Assembly Line Data Link.
ALDL Siglas para una trasmisión de datos de la planta de fabricación.

Alignment An adjustment to a line or to bring into a line.
Alineación Un ajuste que se efectúa en una linea o alinear.

Ammeter A test instrument used to measure electrical current.
Amperímetro Un aparato que mide corriente eléctrica.

Amplitude The height of a waveform is called its amplitude.
Amplitud La altura de onda eléctrica se llama su amplitud.

Antifriction bearing A bearing designed to reduce friction. This type of bearing normally uses ball or roller inserts to reduce the friction.
Cojinetes de antifricción Un cojinete diseñado con el fin de disminuir la fricción. Este tipo de cojinete suele incorporar una pieza inserta esférica o de rodillos para disminuir la fricción.

Antiseize Thread compound designed to keep threaded connections from damage due to rust or corrosion.
Antiagarrotamiento Un compuesto para filetes diseñado para protejer a las conecciones fileteados de los daños de la oxidación o la corrosión.

Apply devices Devices that hold or drive members of a planetary gearset. They may be hydraulically or mechanically applied.
Dispositivos de aplicación Los dispositivos que sujeten o manejan los miembros de un engranaje planetario. Se pueden aplicar mecánicamente o hidráulicamente.

Arbor press A small, hand-operated shop press used when only a light force is required against a bearing, shaft, or other part.
Prensa para calar Una prensa de mano pequeña del taller que se puede usar en casos que requieren una fuerza ligera contra un cojinete, una flecha u otra parte.

ATF Automatic transmission fluid.
ATF Fluido de transmisión automática.

Automatic transmission A transmission in which gear or ratio changes are self-activated, eliminating the necessity of hand-shifting gears.
Transmisión automática Una transmisión en la cual un cambio deengranajes o los cambios en relación son por mando automático, así eliminando la necesad de cambios de velocidades manual.

Automotive Service Excellence (ASE) The National Institute for Automotive Service Excellence (ASE) has established a voluntary certification program for technicians employed in the many related fields of the automotive industry.
Excelencia Automotora Del Servicio El instituto nacional para la excelencia automotora del servicio (ASE) ha establecido un programa voluntario de la certificación para los técnicos empleados en los muchos ramos relacionados de la industria del automóvil.

Axial Parallel to a shaft or bearing bore.
Axial Paralelo a una flecha o al taladro del cojinete.

Axis The centerline of a rotating part, a symmetrical part, or a circular bore.
Eje La linea de quilla de una parte giratoria, una parte simétrica, o un taladro circular.

Axle The shaft or shafts of a machine upon which the wheels are mounted.
Semieje El eje o los ejes de una máquina sobre los cuales se montan las ruedas.

Axle ratio The ratio between the rotational speed (rpm) of the driveshaft and that of the driven wheel; gear reduction through the differential, determined by dividing the number of teeth on the ring gear by the number of teeth on the drive pinion.
Relación del eje La relación entre la velocidad giratorio (rpm) del árbol propulsor y la de la rueda arrastrada; reducción de los engranajes por medio del diferencial, que se determina por dividir el número de dientes de la corona por el número de los dientes en el pinión de ataque.

Axle shaft A shaft on which the road wheels are mounted.
Flecha del semieje Una flecha en la cual se monta las ruedas.

Backlash The amount of clearance or play between two meshed gears.
Juego La cantidad de holgura o juego entre dos engranajes endentados.

Balance Having equal weight distribution. The term is usually used to describe the weight distribution around the circumference and between the front and back sides of a wheel.
Equilibrio Lo que tiene una distribución igual de peso. El término suele usarse para describir la distribución del peso alrededor de la circunferencia y entre los lados delanteros y traseros de una rueda.

Balance valve A regulating valve that controls a pressure of just the right value to balance other forces acting on the valve.
Válvula niveladora Una válvula de reglaje que controla a la presión del valor correcto para mantener el equilibrio contra las otras fuerzas que afectan a la válvula.

Ball bearing An antifriction bearing consisting of a hardened inner and outer race with hardened steel balls that roll between the two races, and supports the load of the shaft.

Rodamiento de bolas Un cojinete de antifricción que consiste de una pista endurecida interior e exterior y contiene bolas de acero endurecidos que ruedan entre las dos pistas, y sostiene la carga de la flecha.

Ball joint A suspension component that attaches the control arm to the steering knuckle and serves as the lower pivot point for the steering knuckle. The ball joint gets its name from its ball-and-socket design. It allows both up-and-down motion as well as rotation. In a MacPherson strut FWD suspension system, the two lower ball joints are nonload carrying.

Articulación esférica Un componente de la suspensión que une el brazo de mando a la articulación de la dirección y sirve como un punto pivote inferior de la articulación de la dirección. La articulación esférica derive su nombre de su diseño de bola y casquillo. Permite no sólo el movimiento de arriba y abajo sino también el de rotación. En un sistema de suspensión tipo FWD con poste de MacPherson, las articulaciones esféricas inferiores no soportan el peso.

Ballooning A condition in which the torque converter has been blown up like a balloon caused by excessive pressure in the converter.

Inflación Una condición en el qual el convertidor del esfuerzo de torsión se ha inflado como un baloon, que es una condición causada por la presión excesiva dentro del convertidor.

Band A steel band with an inner lining of frictional material. Device used to hold a clutch drum at certain times during transmission operation.

Banda Una banda de acero que tiene un forro interior de una materia de fricción. Un dispositivo que retiene al tambor del embrague en algunos momentos durante la operación de la transmisión.

Bearing The supporting part that reduces friction between a stationary and rotating part or between two moving parts.

Cojinete La parte portadora que reduce la fricción entre una parte fija y una parte giratoria o entre dos partes que muevan.

Bearing cage A spacer that keeps the balls or rollers in a bearing in proper position between the inner and outer races.

Jaula del cojinete Un espaciador que mantiene a las bolas o a los rodillos del cojinete en la posición correcta entre las pistas interiores e exteriores.

Bearing caps In the differential, caps held in place by bolts or nuts which, in turn, hold bearings in place.

Tapones del cojinete En un diferencial, las tapas que se sujeten en su lugar por pernos o tuercas, los cuales en su turno, retienen y posicionan a los cojinetes.

Bearing cone The inner race, rollers, and cage assembly of a tapered roller bearing. Cones and cups must always be replaced in matched sets.

Cono del cojinete La asamblea de la pista interior, los rodillos, y el jaula de un cojinete de rodillos cónico. Se debe siempre reemplazar a ambos partes de un par de conos del cojinete y los anillos exteriores a la vez.

Bearing cup The outer race of a tapered roller bearing or ball bearing.

Anillo exterior La pista exterior de un cojinete cónico de rodillas o de bolas.

Bearing race The surface upon which the rollers or balls of a bearing rotate. The outer race is the same thing as the cup, and the inner race is the one closest to the axle shaft.

Pista del cojinete La superficie sobre la cual ruedan los rodillos o las bolas de un cojinete. La pista exterior es lo mismo que un anillo exterior, y la pista interior es la más cercana a la flecha del eje.

Belleville spring A tempered spring steel cone-shaped plate used to aid the mechanical force in a pressure plate assembly.

Resorte de tensión Belleville Un plato de resorte del acero revenido en forma cónica que aumenta a la fuerza mecánica de una asamblea del plato opresor.

Bell housing A housing that fits over the clutch components and connects the engine and the transmission.

Concha del embrague Un cárter que encaja a los componentes del embrague y conecta al motor con la transmisión.

Bias voltage Voltage applied across a diode.

Tensión de polarización El voltaje aplicado através de un diodo.

Bolt Head The part of a bolt that the socket or wrench fits over to torque or tighten the bolt.

Cabeza del perno La pieza de un perno que el socket o la llave ajusta sobre la cabeza del perno para apretarlo.

Bolt Shank The smooth area on a bolt from the bottom surface of the head to the start of the threads.

Asta del perno El área lisa entre el fondo de la cabeza y el principio de las roscas se llama la asta del perno.

Bolt torque The turning effort required to offset resistance as the bolt is being tightened.

Torsión del perno El esfuerzo de torsión que se requiere para compensar la resistencia del perno mientras que esté siendo apretado.

Brinnelling Rough lines worn across a bearing race or shaft due to impact loading, vibration, or inadequate lubrication.

Efecto brinel Lineas ásperas que aparecen en las pistas de un cojinete o en las flechas debido al choque de carga, la vibración, o falta de lubricación.

Burnish To smooth or polish by the use of a sliding tool under pressure.

Bruñir Pulir o suavizar por medio de una herramienta deslizando bajo presión.

Burr A feather edge of metal left on a part being cut with a file or other cutting tool.

Rebaba Una lima espada de metal que permanece en una parte que ha sido cortado con una lima u otro herramienta de cortar.

Bus A common connector used as an information source for the vehicle's various control units.

Bus Un conector común que se usa como un fuente de información para los varios aparatos de control del vehículo.

Bushing A cylindrical lining used as a bearing assembly made of steel, brass, bronze, nylon, or plastic.

Buje Un forro cilíndrico que se usa como una asamblea de cojinete que puede ser hecho del acero, del latón, del bronce, del nylon, o del plástico.

Butt-end locking ring A locking ring whose ends are cut to butt up to contact each other once in place. There is no gap between the ends of a butt-end ring.

Anillo de enclavamiento a tope Un anillo de enclavamiento cuyos extremidades son cortadas para toparse o ajustarse una contra la otra en lugar. No hay holgura entre las extremidades de un anillo de retén a tope.

C-clip A C-shaped clip used to retain the drive axles in some rear axle assemblies.

Grapa de C Una grapa en forma de C que retiene a las flechas motrices en algunas asambleas de ejes traseras.

Cage A spacer used to keep the balls or rollers in proper relation to one another. In a constant-velocity joint, the cage is an open metal framework that surrounds the balls to hold them in position.

Jaula Una espaciador que mantiene una relación correcta entre los rodillos o las bolas. En una junta de velocidad constante, la jaula es un armazón abierto de metal que rodea a las bolas para mantenerlas en posición.

Cap An object that fits over an opening to stop flow.

Tapón Un objecto que tapa a una apertura para detener el flujo.

Carbon monoxide Part of the exhaust gas from an engine; an odorless, colorless, and deadly gas.

Óxido de carbono Una parte de los vapores de escape de un motor; es un gas sin olor, sin color y puede causar la muerte.

Case porosity Leaks caused by tiny holes that are formed by trapped air bubbles during the casting process.

Porosidad del cárter Las fugas que se causan por los hoyitos pequeños formados por burbújas de aire entrapados durante el proceso del moldeo.

Chamfer A bevel or taper at the edge of a hole or a gear tooth.

Chaflán Un bisél o cono en el borde de un hoyo o un diente del engranaje.

Chamfer face A beveled surface on a shaft or part that allows for easier assembly. The ends of FWD drive shafts are often chamfered to make installation of the CV joints easier.

Cara achaflanada Una superficie biselada en una flecha o una parte que facilita la asamblea. Los extremos de los árboles de mando de FWD suelen ser achaflandos para facilitar la instalación de las juntas CV.

Chase To straighten up or repair damaged threads.

Embutir Enderezar o reparar a los filetes dañados.

Chasing To clean threads with a tap.

Embutido Limpiar a los filetes con un macho.

Chassis The vehicle frame, suspension, and running gear. On FWD cars, it includes the control arms, struts, springs, trailing arms, sway bars, shocks, steering knuckles, and frame. The drive shafts, constant-velocity joints, and transaxle are not part of the chassis or suspension.

Chasis El armazón de un vehículo, la suspensión, y el engranaje de marcha. En los coches de FWD, incluye los brazos de mando, los postes, los resortes (chapas), los brazos traseros, las estabilizadoras, las articulaciones de la dirección y el armazón. Los árboles de mando, las juntas de velocidad constante, y la flecha impulsora no son partes del chasis ni de la suspensión.

Circlip A split steel snap ring that fits into a groove to hold various parts in place. Circlips are often used on the ends of FWD drive shafts to retain the constant-velocity joints.

Grapa circular Un seguro partido circular de acero que se coloca en una ranura para posicionar a varias partes. Las grapas circulares se suelen usar en las extremidades de los árboles de mando en FWD para retener las juntas de velocidad constante.

Class A fire A fire in which wood, paper, and other ordinary materials are burning.

Fuego de la Clase A Un fuego en el cual la madera, el papel, y otros materiales ordinarios se están quemando se llama un fuego de la Clase A.

Class B fire A fire involving flammable liquids, such as gasoline, diesel fuel, paint, grease, oil, and other similar liquids.

Fuego de la Clase B Un fuego que quema líquidos inflamables, tales como gasolina, el combustible diesel, la pintura, la grasa, el aceite, y otros líquidos similares se llama un fuego de la clase B.

Class C fire An electrical fire.

Fuego de la Clase C Un fuego eléctrico se llama un fuego de la clase C.

Class D fire A unique type of fire because the material burning is a metal.

Fuego de la Clase D Un fuego de la clase D es un fuego inusual en el cual un metal se está quemando.

Clearance The space allowed between two parts, such as between a journal and a bearing.

Holgura El espacio permitido entre dos partes, tal como entre un muñon y un cojinete.

Clutch A device for connecting and disconnecting the engine from the transmission or for a similar purpose in other units.

Embrague Un dispositivo para conectar y desconectar el motor de la transmisión o para tal propósito en otros conjuntos.

Clutch packs A series of clutch discs and plates installed alternately in a housing to act as a driving or driven unit.

Conjuntos de embrague Una seria de discos y platos de embrague que se han instalado alternativamente en un cárter para funcionar como una unedad de propulsión o arrastre.

Clutch slippage A situation in which engine speed increases but increased torque is not transferred through to the driving wheels.

Resbalado del embrague Una situátion en el qual la velocidad del motor aumenta pero la torsión aumentada del motor no se transfere a las ruedas de marcha.

Clutch Volume Index (CVI) This represents the volume of fluid needed to compress a clutch pack or apply a brake band.

Índice del Volumen del Embrague Esto representa el volumen de líquido necesitado para comprimir un paquete del embrague o para aplicar una banda del freno.

Coefficient of friction The ratio of the force resisting motion between two surfaces in contact to the force holding the two surfaces in contact.

Coeficiente de la fricción La relación entre la fuerza que resiste al movimiento entre dos superficies que tocan y la fuerza que mantiene en contacto a éstas dos superficies.

Coil preload springs Coil springs are made of tempered steel rods formed into a spiral that resist compression; located in the pressure plate assembly.

Muelles de embrague Los muelles espirales son fabricadas de varillas de acero revenido y resisten la compresión; se ubican en el conjunto del plato opresor.

Coil spring A heavy wire-like steel coil used to support the vehicle weight while allowing for suspension motions. On FWD cars, the front coil springs are mounted around the MacPherson struts. On the rear suspension, they may be mounted to the rear axle, to trailing arms, or around rear struts.

Muelles de embrague Un resorte espiral hecho de acero en forma de alambre grueso que soporte el peso del vehículo mientras que permite a los movimientos de la suspensión. En los coches de FWD, los muelles de embrague delanteros se montan alrededor de los postes Macpherson. En la suspensión trasera, pueden montarse en el eje trasero, en los brasos traseros, o alrededor de los postes traseros.

Compound A mixture of two or more ingredients.

Compuesto Una combinación de dos ingredientes o más.

Concentric Two or more circles having a common center.

Concéntrico Dos círculos o más que comparten un centro común.

Constant-velocity joint A flexible coupling between two shafts that permits each shaft to maintain the same driving or driven speed regardless of operating angle, allowing for a smooth transfer of power. The constant-velocity joint (also called CV joint) consists of an inner and outer housing with balls in between, or a tripod and yoke assembly.

Junta de velocidad constante Un acoplador flexible entre dos flechas que permite que cada flecha mantenga la velocidad de propulsión o arrastre sin importar el ángulo de operación, efectuando una transferencia lisa del poder. La junta de velociadad constante (también llamado junta CV) consiste de un cárter interior e exterior entre los cuales se encuentran bolas, o de un conjunto de trípode y yugo.

Contraction A reduction in mass or dimension; the opposite of expansion.

Contración Una reducción en la masa o en la dimensión; el opuesto de expansión.

Control arm A suspension component that links the vehicle frame to the steering knuckle or axle housing and acts as a hinge to allow up-and-down wheel motions. The front control arms are attached to the frame with bushings and bolts and are connected to the steering knuckles with ball joints. The rear control arms attach to the frame with bushings and bolts and are welded or bolted to the rear axle or wheel hubs.

Brazo de mando Un componente de la suspención que une el armazón del vehículo al articulación de dirección o al cárter del eje y que se porta como una bisagra para permitir a los movimientos verticales de las ruedas. Los brazos de mando delanteros se conectan al armazón por medio de pernos y bujes y se conectan al articulación de dirección por medio de los articulaciones esféricos. Los brazos de mando traseros se conectan al armazón por medio de pernos y bujes y son soldados o empernados al eje trasero o a los cubos de la rueda.

Corrode To eat away gradually as if by gnawing, especially by chemical action.

Corroer Roído poco a poco, primariamente por acción químico.

Corrosion Chemical action, usually by an acid, that eats away (decomposes) a metal.

Corrosión Un acción químico, por lo regular un ácido, que corroe (descompone) un metal.

Cotter pin A type of fastener, made from soft steel in the form of a split pin, that can be inserted in a drilled hole. The split ends are spread to lock the pin in position.

Pasador de chaveta Un tipo de fijación, hecho de acero blando en forma de una chaveta que se puede insertar en un hueco tallado. Las extremidades partidas se despliegen para asegurar la posición de la chaveta.

Counterclockwise rotation Rotating in the opposite direction of the hands on a clock.

Rotación en sentido inverso Girando en el sentido opuesto de las agujas de un reloj.

Coupling A connecting means for transferring movement from one part to another; may be mechanical, hydraulic, or electrical.

Acoplador Un método de conección que transfere el movimiento de una parte a otra; puede ser mecánico, hidráulico, o eléctrico.

Coupling phase Point in torque converter operation in which the turbine speed is 90% of impeller speed and there is no longer any torque multiplication.

Fase del acoplador El punto de la operación del convertidor de la torsión en el cual la velocidad de la turbina es el 90% de la velocidad del impulsor y no queda ningún multiplicación de la torsión.

Cover plate A stamped steel cover bolted over the service access to the manual transmission.

Cubrejuntas Un cubierto de acero estampado que se emperna en la apertura de servicio de la transmisión manual.

Crocus cloth A very fine polishing paper designed to remove very little metal; therefore it is safe to use on critical surfaces.

Tela de óxido férrico Un papel muy fino para pulir. Fue diseñado para raspar muy poco del metal; por lo tanto, se suele emplear en las superficies críticas.

CRT The acronym for a cathode ray tube. This term normally refers to the display of a computer.

CRT La sigla en inglés por un tubo de rayos catódicos. Este termino suele referirse a la presentación en una computadora.

Cycle One set of changes in a signal that repeats itself several times.

Ciclo Un conjunto de cambios en una señal que se repite varias veces.

DLC The data link connector. This is the connector used to connect into a vehicle's computer system for the purpose of diagnostics. Prior to J1930, this was commonly referred to as the ALDL.

DLC El conectador de enlaces de datos. Este es el conectador que se usa en conectarse al sistema computerizado del vehículo con el propósito de efectuar los diagnósticos. Antes del J1930, esto solía referirse como el ALDL.

Default mode A mode of operation that allows for limited use of the transmission in the case of electronic failure.

Modo del uso limitado Un modo de operación que permite el uso limitado de la transmisión en el caso del incidente electrónico.

Deflection Bending or movement away from normal due to loading.

Desviación Curvación o movimiento fuera de lo normal debido a la carga.

Degree A unit of measurement equal to 1/360th of a circle.

Grado Una uneda de medida que iguala al 1/360 parte de un círculo.

Density Compactness; relative mass of matter in a given volume.

Densidad La firmeza; una cantidad relativa de la materia que ocupa a un volumen dado.

Detent A small depression in a shaft, rail, or rod into which a pawl or ball drops when the shaft, rail, or rod is moved. This provides a locking effect.

Detención Un pequeño hueco en una flecha, una barra o una varilla en el cual cae una bola o un linguete al moverse la flecha, la barra o la varilla. Esto provee un efecto de enclavamiento.

Detent mechanism A shifting control designed to hold the manual transmission in the gear range selected.

Aparato de detención Un control de desplazamiento diseñado a sujetar a la transmisión manual en la velocidad selecionada.

Diagnosis A systematic study of a machine or machine parts to determine the cause of improper performance or failure.

Diagnóstico Un estudio sistemático de una máquina o las partes de una máquina con el fín de determinar la causa de una falla o de un operación irregular.

Dial indicator A measuring instrument with the readings indicated on a dial rather than on a thimble as on a micrometer.

Indicador de carátula Un instrumento de medida cuyo indicador es en forma de muestra en contraste al casquillo de un micrómetro.

Differential A mechanism between drive axles that permits one wheel to run at a different speed than the other while turning.

Diferencial Un mecanismo entre dos semiejes que permite que una rueda gira a una velocidad distincta que la otra en una curva.

Differential action An operational situation in which one driving wheel rotates at a slower speed than the opposite driving wheel.

Acción del diferencial Una situación durante la operación en la cual una rueda propulsora gira con una velocidad más lenta que la rueda propulsora opuesta.

Differential case The metal unit that encases the differential side gears and pinion gears, and to which the ring gear is attached.

Caja de satélites La unedad metálica que encaja a los engranajes planetarios (laterales) y a los satélites del diferencial, y a la cual se conecta la corona.

Differential drive gear A large circular helical gear that is driven by the transaxle pinion gear and shaft and drives the differential assembly.

Corona Un engranaje helicoidal grande circular que es arrastrado por el piñon de la flecha de transmisión y la flecha y propela al conjunto del diferencial.

Differential housing Cast-iron assembly that houses the differential unit and the drive axles. Also called the rear axle housing.

Cárter del diferencial Una asamblea de acero vaciado que encaja a la unedad del diferencial y los semiejes. También se llama el cárter del eje trasero.

Differential pinion gears Small beveled gears located on the differential pinion shaft.

Satélites Engranajes pequeños biselados que se ubican en la flecha del piñon del diferencial.

Differential pinion shaft A short shaft locked to the differential case. This shaft supports the differential pinion gears.

Flecha del piñon del diferencial Una flecha corta clavada en la caja de satélites. Esta flecha sostiene a los satélites.

Differential ring gear A large circular hypoid-type gear enmeshed with the hypoid drive pinion gear.

Corona Un engranaje helicoidal grande circular endentado con el piñon de ataque hipoide.

Differential side gears The gears inside the differential case that are internally splined to the axle shafts, and which are driven by the differential pinion gears.

Planetarios (laterales) Los engranajes adentro de la caja de satélites que son acanalados a los semiejes desde el interior, y que se arrastran por los satélites.

Dipstick A metal rod used to measure the fluid in an engine or transmission.

Varilla de medida Una varilla de metal que se usa para medir el nivel de flúido en un motor o en una transmisión.

Direct drive One turn of the input driving member compared to one complete turn of the driven member, such as when there is direct engagement between the engine and driveshaft in which the engine crankshaft and the driveshaft turn at the same rpm.

Mando directo Una vuelta del miembro de ataque o propulsión que se compara a una vuelta completa del miembro de arrastre, tal como cuando hay un enganchamiento directo entre el motor y el árbol de transmisión en el qual el cigueñal y el árbol de transmisión giran al mismo rpm.

Disengage When the operator moves the clutch pedal toward the floor to disconnect the driven clutch disc from the driving flywheel and pressure plate assembly.

Desembragar Cuando el operador mueva el pedal de embrague hacia el piso para desconectar el disco de embrague del volante impulsor y del conjunto del plato opresor.

Distortion A warpage or change in form from the original shape.

Distorción Abarquillamiento o un cambio en la forma original.

DLC The data link connector. This is the connector used to connect into a vehicle's computer system for the purpose of diagnostics. Prior to J1930, this was commonly referred to as the ALDL.

DLC El conectador de enlaces de datos. Este es el conectador que se usa en conectarse al sistema computerizado del vehículo con el propósito de efectuar los diagnósticos. Antes del J1930, esto solía referirse como el ALDL.

DMM The acronym for a digital multimeter.

DMM La sigla en inglés por un multímeter digital.

Dowel A metal pin attached to one object which, when inserted into a hole in another object, ensures proper alignment.

Espiga Una clavija de metal que se fija a un objeto, que al insertarla en el hoyo de otro objeto, asegura una alineación correcta.

Dowel pin A pin inserted in matching holes in two parts to maintain those parts in fixed relation one to another.

Clavija de espiga Una clavija que se inserte en los hoyos alineados en dos partes para mantener ésos dos partes en una relación fijada el uno al otro.

Downshift To shift a transmission into a lower gear.

Cambio descendente Cambiar la velocidad de una transmision a una velocidad más baja.

Driveline torque Relates to rear-wheel driveline and is the transfer of torque between the transmission and the driving axle assembly.

Potencia de la flecha motríz Se relaciona a la flecha motríz de las ruedas traseras y transfere la potencia de la torsión entre la transmisión y el conjunto del eje trasero.

Driven gear The gear meshed directly with the driving gear to provide torque multiplication, reduction, or a change of direction.

Engranaje de arrastre El engranaje endentado directamente al engranaje de ataque para proporcionar la multiplicación, la reducción, o los cambios de dirección de la potencia.

Drive pinion gear One of the two main driving gears located within the transaxle or rear driving axle housing. Together the two gears multiply engine torque.

Engranaje de piñon de ataque Uno de dos engranajes de ataque principales que se ubican adentro de la flecha de transmisión o en el cárter del eje de propulsión. Los dos engranajes trabajan juntos para multiplicar la potencia.

Drive shaft An assembly of one or two universal joints connected to a shaft or tube; used to transmit power from the transmission to the differential. Also called the propeller shaft.

Árbol de mando Una asamblea de uno o dos uniones universales que se conectan a un árbol o un tubo; se usa para transferir la potencia desde la transmisión al diferencial. También se le refiere como el árbol de propulsión.

Dry friction The friction between two dry solids.
Fricción seca Fricción entre dos s-lidos secos.

DSO A common acronym for a Digital Storage Oscilloscope.
DSO Siglas comunes para un osciloscopio con memoria digital.

DTC The acronym for diagnostic trouble code.
DTC La sigla en inglés por un código diagnóstico de averías.

Dynamic In motion.
Dinámico En movimiento.

Eccentric One circle within another circle wherein both circles do not have the same center or a circle mounted off center. On FWD cars, front-end camber adjustments are accomplished by turning an eccentric cam bolt that mounts the strut to the steering knuckle.
Excéntrico Se dice de dos círculos, el uno dentro del otro, que no comparten el mismo centro o de un círculo ubicado descentrado. En los coches FWD, los ajustes de la inclinación se efectuan por medio de un perno excéntrico que fija el poste sobre el articulación de dirección.

EEPROM An electrically erasable programmable read only memory chip.
EEPROM Una placa de memoria, leer solamente, erasible y programable.

Elastomer Any rubber-like plastic or synthetic material used to make bellows, bushings, and seals.
Elastómero Cualquiera materia plást parecida al hule o una materia sintética que se utiliza para fabricar a los fuelles, los bujes y las juntas.

End clearance Distance between a set of gears and their cover, commonly measured on oil pumps.
Holgura del extremo La distancia entre un conjunto de engranajes y su placa de recubrimiento, suele medirse en las bombas de aceite.

Endplay The amount of axial or end-to-end movement in a shaft due to clearance in the bearings.
Juego de las extremidades La cantidad del movimiento axial o del movimiento de extremidad a extremidad en una flecha debido a la holgura que se deja en los cojinetes.

Engage When the vehicle operator moves the clutch pedal up from the floor, this engages the driving flywheel and pressure plate to rotate and drive the driven disc.
Accionar Cuando el operador del vehículo deja subir el pedal del embrague del piso, ésto acciona la volante de ataque y el plato opresor para impulsar al disco de arrastre.

Engagement chatter A shaking, shuddering action that takes place as the driven disc makes contact with the driving members. Chatter is caused by a rapid grip and slip action.
Chasquido de enganchamiento Un movimiento de sacudo o temblor que resulta cuando el disco de ataque viene en contacto con los miembros de propulsión. El chasquido se causa por una acción rápida de agarrar y deslizar.

Engine torque A turning or twisting action developed by the engine, measured in foot-pounds or kilogram meters.
Torsión del motor Una acción de girar o torcer que crea el motor, ésta se mide en librasópie o kilosómetros.

Essential tool kit A set of special tools designed for a particular model of car or truck.

Estuche de herramientas principales Un conjunto de herramientas especiales diseñadas para un modelo específico de coche o camión.

Etching A discoloration or removal of some material caused by corrosion or some other chemical reaction.
Grabado por ácido Una descoloración o remueva de una materia que se efectua por medio de la corrosión u otra reacción química.

Extension housing An aluminum or iron casting of various lengths that encloses the transmission output shaft and supporting bearings.
Cubierta de extensión Una pieza moldeada de aluminio o acero que puede ser de varias longitudes que encierre a la flecha de salida de la transmisión y a los cojinetes de soporte.

External gear A gear with teeth across the outside surface.
Engranaje exterior Un engranaje cuyos dientes estan en la superficie exterior.

Externally tabbed clutch plates Clutch plates that are designed with tabs around the outside periphery to fit into grooves in a housing or drum.
Placas de embrague de orejas externas Las placas de embrague que se diseñan de un modo para que las orejas periféricas de la superficie se acomoden en una ranura alrededor de un cárter o un tambor.

Extreme-pressure lubricant A special lubricant for use in hypoid-gear differentials; needed because of the heavy wiping loads imposed on the gear teeth.
Lubricante de presión extrema Un lubricante especial que se usa en las diferenciales de tipo engranaje hipóide; se requiere por la carga de transmisión de materia pesada que se imponen en los dientes del engranaje.

Face The front surface of an object.
Cara La superficie delantera de un objeto.

Fatigue The buildup of natural stress forces in a metal part that eventually causes it to break. Stress results from bending and loading the material.
Fatiga El incremento de tensiones y esfuerzos normales en una parte de metal que eventualmente causen una quebradura. Los esfuerzos resultan de la carga impuesta y el doblamiento de la materia.

Feeler gauge A metal strip or blade finished accurately with regard to thickness used for measuring the clearance between two parts; such gauges ordinarily come in a set of different blades graduated in thickness by increments of 0.001 inch.
Calibrador de laminillas Una lámina o hoja de metal que ha sido acabado precisamente con respecto a su espesor que se usa para medir la holgura entre dos partes; estas galgas típicamente vienen en un conjunto de varias espesores graduados desde el 0.001 de una pulgada.

Fillet The small, smooth curve on a bolt, where the shank flows into the bolt head.
Filete La curva pequeña, lisa en un perno, adonde la asta fluye en la cabeza del perno se llama un filete.

Final drive ratio The ratio between the drive pinion and ring gear.
Relación del mando final La relación entre el piñon de ataque y la corona.

Fit The contact between two machined surfaces.
Ajuste El contacto entre dos superficies maquinadas.

Fixed-type constant-velocity joint A joint that cannot telescope or plunge to compensate for suspension travel. Fixed joints are always found on the outer ends of the drive shafts of FWD cars. A fixed joint may be of either Rzeppa or tripod type.

Junta tipo fijo de velocidad constante Una junta que no tiene la capacidad de los movimientos telescópicos o repentinos que sirven para compensar en los viajes de suspensión. Las juntas fijas siempre se ubican en las extremidades exteriores de los árboles de mando en los coches de FWD. Una junta tipo fijo puede ser de un tipo Rzeppa o de trípode.

Flange   A projecting rim or collar on an object for keeping it in place.

Reborde Una orilla o un collar sobresaliente de un objeto cuyo función es de mantenerlo en lugar.

Flat rate   A pay system in which a technician is paid for the amount of work he or she does. Each job has a flat rate time.

Tarifa fija Un sistema de la paga en el cual un técnico es pagado para la cantidad de trabajo él o ella lo hace. Cada trabajo tiene un rato de la tarifa fija.

Flexplate   A lightweight flywheel used only on engines equipped with an automatic transmission. The flexplate is equipped with a starter ring gear around its outside diameter and also serves as the attachment point for the torque converter.

Placa articulada Un volante ligera que se usa solamente en los motores que se equipan con una transmisión automática. El diámetro exterior de la placa articulada viene equipado con un anillo de engranajes para arrancar y también sirve como punto de conección del convertidor de la torsión.

Fluid coupling   A device in the powertrain consisting of two rotating members; transmits power from the engine, through a fluid, to the transmission.

Acoplamiento de fluido Un dispositivo en el tren de potencia que consiste de dos miembros rotativos; transmite la potencia del motor, por medio de un fluido, a la transmisión.

Fluid drive   A drive in which there is no mechanical connection between the input and output shafts, and power is transmitted by moving oil.

Dirección fluido Una dirección en la cual no hay conecciones mecánicas entre las flechas de entrada o salida, y la potencia se transmite por medio del aceite en movimiento.

Flywheel   A heavy metal wheel that is attached to the crankshaft and rotates with it; helps smooth out the power surges from the engine power strokes; also serves as part of the clutch and engine-cranking system.

Volante Una rueda pesada de metal que se fija al cigüeñal y gira con ésta; nivela a los sacudos que provienen de la carrera de fuerza del motor; también sirve como parte del embrague y del sistema de arranque.

Flywheel ring gear   A gear fitted around the flywheel that is engaged by teeth on the starting motor drive to crank the engine.

Engranaje anular del volante Un engranaje, colocado alrededor del volante que se acciona por los dientes en el propulsor del motor de arranque y arranca al motor.

Foot-pound (ft.-lb.)   A measure of the amount of energy or work required to lift 1 pound a distance of 1 foot.

Pie libra Una medida de la cantidad de energía o fuerza que requiere mover una libra a una distancia de un pie.

Force   Any push or pull exerted on an object; measured in pounds and ounces, or in newtons (N) in the metric system.

Fuerza Cualquier acción empujado o jalado que se efectua en un objeto; se mide en pies y onzas, o en newtones (N) en el sistema métrico.

Four-wheel drive   On a vehicle, driving axles at both front and rear, so that all four wheels can be driven.

Tracción a cuatro ruedas En un vehículo, se trata de los ejes de dirección fronteras y traseras, para que cada uno de las ruedas puede impulsar.

Frame   The main understructure of the vehicle to which everything else is attached. Most FWD cars have only a subframe for the front suspension and drivetrain. The body serves as the frame for the rear suspension.

Armazón La estructura principal del vehículo al cual todo se conecta. La mayoría de los coches FWD sólo tiene un bastidor auxiliar para la suspensión delantera y el tren de propulsión. El carrocería del coche sirve de chassis par la suspensión trasera.

Freewheel   To turn freely and not transmit power.

Volante libre Da vueltas libremente sin transferir la potencia.

Freewheeling clutch   A mechanical device that will engage the driving member to impart motion to a driven member in one direction but not the other. Also known as an "overrunning clutch."

Embrague de volante libre Un dispositivo mecánico que acciona el miembro de tracción y da movimiento al miembro de tracción en una dirección pero no en la otra. También se conoce bajo el nombre de un "embrague de sobremarcha."

Friction   The resistance to motion between two bodies in contact with each other.

Fricción La resistencia al movimiento entre dos cuerpos que estan en contacto.

Friction bearing   A bearing in which there is sliding contact between the moving surfaces. Sleeve bearings, such as those used in connecting rods, are friction bearings.

Rodamientos de fricción Un cojinete en el cual hay un contacto deslizante entre las superficies en movimiento. Los rodamientos de manguitos, como los que se usan en las bielas, son rodamientos de fricción.

Friction disc   In the clutch, a flat disc, faced on both sides with frictional material and splined to the clutch shaft. It is positioned between the clutch pressure plate and the engine flywheel. Also called the clutch disc or driven disc.

Disco de fricción En el embrague, un disco plano al cual se ha cubierto ambos lados con una materia de fricción y que ha sido estriado a la flecha del embrague. Se posiciona entre el plato opresor del embrague y el volante del motor. También se llama el disco del embrague o el disco de arrastre.

Friction facings   A hard-molded or woven asbestos or paper material that is riveted or bonded to the clutch driven disc.

Superficie de fricción Un recubrimiento remachado o aglomerado al disco de arrastre del embrague que puede ser hecho del amianto moldeado o tejido o de una materia de papel.

Front pump   Pump located at the front of the transmission. It is driven by the engine through two dogs on the torque converter housing. It supplies fluid whenever the engine is running.

Bomba delantera Una bomba ubicado en la parte delantera de la transmisión. Se arrastre por el motor al través de dos álabes en el cárter del convertidor de la torsión. Provee el fluido mientras que funciona el motor.

Front-wheel drive (FWD) The vehicle has all drivetrain components located at the front.

Tracción de las ruedas delanteras (FWD) El vehículo tiene todos los componentes del tren de propulsión en la parte delantera.

FWD Abbreviation for front-wheel drive.

FWD Abreviación de tracción de las ruedas delanteras.

Galling Wear caused by metal-to-metal contact in the absence of adequate lubrication. Metal is transferred from one surface to the other, leaving behind a pitted or scaled appearance.

Desgaste por fricción El desgaste causado por el contacto de metal a metal en la ausencia de lubricación adecuada. El metal se transfere de una superficie a la otra, causando una aparencia agujerado o con depósitos.

Gasket A layer of material, usually made of cork, paper, plastic, composition, or metal, or a combination of these, placed between two parts to make a tight seal.

Empaque Una capa de una materia, normalmente hecho del corcho, del papel, del plástico, de la materia compuesta o del metal, o de cualquier combinación de éstos, que se coloca entre dos partes para formar un sello impermeable.

Gasket cement A liquid adhesive material or sealer used to install gaskets.

Mastique para empaques Una substancia líquida adhesiva, o una substancia impermeable, que se usa para instalar a los empaques.

Gear A wheel with external or internal teeth that serves to transmit or change motion.

Engranaje Una rueda que tiene dientes interiores o exteriores que sirve para transferir o cambiar el movimiento.

Gear lubricant A type of grease or oil blended especially to lubricate gears.

Lubricante para engranaje Un tipo de grasa o aceite que ha sido mezclado específicamente para la lubricación de los engranajes.

Gear ratio The number of revolutions of a driving gear required to turn a driven gear through one complete revolution. For a pair of gears, the ratio is found by dividing the number of teeth on the driven gear by the number of teeth on the driving gear.

Relación de los engranajes El número de las revoluciones requeridas del engranaje de propulsión para dar una vuelta completa al engranaje arrastrado. En una pareja de engranajes, la relación se calcula al dividir el número de los dientes en el engranaje de arrastre por el número de los dientes en el engranaje de propulsión.

Gear reduction When a small gear drives a large gear, there is an output speed reduction and a torque increase that results in a gear reduction.

Velocidad descendente Cuando un engranaje pequeño impulsa a un engranaje grande, hay una reducción en la velocidad de salida y un incremento en la torsión que resulta en una cambio descendente de los velocidades.

Gearshift A linkage-type mechanism by which the gears in an automobile transmission are engaged and disengaged.

Varillaje de cambios Un mecanismo tipo eslabón que acciona y desembraga a los engranajes de la transmisión.

Gear whine A high-pitched sound developed by some types of meshing gears.

Ruido del engranaje Un sonido agudo que proviene de algunos tipos de engranajes endentados.

Glitches Abnormal, slight movements of a waveform on a lab scope. These can be caused by circuit problems or noise in the circuit.

Irregularidades espontáneas ("glitch") Los movimientos pequeños e abnormales en una onda en una pantalla de laboratorio. Estos pueden ser causados por los problemas de los circuitos o el ruido dentro del circuito.

Governor pressure The transmission's hydraulic pressure which is directly related to output shaft speed. It is used to control shift points.

Regulador de presión La presión hidráulica de una transmisión se relaciona directamente a la velocidad de la flecha de salida. Se usa para controlar los puntos de cambios de velocidad.

Governor valve A device used to sense vehicle speed. The governor valve is attached to the output shaft.

Válvula reguladora Un dispositivo que se usa para determinar la velocidad de un vehículo. La válvula reguladora se monta en la flecha de salida.

Grade marks Marks on fasteners that indicate strength.

Marcas del grado Las marcas en los sujetadores que indican fuerza se llaman las marcas del grado.

Hazardous waste Any used material or any by-product that can be classified as potentially hazardous to one's health and/or the environment.

Desechos Peligrosos Cualquier material usado, o cualquier subproducto, que se pueda clasificar como potencialmente dañoso al medio ambiente o parjudicial para salud.

Heat exchanger A heat exchanger may also be called an intercooler and is used to transfer heat from one object to another. Heat is exchanged because of a law of nature in that the heat of an object will always attempt to heat a cooler object.

Un cambiador de calor Un cambiador de calor se puede también llamar un "intercooler" y se utiliza transferir calor a partir de un objeto a otro. El calor se intercambia debido a una ley de la naturaleza en que el calor de un objeto procurará siempre calentar un objeto más fresco.

Hub The center part of a wheel, to which the wheel is attached.

Cubo La parte central de una rueda, a la cual se monta la rueda.

Hydraulic press A piece of shop equipment that develops a heavy force by use of a hydraulic piston-and-jack assembly.

Prensa hidráulica Una herramienta del taller que provee una fuerza grande por medio de una asamblea de gato con un pistón hidráulico.

Hydraulic pressure Pressure exerted through the medium of a liquid.

Presión hidráulica La presión esforzada por medio de un líquido.

ID Inside diameter.

DI Diámetro interior.

Idle Engine speed when the accelerator pedal is fully released and there is no load on the engine.

Marcha lenta La velocidad del motor cuando el pedal accelerador esta completamente desembragada y no hay carga en el motor.

Impedance The operating resistance of an electrical device.

Impedancia La resistencia operativa de un dispositivo eléctrico.

Impeller The pump or driving member in a torque converter.

Impulsor La bomba o el miembro impulsor en un convertidor de torsión.

Increments Series of regular additions from small to large.

Incrementos Una serie de incrementos regulares que va de pequeño a grande.

Index To orient two parts by marking them. During reassembly the parts are arranged so the index marks are next to each other. Used to preserve the orientation between balanced parts.

Índice Orientar a dos partes marcándolas. Al montarlas, las partes se colocan para que las marcas de índice estén alinieadas. Se usan los índices para preservar la orientación de las partes balanceadas.

Input shaft The shaft carrying the driving gear by which the power is applied, as to the transmission.

Flecha de entrada La flecha que porta el engranaje propulsor por el cual se aplica la potencia, como a la transmisión.

Inspection cover A removable cover that permits entrance for inspection and service work.

Cubierta de inspección Una cubierta desmontable que permite a la entrada para inspeccionar y mantenimiento.

Integral Built into, as part of the whole.

Integral Incorporado, una parte de la totalidad.

Internal gear A gear with teeth pointing inward, toward the hollow center of the gear.

Engranaje internal Un engranaje cuyos dientes apuntan hacia el interior, al hueco central del engranaje.

International System (SI) The International System of weights and measures. The system is normally called the metric system.

Sistema Internacional (SI) El sistema internacional de pesos y de medidas. Este sistema normalmente se llama la sistema métrico.

Jam nut A second nut tightened against a primary nut to prevent it from working loose. Used on inner and outer tie-rod adjustment nuts and on many pinion-bearing adjustment nuts.

Contra tuerca Una tuerca secundaria que se aprieta contra una tuerca primaria para prevenir que ésta se afloja. Se emplean en las tuercas de ajustes interiores e exteriores para las barras de acoplamiento y también en muchas de las tuercas de ajuste de portapiñones.

Journal A bearing with a hole in it for a shaft.

Manga de flecha Un cojinete que tiene un hoyo para una flecha.

Key A small block inserted between the shaft and hub to prevent circumferential movement.

Chaveta Un tope pequeño que se meta entre la flecha y el cubo para prevenir un movimiento circunferencial.

Keyway A groove or slot cut to permit the insertion of a key.

Ranura de chaveta Un corte de ranura o mortaja que permite insertar una chaveta.

Knock A heavy metallic sound usually caused by a loose or worn bearing.

Golpe Un sonido metálico fuerte que suele ser causado por un cojinete suelto o gastado.

Knurl To indent or roughen a finished surface.

Moletear Indentar o desbastar a una superficie acabada.

Lapping The process of fitting one surface to another by rubbing them together with an abrasive material between the two surfaces.

Pulido El proceso de ajustar a una superficie con otra por frotarlas juntas con una materia abrasiva entre las dos superficies.

Lash The amount of free motion in a geartrain, between gears, or in a mechanical assembly, such as the lash in a valve train.

Juego La cantidad del movimiento libre en un tren de engranajes, entre los engranajes o en una asamblea mecánica, tal como el juego en un tren de vávulas.

LED The common acronym for a light-emitting diode.

LED La sigla común en inglés por un diódo emisor de luz.

Linkage Any series of rods, yokes, levers, and so on used to transmit motion from one unit to another.

Biela Cualquiera serie de barras, yugos, palancas, y todo lo demás, que se usa para transferir los movimientos de una unedad a otra.

Locking ring A type of sealing ring that has ends that meet or lock together during installation. There is no gap between the ends when the ring is installed.

Anillo de enclavamiento Un tipo de anillo obturador que tiene las extremidades que se tocan o se enclavan durante la instalación. No hay holgura entre las extremidades cuando se ha instalado el anillo.

Locknut A second nut turned down on a holding nut to prevent loosening.

Contra tuerca Una tuerca segundaria apretada contra una tuerca de sostén para prevenir que ésta se afloja.

Lock pin Used in some ball sockets (inner tie-rod end) to keep the connecting nuts from working loose. Also used on some lower ball joints to hold the tapered stud in the steering knuckle.

Clavija de cerrojo Se usan en algunas rótulas (las extremidades interiores de la barra de acoplamiento) para prevenir que se aflojan las tuercas de conexión. También se emplean en algunas juntas esféricas inferiores para retener al perno cónico en la articulación de dirección.

Lockplates Metal tabs bent around nuts or bolt heads.

Placa de cerrojo Chavetas de metal que se doblan alrededor de las tuercas o las cabezas de los pernos.

Lockwasher A type of washer which, when placed under the head of a bolt or nut, prevents the bolt or nut from working loose.

Arandela de freno Un tipo de arandela que, al colocarse bajo la cabeza de un perno, previene que el perno o la tuerca se aflojan.

Low speed The gearing that produces the highest torque and lowest speed of the wheels.

Velocidad baja La velocidad que produce la torsión más alta y la velocidad más baja a las ruedas.

Lubricant Any material, usually a petroleum product such as grease or oil, that is placed between two moving parts to reduce friction.

Lubricante Cualquier substancia, normalmente un producto de petróleo como la grasa o el aciete, que se coloca entre dos partes en movimiento para reducir la fricción.

Mainline pressure The hydraulic pressure that operates apply devices and is the source of all other pressures in an automatic transmission. It is developed by pump pressure and regulated by the pressure regulator.

Línea de presión La presión hidráulica que opera a los dispositivos de aplicación y es el orígen de todas las presiones en la transmisión automática. Proviene de la bomba de presión y es regulada por el regulador de presión.

Main oil pressure regulator valve Regulates the line pressure in a transmission.

Válvula reguladora de la linea de presión Regula la presión en la linea de una transmisión.

Manifold absolute pressure (MAP) Sensor The sensor that measures changes in the intake manifold pressure that result from changes in engine load and speed.

Sensor de presión absoluta del colector de escape Un sensor que mide los cambios en la presión del colector de escape que resultan de cambios en carga y velocidad del motor.

Manual control valve A valve used to manually select the operating mode of the transmission. It is moved by the gearshift linkage.

Válvula de control manual Una válvula que se usa para escoger a una velocidad de la transmisión por mano. Se mueva por la biela de velocidades.

Meshing The mating, or engaging, of the teeth of two gears.

Engrane Embragar o endentar a los dientes de dos engranajes.

Meter 1/10,000,000 of the distance from the North Pole to the Equator, or 39.37 inches.

Metro Un 1/10,000,000 de la distancia del polo del norte al ecuador.

Micrometer A precision measuring device used to measure small bores, diameters, and thicknesses. Also called a mike.

Micrómetro Un dispositivo de medida precisa que se emplea a medir a los taladros pequeños y a los espesores. También se llama un mike (mayk).

MIL The malfunction indicator lamp for a computer control system. Prior to J1930, the MIL was commonly called a Check Engine or Service Engine Soon lamp.

MIL La lámpara de indicación de averías para un sistema de control de computadora. Antes del J1930, la MIL se llamaba la lámpara de Revise Motor o Servicia el Motor Pronto.

Millimeter (mm) The third and most commonly used division of a meter.

Milímetro (mm) Un milímetro es el tercero y la división lo más comúnmente posible usada de un metro, una medida lineal en la sistema métrico, conteniendo la milésima pieza de un metro; igual al 0.03934 de una pulgada.

Mineral spirits An alcohol-based liquid that leaves no surface residue after it dries.

Alcoholes mineral Un líquido a base de alcohol que no sale de ningún residuo superficial después de que se seque.

Misalignment When bearings are not on the same centerline.

Desalineamineto Cuando los cojinetes no comparten la misma linea central.

Modulator A vacuum diaphragm device connected to a source of engine vacuum. It provides an engine load signal to the transmission.

Modulador Un dispositivo de diafragma de vacío que se conecta a un orígen de vacío en el motor. Provee un señal de carga del motor a la transmisión.

Mounts Made of rubber to insulate vibrations and noise while they support a powertrain part, such as engine or transmission mounts.

Monturas Hecho de hule para insular a las vibraciones y a los ruidos mientras que sujetan una parte del tren de propulsión, tal como las monturas del motor o las monturas de la transmisión.

Multimeter A tool that combines the voltmeter, ohmmeter, and ammeter together in a diagnostic instrument.

Multímetro Una herramienta que combina el voltímetro, el ohmímetro, y el amperímetro junta en uno instrumento de diagnóstico.

Multiple disc A clutch with a number of driving and driven discs as compared to a single plate clutch.

Discos múltiples Un embrague que tiene varios discos de propulsión o de arraste al contraste con un embrague de un sólo plato.

Needle bearing An antifriction bearing using a great number of long, small-diameter rollers. Also known as a quill bearing.

Rodamiento de agujas Un rodamiento (cojinete) antifricativo que emplea un gran cantidad de rodillos largos y de diámetro muy pequeños.

Needle deflection Distance of travel from zero of the needle on a dial gauge.

Desviación de la aguja La distancia del cero que viaja una aguja de un indicador.

Neoprene A synthetic rubber that is not affected by the various chemicals that are harmful to natural rubber.

Neoprene Un hule sintético que no se afecta por los varios productos químicos que pueden dañar al hule natural.

Neutral In a transmission, the setting in which all gears are disengaged and the output shaft is disconnected from the drive wheels.

Neutral En una transmisión, la velocidad en la cual todos los engranajes estan desembragados y el árbol de salida esta desconectada de las ruedas de propulsión.

Neutral-start switch A switch wired into the ignition switch to prevent engine cranking unless the transmission shift lever is in neutral or the clutch pedal is depressed.

Interruptor de arranque en neutral Un interruptor eléctrico instalado en el interruptor de encendido que previene el arranque del motor al menos de que la palanca de cambio de velocidad esté en una posición neutral o que se pisa en el embrague.

Newton-meter (Nm) Metric measurement of torque or twisting force.

Metro newton (Nm) Una medida métrica de la fuerza de torsión.

Nominal shim A shim with a designated thickness.

Laminilla fina Una cuña de un espesor especificado.

Nonhardening A gasket sealer that never hardens.

Sinfragua Un cemento de empaque que no endurece.

Nut A removable fastener used with a bolt to lock pieces together; made by threading a hole through the center of a piece of metal that has been shaped to a standard size.

Tuerca Un retén removable que se usa con un perno o tuerca para unir a dos piezas; se fabrica al filetear un hoyo taladrado en un pedazo de metal que se ha formado a un tamaño especificado.

Occupational Safety and Health Administration (OSHA) The government agency charged with ensuring safe work environments for all workers.

Administraion de la seguridad y de la salud ocupacionales (OSHA) La agencia de estatal cargada con asegurar los ambientes seguros del trabajo para todos los trabajadores.

OD Outside diameter.

DE Diámetro exterior.

Ohmmeter A test meter used to measure electrical resistance and/or continuity.

Contador de ohmios Un aparato que mide resistencia y/o continuidad eléctrica.

Oil seal A seal placed around a rotating shaft or other moving part to prevent leakage of oil.

Empaque de aciete Un empaque que se coloca alrededor de una flecha giratoria para prevenir el goteo de aceite.

One-way clutch See Sprag clutch.

Embrague de una via Vea Sprag clutch.

Open An open is a break in the circuit that stops current flow.

Abierto Una rotura en el circuito que para el flujo de la corriente eléctrica.

O-ring A type of sealing ring, usually made of rubber or a rubber-like material. In use, the O-ring is compressed into a groove to provide the sealing action.

Anillo en O   Un tipo de sello anular, suele ser hecho de hule o de una materia parecida al hule. Al usarse, el anillo en O se comprime en una ranura para proveer un sello.

Oscillate   To swing back and forth like a pendulum.

Oscilar   Moverse alternativamente en dos sentidos contrarios como un péndulo.

Outer bearing race   The outer part of a bearing assembly on which the balls or rollers rotate.

Pista exterior de un cojinete   La parte exterior de una asamblea de cojinetes en la cual ruedan las bolas o los rodillos.

Out-of-round   Wear of a round hole or shaft which, when viewed from an end, will appear egg-shaped.

Defecto de circularidad   Desgaste de un taladro o de una flecha circular, que al verse de una extremidad, tendrá una forma asimétrica, como la de un huevo.

Output shaft   The shaft or gear that delivers the power from a device, such as a transmission.

Flecha de salida   La flecha o la velocidad que transmite la potencia de un dispositivo, tal como una transmisión.

Overall ratio   The product of the transmission gear ratio multiplied by the final drive or rear axle ratio.

Relación global   El producto de multiplicar la relación de los engranajes de la transmisión por la relación del impulso final o por la relación del eje trasero.

Overdrive   Any arrangement of gearing that produces more revolutions of the driven shaft than of the driving shaft.

Sobremultiplicación   Un arreglo de los engranajes que produce más revoluciones de la flecha de arrastre que los de la flecha de propulsión.

Overdrive ratio   Identified by the decimal point indicating less than one driving input revolution compared to one output revolution of a shaft.

Relación del sobremultiplicación   Se identifica por el punto decimal que indica menos de una revolución del motor comparado a una revolución de una flecha de salida.

Overrun coupling   A freewheeling device to permit rotation in one direction but not in the other.

Acoplamiento de sobremarcha   Un dispositivo de marcha de rueda libre que permite las giraciones en una dirección, pero no en la otra dirección.

Overrunning clutch   A device consisting of a shaft or housing linked together by rollers or sprags operating between movable and fixed races. As the shaft rotates, the rollers or sprags jam between the movable and fixed races. This jamming action locks together the shaft and housing. If the fixed race should be driven at a speed greater than the movable race, the rollers or sprags will disconnect the shaft.

Embrague de sobremarcha   Un dispositivo que consiste en una flecha o un cárter eslabonados por medio de rodillos o palancas de detención que operan entre pistas fijas y movibles. Al girar la flecha, los rodillos o palancas de detención se aprietan entre las pistas fijas y movibles. Este acción de apretarse enclava el cárter con la flecha. Si la pista fija se arrastra en una velocidad más alta que la pista movible, los rodillos o palancas de detención desconectarán a la flecha.

Oxidation   Burning or combustion; the combining of a material with oxygen. Rusting is slow oxidation, and combustion is rapid oxidation.

Oxidación   Quemando o la combustión; la combinación de una materia con el oxígeno. El orín es una oxidación lenta, la combustión es la oxidación rápida.

Pascal's Law   The law of fluid motion.

Ley de pascal   La ley del movimiento del fluido.

Parallel   The quality of two items being the same distance from each other at all points; usually applied to lines and, in automotive work, to machined surfaces.

Paralelo   La calidad de dos artículos que mantienen la misma distancia el uno del otro en cada punto; suele aplicarse a las líneas y, en el trabajo automotívo, a las superficies acabadas a máquina.

Pawl   A lever that pivots on a shaft. When lifted, it swings freely and when lowered, it locates in a detent or notch to hold a mechanism stationary.

Trinquete   Una palanca que gira en una flecha. Levantado, mueve sín restricción, bajado, se coloca en una endentación o una muesca para mantener sín movimiento a un mecanismo.

PCM   The powertrain control module of a computer control system. Prior to J1930, the PCM was commonly called a ECA, ECM, or one of many acronyms used by the various manufacturers.

PCM   El módulo de control del sistema de transmisión de fuerza de un sistema de control de una computadora. Antes d3l J1930, el PCM se llamaba un ECA, un ECM, o una de varias siglas usadas por los varios fabricantes.

Pitch   The number of threads per inch on any threaded part.

Paso   El número de filetes por pulgada de cualquier parte fileteada.

Pivot   A pin or shaft upon which another part rests or turns.

Pivote   Una chaveta o una flecha que soporta a otra parte o sirve como un punto para girar.

Planetary gearset   A system of gearing that is modeled after the solar system. A pinion is surrounded by an internal ring gear and planet gears are in mesh between the ring gear and pinion around which all revolve.

Conjunto de engranajes planetarios   Un sistema de engranaje cuyo patrón es el sistema solar. Un engranaje propulsor (la corona interior) rodea al piñon de ataque y los engranajes satélites y planetas se endentan entre la corona y el piñon alrededor del cual todo gira.

Planet carrier   The carrier or bracket in a planetary gear system that contains the shafts upon which the pinions or planet gears turn.

Perno de arrastre planetario   El soporte o la abrazadera que contiene las flechas en las cuales giran los engranajes planetarios o los piñones.

Planet gears   The gears in a planetary gearset that connect the sun gear to the ring gear.

Engranajes planetarios   Los engranajes en un conjunto de engranajes planetario que connectan al engranaje propulsor interior (el engranaje sol) con la corona.

Planet pinions   In a planetary gear system, the gears that mesh with, and revolve about, the sun gear; they also mesh with the ring gear.

Piñones planetarios   En un sistema de engranajes planetarios, los engranajes que se endentan con, y giran alrededor, el engranaje propulsor (sol); también se endentan con la corona.

Plug   Anything that will fit into an opening to stop fluid or airflow.

Tapón   Cualquier cosa que se ajuste en una apertura para prevenir el goteo o el escape de un corriente del aire.

Pneumatic tools   Power tools that rely on compressed air for power.

Herramientas neumáticas   Las herramientas de motor cuyo energía proviene del aire comprimido.

Porosity   A statement of how porous or permeable to liquids a material is.

Porosidad   Una expresión de lo poroso o permeable a los líquidos es una materia.

Powertrain   The mechanisms that carry the power from the engine crankshaft to the drive wheels; these include the clutch, transmission, driveline, differential, and axles.

Tren impulsor   Los mecanismos que transferen la potencia desde el cigüeñal del motor a las ruedas de propulsión; éstos incluyen el embrague, la transmisión, la flecha motríz, el diferencial y los semiejes.

Preload   A load applied to a part during assembly so as to maintain critical tolerances when the operating load is applied later.

Carga previa   Una carga aplicada a una parte durante la asamblea para asegurar sus tolerancias críticas antes de que se le aplica la carga de la operación.

Press-fit   Forcing a part into an opening that is slightly smaller than the part itself to make a solid fit.

Ajustamiento a presión   Forzar a una parte en una apertura que es de un tamaño más pequeño de la parte para asegurar un ajustamiento sólido.

Pressure   Force per unit area, or force divided by area. Usually measured in pounds per square inch (psi) or in kilopascals (kPa) in the metric system.

Presión   La fuerza por unedad de una area, o la fuerza divida por la area. Suele medirse en libras por pulgada cuadrada (lb/pulg2) o en kilopascales (kPa) en el sistema métrico.

Pressure plate   That part of the clutch that exerts force against the friction disc; it is mounted on and rotates with the flywheel.

Plato opresor   Una parte del embrague que aplica la fuerza en el disco de fricción; se monta sobre el volante, y gira con éste.

Propeller shaft   See Drive shaft.

Flecha de Propulsion   Vea Flecha motríz.

Prussian blue   A blue pigment; in solution, useful in determining the area of contact between two surfaces.

Azul de Prusia   Un pigmento azul; en forma líquida, ayuda en determinar la area de contacto entre dos superficies.

PSI   Abbreviation for pounds per square inch, a measurement of pressure.

Lb/pulg2   Una abreviación de libras por pulgada cuadrada, una medida de la presión.

Puller   Generally, a shop tool used to separate two closely fitted parts without damage. Often contains one screw, or several screws that can be turned to apply a gradual force.

Extractor   Generalmente, una herramienta del taller que sirve para separar a dos partes apretadas sin incurrir daños. Suele tener una tuerca o varias tuercas, que se pueden girar para aplicar la fuerza gradualmente.

Pulsation   To move or beat with rhythmic impulses.

Pulsación   Moverse o batir con impulsos rítmicos.

Race   A channel in the inner or outer ring of an antifriction bearing in which the balls or rollers roll.

Pista   Un canal en el anillo interior o exterior de un cojinete antifricción en el cual ruedan las bolas o los rodillos.

Radial   The direction moving straight out from the center of a circle. Perpendicular to the shaft or bearing bore.

Radial   La dirección al moverse directamente del centro de un círculo. Perpendicular a la flecha o al taladro del cojinete.

Radial clearance   Clearance within the bearing and between balls and races perpendicular to the shaft. Also called radial displacement.

Holgura radial   La holgura en un cojinete entre las bolas y las pistas que son perpendiculares a la flecha. También se llama un desplazamiento radial.

Radial load   A force perpendicular to the axis of rotation.

Carga radial   Una fuerza perpendicular al centro de rotación.

Radio frequency interference (RFI)   An unwanted voltage signal that rides on another voltage signal.

Interferencia de la radiofrecuencia   Una señal indeseada del voltaje que interfiere con otra señal del voltaje.

Ratio   The relation or proportion that one number bears to another.

Relación   La correlación o proporción de un número con respeto a otro.

Rear-wheel drive   A term associated with a vehicle in which the engine is mounted at the front and the driving axle and driving wheels are at the rear of the vehicle.

Tracción trasera   Un término que se asocia con un vehículo en el cual el motor se ubica en la parte delantera y el eje propulsor y las ruedas propulsores se encuentran en la parte trasera del vehículo.

Relief valve   A valve used to protect against excessive pressure in the case of a malfunctioning pressure regulator.

Válvula de seguridad   Una válvula que se usa para guardar contra una presión excesiva en caso de que malfulciona el regulador de presión.

Retaining ring   A removable fastener used as a shoulder to retain and position a round bearing in a hole.

Anillo de retén   Un seguro removible que sirve de collarín para sujetar y posicionar a un cojinete en un agujero.

RFI   Radio frequency interference. This acronym is used to describe a type of electrical interference that may affect voltage signals in a computerized system.

RFI   Interferencia de frecuencias de radio. Esta sigla se usa en describir un tipo de interferencia eléctrica que puede afectar los señales de voltaje en un sistema computerizado.

Right-to-Know Laws   Laws requiring employers to provide employees with a safe work place as it relates to hazardous materials, and information about any and all hazards the employees face while performing their job.

Prerrogativa a saber leyes   Los leyes que requieren a patrones proveer de empleados un lugar de trabajo seguro, relacionada con los materiales peligrosos y la información sobre cualesquiera y todos todos los peligros que los empleados pudieron encontrar mientras que realizaban su trabajo.

Rivet   A headed pin used for uniting two or more pieces by passing the shank through a hole in each piece and securing it by forming a head on the opposite end.

Remache   Una clavija con cabeza que sirve para unir a dos piezas o más al pasar el vástago por un hoyo en cada pieza y asegurarlo por formar una cabeza en el extremo opuesto.

Roller bearing   An inner and outer race upon which hardened steel rollers operate.

Cojinete de rodillos   Una pista interior y exterior en la cual operan los rodillos hecho de acero endurecido.

Rollers   Round steel bearings that can be used as the locking element in an overrunning clutch or as the rolling element in an antifriction bearing.

Rodillos   Articulaciones redondos de acero que pueden servir como un elemento de enclavamiento en un embrague de sobremarcha o como el elemento que rueda en un cojinete antifricción.

Rotary flow A fluid force generated in the torque converter that is related to vortex flow. The vortex flow leaving the impeller is not only flowing out of the impeller at high speed but is also rotating faster than the turbine. The rotating fluid striking the slower turning turbine exerts a force against the turbine that is defined as rotary flow.

Flujo rotativo Una fuerza fluida producida en el convertidor de torsión que se relaciona al flujo torbellino. El flujo torbellino saliendo del rotor no sólo viaja en una alta velocidad sino también gira más rápidamente que el turbino. El fluido rotativo chocando contra el turbino que gira más lentamente, impone una fuerza contra el turbino que se define como flujo rotativo.

RPM Abbreviation for revolutions per minute, a measure of rotational speed.

RPM Abreviación de revoluciones por minuto, una medida de la velocidad rotativa.

RTV sealer Room-temperature vulcanizing gasket material that cures at room temperature; a plastic paste squeezed from a tube to form a gasket of any shape.

Sellador RTV Una materia vulcanizante de empaque que cura en temperaturas del ambiente; una pasta plástica exprimida de un tubo para formar un empaque de cualquiera forma.

Runout Deviation of the specified normal travel of an object. The amount of deviation or wobble a shaft or wheel has as it rotates. Runout is measured with a dial indicator.

Corrimiento Una desviación de la carrera normal e especificada de un objeto. La cantidad de desviación o vacilación de una flecha o una rueda mientras que gira. El corrimiento se mide con un indicador de carátula.

RWD Abbreviation for rear-wheel drive.

RWD Abreviación de tracción trasera.

SAE Society of Automotive Engineers.

SAE La Sociedad de Ingenieros Automotrices.

Scan tool A microprocessor designed to communicate with a vehicle's on-board computer in order to perform diagnosis and troubleshooting.

Herramienta de la exploración Una herramienta con un microprocesador diseñado para comunicarse con el ordenador a bordo de un vehículo para realizar diagnosis y la localización de averías.

Score A scratch, ridge, or groove marring a finished surface.

Entalladura Una raya, una arruga o una ranura que desfigure a una superficie acabada.

Scuffing A type of wear in which there is a transfer of material between parts moving against each other; shows up as pits or grooves in the mating surfaces.

Erosión Un tipo de desgaste en el cual hay una tranferencia de una materia entre las partes que estan en contacto mientras que muevan; se manifesta como hoyitos o muescas en las superficies apareadas.

Seal A material shaped around a shaft, used to close off the operating compartment of the shaft, preventing oil leakage.

Sello Una materia, formado alrededor de una flecha, que sella el compartimiento operativo de la flecha, previniendo el goteo de aceite.

Sealer A thick, tacky compound, usually spread with a brush, that may be used as a gasket or sealant to seal small openings or surface irregularities.

Sellador Un compuesto pegajoso y espeso, comœnmente aplicado con una brocha, que puede usarse como un empaque o un obturador para sellar a las aperturas pequeñas o a las irregularidades de la superficie.

Seat A surface, usually machined, upon which another part rests or seats; for example, the surface upon which a valve face rests.

Asiento Una superficie, comúnmente maquinada, sobre la cual yace o se asienta otra parte; por ejemplo, la superficie sobre la cual yace la cara de la válvula.

Selective washer This type of washer is available in different thicknesses to provide the correct endplay or clearance.

Arandela selectiva Este tipo de arandela está disponible en diversos espesores, para proporcionar a la separación correcta o amplitud correcta.

Serial data The communications to and from the computer are commonly referred to as the system's serial data.

Datos seriales del sistema Las comunicaciones al ordenador y del ordenador se llama comúnmente los datos seriales del sistema.

Servo A device that converts hydraulic pressure into mechanical movement, often multiplying it. Used to apply the bands of a transmission.

Servo Un dispositivo que convierte la presión hidráulica al movimiento mecánico, frecuentemente multiplicándola. Se usa en la aplicación de las bandas de una transmisión.

Shift lever The lever used to change gears in a transmission. Also the lever on the starting motor that moves the drive pinion into or out of mesh with the flywheel teeth.

Palanca del cambiador La palanca que sirve para cambiar a las velocidades de una transmisión. También es la palanca del motor de arranque que mueva al piñon de ataque para engranarse o desegranarse con los dientes del volante.

Shift valve A valve that controls the shifting of the gears in an automatic transmission.

Válvula de cambios Una válvula que controla a los cambios de las velocidades en una transmisión automática.

Shim Thin sheets used as spacers between two parts, such as the two halves of a journal bearing.

Laminilla de relleno Hojas delgadas que sirven de espaciadores entre dos partes, tal como las dos partes de un muñón.

Shim stock Sheets of metal of accurately known thickness that can be cut into strips and used to measure or correct clearances.

Materia de laminillas Las hojas de metal cuyo espesor se conoce precisamente que pueden cortarse en tiras y usarse para medir o correjir a las holguras.

Short A short is best described as an unwanted path for current.

Cortocircuito Un cortocircuito se describe lo más mejor posible como un camino indeseado para la corriente eléctrica.

Side clearance The clearance between the sides of moving parts when the sides do not serve as load-carrying surfaces.

Holgura lateral La holgura entre los lados de las partes en movimiento mientras que los lados no funcionan como las superficies de carga.

Sinusoidal The term sinusoidal means the wave is shaped like a sine wave.

Sinusoidal El término sinusoidal quiere decir que la onda tiena una formaparecida a una onda senoidal.

Sliding-fit Where sufficient clearance has been allowed between the shaft and journal to allow freerunning without overheating.

Ajuste corredera Donde se ha dejado una holgura suficiente entre la flecha y el muñón para permitir una marcha libre sin sobrecalentamiento.

Snap ring Split spring-type ring located in an internal or external groove to retain a part.

Anillo de seguridad Un anillo partido tipo resorte que se coloca en una muesca interior o exterior para retener a una parte.

Spalling A condition in which the material of a bearing surface breaks away from the base metal.

Escamación Una condición en la cual una materia de la superficie de un rodamiento se separa del metal base.

Spindle The shaft on which the wheels and wheel bearings mount.

Husillo La flecha en la cual se montan las ruedas y el conjunto del cojinete de las ruedas.

Spline Slot or groove cut in a shaft or bore; a splined shaft onto which a hub, wheel, gear, and so on, with matching splines in its bore is assembled so that the two must turn together.

Acanaladura (espárrago) Una muesca o ranura cortada en una flecha o en un taladro; una flecha acanalada en la cual se asamble un cubo, una rueda, un engranaje, y todo lo demás que tiene un acanaladura pareja en el taladro de manera de que las dos deben girar juntos.

Split lip seal Typically, a rope seal sometimes used to denote any two-part oil seal.

Sello hendido Típicamente, un sello de cuerda que se usa a veces para demarcar cualquier sello de aceite de dos partes

Split pin A round split spring steel tubular pin used for locking purposes; for example, locking a gear to a shaft.

Chaveta hendida Una chaveta partida redonda y tubular hecho de acero para resorte que sirve para el enclavamiento; por ejemplo, para enclavar un engranaje a una flecha.

Spool valve A cylindrically shaped valve with two or more valleys between the lands. Spool valves are used to direct fluid flow.

Válvula de carrete Una válvula de forma cilíndrica que tiene dos acanaladuras de cañón o más entre las partes planas. Las válvulas de carrete sirven para dirigir el flujo del fluido.

Sprag clutch A member of the overrunning clutch family using a sprag to jam between the inner and outer races used for holding or driving action.

Embrague de puntal Un miembro de la familia de embragues de sobremarcha que usa a una palanca de detención trabada entre las pistas interiores e exteriores para realizar una acción de asir o marchar.

Spring A device that changes shape when it is stretched or compressed, but returns to its original shape when the force is removed; the component of the automotive suspension system that absorbs road shocks by flexing and twisting.

Resorte Un dispositivo que cambia de forma al ser estirado o comprimido, pero que recupera su forma original al levantarse la fuerza; es un componente del sistema de suspensión automotívo que absorba los choques del camino al doblarse y torcerse.

Spring retainer A steel plate designed to hold a coil or several coil springs in place.

Retén de resorte Una chapa de acero diseñado a sostener en su posición a un resorte helicoidal o más.

Squeak A high-pitched noise of short duration.

Chillido Un ruido agudo de poca duración.

Squeal A continuous high-pitched noise.

Alarido Un ruido agudo continuo.

Stall A condition in which the engine is operating and the transmission is in gear, but the drive wheels are not turning because the turbine of the torque converter is not moving.

Paro Una condición en la cual opera el motor y la transmisión esta embragada pero las ruedas de impulso no giran porque no mueva el turbino del convertidor de la torsión.

Stall test A test of the one-way clutch in a torque converter.

Prueba de paro Una prueba del embrague de una vía en un convertidor de la torsión.

Static A form of electricity caused by friction.

Estático Una forma de la electridad causada por la fricción.

Stress The force to which a material, mechanism, or component is subjected.

Esfuerzo La fuerza a la cual se somete a una materia, un mecanísmo o un componente.

Sun gear The central gear in a planetary gear system around which the rest of the gears rotate. The innermost gear of the planetary gearset.

Engranaje principal (sol) El engranaje central en un sistema de engranajes planetarios alrededor del cual giran los otros engranajes. El engranaje más interno del conjunto de los engranajes planetarios.

Tap To cut threads in a hole with a tapered, fluted, threaded tool.

Roscar con macho Cortar las roscas en un agujero con una herramienta cónica, acanalada y fileteada.

Teardown A term often used to describe the process of disassembling a transmission.

Desmontaje Un término común que describe el proceso de desarmar a una transmisión.

Temper To change the physical characteristics of a metal by applying heat.

Templar Cambiar las características físicas de un metal mediante una aplicación del calor.

Tension Effort that elongates or "stretches" a material.

Tensión Un esfuerzo que alarga o "estira" a una materia.

Thickness gauge Strips of metal made to an exact thickness, used to measure clearances between parts.

Calibre de espesores Las tiras del metal que se han fabricado a un espesor exacto, sirven para medir las holguras entre las partes.

Thread chaser A device similar to a die that is used to clean threads.

Peine de roscar Un dispositivo parecido a una terraja que sirve para limpiar a las roscas.

Threaded insert A threaded coil that is used to restore the original thread size to a hole with damaged threads.

Pieza inserta roscada Una bobina roscada que sirve para restaurar a su tamaño original una rosca dañada.

Throttle position (TP) sensor A potentiometer used to monitor changes in throttle plate opening. The position of the throttle plate determines the voltage drop at the sensor's resistor and the resultant voltage signal is sent to a computer system.

Sensor de posición de la válvula reguladora Un potenciómetro usado para vigilar cambios en la apertura de la placa de la válvula reguladora. La posición de la placa de la válvula reguladora determina la caída de voltaje en el resistor del sensor y la señal resultante del voltaje se envía a un sistema informático.

Thrust bearing A bearing designed to resist or contain side or end motion as well as reduce friction.

Cojinete de empuje Un cojinete diseñado a detener o reprimir a los movimientos laterales o de las extremidades y también reducir la fricción.

Thrust load A load that pushes or reacts through the bearing in a direction parallel to the shaft.

Carga de empuje Una carga que empuja o reacciona por el cojinete en una dirección paralelo a la flecha.

Thrust washer A washer designed to take up end thrust and prevent excessive endplay.

Arandela de empuje Una arandela diseñada para rellenar a la holgura de la extremidad y prevenir demasiado juego en la extremidad.

Tolerance A permissible variation between the two extremes of a specification or dimension.

Tolerancia Una variación permisible entre dos extremos de una especificación o de un dimensión.

Torque A twisting motion, usually measured in ft.-lb. (Nm).

Torsión Un movimiento giratorio, suele medirse en pies/libra (Nm).

Torque capacity The ability of a converter clutch to hold torque.

Capacidad de la torsión La abilidad de un convertidor de embraque a sostener a la torsión.

Torque converter A turbine device utilizing a rotary pump, one or more reactors (stators), and a driven circular turbine or vane, whereby power is transmitted from a driving to a driven member by hydraulic action. It provides varying drive ratios; with a speed reduction, it increases torque.

Convertidor de la torsión Un dispositivo de turbino que utilisa a una bomba rotativa, a un reactor o más, y un molinete o turbino circular impulsado, por cual se transmite la energía de un miembro de impulso a otro arrastrado mediante la acción hidráulica. Provee varias relaciones de impulso; al descender la velocidad, aumenta la torsión.

Torque curve A line plotted on a chart to illustrate the torque personality of an engine. When the engine operates on its torque curve, it is producing the most torque for the quantity of fuel being burned.

Curva de la torsión Una linea delineada en una carta para ilustrar las características de la torsión del motor. Al operar un motor en su curva de la torsión, produce la torsión óptima para la cantidad del combustible que se consuma.

Torque multiplication The result of meshing a small driving gear and a large driven gear to reduce speed and increase output torque.

Multiplicación de la torsión El resultado de engranar a un engranaje pequeño de ataque con un engranaje más grande arrastrado para reducir la velocidad y incrementar la torsión de salida.

Torque steer An action felt in the steering wheel as the result of increased torque.

Dirección la torsión Una acción que se nota en el volante de dirección como resultado de un aumento de la torsión.

Traction The gripping action between the tire tread and the road's surface.

Tracción La acción de agarrar entre la cara de la rueda y la superficie del camino.

Transaxle Type of construction in which the transmission and differential are combined in one unit.

Flecha de transmisión Un tipo de construcción en el cual la transmisión y el diferencial se combinan en una unidad.

Transaxle assembly A compact housing most often used in front-wheel-drive vehicles that houses the manual transmission, final drive gears, and differential assembly.

Asamblea de la flecha de transmisión Un cárter compacto que se usa normalmente en los vehículos de tracción delantera que contiene la transmisión manual, los engranajes de propulsión, y la asamblea del diferencial.

Transfer case An auxiliary transmission mounted behind the main transmission. Used to divide engine power and transfer it to both front and rear differentials, either full-time or part-time.

Cárter de la transferencia Una transmisión auxiliar montada detrás de la transmisión principal. Sirve para dividir la potencia del motor y transferirla a ambos diferenciales delanteras y traseras todo el tiempo o la mitad del tiempo.

Transmission The device in the powertrain that provides different gear ratios between the engine and drive wheels as well as reverse.

Transmisión El dispositivo en el trén de potencia que provee las relaciones diferentes de engranaje entre el motor y las ruedas de impulso y también la marcha de reversa.

Transverse Powertrain layout in a front-wheel-drive automobile extending from side to side.

Transversal Una esquema del tren de potencia en un automóvil de tracción delantera que se extiende de un lado a otro.

U-joint A four-point cross connected to two U-shaped yokes that serves as a flexible coupling between shafts.

Junta de U Una cruceta de cuatro puntos que se conecta a dos yugos en forma de U que sirven de acoplamientos flexibles entre las flechas.

Underdrive The condition that exists when there is speed reduction through the gears.

Bajamarcha Bajamarcha es la condición que existe cuando hay reducción de velocidad a través de los engranajes.

United States Customary System (USCS) A system of weights and measures used in places that do not use the metric system, such as the United States. The USCS system is also known as the English or inch system.

Sistema acostumbrado de Estados Unidos (USCS) Un sistema de los pesos y de las medidas usados en los países que no utilizan la sistema métrico, tal como los Estados Unidos. El sistema de USCS también se conoce como el sistema del inglés o de la pulgada.

Universal joint A mechanical device that transmits rotary motion from one shaft to another shaft at varying angles.

Junta Universal Un dispositivo mecánico que transmite el movimiento giratorio desde una flecha a otra flecha en varios ángulos.

Upshift To shift a transmission into a higher gear.

Cambio ascendente Cambiar a la velocidad de una transmisión a una más alta.

Valve body Main hydraulic control assembly of a transmission containing the components necessary to control the distribution of pressurized transmission fluid throughout the transmission.

Cuerpo de la válvula Asamblea principal del control hidráulico de una transmisión que contiene los componentes necesarios para controlar a la distribución del fluido de la transmisión bajo presión por toda la transmisión.

Vehicle identification number (VIN) The number assigned to each vehicle by its manufacturer, primarily for registration and identification purposes.

Número de identificacíon del vehículo El número asignado a cada vehículo por su fabricante, primariamente con el propósito de la registración y la identificación.

Vehicle speed sensor (VSS) A sensor used to track the current speed and the total miles traveled by a vehicle. This input is used by many computer systems in the vehicle.

Sensor de velocidad del vehículo Un sensor utilizado para medir la velocidad actual y las millas totales viajaron por un vehículo. Muchos de los sistemas informáticos del vehículo utilizan estos datos como entrada de información.

Vibration A quivering, trembling motion felt in the vehicle at different speed ranges.

Vibración Un movimiento de estremecer o temblar que se siente en el vehículo en varios intervalos de velocidad.

Viscosity The resistance to flow exhibited by a liquid. A thick oil has greater viscosity than a thin oil.

Viscosidad La resistencia al flujo que manifiesta un líquido. Un aceite espeso tiene una viscosidad mayor que un aceite ligero.

Voltmeter A test instrument used to measure voltage.

Voltímetro Un aparato utilizado para medir voltaje.

Vortex Path of fluid flow in a torque converter. The vortex may be high, low, or zero depending on the relative speed between the pump and turbine.

Vórtice La vía del flujo de los fluidos en un convertido de torsión. El vórtice puede ser alto, bajo, o cero depende de la velocidad relativa entre la bomba y la turbina.

Vortex flow Recirculating flow between the converter impeller and turbine that causes torque multiplication.

Flujo del vórtice El fluyo recirculante entre el impulsor del convertidor y la turbina que causa la multiplicación de la torsión.

Wet-disc clutch A clutch in which the friction disc (or discs) is operated in a bath of oil.

Embrague de disco flotante Un embrague en el cual el disco (o los discos) de fricción opera en un baño de aceite.

Wheel A disc or spokes with a hub at the center that revolves around an axle, and a rim around the outside on which the tire is mounted.

Rueda Un disco o rayo que tiene en su centro un cubo que gira alrededor de un eje, y tiene un rim alrededor de su exterior en la cual se monta el neumático.

Yoke In a universal joint, the drivable torque-and-motion input and output member attached to a shaft or tube.

Yugo En una junta universal, el miembro de la entrada y la salida que transfere a la torsión y al movimiento, que se conecta a una flecha o a un tubo.

INDEX